PHYSICAL CHEMISTRY

An Advanced Treatise

Volume I / Thermodynamics

PHYSICAL CHEMISTRY

An Advanced Treatise

Edited by

HENRY EYRING
Departments of Chemistry and Metallurgy
University of Utah
Salt Lake City, Utah

DOUGLAS HENDERSON
IBM Research Laboratories
San Jose, California

WILHELM JOST
Institut für Physikalische Chemie der Universität Göttingen
Göttingen, Germany

PHYSICAL CHEMISTRY

An Advanced Treatise

VOLUME I / Thermodynamics

Edited by

WILHELM JOST
Institut für Physikalische
Chemie der Universität Göttingen
Göttingen, Germany

1971

ACADEMIC PRESS NEW YORK / LONDON

ACADEMIC PRESS, INC.
111 Fifth Avenue, New York, New York 10003

United Kingdom Edition published by
ACADEMIC PRESS, INC. (LONDON) LTD.
Berkeley Square House, London W1X 6BA

LIBRARY OF CONGRESS CATALOG CARD NUMBER: 73-117081

PRINTED IN THE UNITED STATES OF AMERICA

Contents

Chapter 1 / Survey of Fundamental Laws

R. Haase

Chapter 2A / Equilibrium, Stability, and Displacements

A. Sanfeld

Chapter 2B / Irreversible Processes

A. Sanfeld

Chapter 2C / Thermodynamics of Surfaces

A. Sanfeld

Chapter 3 / Thermodynamic Properties of Gases, Liquids, and Solids

R. Haase

Chapter 4 / Gas–Liquid and Gas–Solid Equilibria at High Pressure, Critical Curves, and Miscibility Gaps

E. U. Franck

Chapter 5 / Thermodynamics of Matter in Gravitational, Electric, and Magnetic Fields

Herbert Stenschke

Chapter 6 / The Third Law of Thermodynamics

J. Wilks

Chapter 7 / Practical Treatment of Coupled Gas Equilibrium

Max Klein

Chapter 8 / Equilibria at Very High Temperatures

H. Krempl

Chapter 9 / High Pressure Phenomena

Robert H. Wentorf, Jr.

Chapter 10 / Carathéodory's Formulation of the Second Law

S. M. Blinder

List of Contributors

Numbers in parentheses indicate the pages on which the authors' contributions begin.

S. M. Blinder, Chemistry Department, University of Michigan, Ann Arbor, Michigan (613)

E. U. Franck, Institut für Physikalische Chemie, Universität Karlsruhe, Karlsruhe, Germany (367)

R. Haase, Institut für Physikalische Chemie, Technische Universität Aachen, Aachen, Germany (1, 239)

Max Klein, Equation of State Section, Heat Division, National Bureau of Standards, Washington, D.C. (489)

H. Krempl, Institut für Physikalische Chemie und Electrochemie der Technischen Universität München, Munich, Germany (545)

A. Sanfeld, Faculte des Sciences, Chimie Physique 2, Université Libre de Bruxelles, Brussels, Belgium (99, 217, 245)

Herbert Stenschke, Batelle Memorial Institute, Columbus, Ohio (403)

Robert H. Wentorf, Jr., General Electric Research and Development Center, Schenectady, New York (571)

J. Wilks, Clarendon Laboratory, University of Oxford, Oxford, England (437)

Foreword

In recent years there has been a tremendous expansion in the development of the techniques and principles of physical chemistry. As a result most physical chemists find it difficult to maintain an understanding of the entire field.

The purpose of this treatise is to present a comprehensive treatment of physical chemistry for advanced students and investigators in a reasonably small number of volumes. We have attempted to include all important topics in physical chemistry together with borderline subjects which are of particular interest and importance. The treatment is at an advanced level. However, elementary theory and facts have not been excluded but are presented in a concise form with emphasis on laws which have general importance. No attempt has been made to be encyclopedic. However, the reader should be able to find helpful references to uncommon facts or theories in the index and bibliographies.

Since no single physical chemist could write authoritatively in all the areas of physical chemistry, distinguished investigators have been invited to contribute chapters in the field of their special competence.

If these volumes are even partially successful in meeting these goals we will feel rewarded for our efforts.

We would like to thank the authors for their contributions and to thank the staff of Academic Press for their assistance.

HENRY EYRING
DOUGLAS HENDERSON
WILHELM JOST

Preface

Thermodynamics, complete in its fundamental development, is a necessary tool for the chemist in almost all fields, and new lines are generally induced by the requirements of applications. Thus this volume starts with a survey of basic laws, as a preparation for the several topics to be covered in later chapters. Emphasis is placed on questions of stability, irreversible processes, surfaces, the third law, and a short introduction to Carathéodory's axiomatic foundation.

This volume deals with the applications of thermodynamics to mixtures, to fluids and solid systems at high pressures and temperatures, to critical phenomena, to the practical handling of coupled gas equilibria, and to matter in electric, magnetic, and gravitational fields.

The editor had been fortunate in having had the cooperation of A. Michels, Amsterdam, with his superior experimental and theoretical knowledge in the field of equation of state. It is a great loss for this volume that this outstanding man was taken from us, still in the middle of his work, though of a biblical age.

WILHELM JOST

Contents of Previous and Future Volumes

VOLUME IV

VOLUME V

VOLUME IXA

PHYSICAL CHEMISTRY

An Advanced Treatise

Volume I / Thermodynamics

Chapter 1

Survey of Fundamental Laws

R. HAASE

I. Introduction

Thermodynamics, in its most general form, is the theory of those aspects of the macroscopic behavior of matter not covered by mechanics and electrodynamics. We shall assume the basic concepts of mechanics and electrodynamics to be known, but we shall rigorously define and explain all concepts that are specific to thermodynamics, such as temperature, heat, entropy, and so on.

This survey of fundamental laws applies not only to classical thermodynamics but also to nonequilibrium thermodynamics (thermodynamics of irreversible processes); it will become clear that both these branches derive from the same root.

We shall give examples if they are needed for the sake of clarity, but we shall not go into details of application since this is the subject of later chapters and of other volumes.

From the very beginning, the discussion will not be restricted to equilibrium or to isotropic systems in the absence of external fields. Such a restriction made by most textbooks, at least initially, masks the very general character of modern thermodynamics.

We adopt a purely macroscopic point of view, that is, we start from some fundamental empirical laws and then develop the consequences mathematically, step by step. We do not imply that this is the only approach or even the most satisfactory one. But we stress that this procedure is consistent and logical since it can be traced back to experience.

Statistical mechanics, the microscopic counterpart of thermodynamics, has been dealt with in Volume II.

II. Some Basic Concepts

Let us call a *system* any macroscopic portion of matter on which we fix our attention. Then the rest of the physical world as far as it may interact with the given system is said to be the *surroundings*.

If the system has the same properties throughout its extension, it is called a *homogeneous system*. If the system is composed of a number of homogeneous parts called *phases*, then it is said to be a *heterogeneous* or *discontinuous system*. Finally, if properties such as pressure or composition vary continuously over macroscopic parts of the system, then we speak of a *continuous system*. A quantity of water or a piece of ice under ordinary conditions is an example of a homogeneous system. A quantity of water together with a piece of ice is an example of a heterogeneous system. A high column of air in the Earth's gravitational field or a liquid in a centrifuge is an example of a continuous system.

Let a *region* of a system be either a single phase (homogeneous system or homogeneous part of a heterogeneous system) or a volume element of a continuous system. The volume element is considered to be macroscopic in the sense of containing a large number of particles (atoms, molecules, ions, electrons, and so on) but small enough to be referred to as an "element" of macroscopic matter in space.

A system or a region is said to be *open* if it exchanges matter with the surroundings. If there is no such exchange, the system or region is called *closed*. An *isolated system* shows no interactions whatsoever with the surroundings, thus being a special case of a closed system. A more detailed classification of systems will be given in a later section.

The macroscopic variables that describe the behavior of a system or region are called *state variables*. They may be either *external coordinates*, such as macroscopic velocities or position coordinates in an external field, or *internal state variables*, such as pressure, volume, or masses of all the substances. A definite set of values of state variables fixes a definite *state* of the system or region.

An *intensive property* is an internal state variable that does not depend on the quantity of the region considered. It has a definite value for each point in a system. In particular, it is constant throughout a homogeneous system. Examples are: pressure, density, electric field strength.

An *extensive property* is an internal state variable that depends on the quantity of the region. If the masses of all the substances contained in the region increase by a certain factor, then, for given values of the intensive properties, all the extensive properties increase by the same factor. Thus, any extensive property is a homogeneous function of first degree in the masses. Furthermore, an extensive property of any system is equal to the sum of the extensive properties of the macroscopic subsystems into which the given system may be divided. Examples of extensive properties are: mass, volume, total electric moment.

Let us now consider an arbitrary system and let us suppose that the intensive properties do not change in time. This situation may arise in two cases, which can be distinguished by cutting off the system from all external influences, except external force fields possibly present (in the absence of external fields, we then have an isolated system): If, after this insulation, nothing happens within the system, then the system is said to be in *equilibrium*; if, on the other hand, changes of state occur within the system after the insulation, then the system was in a *stationary state* (or "steady state"). Thus, a stationary state is a nonequilibrium state in which the intensive properties happen to be independent of time by virtue of external agencies, while an equilibrium state is produced by internal causes.

If we fix our attention on a certain subsystem of a given system and find that the subsystem is in equilibrium while there is not necessarily equilibrium within the whole system, then the subsystem is said to be in *internal equilibrium*. For example, there may be chemical equilibrium within a given phase of a heterogeneous system while there is no equilibrium at all between the constituent phases.

If, within a given system, a change of state occurs, then we say that a *process* takes place.

A process is called *quasistatic* if a system subject to external influences passes through a continuous series of equilibrium states and if all actions produced by the system are exactly counterbalanced by external actions. A simple example is the expansion or compression of a gas acting on a moving piston by means of the gas pressure opposed by an equal pressure exerted by the piston so that effects such as friction are to be excluded. Obviously, such a process is fictitious. A process that actually occurs in nature is a *natural* process; it proceeds in a direction toward equilibrium unless there is periodic behavior. Quasistatic processes are limiting cases of natural processes.

A process is said to be *reversible* if the system can return from its final state to its initial state without any change in the surroundings. A process is called *irreversible* if the initial state of the system cannot be restored without changes in the surroundings.

It will be proved later (Section XI) that all quasistatic processes are reversible and that all natural processes are irreversible. Meanwhile, the distinction will be maintained.

III. Mass, Amount of Substance, and Composition

The *chemical content* of a region is determined by the quantities of all the constituent substances. Without leaving the province of thermodynamics, we may introduce the *mass* m_k of each substance k as a measure of the content of the region.* The total mass m of the region is then given by

$$m = \sum_k m_k, \tag{3.1}$$

the summation being extended over all constituents present. The units usually chosen for m_k or m are the gram (symbol: g) and the kilogram (symbol: kg).

But it turns out to be expedient to use another variable for describing the quantity of matter. We introduce the concept of "particles" (atoms, molecules, ions, electrons, etc.) and the quantity N_k, the *number of particles* of kind k, thus anticipating a simple result of atomistics. The quantity N_k is a pure number (a "dimensionless" quantity).

For practical purposes, however, N_k is too large. Therefore, we define another quantity n_k called the *amount of substance* of the chemical species k:

$$n_k \equiv N_k/L, \tag{3.2}$$

where L is a universal constant called the *Avogadro constant*. The value of L depends on the unit to be chosen for n_k.

We adopt the following convention: 1 mol is an amount of substance of a chemical species containing the same number of particles as there are atoms in 12 g (exactly) of the pure carbon nuclide ^{12}C.

If we measure n_k in mol or kmol, the unit to be used for L is mol^{-1} or $kmol^{-1}$, respectively. The numerical value of n_k expressed in mol may be called "number of moles." In a similar way, the numerical value of L expressed in mol^{-1} may be called the "Avogadro number."

Let us denote the mass of a single particle of kind k by $\mathfrak{m}_k$, so that $m_k = N_k \mathfrak{m}_k$. Then we have, according to Eq. (3.2),

$$m_k = L\mathfrak{m}_k n_k. \tag{3.3}$$

Applying Eq. (3.3) to ^{12}C and taking account of the definition of the mole, we find

$$12m^\dagger = L\mathfrak{m}_C n^\dagger, \tag{3.4}$$

where

$$m^\dagger \equiv 1 \text{ g}, \qquad n^\dagger \equiv 1 \text{ mol} \tag{3.5}$$

and $\mathfrak{m}_C$ is the mass of an atom ^{12}C. By means of Eq. (3.4), we may, in

* Major symbols used in this chapter are defined in the section on Nomenclature.

principle, derive the Avogadro constant L from experimental data on $\mathfrak{m}_C$. The approximate value is

$$L = 6.023 \times 10^{23} \quad \text{mol}^{-1}. \tag{3.6}$$

The total amount of substance n of the region is given by

$$n = \sum_k n_k, \tag{3.7}$$

a relation completely analogous to Eq. (3.1).

The *molar mass* M_k of substance k is defined by

$$M_k \equiv m_k/n_k. \tag{3.8}$$

The units usually chosen for M_k are g mol^{-1} = kg kmol^{-1}, and sometimes kg mol^{-1}.

The *composition* of a region is measured in a number of ways. For the present purpose, it will be sufficient to choose the following composition variables:

mass fraction	$\chi_k \equiv m_k/m,$	(3.9)
mole fraction	$x_k \equiv n_k/n,$	(3.10)
mass concentration	$\varrho_k \equiv m_k/V,$	(3.11)
molarity	$c_k \equiv n_k/V$	(3.12)

where V is the volume of the region considered. As is obvious from Eqs. (3.1) and (3.7), the following identities hold:

$$\sum_k \chi_k = 1, \qquad \sum_k x_k = 1, \tag{3.13}$$

$$\sum_k \varrho_k = m/V = \varrho, \tag{3.14}$$

$$\sum_k c_k = n/V. \tag{3.15}$$

Here ϱ denotes the density of the region and n/V is the reciprocal value of a quantity called molar volume (Section VI).

The quantities χ_k and x_k are dimensionless, while ϱ_k has the dimension of mass divided by volume and c_k that of amount of substance divided by volume.

It should be noted that all composition variables are intensive properties, while m_k, m, n_k, n, and V clearly belong to the class of extensive properties.

IV. Zeroth Law of Thermodynamics (Empirical Temperature)

Everyday experience tells us that the properties of a system (or of a region) depend on its thermal state, that is, on whether it is hot or cold. Thus, the volume of a gas or liquid is not only determined by its pressure and chemical content, but also by its thermal state. Alternatively, the volume of an isotropic region polarized by an electrostatic field is not fixed by pressure, electric field strength, and chemical content, but also depends on the thermal state of the region. In other words, prethermodynamic quantities such as volume, pressure, electric field strength, and chemical content do not suffice to describe the real behavior of systems. It is the purpose of this section to replace the vague term "thermal state" by a quantity defined rigorously (cf. Guggenheim (1967)).

Any boundary or enclosure of a system that is impermeable to matter is called a *wall*. If each part of the wall is fixed in space, the wall is said to be *rigid*.

We now introduce a *thermally conducting wall*. If two systems each in internal equilibrium are brought into contact through a certain type of rigid wall, then, in general, the two systems will not be in mutual equilibrium, but there will be changes of state in each of the systems until there is no further change. The type of rigid wall thus defined is said to be a thermally conducting wall. The final state of mutual equilibrium reached after this "thermal contact" is referred to as a state of *thermal equilibrium*.

Since the thermally conducting wall is by definition rigid and impermeable to matter and both systems are in internal equilibrium, the processes that occur and the mutual equilibrium eventually established have nothing to do with equalization of pressures, passage of matter, or chemical reactions. On the contrary, it can be checked, at least qualitatively, that changes take place if the two systems have different degrees of "hotness" and that thermal equilibrium is reached when the two systems appear to have the same degree of "hotness." Therefore, thermal equilibrium and the processes preceding it are related to what we vaguely called the "thermal state."

Let us consider two systems in internal equilibrium not surrounded by rigid walls except at the surfaces of thermal contact. Then, for example, by measuring the volumes at fixed pressure and fixed content, we can decide whether or not the two systems are in thermal equilibrium with each other. By making experiments on a number of such systems, we conclude that there is an important general principle:

If two systems are both in thermal equilibrium with a third system, then they are in thermal equilibrium with each other.

This will be called the *zeroth law of thermodynamics* (Carathéodory, 1909; Fowler and Guggenheim, 1952).

According to this law, all systems in thermal equilibrium with one another have a property in common. They are said to have the same "temperature." Systems not in thermal equilibrium are said to have different "temperatures." The property thus defined is called *empirical temperature*. The definition is still incomplete,* since we have not yet given a description of how to measure the temperature quantitatively.

The next step consists in constructing a *thermometer*, i.e., a system to be brought into thermal contact with, and very small in comparison to, the system whose temperature is to be investigated. The thermometer must have one property that can be easily and reproducibly recorded, for example, the volume at fixed pressure and fixed content. As we can deduce from the statements given above, this property may be taken to be a function of temperature. We are now able to say whether or not two separate systems have equal temperatures.

Let us consider a heterogeneous system consisting of pure water and composed of three phases in mutual equilibrium: ice, liquid water, and steam. Using our provisional thermometer, we find that all samples of systems of this kind have the same temperature, regardless of the masses of the three phases. Since the heterogeneous system mentioned is easily reproducible, it will be our reference system. The equilibrium temperature in this three-phase system recorded by the thermometer will be referred to as the "triple point of water."

We now measure the product of pressure P and volume V of a gas of fixed content for various values of P and V. We then extrapolate the experimental values of PV to $P = O$, thus obtaining a limiting value A of PV for vanishing pressure. We carry out these experiments at the

* For a fuller discussion of this first step in defining temperature, see Landsberg (1961).

triple point of water (subscript tr) and at any other temperature as recorded by the provisional thermometer. We then find

$$\lim_{P\to 0} (PV)_{tr} = A_{tr}, \tag{4.1}$$

$$\lim_{P\to 0} PV = A. \tag{4.2}$$

Experience shows that the ratio A/A_{tr} is independent of the nature of the gas considered. (This would not be true if we had chosen a liquid or a solid.) Therefore, it is reasonable to define the empirical temperature Θ by

$$\Theta/\Theta_{tr} \equiv A/A_{tr}. \tag{4.3}$$

The quantity Θ thus defined is called the *empirical Kelvin temperature*. As can be seen from Eq. (4.3), it is an intensive property.

It was decided in 1954 to fix the unit of Θ by assigning a definite value (numerical value: 273.16) to the empirical Kelvin temperature Θ_{tr} of the triple point of water. This unit was formerly denoted as the degree Kelvin (symbol: °K) and is now called the kelvin (symbol: K). We accordingly write

$$\Theta_{tr} \equiv 273.16\text{K}. \tag{4.4}$$

Since the right-hand side of Eq. (4.3) is measurable, the value of Θ is thus uniquely determined. The empirical Kelvin temperature of any system may now be found by using a fixed quantity of gas as a thermometer and applying Eqs. (4.3) and (4.4).

If we measure the values Θ_0 and Θ_1 of the empirical Kelvin temperature for the normal freezing point and normal boiling point of water, respectively, then we obtain

$$\Theta_0 = 273.15\text{K}, \qquad \Theta_1 = 373.15\text{K}. \tag{4.5}$$

These figures are only approximately true, while Eq. (4.4) holds exactly, by definition. Originally, a "temperature scale" was defined by fixing the *two* values given by Eq. (4.5), and this is why the difference $\Theta_1 - \Theta_0$ is just 100K, within present experimental accuracy.

Other empirical temperatures still used in practice have a similar basis. Any such temperature ϑ may be defined by

$$\vartheta \equiv a[(\Theta/\Theta^{\dagger}) - 273.15] + b, \tag{4.6}$$

where a and b are constants and

$$\Theta^{\dagger} \equiv 1\text{K}. \tag{4.7}$$

If, for instance, we wish to express ϑ in the unit degree Celsius (°C) or in the unit degree Fahrenheit (°F), respectively, we have to insert the following constants into Eq. (4.6):

$$a = 1°\text{C}, \qquad b = 0 \qquad \text{(Celsius temperature)}, \tag{4.8}$$

$$a = \tfrac{9}{5}°\text{F}, \qquad b = 32°\text{F} \qquad \text{(Fahrenheit temperature)}. \tag{4.9}$$

No further use will be made of these temperature scales, except in numerical examples where the Celsius temperature will be occasionally employed.

We shall prove later that the so-called *thermodynamic temperature* T, when properly defined, coincides with the empirical Kelvin temperature:

$$T = \Theta. \tag{4.10}$$

We anticipate this result until we reach the sections concerning the second law, where the distinction will again be relevant. Meanwhile, we refer to Θ or T as *the* temperature.

This may be the appropriate place to say a few words about dimensions and units.

As far as dimensions are concerned, we have the following *basic quantities*: length, time, mass, amount of substance, charge, temperature, and luminous intensity. According to all recent international recommendations, the corresponding *basic units* are: the meter (m), the second (s), the kilogram (kg), the mole (mol), the coulomb (C), the kelvin (K), and the candela (cd). For practical reasons, the unit of electric current strength, the ampere (A), is considered to be a basic unit, so that the coulomb appears as a derived unit defined by C $\equiv$ As. The meter, the second, the kilogram, the mole, the ampere, the kelvin, and the candela are the basic units of the International System of Units (SI units).

The quantities considered in this section, besides temperature, are volume and pressure. They are both derived quantities, having the dimensions $(\text{length})^3$ and $\text{force}/(\text{length})^2$, respectively, where the force is again a derived quantity of the dimension mass $\times$ $\text{length}/(\text{time})^2$. Hence, the SI unit of volume is the cubic meter (m^3) and that of force is the newton (N) defined by $\text{N} \equiv \text{kg m s}^{-2}$. The practical unit of pressure is the bar ($\text{bar} \equiv 10^5\ \text{N m}^{-2}$).

Sometimes it is expedient to make use of the centimeter (cm) and the gram (g). The corresponding units of volume, force, and pressure are: the cubic centimeter (cm^3), the dyne ($dyn \equiv g\ cm\ s^{-2}$), and the dyne divided by the square centimeter ($dyn\ cm^{-2}$). We have the relations:

$$m^3 = 10^6 \quad cm^3, \qquad N = 10^5 \quad dyn, \qquad bar = 10^6 \quad dyn\ cm^{-2}.$$

The units liter (l), atmosphere (atm), millimeter of mercury (mm Hg), and torr (Torr) are no longer recommended. Nevertheless, we shall occasionally use the atmosphere (atm) defined by

$$atm \equiv 1.01325 \quad bar.$$

V. Equation of State

We first consider a single gas. We want to investigate the relation connecting volume V, pressure P, temperature T, and amount of substance n.

The product PV is an extensive property. Thus, for a one-component system,* it must be proportional to n. If we take into account Eqs. (4.2), (4.3), and (4.10), we obtain for any gas

$$\lim_{P \to 0} PV = nRT. \tag{5.1}$$

Here R is a universal constant called the *gas constant*. Its approximate value is

$$R = 8.314 \times 10^{-5} \quad bar\ m^3\ K^{-1}\ mol^{-1} = 8.314 \quad N\ m\ K^{-1}\ mol^{-1}. \tag{5.2}$$

At the outset we only know the mass m, but not the amount of substance n of the gas considered. Thus we conclude from Eqs. (4.2), (4.3), and (4.10)

$$\lim_{P \to 0} PV = mR'T \tag{5.1a}$$

* For a multicomponent system, an extensive property is still a homogeneous function of first degree in the amounts of substance $n_1, n_2, \ldots$, but this is not necessarily a linear relation. Thus, the function $PV = \text{const}(n_1^2 + n_2^2 + \cdots)^{1/2}$ is homogeneous of first degree in $n_1, n_2, \ldots$. Obviously, this expression reduces to a linear relation for $n_1 = n$, $n_2 = \cdots = 0$.

where

$$R' = A_{\text{tr}}/mT_{\text{tr}}$$

is a constant dependent on the nature of the gas as borne out by experiment. We define a constant M characteristic of each gas

$$M \equiv R/R', \tag{5.1b}$$

where R is a universal constant. Introducing the density $\varrho = m/V$ of the gas, we obtain from Eqs. (5.1a) and (5.1b)

$$M = RT \lim_{P\to 0}(\varrho/P). \tag{5.1c}$$

If we assign the value given by (5.2) to R, then, from comparison to the results of molecular theory, we find that the measurable quantity M is the molar mass as defined by Eq. (3.8):

$$M = m/n. \tag{5.1d}$$

Combining Eqs. (5.1a), (5.1b), and (5.1d), we recover Eq. (5.1).

Any gas under a pressure low enough to render indistinguishable its properties from those under zero pressure is called a *perfect gas*. Thus, we have, according to Eq. (5.1),

$$PV = nRT \qquad \text{(perfect gas)}. \tag{5.3}$$

If we define a perfect gaseous mixture in a similar way, we find that Eq. (5.3) continues to hold, the quantity n now being the total amount of substance given by Eq. (3.7). The relation (5.3) is the simplest example of an *equation of state*.

Any gas which does not obey Eq. (5.3) is said to be an *imperfect gas*. A number of more complicated equations of state have been developed in order to describe the behavior of imperfect gases, such as the equations proposed by van der Waals (1881), by Beattie and Bridgeman (1927), and by Benedict *et al.* (1940).

Except for liquids near the critical region, where the equations of state for imperfect gases are valid, at least approximately, there are no explicit relations of a similar type for liquids and solids. But we may state that there always exists a function ($n_1, n_2, \ldots$ being the amounts of substance of species 1, 2, ...)

$$V = V(T, P, n_1, n_2, \ldots) \tag{5.4}$$

for any gas, liquid, or isotropic solid. This also holds for a region within a continuous isotropic system, e.g., for a volume element of a gas or liquid in a gravitational or centrifugal field.

In Eq. (5.4), it has been taken for granted that the region considered is isotropic and not subject to any influence by electromagnetic fields (which might lead to polarization of matter) or by surface effects. We shall call such a region a *simple region*, since not only the equation of state, but many other thermodynamic relations take a particularly simple form in this case. A *simple system* is any system composed of simple regions, for instance, a heterogeneous system consisting of isotropic, nonpolarized phases not subject to surface effects.

Equation (5.4) is the unspecified form of an equation of state for a simple region. We shall now generalize the concept "equation of state" so as to include more complicated cases.

If we study an *anisotropic region*, e.g., a volume element of a non-cubic crystal, we discover that its mechanical behavior can no longer be described by two variables such as volume and pressure. Mechanics tells us indeed that here in general the state of deformation is given by a set of six variables $e_1, e_2, \ldots, e_6$ called "strain components" and that the state of stress is given by another set of six variables $\tau_1, \tau_2, \ldots, \tau_6$ called "stress components."* The quantities analogous to the volume of an isotropic region are the extensive properties $V_0 e_i$ ($i = 1, 2, \ldots, 6$), where V_0 denotes the volume of the reference state. On the other hand, the stress components correspond to the negative pressure and are intensive properties. There exists a "stress–strain relation"

$$e_i = e_i(T, \tau_1, \tau_2, \ldots, \tau_6, x_1, x_2, \ldots) \qquad (i = 1, 2, \ldots, 6). \quad (5.5)$$

Here the mole fractions $x_1, x_2, \ldots$ of the various species have been introduced since the strain components, being intensive properties, do not depend on the amount of the volume element considered. Equation (5.5) may again be called an equation of state. The simplest relation of this kind is the "generalized Hooke law" which holds for small deformations (see Section XVI).

Let us now consider a *polarized region*, that is, a phase or a volume element of a system polarized by an electromagnetic field. To describe

* If we denote the components of the strain tensor and of the stress tensor by e_{xx}, $e_{xy}, \ldots$, and $\tau_{xx}, \tau_{xy}, \ldots$, respectively, then we have: $e_1 = e_{xx}$, $e_2 = e_{yy}$, $e_3 = e_{zz}$, $e_4 = 2e_{yz}$, $e_5 = 2e_{xz}$, $e_6 = 2e_{xy}$, $\tau_1 = \tau_{xx}$, $\tau_2 = \tau_{yy}$, $\tau_3 = \tau_{zz}$, $\tau_4 = \tau_{yz}$, $\tau_5 = \tau_{xz}$, $\tau_6 = \tau_{xy}$.

the internal state of an *isotropic* polarized region, we have to use not only the volume V, the pressure P, the temperature T, and the amounts of substance $n_1, n_2, \ldots$, but we must also introduce the electric field strength $\mathfrak{E}$, the electric polarization $\mathfrak{P}$, the magnetic field strength $\mathfrak{H}$, and the magnetic polarization $\mathfrak{J}$, all these quantities being vectors defined by electrodynamics. The quantities analogous to the negative pressure are the intensive properties $\mathfrak{E}$ and $\mathfrak{H}$, while the analogs of volume are the extensive properties $\mathfrak{P}V$ and $\mathfrak{J}V$ called the total electric moment and the total magnetic moment, respectively.* There are then equations of state of the following types:

$$V = V(T, P, \mathfrak{E}, n_1, n_2, \ldots), \qquad \mathfrak{P} = \mathfrak{P}(T, P, \mathfrak{E}, x_1, x_2, \ldots) \tag{5.6}$$

and

$$V = V(T, P, \mathfrak{H}, n_1, n_2, \ldots), \qquad \mathfrak{J} = \mathfrak{J}(T, P, \mathfrak{H}, x_1, x_2, \ldots). \tag{5.7}$$

In *anisotropic* polarized regions, we have to introduce the strain and stress components, the temperature, the amounts of substance, and the components $\mathfrak{E}_1, \mathfrak{E}_2, \mathfrak{E}_3, \ldots$ of $\mathfrak{E}$, $\mathfrak{P}$, $\mathfrak{H}$, $\mathfrak{J}$ (three of them for each vector). The extensive properties which take the place of $\mathfrak{P}V$ and $\mathfrak{J}V$ are the components $\mathfrak{P}_1 V, \ldots, \mathfrak{J}_3 V$ of the total electric and magnetic moment, respectively. Here, V is the actual volume of the strained element: $V = V_0(1 + e_1 + e_2 + e_3)$.

A peculiar kind of region is an *interfacial layer*. It is defined as that part of the boundary zone between two fluid phases which includes all inhomogeneities. The extent of the interfacial layer is to a certain degree arbitrary. The quantities that take the place of the volume of a simple region are the volume V and the surface area ω of the layer. The quantities to be introduced instead of the negative pressure in a simple region are the negative pressure $-P$ and the interfacial tension σ. For curved interfaces, P is to be interpreted as any value between the two pressures in the two adjacent bulk phases. The area ω of the interfacial layer is an extensive property in the following sense: if one increases the amount of the layer by a certain factor, keeping the thickness, the curvature of the surface, the temperature, and the composition constant, then the area increases by the same factor. It must be admitted, however, that there

* Here, V is the volume of a homogeneous system or of a volume element of a continuous system. The total electric moment of a whole continuous system of volume V is, of course, given by $\int_V \mathfrak{P}\, dV$, as implied by Eq. (6.21) in Section VI.

are difficulties in treating the interfacial layer as a "region": in general, the properties of the layer are not uniquely determined by the internal state variables of the layer itself, but also depend on the state of the two adjacent bulk phases. It is only for equilibrium and for states near equilibrium that the interfacial layer is "autonomous." We shall restrict our treatment to this case.

As we infer from the preceding discussion, there are, for a region of given temperature and chemical content, certain variables describing the mechanical and electromagnetic behavior. They may be divided into two classes. The first class contains quantities analogous to the negative pressure, such as stress components, electric field strength, and interfacial tension. These are intensive properties and will be designated by the symbol λ_i. The second class comprises quantities analogous to the volume, such as strain components multiplied by reference volume, total electric moment, and surface area. These are extensive properties and will be denoted by the symbol l_i. For reasons to be explained later (Section VII), the λ_i and l_i are called *work coefficients* and *work coordinates*, respectively.

In Table I we summarize the work coefficients and work coordinates to be used for the different types of region considered. It is important to note that there is for each work coefficient a conjugate work coordinate. Such pairs of conjugate variables are $-P$ and V, τ_1 and $V_0 e_1$, σ and ω, $\mathfrak{E}_1$ and $\mathfrak{P}_1 V$, and so on.

TABLE I

WORK COEFFICIENTS λ_i AND WORK COORDINATES l_i FOR SPECIAL CASES

Kind of region	λ_i	l_i
Simple region	$-P$	V
Anisotropic region	$\tau_1, \ldots, \tau_6$	$V_0 e_1, \ldots, V_0 e_6$
Interfacial layer	$-P, \sigma$	V, ω
Isotropic region polarized by electric field	$-P, \mathfrak{E}$	$V, \mathfrak{P}V$
Isotropic region polarized by magnetic field	$-P, \mathfrak{H}$	$V, \mathfrak{I}V$
Anisotropic region polarized by electromagnetic field	$\tau_1, \ldots, \tau_6$ $\mathfrak{E}_1, \mathfrak{E}_2, \mathfrak{E}_3$, $\mathfrak{H}_1, \mathfrak{H}_2, \mathfrak{H}_3$	$V_0 e_1, \ldots, V_0 e_6$, $\mathfrak{P}_1 V, \mathfrak{P}_2 V, \mathfrak{P}_3 V$, $\mathfrak{I}_1 V, \mathfrak{I}_2 V, \mathfrak{I}_3 V$

There may be situations which can no longer be described by macroscopic variables alone. Thus, a region with turbulent flow or a reaction zone in a burning gas mixture will possibly be outside the scheme of description given above. In this case, there will be no definite equation of state, and even the definitions of quantities such as temperature will become difficult or impossible. Then, of course, we leave the province of thermodynamics.

VI. Specific, Molar, and Partial Molar Quantities, and Generalized Densities

Let us consider a single region of mass m, amount of substance n, and volume V. Let us denote any extensive property by Z. We then define

$$\text{specific quantity:} \qquad \tilde{Z} \equiv Z/m, \tag{6.1}$$

$$\text{molar quantity:} \qquad \bar{Z} \equiv Z/n, \tag{6.2}$$

$$\text{generalized density:} \qquad \hat{Z} \equiv Z/V. \tag{6.3}$$

Examples of specific quantities are: mass fraction χ_k of substance k, specific volume $\tilde{V}$, and specific electric moment $\mathfrak{P}\tilde{V}$, the two last expressions being specific work coordinates. As examples of molar quantities, we mention: mole fraction x_k of substance k, molar volume $\bar{V}$, and molar electric moment $\mathfrak{P}\bar{V}$, where the two last quantities are molar work coordinates. Finally, the following examples of generalized densities should be noted: density ϱ, mass concentration ϱ_k or molarity c_k of substance k, and electric polarization $\mathfrak{P}$, which represents a density of a work coordinate.

It is obvious that all these quantities are intensive properties as defined in Section II. But it should be realized that in many thermodynamic relations specific and molar quantities as well as generalized densities behave like the extensive properties to which they are connected. Thus, it may be expedient occasionally to group together the above-mentioned quantities and the extensive properties, while the remaining "proper" intensive properties are quantities such as pressure, temperature, and electric field strength (see Ulich and Jost (1970)).

Let us write the extensive property Z as a function of temperature T, work coefficients λ_i, and amounts of substance n_k:

$$Z = Z(T, \lambda_i, n_k). \tag{6.4}$$

We define

$$Z_k \equiv (\partial Z/\partial n_k)_{T,\lambda_i,n_j}, \tag{6.5}$$

where the subscript n_j means that all amounts of substance are to be held constant except n_k. The intensive property Z_k is called the *partial molar quantity* of substance k in the region considered. The quantity Z_k, as well as $\tilde{Z}$, $\bar{Z}$, and $\hat{Z}$, depends on T, λ_i, and composition (e.g., mole fractions).

For simple regions, we have

$$Z_k \equiv (\partial Z/\partial n_k)_{T,P,n_j}, \tag{6.6}$$

P being the pressure. The quantity defined by Eq. (6.6) was already introduced by Lewis and Randall (1921). In simple regions, Z_k depends on T, P, and composition.

Since the extensive property Z is a homogeneous function of first degree in the amounts of substance n_k (see Section II), we obtain from Euler's theorem and Eq. (6.5)

$$Z = \sum_k n_k Z_k. \tag{6.7}$$

Using Eqs. (3.10), (3.12), (6.2), and (6.3), we find

$$\bar{Z} = \sum_k x_k Z_k, \qquad \hat{Z} = \sum_k c_k Z_k. \tag{6.8}$$

It follows from Eq. (6.8) that for a one-component system the partial molar function coincides with the molar quantity.

In particular, there results for $Z = V$, owing to Eqs. (6.7) and (6.8),

$$\sum_k n_k V_k = V, \qquad \sum_k x_k V_k = \bar{V}, \qquad \sum_k c_k V_k = 1, \tag{6.9}$$

where V_k is the partial molar volume of substance k.

The total differential of the expression (6.4) is

$$dZ = \frac{\partial Z}{\partial T}\, dT + \sum_i \frac{\partial Z}{\partial \lambda_i}\, d\lambda_i + \sum_k Z_k\, dn_k \tag{6.10}$$

where Eq. (6.5) has been used. From Eq. (6.7), we deduce

$$dZ = \sum_k Z_k\, dn_k + \sum_k n_k\, dZ_k. \tag{6.11}$$

Comparing Eqs. (6.10) and (6.11), we obtain

$$\sum_k n_k\, dZ_k = (\partial Z/\partial T)\, dT + \sum_i (\partial Z/\partial \lambda_i)\, d\lambda_i, \tag{6.12}$$

or, as a special case,

$$\sum_k n_k\, dZ_k = 0 \qquad (T,\ \lambda_i \quad \text{constant}). \tag{6.13}$$

Let us apply Eq. (6.13) to a simple region and let Z be the volume V. Dividing by V and using Eq. (3.12), we derive

$$\sum_k c_k\, dV_k = 0 \qquad (T, P \quad \text{constant}). \tag{6.14}$$

From this and the last equation in (6.9), we find

$$\sum_k V_k\, dc_k = 0 \qquad (T, P \quad \text{constant}), \tag{6.15}$$

a useful relation.

We return to the general case and express the partial molar quantities Z_k in terms of the molar property $\bar{Z}$ and its partial derivatives $\partial\bar{Z}/\partial x_k$ with respect to the mole fractions x_k at fixed values of the temperature and of the work coefficients. If the region contains N substances, there are $N - 1$ independent mole fractions, which will be arbitrarily chosen to be $x_2, x_3, \ldots, x_N$. According to Eq. (3.13), we have

$$x_1 = 1 - x_2 \cdots - x_i \cdots - x_N. \tag{6.16}$$

From Eqs. (6.8) and (6.16), we obtain

$$(\partial\bar{Z}/\partial x_i)_{T,\lambda,x_j} = \sum_{k=1}^{N} x_k(\partial Z_k/\partial x_i)_{T,\lambda,x_j} + Z_i - Z_1 \quad (i = 2, 3, \ldots, N), \tag{6.17}$$

where λ refers to all work coefficients and x_j to all independent mole fractions except x_i. On the other hand, there follows from Eqs. (3.10) and (6.13) that the sum in Eq. (6.17) disappears. Thus, we find

$$(\partial\bar{Z}/\partial x_i)_{T,\lambda,x_j} = Z_i - Z_1 \qquad (i = 2, 3, \ldots, N). \tag{6.18}$$

On substitution of Eqs. (6.16) and (6.18) into Eq. (6.8), we derive

$$Z_1 = \bar{Z} - \sum_{i=2}^{N} x_i(\partial\bar{Z}/\partial x_i)_{T,\lambda,x_j}. \tag{6.19}$$

Since the component 1 can be arbitrarily chosen, there are N relations of the type (6.19). This set of equations, together with Eq. (6.18), is important for certain geometrical constructions and for the derivation of partial molar quantities from experimental values of molar quantities (Haase, 1948).

If Z is an extensive property of a heterogeneous (discontinuous) system composed of several phases, then we have

$$Z = \sum_{\alpha} Z^{\alpha} \tag{6.20}$$

where Z^{α} is the extensive property of the phase α.

For a continuous system, it is expedient to introduce the generalized density $\hat{Z}$ for each volume element.* If we multiply $\hat{Z}$ by the infinitesimal volume dV of the element considered and integrate over the total volume V, we obtain the extensive property Z of the whole system:

$$Z = \int_V \hat{Z}\, dV. \tag{6.21}$$

Here, $\hat{Z}$ may be called a "local generalized density." Examples are: local density, local mass concentration, local molarity, and local electric polarization. Such quantities are of fundamental importance for the description of continuous systems.

VII. Work

Up to this point, all definitions and derivations referred to quantities uniquely related to the state of the system considered. Thus, properties such as volume, pressure, and temperature are said to be functions of state.† The infinitesimal dy of any function of state y is an exact differential. Furthermore, in all the equations studied, it was immaterial whether the system was closed or open: the change of a function of state does not depend on the path of the process considered, and, thus, in particular, is independent of how changes in amounts are brought about.

* The rigorous definition of $\hat{Z}$ is: $\hat{Z} = \lim_{V\to 0} (Z/V)$, all the intensive properties of the system being left unaltered when proceeding to the limit.

† If, for instance, temperature T and pressure P are chosen to be the independent variables, then the volume $V(T, P)$ is said to be a function of state. But of course $P(T, V)$ and $T(P, V)$ may also be called functions of state.

In this section, we shall fix our attention on a quantity called *work* and denoted by W. This quantity, as will be evident on further inspection, is *not* a function of state* and is *not* defined for open systems.† Thus, an element of work done on the system during an infinitesimal change of state will be an inexact differential represented by $đW$, and the discussion will be restricted to closed systems.

If a system or part of a system undergoes a displacement under the action of a force, the work done on the system is the integral over the product of the force and the component of the infinitesimal displacement parallel to the force. This definition, already familiar in mechanics and electrodynamics, will be used here, too. But it should be noted that only work involving an interaction between the system and its surroundings is significant in thermodynamics. Thus, the forces to be considered are macroscopic and do not include forces acting internally between different parts of the system. The forces may be of mechanical, electric, or magnetic origin.

The sign convention is such that work done on the system is positive and work done by the system is negative.

We first discuss *quasistatic work*. This means that the processes related to the work are quasistatic in the sense defined in Section II. Hence, if a system undergoes a deformation, there must be equilibrium between the acting forces and the forces opposing the action. There is not necessarily equilibrium with regard to other processes occurring inside the system. For example, the quasistatic compression or expansion of a gas only implies equality of the pressure exerted by the gas and of the pressure opposing it (thus ruling out effects such as friction), but it does not imply that there is equilibrium with respect to mass transfer or chemical reactions in the interior of the system; in other words: the processes related to the change of volume are quasistatic, while all other changes may be produced by natural processes.

Only quasistatic work that leads to an internal change of state of the system is relevant to thermodynamics. As far as a single region is concerned, it is either work of quasistatic deformation or work of quasistatic polarization in electromagnetic fields. We shall denote an infinitesimal work of this type by $đW_l$.

The best-known example is the quasistatic work of compression or

* This will be demonstrated in general in Section VIII. But for a special case, the direct proof will be given below.

† See Section X.

expansion done on a simple region of pressure P and volume V. From elementary considerations, we may derive that for an infinitesimal change of volume

$$đW_l = -P\,dV. \tag{7.1}$$

This also holds in principle for a volume element of a continuous simple system in a gravitational or centrifugal field, though in this case a mathematically more convenient formulation would be used.

While dV and $d(PV)$ are exact differentials, $P\,dV$ is an inexact differential. This can be shown as follows. Suppose the content of the region to be fixed. (This implies absence of chemical reactions, since open regions were excluded from the very beginning.) Then the volume V is a function of temperature T and pressure P only. Hence,

$$P\,dV = P(\partial V/\partial T)\,dT + P(\partial V/\partial P)\,dP.$$

If $P\,dV$ were an exact differential, the relation

$$\frac{\partial}{\partial P}\left(P\frac{\partial V}{\partial T}\right) = \frac{\partial}{\partial T}\left(P\frac{\partial V}{\partial P}\right)$$

or

$$\left(\frac{\partial V}{\partial T}\right)_P + P\frac{\partial^2 V}{\partial T\,\partial P} = P\frac{\partial^2 V}{\partial P\,\partial T}$$

would hold. But we have

$$\partial^2 V/\partial T\,\partial P = \partial^2 V/\partial P\,\partial T,$$

since V is a function of state. Therefore, the condition can only be fulfilled if in general $(\partial V/\partial T)_P = 0$, and this is not true. Thus, $P\,dV$ is an inexact differential.

Explicit expressions for $đW_l$ may also be obtained in more complicated cases, such as interfacial layers (Guggenheim, 1967), anisotropic regions (Callen, 1961) and polarized regions (Callen, 1961). The derivations of these expressions are lengthy and fall within the scope of mechanics and electrodynamics, and thus will not be given here. The results, however, are remarkably simple. They are analogous to Eq. (7.1) and can be written in the general form

$$đW_l = \sum_i \lambda_i\,dl_i \tag{7.2}$$

where the λ_i and the l_i are the quantities introduced in Section V. It is now evident why we called λ_i and l_i work coefficients and work coordinates, respectively.* Each term in Eq. (7.2) is a work coefficient λ_i multiplied by the differential of the conjugate work coordinate l_i. The appropriate quantities to be inserted follow from Table I. Equation (7.1) is a special case of Eq. (7.2).

During a real change of state (a natural process), the infinitesimal work $đW$ done on the region is not necessarily equal to $đW_l$ given by Eq. (7.2). The compression of a fluid, for example, requires an external pressure exceeding the pressure in the interior of the fluid, and at the same time friction forces must be overcome. Thus, the real work done exceeds the quasistatic work given by Eq. (7.1). There are similar effects in the deformation of anisotropic bodies, in the polarization of matter by electromagnetic fields, and so on. Such phenomena are called "dissipative effects." They include the flow of electric current† from external sources into the region considered, since this may lead to a change of the internal state variables (temperature, etc.). Therefore, we define a quantity W_{diss} called *dissipated work* as the difference between the real work and the quasistatic work done on the region during a change of the *internal* state. If there is, for example, an electric potential difference E involving an electric current of strength I, then the electric work done on the region during the time element dt is $EI\,dt = đW_{\text{diss}}$. Furthermore, there may be changes of the *external* coordinates caused by forces accelerating the region or opposing external fields, such as forces acting against the gravitational field to which the region is exposed. The work due to these forces is said to be *external work* W_a. Thus, we have for any infinitesimal change of state in the region considered

$$đW = đW_l + đW_{\text{diss}} + đW_a. \tag{7.3}$$

Here, $đW_l$ is explicitly given by Eq. (7.2).

* The terms "generalized forces" and "generalized coordinates" often used should not be applied, since these have a different meaning in nonequilibrium thermodynamics and in statistical mechanics.

† Here, a passage of electrons through the system takes place. This may be considered to be an exchange of matter with the surroundings, which is, at first sight, incompatible with a closed system. But the mass of an electron is negligible as compared with the mass of an ion, atom, or molecule, and, furthermore, for a given time interval, the number of electrons entering the system is equal to the number of electrons leaving it. Thus, it is *practically* consistent to talk of a closed system through which electricity passes from external sources.

For a system composed of several regions, the work done on the whole system is not necessarily equal to the sum obtained from Eq. (7.3). Thus, for a heterogeneous system consisting of two phases (volumes V' and V'') under different pressures (P' and P''), the element of quasistatic work of compression or expansion done on the whole system need not equal $-P'\,dV' - P''\,dV''$. This is easily seen if we impose the condition of fixed total volume ($V' + V'' =$ const) and let the volumes of both phases be variable; we then find

$$\int (P'' - P')\,dV' \neq 0$$

while the work done on the whole system is zero. This case, though exceptional, must be taken into account in principle. On the other hand, there may be work done on a heterogeneous system which does not appear in the single terms referring to the constituent phases. An important example is the electric work done on a galvanic cell.*

An extreme case is the experiment of Gay-Lussac and Joule (1807). Here, a gas is allowed to expand suddenly into a vacuum, the whole experimental setup being surrounded by rigid walls. In such a process, there is no work done on or by the gas (as far as the interaction of the system with the surroundings is concerned) though the volume of the gas changes. Obviously, this process cannot be described by differential equations, since only the initial and the final situations are completely specified states, while the intermediate situations are ill-defined (cf. Section V).

Let us write for the work done on a system composed of several regions during an infinitesimal change of state

$$đW = đW_l + đW_a + đW^*. \tag{7.4}$$

The first term, $đW_l$, is the quasistatic work of deformation or electric and magnetic polarization done on the whole system. For a heterogeneous system, this expression can be derived from Eq. (7.2) by a summation over all phases (superscript α):

$$đW_l = \sum_{\alpha} \sum_{i} \lambda_i^{\alpha}\, dl_i^{\alpha} \tag{7.5}$$

provided the dl_i^{α} refer to those changes in the work coordinates of the several phases that contribute to the work on the whole system. (On

* For a reversible galvanic cell, i.e., for a cell in which nothing happens at zero electric current, the work done on the cell may even be quasi static (see Section XXI).

the other hand, if we suppose $l_i^\alpha = \text{const}$ for each phase, then the term $đW_l$ will always disappear.) The second term, $đW_a$, denotes the external work done on the whole system. The last term, $đW^*$, refers to any other kind of work, such as work of friction and electric work done on the whole system. We briefly call W^* the *excess work*.

We stress that, according to our definitions, the work coefficients λ_i and work coordinates l_i do not include quantities such as external forces and external coordinates. Consequently, there may be work done on a system though all work coordinates have fixed values. This situation, strange as it may appear at first sight, is a logical consequence of our definitions. The reason for this procedure is the fact that, as will be obvious later, the λ_i and l_i occur in important thermodynamic functions while quantities such as external forces and external coordinates do not.

We repeat that all the formulas given in this section refer to *closed* regions or systems.

VIII. First Law of Thermodynamics (Energy and Heat)

Let us consider a system in internal equilibrium completely surrounded by a wall. If no change can be produced in the system except by work done on the system, then the wall is called a *thermally insulating wall.* Any system enclosed by such a wall is said to be *thermally insulated* and any change of state in the system is called an *adiabatic process.*

We can measure the work W_{ad} required to bring a system from any initial state to any final state in an adiabatic path. Experience then tells us that there is an important general principle:

The work done on a thermally insulated system is independent of the source of work and of the path through which the system passes from the initial to the final state.

This statement is the first part of the *first law of thermodynamics.* The experimental basis on which it rests is largely due to Joule (1840–1845).

We can now introduce a function of state called *energy E*. The increase ΔE of the energy due to a change of state I $\rightarrow$ II is given by

$$\Delta E = E_{\text{II}} - E_{\text{I}} \equiv W_{\text{ad}}, \tag{8.1}$$

E_{I} and E_{II} being the energies of the system in the initial state I and in the final state II, respectively. Since any change of state can be made to occur through an adiabatic path, at least in one direction, the change of energy can be determined for any process. Thus, the energy of a system is defined, apart from an arbitratry additive constant without physical significance.

These rigorous and clear formulations are due to Carathéodory (1909).

We now consider a nonadiabatic process in a closed system, that is, any change of state in a system surrounded by a wall which only prevents the passage of matter. If we measure the work W done on the system and compare W to the increase of energy ΔE given by Eq. (8.1), we discover another important general principle:

The work done on an arbitrary closed system is not necessarily equal to the increase of the energy of the system. In particular, there may be a change of energy without any work, even for a system in internal equilibrium initially.

This statement is the second part of the *first law of thermodynamics*, implicititly contained in the investigations of Joule and others.

Hence, there are external agencies different from work which lead to a change of state and thus to a change of energy in the system considered. Therefore, we define a quantity

$$Q \equiv \Delta E - W \qquad \text{(closed system)} \tag{8.2}$$

called the *heat* supplied to the system or absorbed (received) by the system. If $Q < 0$, then we say that heat is given off by the system. The case $Q \neq 0$ may be referred to as a situation in which there is a "flow of heat" or an "exchange of heat" between the system and its surroundings. The case $Q = 0$ is that of an adiabatic process.

This unambiguous definition of heat is due to Born (1921).

According to the previous discussions, we may have, for a given change of state, $Q = 0$, $W \neq 0$ or $Q \neq 0$, $W = 0$, or anything between. It follows that neither Q nor W are functions of state.* Thus, the infinitesimals $đQ$ and $đW$ are inexact differentials, while dE is an exact differential. So we may write for an infinitesimal change of state

$$dE = đQ + đW \qquad \text{(closed system)}. \tag{8.3}$$

For a noninfinitesimal process, we obtain

$$\Delta E = Q + W \qquad \text{(closed system)}, \tag{8.4}$$

in agreement with Eq. (8.2).

* There is no such thing as the "heat of a system" or the "work of a system."

A process may be either due to a change of external coordinates or to a change of internal state variables (see Section II). In the first case, there are two alternatives again: a change of macroscopic velocities or of position coordinates in external fields. If only the macroscopic velocities alter, then the change of energy of the system defined by Eq. (8.1) is called a change of *kinetic energy* E_{kin}. If, on the other hand, there is an energy change due to an alteration of the position coordinates with respect to an external field, then we refer to this as a change of *potential energy* E_{pot}. It can be easily verified that the quantities E_{kin} and E_{pot} thus defined coincide with the corresponding quantities used in classical mechanics. Finally, the energy change due to alterations of the internal state of the system is said to be a change of *internal energy* U. Experience tells us that it is always possible to write

$$E = E_{\text{kin}} + E_{\text{pot}} + U \tag{8.5}$$

provided the external field is a conservative field.* Here, E_{kin} depends on the macroscopic velocities, E_{pot} on the position coordinates in the external field, and U on the internal state variables. As can be verified from experiments made on a number of systems of varying chemical content, each term in Eq. (8.5) is additive in the sense that the energy of the whole system is the sum of, or the integral over, all the energies of the constituent phases or volume elements, so that Eq. (6.20) or (6.21) is applicable. In particular, the internal energy is an extensive property. Thus, for a given region, we may introduce the molar internal energy $\bar{U}$ and the partial molar internal energy of substance k, U_k.

Equations (8.4) and (8.5) combine to give

$$\Delta E = \Delta E_{\text{kin}} + \Delta E_{\text{pot}} + \Delta U = Q + W \qquad \text{(closed system).} \tag{8.6}$$

For the purely "mechanical" case, we have $\Delta U = 0$, $Q = 0$, $W = W_a$ (external work). On the other hand, a change of the internal state of the system is given by the conditions $\Delta E_{\text{kin}} = 0$, $\Delta E_{\text{pot}} = 0$, $W_a = 0$. Thus, there results from Eqs. (7.4) and (8.6)

$$\Delta U = Q + W = Q + W_i + W^* \qquad \text{(closed system, internal change).} \tag{8.7}$$

Here, W_i is the quasistatic work of deformation and polarization done

* A field is said to be conservative if, for the force **K** acting in any volume element, the condition curl **K** = 0 holds.

on the system and W^* the excess work. If all the work coordinates (l_i^α) have fixed values ($W_l = 0$), we obtain

$$\Delta U = Q + W^* \qquad \text{(closed system, internal change, } l_i^\alpha \text{ constant).} \qquad (8.8)$$

In particular, for a simple phase of pressure P and volume V, the condition of fixed work coordinates is equivalent to the statement $V = \text{const}$ ("isochoric process").

An *isolated system* has no interactions with the surroundings. Therefore, such a system may be described as a closed system with no work done ($W = 0$) and no heat supplied ($Q = 0$). A classification of systems by means of the concepts of "work" and "heat" is shown in Table II. (We shall revert to the subject of open systems in Section X.)

TABLE II

CLASSIFICATION OF SYSTEMS ACCORDING TO THE TYPE OF INTERACTION WITH THE SURROUNDINGS

Kind of system	Exchange of matter	Flow of heat	Work
Open	+	+	+
Closed	−	+	+
Thermally insulated	−	−	+
Isolated	−	−	−

For an isolated system, there follows from Eq. (8.6)

$$E = E_{\text{kin}} + E_{\text{pot}} + U = \text{const} \qquad \text{(isolated system).} \qquad (8.9)$$

This formulation is known as the "law of conservation of energy." Early investigations on the first law of thermodynamics usually referred to this statement (Carnot, 1832; Helmholtz, 1847; Clausius, 1850).

The experiment of Gay-Lussac and Joule mentioned on p. 23 is carried through adiabatically. Since no work is done in this experiment, it belongs to the class of internal changes in isolated systems where, in view of Eq. (8.9), the internal energy U remains constant. Such a process is called "isoenergetic."

Work, heat, and energy have the same dimension by virtue of the definitions (8.1) and (8.2). The SI unit is the joule (J) defined by J $\equiv$ N m and equal to the product of the coulomb (C) and the volt (V). Many authors, however, still prefer the erg (erg $\equiv$ dyn cm) or the thermochemical calorie (cal). The units mentioned are interrelated by

$$\mathrm{J} = \mathrm{N\,m} = \mathrm{C\,V} = 10^7 \quad \mathrm{erg}, \tag{8.10}$$

$$\mathrm{cal} = 4.184 \quad \mathrm{J}. \tag{8.11}$$

The unit to be used preferably for the molar and partial molar internal energy is the joule divided by the mole (J $\mathrm{mol^{-1}}$).

The unit to be recommended for the gas constant R (see Section V) is the joule divided by the product of the kelvin and the mole (J $\mathrm{K^{-1}}$ $\mathrm{mol^{-1}}$). From Eqs. (5.2) and (8.10), we derive the approximate value of R expressed in this unit:

$$R = 8.314 \quad \mathrm{J\,K^{-1}\,mol^{-1}}. \tag{8.12}$$

IX. Enthalpy

The *enthalpy* H^α of a region (superscript α) is defined by

$$H^\alpha \equiv U^\alpha - \sum_i \lambda_i^\alpha l_i^\alpha \tag{9.1}$$

where U^α represents the internal energy and the λ_i^α and l_i^α are the work coefficients and work coordinates, respectively. The enthalpy H of a heterogeneous system is given by

$$H \equiv \sum_\alpha H^\alpha. \tag{9.2}$$

The enthalpy is also called "heat content" or "heat function."

For the sake of brevity, we shall either discuss single regions or heterogeneous systems, but we shall revert to the subject of continuous systems at the end of the section.

The enthalpy, like the internal energy, is only defined apart from an arbitrary additive constant. This indefiniteness is removed by assigning the value zero to the enthalpy of each element in its most stable form at 25°C and 1 atm. By this arbitrary convention, the value of the enthalpy, and incidentally, that of the internal energy, of any other state of matter is uniquely fixed.

Enthalpy is an extensive property having the dimension of energy. Thus we may introduce the molar enthalpy of a region ($\bar{H}^\alpha$) and the partial molar enthalpy of substance k in the region considered (H_k^α). Both quantities have the dimension of molar energy.

We consider two special cases of Eq. (9.1), using Table I (p. 15). The first example refers to a simple region (nonpolarized isotropic region) of pressure P^α and volume V^α:

$$H^\alpha = U^\alpha + P^\alpha V^\alpha. \tag{9.3}$$

The second example concerns a nonpolarized anisotropic region:

$$H^\alpha = U^\alpha - V_0^\alpha \sum_{i=1}^{6} \tau_i^\alpha e_i^\alpha. \tag{9.4}$$

Here, V_0^α, τ_i^α, and e_i^α denote the reference volume, the stress components, and the strain components, respectively.

Equation (9.3) is the familiar definition of enthalpy as introduced by Gibbs in 1875 ("heat function for constant pressure"). Obviously, Eq. (9.3) cannot be applied to anisotropic bodies, since here the concept of a single pressure P is meaningless. Furthermore, the general definition (9.1) has the advantage of leading to simple physical behavior of the function H.

From Eqs. (9.1) and (9.2), we obtain for a heterogeneous system

$$H = U - \sum_\alpha \sum_i \lambda_i^\alpha l_i^\alpha, \tag{9.5}$$

where

$$U = \sum_\alpha U^\alpha \tag{9.6}$$

is the internal energy of the whole system.

In particular, for a simple heterogeneous system of uniform pressure P, we conclude from Eqs. (9.2), (9.3), and (9.6)

$$H = U + PV, \tag{9.7}$$

where

$$V = \sum_\alpha V^\alpha \tag{9.8}$$

is the total volume of the system.

For any infinitesimal change of the internal state of a heterogeneous system, we have on account of Eq. (9.5)

$$dH = dU - \sum_\alpha \sum_i (\lambda_i^\alpha \, dl_i^\alpha + l_i^\alpha \, d\lambda_i^\alpha). \tag{9.9}$$

If the system is closed,[†] we find[‡] from Eqs. (7.5), (8.7), and (9.9)

$$dH = đQ + đW^* - \sum_\alpha \sum_i l_i^\alpha \, d\lambda_i^\alpha \quad \text{(closed system, internal change).} \quad (9.10)$$

Here, $đQ$ and $đW^*$ are the heat received by the system and the excess work done on the system, respectively. When all the work coefficients λ_i^α have fixed values, we derive from Eq. (9.10)

$$\Delta H = Q + W^* \quad \text{(closed system, internal change, } \lambda_i^\alpha \text{ constant),} \quad (9.11)$$

in complete analogy to Eq. (8.8).

If we consider a simple system of uniform pressure and apply Eq. (9.7), then Eqs. (9.9) and (9.10) reduce to the following relations:

$$dH = dU + P\,dV + V\,dP \quad \text{(simple system),} \quad (9.12)$$

$$dH = đQ + đW^* + V\,dP \quad \text{(closed simple system, internal change)} \quad (9.13)$$

while Eq. (9.11) is left unchanged except that the condition $\lambda_i^\alpha = \text{const}$ now takes the simple form $P = \text{const}$ ("isobaric process").

For adiabatic changes ($Q = 0$), it follows immediately from Eq. (9.11) that

$$\Delta H = W^* \quad \text{(adiabatic change, } \lambda_i^\alpha \text{ constant).} \quad (9.14)$$

This formula is the basis for *calorimetry*. Let us illustrate this by two examples.

We first consider a system of fixed chemical content and of fixed work coefficients. We want to know the increase of enthalpy $\Delta H = H_{\text{II}} - H_{\text{I}}$ due to a change of the (empirical) temperature from the initial value Θ_{I} to the final value Θ_{II} (which are measurable quantities according to Section IV). We then simply determine the electric work W^* required to raise the temperature from Θ_{I} to Θ_{II}, the system being thermally insulated, and apply Eq. (9.14). We thus obtain experimental values of $H_{\text{II}} - H_{\text{I}} = H(\Theta_{\text{II}}) - H(\Theta_{\text{I}})$.

The second example refers to the calorimetric determination of a

[†] It will be shown in the next section that the definitions of energy and enthalpy continue to hold for open systems. Thus, the relations (9.1)–(9.9), (9.15), and (9.16) are valid for both closed and open systems (or regions).

[‡] Equations (9.10) and (9.11) strictly hold only if all the changes of the work coordinates of the constituent phases contribute to the work done on the whole system. But the cases for which this is not true are rather exceptional (see Section VII).

"heat of reaction," that is, of the increase of enthalpy $H_{II}(\Theta) - H_I(\Theta)$ in a simple system in which a chemical reaction takes place at a given temperature (Θ) and a given pressure. Let us suppose that when the isobaric reaction occurs in a thermally insulated system the temperature diminishes from Θ to Θ', the enthalpy changing from its initial value H_I to the intermediate value H_{III}. In this first step of the experiment, there is no excess work ($W^* = 0$). Thus, we find from Eq. (9.14)

$$H_{III}(\Theta') - H_I(\Theta) = 0.$$

The second step of the calorimetric experiment consists in raising the temperature of the system from Θ' to the initial temperature Θ by electric work W^* done isobarically and adiabatically, thus leading to the final state of the reacting system. Hence, by virtue of Eq. (9.14)

$$H_{II}(\Theta) - H_{III}(\Theta') = W^*.$$

Combining the two relations, we obtain

$$H_{II}(\Theta) - H_I(\Theta) = W^*.$$

Therefore, the "heat of reaction" is measurable.

Let us now discuss an experiment carried out by Joule and Thomson (Lord Kelvin) (1852), often called the *Joule–Thomson throttling experiment*. Here, a stream of gas in a thermally insulated vessel is forced through a plug and the empirical temperatures and the pressures on both sides of the plug are measured. Let us suppose the system to be in a stationary state such that, in a given time, the amount of gas entering the plug (volume V_I, pressure P_I, empirical temperature Θ_I, internal energy U_I) equals the amount of gas leaving the plug (V_{II}, P_{II}, Θ_{II}, U_{II}). Owing to Eq. (8.7), the work done on the system during an adiabatic internal change in a fixed amount of gas is equal to the increase of internal energy $U_{II} - U_I$ and thus independent of the path. Hence, the work done on the gas may be simply computed for the limiting case of quasi-static deformation. Therefore, we have, in view of Eqs. (7.1) and (9.3),

$$U_{II} - U_I = P_I V_I - P_{II} V_{II},$$

$$H_{II} = H_I,$$

or

$$H(\Theta_I, P_I) = H(\Theta_{II}, P_{II}).$$

Since we have already shown how to measure the difference $H(\Theta_{II}) - H(\Theta_{I})$ at fixed pressure, we can now determine the experimental value of

$$H(\Theta, P_{II}) - H(\Theta, P_{I})$$

for a given value of the empirical temperature Θ. The general result of measurements of this kind is

$$H(\Theta, P_{II}) - H(\Theta, P_{I}) = \text{const}(P_{II} - P_{I}) + \text{higher terms},$$

where the "higher terms" refer to quadratic and cubic expressions in the pressures. Thus, terms such as $\ln(P_{II}/P_{I})$ are safely excluded and the linear term is the leading one. This result will be used in Section XVII.

We now turn to *isolated systems*. If we only consider internal changes, then the internal energy is constant. But there is no "law of conservation of enthalpy," for we cannot leave unchanged the work coefficients in a process occurring in the interior of an isolated system, and thus Eq. (9.14), which would lead to $\Delta H = 0$ for $W^* = 0$, is not applicable. A simple isolated system, for example, must have a fixed total volume V and variable pressures if processes take place inside the system, e.g., chemical reactions or phase transitions. Let us suppose the pressure P to be uniform. Then, using Eq. (9.13) and the conditions $V = \text{const}$, $Q = 0$, $W^* = 0$, we obtain

$$\Delta H = V \, \Delta P,$$

where ΔP is the increase of pressure due to the change of state.

For a *continuous system*, the sums over the phases occurring in Eqs. (9.2), (9.5), (9.6), and (9.8)–(9.10) are to be replaced by integrals of the type (6.21). In these integrals, there appears the local enthalpy density defined by Eqs. (6.3) and (9.1):

$$\hat{H} = \hat{U} - \sum_i \lambda_i \hat{l}_i. \tag{9.15}$$

The symbols $\hat{U}$ and $\hat{l}_i$ denote the local densities of internal energy and of the work coordinates, respectively. For a simple volume element (i.e., a volume element of a nonpolarized, isotropic, continuous system), Eq. (9.15) reduces to

$$\hat{H} = \hat{U} + P. \tag{9.16}$$

X. Open Systems

In an open system, there is an exchange of matter with the surroundings. We shall treat both homogeneous and heterogeneous open systems. Continuous systems will not be dealt with, since the mathematics required are much more complicated, though the results obtained do not differ in principle from those to be derived here (cf. Haase (1969)).

Let us consider a single open phase (superscript α) which is either a homogeneous system or part of a heterogeneous system. Then the amount of substance n_k^α of species k may vary as a consequence of (a) chemical reactions in the phase (subscript r), (b) transport of matter to or from other phases of the system (subscript t), (c) exchange of matter with the surroundings of the whole system (subscript e).

Thus, we may write for an infinitesimal change of state

$$dn_k^\alpha = d_\mathrm{r}n_k^\alpha + d_\mathrm{t}n_k^\alpha + d_\mathrm{e}n_k^\alpha \tag{10.1}$$

or

$$dn_k^\alpha = d_\mathrm{r}n_k^\alpha + d_\mathrm{a}n_k^\alpha, \tag{10.2}$$

where

$$d_\mathrm{a}n_k^\alpha \equiv d_\mathrm{t}n_k^\alpha + d_\mathrm{e}n_k^\alpha \tag{10.3}$$

or

$$dn_k^\alpha = d_\mathrm{i}n_k^\alpha + d_\mathrm{e}n_k^\alpha, \tag{10.4}$$

where

$$d_\mathrm{i}n_k^\alpha \equiv d_\mathrm{r}n_k^\alpha + d_\mathrm{t}n_k^\alpha. \tag{10.5}$$

Here, only dn_k^α is an exact differential.* The subscript a refers to "external" changes as far as the phase α is concerned. The subscript i relates to "internal" changes as far as the whole system is concerned. For a homogeneous system, we may omit the superscript α, and we have $d_\mathrm{t}n_k = 0$.

The amount of substance of any species k present in the whole system does not change by transfer of matter between neighboring phases of the system:

$$\sum_\alpha d_\mathrm{t}n_k^\alpha = 0. \tag{10.6}$$

* It is unnecessary to write $đ_\mathrm{r}n_k^\alpha$, $đ_\mathrm{t}n_k^\alpha$, and so on, since subscripts such as r and t already indicate that the infinitesimal is an inexact differential. It is obvious that a given change of n_k^α can be brought about by, say, chemical reactions or by exchange of matter with the surroundings or by a combination of both.

Here, the sum is to be extended over all phases. Multiplying Eq. (10.6) by the molar mass M_k of substance k and summing over all species, we obtain

$$\sum_\alpha \sum_k M_k \, d_t n_k^\alpha = 0. \tag{10.7}$$

This is analogous to the relation

$$\sum_k M_k \, d_r n_k^\alpha = 0, \tag{10.8}$$

which expresses the fact that the total mass of a phase cannot be changed by chemical reactions within this phase. (A heterogeneous reaction may always be regarded as a combination of a homogeneous reaction and a transport of matter between different phases.) Multiplying Eq. (10.1) by M_k, summing over all species and phases, and comparing with Eqs. (10.7) and (10.8), we find

$$dm = \sum_\alpha \sum_k M_k \, dn_k^\alpha = \sum_\alpha \sum_k M_k \, d_e n_k^\alpha, \tag{10.9}$$

where m is the total mass of the system. Since for a closed system (whose constituent phases may still be open)

$$d_e n_k^\alpha = 0 \qquad \text{(closed system)}, \tag{10.10}$$

we derive from Eq. (10.9)

$$dm = 0 \qquad \text{(closed system)}. \tag{10.11}$$

Equation (10.11) may be said to express the "law of conservation of mass."

As explained in the beginning of Section VII, any function of state has a definite meaning both for closed and open systems. In particular, the change of energy in an open system can be determined by means of Eq. (8.1) by measurements on a number of systems of different chemical content. Thus, the change of enthalpy is equally well defined. Therefore, we may unambiguously refer to the partial molar enthalpy H_k^α of substance k in an open phase α. (This quantity will be used later.)

On the other hand, work and heat are not functions of state; they essentially relate to interactions of the system with its surroundings. Hence, these quantities as defined heretofore have no definite meaning for open systems (cf. Defay (1929). See also Haase (1956a).)

There are, however, important exceptions: the external work W_a and the dissipated work W_{diss} can always be calculated from the external actions. If, for instance, there is a flow of electricity coming from external sources, then the work done on the open system is still the product of the electrical potential difference and the electric charge passing the system during the time interval considered.

But the total work done on the open system remains indefinite. We shall not try to extend the general concept of work to open systems, since such an extension would not serve any useful purpose. The only thing we shall need is the work w^α that *would* be done on an open phase α if the phase *were* closed. This quantity is explicitly given by Eqs. (7.2) and (7.3):

$$đw^\alpha = đW_a{}^\alpha + đW_{\text{diss}}^\alpha + \sum_i \lambda_i{}^\alpha \, dl_i{}^\alpha. \qquad (10.12)$$

Here, the work coefficients λ_i^α and work coordinates l_i^α have a definite meaning, though the sum in Eq. (10.12) is no longer to be interpreted as the actual quasistatic work of deformation and polarization except in the trivial case of a closed phase.

We shall now redefine the "heat" for open systems, since this concept is required, especially in nonequilibrium thermodynamics. (The simultaneous occurrence of heat flow and flow of matter is a familiar problem in the theory of transport processes.)

We call the quantity

$$đQ^\alpha \equiv dE^\alpha - đw^\alpha - \sum_k H_k{}^\alpha \, d_a n_k{}^\alpha \qquad (10.13)$$

the *heat* supplied to the open phase α during an infinitesimal change of state. The quantity E^α is the energy of the phase and $đw^\alpha$ is defined by Eq. (10.12).

For $d_a n_k{}^\alpha = 0$ (closed phase), w^α becomes the work actually done, and we recover Eq. (8.2), the definition of heat for a closed system.

For an internal change of state, we derive from Eqs. (10.12) and (10.13)

$$đQ^\alpha = dU^\alpha - đW_{\text{diss}}^\alpha - \sum_i \lambda_i{}^\alpha \, dl_i{}^\alpha - \sum_k H_k{}^\alpha \, d_a n_k{}^\alpha \qquad (10.14)$$

(internal change)

where U^α is the internal energy of the phase.

It is instructive to consider the enthalpy H^α of an open phase. Owing to Eqs. (6.5), (6.10), and (10.2), we may write

$$dH^\alpha = \frac{\partial H^\alpha}{\partial T^\alpha} dT^\alpha + \sum_i \frac{\partial H^\alpha}{\partial \lambda_i^\alpha} d\lambda_i^\alpha + \sum_k H_k^\alpha (d_r n_k^\alpha + d_a n_k^\alpha) \quad (10.14a)$$

where T^α is the temperature of the phase. For an internal change in a closed phase, we obtain with the help of Eqs. (7.2), (7.3), (8.3), and (9.1)

$$dH^\alpha = đQ^\alpha + đW^\alpha_{\text{diss}} - \sum_i l_i^\alpha \, d\lambda_i^\alpha \quad (10.14b)$$

(closed phase, internal change).

For the closed phase, the term $\sum_k H_k^\alpha \, d_a n_k^\alpha$ will disappear in Eq. (10.14a). We thus conjecture that it is reasonable to add this term to the right-hand side of Eq. (10.14b) in order to extend this relation to open phases:

$$dH^\alpha = đQ^\alpha + đW^\alpha_{\text{diss}} - \sum_i l_i^\alpha \, d\lambda_i^\alpha + \sum_k H_k^\alpha \, d_a n_k^\alpha \quad (10.14c)$$

(open phase, internal change).

If we insert Eq. (9.1) into Eq. (10.14c), we recover Eq. (10.14). Hence, the heat implicitly defined by Eq. (10.14c) coincides with the heat in Eq. (10.14).

We now consider a heterogeneous open system. Here, the heat $đQ^\alpha$ absorbed by a phase is still given by Eqs. (10.12) and (10.13). We split it into two terms:

$$đQ^\alpha = d_i Q^\alpha + d_e Q^\alpha. \quad (10.15)$$

The first term refers to the heat received from neighboring phases and the second term to the heat absorbed from the surroundings of the whole system [cf. Eq. (10.4)]. This decomposition is uniquely determined by the further convention

$$đQ = \sum_\alpha d_e Q^\alpha, \quad (10.16)$$

where $đQ$ is the heat supplied to the whole system from the surroundings.

Combining Eqs. (10.12), (10.13), (10.15), and (10.16), we find

$$đQ + \sum_\alpha d_i Q^\alpha = dE - đW_a - đW_{\text{diss}} - \sum_\alpha \sum_i \lambda_i^\alpha \, dl_i^\alpha - \sum_\alpha \sum_k H_k^\alpha \, d_a n_k^\alpha. \quad (10.17)$$

Here, E denotes the total energy of the system, while W_a and W_{diss} denote the external and dissipated work, respectively, done on the whole system.

If we stipulate that the heterogeneous system is closed (the single phases still remaining open), then we conclude from Eqs (7.3)–(7.5) and (8.3)

$$đQ = dE - \sum_{\alpha} \sum_{i} \lambda_i^\alpha \, dl_i^\alpha - đW_a - đW_{\text{diss}} \qquad \text{(closed system)} \qquad (10.18)$$

provided we exclude all cases where the work done on the total system cannot be decomposed into contributions due to single phases (see Section VII). Remembering that, according to Eqs. (10.3) and (10.10) for a closed system,

$$d_a n_k^\alpha = d_t n_k^\alpha \qquad \text{(closed system)} \qquad (10.19)$$

and inserting Eq. (10.18) into Eq. (10.17), we obtain

$$\sum_{\alpha} d_i Q^\alpha + \sum_{\alpha} \sum_{k} H_k^\alpha \, d_t n_k^\alpha = 0 \qquad \text{(closed system)}, \qquad (10.20)$$

where the condition (10.6) is to be imposed. In particular, if there are only two phases (superscripts $'$ and $''$), there results from Eqs. (10.6) and (10.20)

$$d_i Q' + d_i Q'' = \sum_{k} (H_k'' - H_k') \, d_t n_k' \qquad (10.21)$$

$$\text{(two open phases in a closed system).}$$

If each of the phases is closed ($d_t n_k' = 0$), there follows from Eq. (10.21)

$$d_i Q' + d_i Q'' = 0 \qquad \text{(two closed phases).} \qquad (10.22)$$

Thus, during the flow of heat between two closed phases, the heat received by one phase is equal to the heat given off by the other phase, irrespective of the heat absorbed by the two phases from the surroundings of the system. The result (10.22), of course, could have been obtained directly by applying Eqs. (8.3), (10.15), and (10.16) to a system consisting of two closed phases.

Finally, it should be mentioned that there are several other definitions of "heat" for open systems, given implicitly or explicitly in the literature. These definitions differ from the convention (10.13) by the replacement of H_k^α by another intensive property including the value zero. Thermodynamics of irreversible processes can be developed consistently on the basis of these definitions, too. But many simple properties shown by our quantity $đQ^\alpha$ are then lost (Haase, 1969).

XI. Second Law of Thermodynamics (Entropy and Thermodynamic Temperature)

By the zeroth and first laws of thermodynamics, two functions of state were introduced, the empirical temperature and the energy, respectively. We are now ready to formulate the *second law of thermodynamics*. This principle leads to two new fundamental quantities (entropy and thermodynamic temperature).

The second law is to be justified ultimately by comparison of its consequences with experimental facts. In this respect, it is similar to the other thermodynamic laws. But the connection between experience and mathematical statements is not as obvious as it is in the other principles. Thus, the following formulation will appear rather abstract at first sight, but it contains the necessary and sufficient number of statements:

There exists a function of state called *entropy* having the following properties:

(a) The entropy S^α of any region α depends on all the variables describing the internal state, such as internal energy U^α (or empirical temperature), work coordinates l_i^α (or work coefficients λ_i^α), and amounts (masses or amounts of substance) of the species present.

(b) For any infinitesimal change in a region of fixed chemical content, the relation

$$T^\alpha \, dS^\alpha = dU^\alpha - \sum_i \lambda_i^\alpha \, dl_i^\alpha \tag{11.1}$$

holds, where T^α is a positive intensive property called *thermodynamic temperature* which is a universal function of the empirical temperature of the region.

(c) The entropy is an extensive property.

(d) The increase ΔS of entropy during a change of state in any system can be decomposed into two parts:

$$\Delta S = \Delta_e S + \Delta_i S, \tag{11.2}$$

the decomposition being defined by the requirement

$$\Delta_e S = 0 \qquad \text{(adiabatic process).} \tag{11.3}$$

(e) The term $\Delta_i S$ due to changes occurring inside the system is zero for quasi-static processes, is positive for natural processes, and is never negative:

$$\Delta_i S = 0 \qquad \text{(quasi-static process),} \tag{11.4}$$

$$\Delta_i S > 0 \qquad \text{(natural process).} \tag{11.5}$$

The entropy is indefinite to the extent of an arbitrary additive constant. In this it resembles energy (Section VIII) and enthalpy (Section IX). In Section XXIII it will be shown how the indefiniteness of the entropy can be removed.

It is important to note that we postulated the existence of the entropy as a function of the masses or amounts of substance. This will lead us to the concept of "chemical potential" and to an extension of Eq. (11.1) valid for regions of variable content (Section XIII). The connection between empirical and thermodynamic temperature will also be discussed later (Section XVII).

Since entropy is an extensive property, we may define quantities such as specific, molar, and partial molar entropies, and entropy density for each region of a system. According to Eq. (6.20), the entropy S of a heterogeneous (discontinuous) system is given by

$$S = \sum_{\alpha} S^{\alpha} \tag{11.6}$$

where S^{α} is the entropy of the phase α. Owing to Eq. (6.21), we may write for the entropy S of a continuous system of volume V

$$S = \int_V \hat{S}\, dV, \tag{11.7}$$

where $\hat{S}$ is the local entropy density and dV an element of volume.

On account of Eqs. (11.2)–(11.5), $\Delta_e S$ denotes the part of the increase ΔS of the entropy of the system due to exchange of matter and heat with the surroundings, while $\Delta_i S$ is the part of this increase due to processes occurring inside the system. The quantity $\Delta_i S$ is never negative, while $\Delta_e S$ and ΔS may be positive, negative, or zero. Explicit expressions for $\Delta_e S$ and $\Delta_i S$ will be given in Section XIV.

For adiabatic processes, i.e., for changes in thermally insulated systems, we derive from Eqs. (11.2)–(11.5)

$$\Delta S = 0 \qquad \text{(quasi-static adiabatic process),} \tag{11.8}$$

$$\Delta S > 0 \qquad \text{(natural adiabatic process).} \tag{11.9}$$

Since isolated systems are special cases of thermally insulated systems, these relations hold for processes in isolated systems, too.

Let us consider a system A and its surroundings B. If inside A there occurs a process which makes the system pass from its initial state to its final state, we can restore the initial state by a second process, and we may ask if there is any change in the surroundings B after these processes have taken place.

If the first process is *quasi static*, the second process can be chosen to be quasi static, too, since this only means that the system passes through a continuous series of equilibrium states in either direction (see Section II). Furthermore, all actions produced by the system are exactly counterbalanced by external actions. Thus, a reversal of a quasistatic process implies a reversal of all changes in the surroundings. We may also prove formally that the entropy of the surroundings does not change. Evidently, A and B, taken together, constitute an isolated system. Therefore, Eq. (11.8) holds, telling us that there is no change of entropy in the composite system $A + B$. But the system A is in its initial state again, and hence its entropy has been left unchanged. Thus, there cannot be any change of entropy in the surroundings B. We conclude that any quasistatic process is *reversible* in the sense defined in Section II.

If the first process is a *natural* process, then the second process may be either quasistatic or natural. In any case, the entropy of the composite system $A + B$ now increases, due to the inequality (11.9). Since the system A has returned to its initial state and thus its entropy has recovered its initial value, there must have been an increase of entropy and thus a change in the surroundings B. In other words: Any natural process is *irreversible* in the sense defined in Section II.

This "principle of irreversibility" or equivalent statements have frequently been used as starting points for the development of the second law. But one needs a number of additional empirical facts in order to derive all the equations and inequalities given above. Thus, the set of theorems considered to be "basic" is at least as complex as the final formulation. Therefore, the older approaches, though ingenious, are nowadays mainly of historical interest. But we shall revert to Carathéodory's axiomatic treatment in the next section.

The second law has been gradually developed from the earliest studies on reversible cycles (Carnot, 1824) through more powerful statements (Clausius, 1850; Thomson (later Lord Kelvin), 1851) into a general principle of axiomatic character (Gibbs, 1875; Planck, 1887, 1926; Carathéodory, 1909, 1925; Born, 1921). The concept of thermodynamic temperature was conceived by Thomson (1848). Entropy was introduced by Clausius (1865). The modern form of the decomposition of the entropy change into an "external" and an "internal" part is due to De Donder (1920, 1927).

According to Eq. (11.1), the product TS has the dimension of energy. The unit recommended for the entropy is the joule divided by the kelvin (J K^{-1}).

XII. Carathéodory's Approach

Carathéodory's axiomatic approach to the second law of thermodynamics deserves a fuller discussion. An outline of the essential features will be given here.*

If we compare the right-hand side of Eq. (11.1),

$$dU^\alpha - \sum_i \lambda_i^\alpha \, dl_i^\alpha, \tag{12.1}$$

to Eq. (10.14), we see that it is the heat absorbed by a closed region α during an infinitesimal change of internal state without dissipative effects. Since Eq. (11.1) only refers to regions of fixed content, chemical reactions are to be excluded, too. Thus, the expression (12.1) is the heat supplied to a region of given chemical content during an infinitesimal reversible (quasistatic) process; it is an inexact differential, a so-called Pfaffian expression. If there are $\nu - 1$ work coefficients λ_i^α (and an equal number of work coordinates l_i^α, of course), we may choose these quantities and the internal energy U^α (or the empirical temperature Θ^α) to be the ν independent variables describing the internal state of the region. The decisive question is now: What are the conditions for the existence of an integrating factor for the inexact differential (12.1) containing ν terms? If such a facter existed, then the Pfaffian expression would be transformed into an exact differential, that is, into the differential of a function of state, as postulated by Eq. (11.1).

Carathéodory starts from the following principle, which is accepted as a physical "axiom:" In the vicinity of any state of a system, there are states which are adiabatically inaccessible (i.e., which cannot be reached by adiabatic processes).

It can be demonstrated that Carathéodory's principle is closely related, though not entirely equivalent, to the principle of irreversibility mentioned earlier. This is most easily shown by discussing an example.

Let us consider a thermally insulated region with fixed values of all

* The original papers are in Carathéodory (1909, 1925). See also Born (1921, 1949).

work coordinates (volume, and so on). The internal state of the region can only be changed by dissipative effects, such as friction or flow of electricity from external sources. As follows from Eq. (10.14), the increase of internal energy of the region equals the dissipated work done on the region. Now suppose there were "antidissipative effects" which would lead to a state of lower internal energy of the region and to a corresponding amount of work done by the system. Then the original state of both the region and the surroundings could be restored and the dissipative effects would be reversible processes. But the dissipative effects are natural processes, and thus, according to the principle of irreversibility, they are irreversible. Therefore, antidissipative effects do not exist, and states of lower internal energy are adiabatically inaccessible for the region considered. Thus, in this case at least, Carathéodory's principle follows directly from the principle of irreversibility. It should be noted that, if there are states inaccessible by adiabatic paths, there must necessarily be states which cannot be reached by reversible adiabatic processes.

According to our previous statements about the expression (12.1), the equation

$$dU^\alpha - \sum_i \lambda_i^\alpha \, dl_i^\alpha = 0 \tag{12.2}$$

describes reversible adiabatic changes in a region of fixed chemical content. If we apply Carathéodory's principle to this class of changes, we conclude: In the vicinity of any point in the ν-dimensional space representing the state of the region, there are points which cannot be reached from the given point by a path subject to the restriction (12.2). By a purely mathematical reasoning,* it can be proved that this statement is precisely the condition for the existence of an integrating factor for the Pfaffian expression (12.1). Thus, we may write

$$\left\{dU^\alpha - \sum_i \lambda_i^\alpha \, dl_i^\alpha\right\} \Big/ T^\alpha = dS^\alpha, \tag{12.3}$$

where $1/T^\alpha$ is an integrating factor (any of an infinite number of integrating factors) and dS^α denotes the exact differential of a function S^α. Both T^α and S^α depend on the internal state variables of the region which may be chosen to be the empirical temperature Θ^α and the $\nu - 1$ work coordinates l_i^α.

* This point in Carathéodory's arguments is explained very clearly by Pippard (1957).

Let us consider a system consisting of two regions (superscripts $'$ and $''$), each of fixed content. Then, according to Eqs. (10.17) and (10.22), the relation

$$dU' + dU'' - \sum_i \lambda_i' \, dl_i' - \sum_i \lambda_i'' \, dl_i'' = 0 \tag{12.4}$$

describes reversible adiabatic changes of the internal state of the composite system. Now Carathéodory's argument can again be applied provided all independent variables except one can be arbitrarily chosen. This implies that the system has only one empirical temperature Θ or, in other words, that there is thermal equilibrium (see Section IV) between the two regions.* Under these circumstances, the left-hand side of Eq. (12.4) may again be transformed into the exact differential dS of a function S:

$$dU' + dU'' - \sum_i \lambda_i' \, dl_i' - \sum_i \lambda_i' \, dl_i'' = T\, dS \tag{12.5}$$

where $1/T$ is the integrating factor. Comparing Eqs. (12.3) and (12.5), we obtain

$$T' \, dS' + T'' \, dS'' = T \, dS. \tag{12.6}$$

Let Θ, S', S'', and $\nu - 2$ work coordinates in each region be the independent variables. Then we have

$$\begin{aligned} &T' = T'(\Theta, S', l_i'), \qquad T'' = T''(\Theta, S'', l_i''), \\ &T = T(\Theta, S', S'', l_i', l_i''), \\ &S = S(\Theta, S', S'', l_i', l_i'') \qquad (i = 1, 2, \ldots, \nu - 2). \end{aligned} \tag{12.7}$$

From Eq. (12.6), we conclude

$$\frac{\partial S}{\partial S'} = \frac{T'}{T}, \quad \frac{\partial S}{\partial S''} = \frac{T''}{T}, \quad \frac{\partial S}{\partial \Theta} = 0, \quad \frac{\partial S}{\partial l_i'} = 0, \quad \frac{\partial S}{\partial l_i''} = 0;$$

hence,

$$\frac{\partial}{\partial x} \frac{T'}{T} = 0, \qquad \frac{\partial}{\partial x} \frac{T''}{T} = 0 \qquad (x = \Theta, l_i', l_i''), \tag{12.8}$$

* If the two regions have different temperatures, there are two variables which cannot be changed arbitrarily. This leads to fundamental difficulties in Carathéodory's approach [see Ehrenfest-Afanassjewa (1925)].

and thus

$$\frac{1}{T'}\frac{\partial T'}{\partial x}=\frac{1}{T''}\frac{\partial T''}{\partial x}=\frac{1}{T}\frac{\partial T}{\partial x}. \tag{12.9}$$

For $x = l_i'$, the derivative $\partial T''/\partial x$ must vanish, and, equally, for $x = l_i''$, the derivative $\partial T'/\partial x$ becomes zero [see Eq. (12.7)]. Therefore, T', T'', and T only depend on Θ. But, owing to Eq. (12.8), the ratios T'/T and T''/T are independent of Θ. Hence, T', T'', and T are universal functions of Θ. Since the choice of the integrating denominators T', T'', T is still arbitrary with regard to the numerical values, we may put $T' = T'' = T$ and from Eq. (12.6) we find $dS' + dS'' = dS$. We may now call T the "thermodynamic temperature" and S', S'', and S the "entropies" of the two regions and of the whole system, respectively. Thus, we have established the validity of Eq. (11.1) and the additivity of the entropy of a composite system. (That it is possible and convenient to use the further convention $\Theta = T$ will be shown in Section XVII.)

We stress that for a simple region where the expression (12.3) reduces to

$$dU^\alpha + P^\alpha\, dV^\alpha = T^\alpha\, dS^\alpha, \tag{12.10}$$

P^α and V^α being pressure and volume, respectively, the existence of an integrating denominator is trivial. But the fact that T^α is a universal function of the empirical temperature is not trivial. This can be seen from the necessity of using Eq. (12.5) which, for two simple regions, contains two pressures and two volumes.

From Eqs. (12.2)–(12.5), we conclude that for any reversible adiabatic change the entropy remains constant. Thus we have verified Eq. (11.8), too.

Let us now consider an irreversible (natural) adiabatic process inside a system. Since we only treat systems of uniform temperature composed of regions of fixed chemical content, the irreversible processes taking place belong to a rather restricted class: they are either dissipative effects in the sense explained earlier or adiabatic changes of volume or of other work coordinates with no work done (isoenergetic processes of the type of the Gay-Lussac–Joule experiment). Therefore, important processes, such as chemical reactions and transport of matter or heat between different regions of the system, are to be excluded. Nevertheless, the following arguments are instructive in principle.

Any irreversible change in which the system passes from a specified initial state I to a specified final state II by adiabatic processes can be subdivided into two steps. The first step consists in changing the work coordinates reversibly and adiabatically until the final values of these variables have been established. The second step consists in approaching the final state of the system by changing its temperature and thus its energy, the work coordinates having fixed values; this can be achieved by adiabatic dissipative effects. The entropy of the system is not changed by the first (reversible) step, but it will alter during the second (irreversible) step. Let us denote the entropy of the system in the initial and in the final states by S_{I} and S_{II}, respectively. If the sign of $S_{\mathrm{II}} - S_{\mathrm{I}}$ were different for different irreversible processes, then any state in the vicinity of the initial state would be accessible by adiabatic paths and this would contradict Carathéodory's principle.* Thus, we have either $S_{\mathrm{II}} - S_{\mathrm{I}} > 0$ or $S_{\mathrm{I}} - S_{\mathrm{II}} > 0$, that is, the entropy always changes in the same direction in irreversible adiabatic processes. We now fix the sign of the temperature T by the convention $T > 0$. We then evaluate a single experiment on an irreversible adiabatic process. Let us first take the adiabatic change of internal energy in a region of fixed work coordinates produced by friction. We know from experience that here the internal energy U increases. Therefore, we conclude from Eq. (12.3) and $T > 0$ that

$$\Delta S = S_{\mathrm{II}} - S_{\mathrm{I}} = \int_{\mathrm{I}}^{\mathrm{II}} (1/T)\, dU > 0. \tag{12.11}$$

Alternatively, we may consider the adiabatic change of volume in a simple region (a gas, for instance) under the condition of fixed internal energy (e.g., in the Gay-Lussac–Joule experiment). Here, we derive from Eq. (12.10), and from the experience that the volume always increases in such a process,

$$\Delta S = S_{\mathrm{II}} - S_{\mathrm{I}} = \int_{\mathrm{I}}^{\mathrm{II}} (P/T)\, dV > 0 \tag{12.12}$$

since $T > 0$, $P > 0$. The relations (12.11) and (12.12) verify the inequality (11.9).

* Take a simple region, the internal state of which can be specified by its volume V and its entropy S. Let the system pass from an initial state V_{I}, S_{I} to a final state V_{II}, S_{II} by an irreversible adiabatic process. If now we have $S_{\mathrm{I}} > S_{\mathrm{II}}$ in one case and $S_{\mathrm{II}} > S_{\mathrm{I}}$ in another case, then, from the state I, the state II can always be reached adiabatically, e.g., by a reversible adiabatic change of volume ($V_{\mathrm{I}} \to V_{\mathrm{II}}$) and an irreversible adiabatic change of entropy ($S_{\mathrm{I}} \to S_{\mathrm{II}}$), the last step being achieved by the appropriate choice of the dissipative effects.

Thus, we conclude that the entropy always increases in irreversible adiabatic processes. It must be admitted, however, that in order to show this we had to go back to experience again or, in other words, we used a second axiom besides Carathéodory's principle. If we had started from the principle of irreversibility or from an equivalent axiom, we should have obtained all the previous statements without a second axiom and, in the mathematical development, we could have employed Carathéodory's elegant methods. It is this way of reasoning which has been carried through in detail by Planck (1926).

Though Carathéodory's methods are debated today for other reasons, too (Falk and Jung, 1959; Landsberg, 1961), it should be stressed that Carathéodory was the first to use an axiomatic approach to the second law of thermodynamics.

As foreshadowed on several occasions, the general formulation of the second law in Section XI is not derivable from Carathéodory's principle or from the principle of irreversibility, nor can it be deduced from any other axiom of comparable simplicity. In a really satisfactory approach, systems of nonuniform temperature and regions of variable chemical content should be included and instead of (11.8) and (11.9) the general statements (11.2)–(11.5) should be obtained. But then, so many "auxiliary axioms" are needed that the usefulness of the procedure becomes doubtful.

XIII. Generalized Gibbs Equation (Chemical Potential)

According to the second law as formulated in Section XI, the entropy S^α of any region α may be considered to be a function of the internal energy U^α, of the work coordinates l_i^α, and of the amounts of substance n_k^α. On account of Eq. (11.1), we may write:

$$T^\alpha(\partial S^\alpha/\partial U^\alpha)_{l_i^\alpha, n_k^\alpha} = 1, \qquad (13.1)$$

$$T^\alpha(\partial S^\alpha/\partial l_i^\alpha)_{U^\alpha, l_j^\alpha, n_k^\alpha} = -\lambda_i^\alpha, \qquad (13.2)$$

where T^α is the thermodynamic temperature, λ_i^α is the work coefficient conjugate to l_i^α, and l_j^α denotes all work coordinates except l_i^α. A differential coefficient analogous to the expression (13.2)

$$T^\alpha(\partial S^\alpha/\partial n_k^\alpha)_{U^\alpha, l_i^\alpha, n_j^\alpha} \equiv -\mu_k^\alpha \qquad (13.3)$$

defines an intensive property μ_k^α called the *chemical potential* of substance k in the region α. Here, n_j^α refers to all amounts of substance except n_k^α.

From Eqs. (13.1)–(13.3), we derive the *generalized Gibbs equation*

$$T^\alpha \, dS^\alpha = dU^\alpha - \sum_i \lambda_i^\alpha \, dl_i^\alpha - \sum_k \mu_k^\alpha \, dn_k^\alpha. \tag{13.4}$$

This relation is an extension of Eq. (11.1) and holds for a region of variable chemical content. Equation (13.4), the simplest form of which is due to Gibbs (see below), is valid for both closed and open regions and includes chemical reactions. It is one of the most fundamental formulas in both classical and nonequilibrium thermodynamics.

All the quantities that occur as differentials in Eq. (13.4) are extensive properties, the remaining quantities (the coefficients of the differentials) being intensive properties. Therefore, we can apply Euler's theorem for homogeneous functions of first degree. We thus find

$$T^\alpha S^\alpha = U^\alpha - \sum_i \lambda_i^\alpha l_i^\alpha - \sum_k \mu_k^\alpha n_k^\alpha. \tag{13.5}$$

Physically, the transition from Eq. (13.4) to (13.5) corresponds to an increase of the quantity of the region without any change of its quality.

Let us denote any function of state by y and let us drop the superscript α. We then have

$$T \, d(S/y) = TS \, d(1/y) + (T/y) \, dS.$$

Hence, we derive by means of Eqs. (13.4) and (13.5)

$$T \, d(S/y) = \left[U - \sum_i \lambda_i l_i - \sum_k \mu_k n_k\right] d(1/y)$$
$$+ (1/y)\left(dU - \sum_i \lambda_i \, dl_i - \sum_k \mu_k \, dn_k\right)$$

or

$$T \, d(S/y) = d(U/y) - \sum_i \lambda_i \, d(l_i/y) - \sum_k \mu_k \, d(n_k/y), \tag{13.6}$$

a useful general relation.

The first example of Eq. (13.6) results for $y = m$ (total mass of the region) on account of Eqs. (3.8), (3.9), and (6.1):

$$T \, d\tilde{S} = d\tilde{U} - \sum_i \lambda_i \, d\tilde{l}_i - \sum_k \tilde{\mu}_k \, d\chi_k \tag{13.7}$$

where

$$\tilde{\mu}_k \equiv \mu_k/M_k. \tag{13.8}$$

Here, $\tilde{S}$ is the specific entropy, $\tilde{U}$ is the specific internal energy, the $\tilde{l}_i$ are the specific work coordinates of the region, and χ_k, M_k, and $\tilde{\mu}_k$ denote the mass fraction, the molar mass, and the "specific chemical potential" of substance k, respectively.

If we put $y = n$ (total amount of substance of the region) and consider Eqs. (3.10) and (6.2), we obtain the second example of Eq. (13.6):

$$T\,d\bar{S} = d\bar{U} - \sum_i \lambda_i\,d\bar{l}_i - \sum_k \mu_k\,dx_k \tag{13.9}$$

where $\bar{S}$, $\bar{U}$, and $\bar{l}_i$ denote the molar entropy, the molar internal energy, and the molar work coordinates, respectively, while x_k is the mole fraction of substance k.

Finally, for $y = V$ (volume of the region), we derive from Eqs. (3.12), (6.3), and (13.6)

$$T\,d\hat{S} = d\hat{U} - \sum_i \lambda_i\,d\hat{l}_i - \sum_k \mu_k\,dc_k. \tag{13.10}$$

Here, $\hat{S}$, $\hat{U}$, and $\hat{l}_i$ are the densities of entropy, internal energy, and of the work coordinates, respectively, and c_k denotes the molarity of substance k. In continuous systems, all these quantities have local character. Thus, Eq. (13.10) is the local form of the generalized Gibbs equation applicable to each volume element of a continuous system.

For a simple region (nonpolarized isotropic region) of pressure P and volume V, the relations (13.4), (13.7), (13.9), and (13.10) reduce to the following equations:

$$T\,dS = dU + P\,dV - \sum_k \mu_k\,dn_k, \tag{13.11}$$

$$T\,d\tilde{S} = d\tilde{U} + P\,d\tilde{V} - \sum_k \tilde{\mu}_k\,d\chi_k, \tag{13.12}$$

$$T\,d\bar{S} = d\bar{U} + P\,d\bar{V} - \sum_k \mu_k\,dx_k, \tag{13.13}$$

$$T\,d\hat{S} = d\hat{U} - \sum_k \mu_k\,dc_k, \tag{13.14}$$

where $\tilde{V}$ and $\bar{V}$ are the specific volume and the molar volume, respectively.

Equation (13.11) is equivalent to a relation given by Gibbs in 1875. By means of this equation, Gibbs introduced the chemical potential μ_k, or rather the quantity $\tilde{\mu}_k$, which he called the "potential" of a substance. Gibbs also established a relation equivalent to Eq. (13.10) for the special case of nonpolarized anisotropic bodies and thus already extended the definition of the chemical potential. But usually Eq. (13.11) is called the "Gibbs equation."

Owing to the definition (13.3), the chemical potential μ_k has the dimension of molar energy. Thus, the preferred unit to be used is the joule divided by the mole (J mol^{-1}).

Let us now define a function of state

$$G^\alpha \equiv U^\alpha - T^\alpha S^\alpha - \sum_i \lambda_i^\alpha l_i^\alpha. \tag{13.15}$$

We postpone the discussion of the properties of this function until a later stage. Here, we only need it formally to simplify the mathematical development. By comparison of the total differential of (13.15) with Eq. (13.4), we find

$$dG^\alpha = - S^\alpha \, dT^\alpha - \sum_i l_i^\alpha \, d\lambda_i^\alpha + \sum_k \mu_k^\alpha \, dn_k^\alpha. \tag{13.16}$$

Thus, we derive

$$\mu_k^\alpha = (\partial G^\alpha / \partial n_k^\alpha)_{T^\alpha, \lambda_i^\alpha, n_j^\alpha}. \tag{13.17}$$

On the other hand, we have on account of Eqs. (9.1) and (13.15)

$$G^\alpha = H^\alpha - T^\alpha S^\alpha, \tag{13.18}$$

where H^α is the enthalpy of the region. From Eqs. (6.5), (13.17), and (13.18), we obtain

$$\mu_k^\alpha = H_k^\alpha - T^\alpha S_k^\alpha. \tag{13.19}$$

Here, H_k^α and S_k^α denote the partial molar enthalpy and the partial molar entropy of substance k, respectively.

XIV. Connection between Entropy and Heat

We have carefully avoided the introduction of the concept "heat" into the relations defining entropy. Thus, the development of the consequences of the second law was not unnecessarily restrictive. In particular, the generalized Gibbs equation (Section XIII) refers to the most general variation of the entropy of any region and, since it only involves functions of state, it holds for both reversible and irreversible changes. Even when the system passes from a specified initial state to a specified final state through a series of ill-defined intermediate situations, e.g., in the Gay-Lussac–Joule experiment, the integrated form of the generalized Gibbs equation can always be applied. As a matter of fact, we already made use of this procedure in the relation (12.12).

We are now ready to derive the connection between the entropy change of a system and the heat supplied to the different parts of the system. This also involves the consideration of transfer of matter, of chemical reactions, and of dissipative effects. Thus, the discussion will lead us to a number of fundamental conclusions.

We consider a discontinuous (heterogeneous), open system of non-uniform temperature and composed of any number of open phases. The discussion of this case, usually not treated in textbooks, is straightforward (Haase, 1959). We shall not deal with continuous systems, but the results obtained for these (Haase, 1969) resemble those to be derived here in all essential features.

We first combine the equations developed for open phases in Section X with the generalized Gibbs equation. We accordingly insert Eqs. (10.1), (10.3), (10.14), and (10.15) into Eq. (13.4) and also use the relationship (13.19) connecting chemical potential μ_k^α, partial molar enthalpy H_k^α, partial molar entropy S_k^α, and thermodynamic temperature T^α:

$$dS^\alpha = \frac{d_e Q^\alpha}{T^\alpha} + \sum_k S_k^\alpha \, d_e n_k^\alpha + \frac{đW_{\text{diss}}^\alpha}{T^\alpha} + \frac{d_i Q^\alpha}{T^\alpha} + \frac{1}{T^\alpha} \sum_k H_k^\alpha \, d_t n_k^\alpha - \frac{1}{T^\alpha} \sum_k \mu_k^\alpha (d_t n_k^\alpha + d_r n_k^\alpha) \quad \text{(internal change)}. \tag{14.1}$$

This is an expression for the infinitesimal increase of the entropy S^α of a phase α in terms of the external and internal agencies producing a change of internal state. The quantities $d_e Q^\alpha$, $d_i Q^\alpha$, and $đW_{\text{diss}}^\alpha$ refer to the heat received from external sources (i.e., from the surroundings of the whole system), to the heat absorbed from neighboring phases of the system, and to the dissipated work done on the phase α, respectively. The symbols $d_e n_k^\alpha$, $d_t n_k^\alpha$, and $d_r n_k^\alpha$ relate to the increase of the amount of substance of species k due to exchange of matter with the surroundings and with neighboring phases, and to chemical reactions within the phase α, respectively.

As we infer from Eqs. (10.15) and (14.1), the general relation between dS^α and the total heat $đQ^\alpha = d_e Q^\alpha + d_i Q^\alpha$ supplied to the phase α is rather complex. But if we consider reversible changes in a phase of fixed content and thus exclude dissipative effects, chemical reactions, and exchange of matter between the phase and its surroundings then we have

$$T^\alpha dS^\alpha = đQ^\alpha \quad \text{(reversible internal change in phase of fixed content)}. \tag{14.2}$$

This formula could have been derived more directly from Eqs. (10.14) and (11.1).

We now turn to the discussion of the total entropy S of a heterogeneous system. From Eqs. (11.6) and (14.1), we obtain

$$dS = d_e S + d_i S, \tag{14.3}$$

where

$$d_e S \equiv \sum_\alpha (d_e Q^\alpha / T^\alpha) + \sum_\alpha \sum_k S_k^\alpha \, d_e n_k^\alpha \tag{14.4}$$

and

$$d_i S \equiv \sum_\alpha \frac{đW_{\text{diss}}^\alpha}{T^\alpha} + \sum_\alpha \frac{d_i Q^\alpha}{T^\alpha} + \sum_\alpha \sum_k \frac{H_k^\alpha}{T^\alpha} \, d_t n_k^\alpha - \sum_\alpha \sum_k \frac{\mu_k^\alpha}{T^\alpha} (d_t n_k^\alpha + d_r n_k^\alpha). \tag{14.5}$$

Here, $d_e S$ refers to the exchange of heat and matter with the surroundings of the system and $d_i S$ to the dissipative effects, to the transport of heat and matter between different phases of the system, and to the chemical reactions. Thus, $d_e S$ vanishes for a thermally insulated system, while $d_i S$ is related to processes taking place inside the system.

Integrating Eqs. (14.3)–(14.5) and comparing the result to Eqs. (11.2) and (11.3), we conclude that the integrals over dS, $d_e S$, and $d_i S$ coincide with ΔS, $\Delta_e S$, and $\Delta_i S$, respectively. Thus, we have in view of the relations (11.4) and (11.5)

$$\Delta S = \Delta_e S + \Delta_i S = \int dS = \int d_e S + \int d_i S, \tag{14.6}$$

$$\Delta_i S = \int d_i S \geq 0. \tag{14.7}$$

Here, the inequality sign refers to irreversible (natural) processes in the interior of the system and the equality sign to the limiting case of reversible (quasistatic) changes.

If we do not integrate the relations (14.3)–(14.5) but divide them by the time element dt and postulate $d_i S/dt \geq 0$, then we obtain the "entropy balance" of nonequilibrium thermodynamics which evaluates the results of the second law in such a way that they hold instantaneously and locally.

Before we continue the general discussion, we want to draw attention to a special corollary of (14.7). We consider an irreversible change in a system composed of two closed phases (superscripts $'$ and $''$) and exclude both dissipative effects and chemical reactions; that is, we treat a system of two phases each of fixed content where nothing happens except a transfer of heat between two phases of different temperatures (T' and T'') and an exchange of heat between the whole system and its surroundings. Then we derive from Eqs. (10.22), (14.5), and (14.7)

$$\int \left(\frac{1}{T'} - \frac{1}{T''} \right) d_i Q' = \int \left(\frac{1}{T''} - \frac{1}{T'} \right) d_i Q'' > 0. \tag{14.8}$$

We conclude that, under the restrictions mentioned, the heat flows from the higher to the lower temperature, irrespective of the heat received by the two phases from the surroundings of the system. This result corroborates the usefulness of our previous formal definitions of heat (Section VIII and X) and of thermodynamic temperature (Section XI).

The explicit expression for the increase ΔS of the entropy of any heterogeneous system follows from the relations (14.4), (14.6), and (14.7):

$$\Delta S \geq \sum_\alpha \int \left(\frac{d_e Q^\alpha}{T^\alpha} + \sum_k S_k^\alpha \, d_e n_k^\alpha \right). \tag{14.9}$$

The plausible form of the integral in (14.9) is one of the reasons for the definition of heat in open systems chosen in Section X. A relation analogous to (14.9) can be shown to hold for continuous systems (Haase, 1969).

For a *closed* system (the single phases still remaining open) we have on account of Eq. (10.10) $d_e n_k^\alpha = 0$. Thus, we obtain from (14.9)

$$\Delta S \geq \sum_\alpha \int \frac{d_e Q^\alpha}{T^\alpha} \qquad \text{(closed system).} \tag{14.10}$$

If the temperature of the closed system is *uniform* ($T^\alpha = T$), we derive from Eqs. (10.16) and (14.10)

$$\Delta S \geq \int \frac{đQ}{T} \qquad \text{(closed system of uniform temperature),} \tag{14.11}$$

where $đQ$ denotes the heat received by the whole system. Here, we can also apply Eq. (10.20) and thus simplify the expression for $\Delta_i S$ following from Eqs. (14.5) and (14.7):

$$\Delta_i S = \int \frac{đW_{\text{diss}}}{T} - \sum_\alpha \int \frac{1}{T} \sum_k \mu_k^\alpha (d_t n_k^\alpha + d_r n_k^\alpha) \geq 0 \tag{14.12}$$
$$\text{(closed system of uniform temperature),}$$

where W_{diss} is the dissipated work done on the whole system.

A useful application of (14.12) refers to the irreversible transfer of matter in a closed system of uniform temperature consisting of two phases (superscripts $'$ and $''$) without dissipative effects and without chemical reactions. Let us further assume that only a single substance k passes from one phase to the other. We then have, using Eqs. (10.6) and (14.12) and the convention $T > 0$,

$$\int (\mu_k'' - \mu_k')\, d_t n_k' = \int (\mu_k' - \mu_k'')\, d_t n_k'' > 0,$$

in close analogy to the inequality (14.8). Thus, under the restrictions mentioned, matter flows from the phase in which μ_k has the higher value to the phase where μ_k has the lower value. This justifies the name chemical potential given to μ_k. An analogous inequality valid for chemical reactions will be derived in Section XX. The common bases of all these inequalities are the formulas (14.5) and (14.7).

To obtain similar inequalities for the work coordinates and work coefficients, we have to start from (14.11) and combine this with the first law and the generalized Gibbs equation. Let us investigate the irreversible changes of the volumes V' and V'' in a closed simple system of uniform temperature containing two phases of pressures P' and P'' each of the phases having fixed content. Let us further suppose the system to be surrounded by rigid walls, so that no work can be done on the system, and the total volume $(V' + V'')$ is constant. Then we conclude from (8.7), (11.6), (13.11), and (14.11)

$$\int (P' - P'')\, dV' = \int (P'' - P')\, dV'' > 0.$$

Thus, the volume of the phase having the higher pressure increases at the expense of the volume of the phase with the lower pressure. This result seems to be obvious from the point of view of mechanics, but it has been derived here from the second law to show the consistency of all our formulas.

Let us return to the general relation (14.11). When the temperature is not only uniform but also constant in time, then the processes occurring in the system are said to be *isothermal*. From Eq. (14.11), we find for $T = \text{const}$

$$\Delta S \geq Q/T \qquad \text{(closed system, isothermal process).} \qquad (14.13)$$

Here, Q denotes the total heat absorbed by the system.

Finally, for an *adiabatic* change of state, we have $d_e Q^\alpha = 0$, $d_e n_k^\alpha = 0$. Thus, we obtain from Eq. (14.9) or Eq. (14.10)

$$\Delta S \geq 0 \qquad \text{(adiabatic process),} \qquad (14.14)$$

in agreement with the relations (11.8) and (11.9).

XV. Helmholtz Function and Gibbs Function

The *Helmholtz function* F^α of a region α is defined by

$$F^\alpha \equiv U^\alpha - T^\alpha S^\alpha, \tag{15.1}$$

where U^α, T^α, and S^α represent internal energy, thermodynamic temperature, and entropy, respectively.

The Helmholtz function F of a heterogeneous or of a continuous system (volume V) is given by

$$F \equiv \sum_\alpha F^\alpha \qquad \text{(heterogeneous system)}, \tag{15.2}$$

$$F \equiv \int_V \hat{F}\, dV \qquad \text{(continuous system)}. \tag{15.3}$$

Here, dV denotes a volume element and $\hat{F}$ is the local density of the Helmholtz function:

$$\hat{F} = \hat{U} - T\hat{S}, \tag{15.4}$$

$\hat{U}$ and $\hat{S}$ being the local densities of the internal energy and of the entropy, respectively.

Thus, the Helmholtz function F is an extensive property having the dimension of energy. This quantity was introduced independently by Gibbs in 1875 and by Helmholtz in 1882 ("Freie Energie"). It is also called "Helmholtz free energy" or "work function."

For the total internal energy U and the total entropy S we have:

$$U = \sum_\alpha U^\alpha, \qquad S = \sum_\alpha S^\alpha \qquad \text{(heterogeneous system)}, \tag{15.5}$$

$$U = \int_V \hat{U}\, dV, \qquad S = \int_V \hat{S}\, dV \qquad \text{(continuous system)}, \tag{15.6}$$

analogous to Eqs. (15.2) and (15.3). When we insert Eq. (15.1) into (15.2) and Eq. (15.4) into (15.3) and compare the results to Eqs. (15.5) and (15.6), respectively, we conclude that for systems of *uniform temperature* ($T^\alpha = T$) there exists the simple relation

$$F = U - TS \qquad (T \quad \text{uniform}). \tag{15.7}$$

If, during a change in the internal state of the system, the temperature T does not alter, then the change or the process is *isothermal*. From Eq. (15.7), we obtain for $T = \text{const}$

$$\Delta F = \Delta U - T\,\Delta S \qquad \text{(isothermal process)} \tag{15.8}$$

where the symbol Δ denotes the increase of a function of state during the change considered.

Restricting the discussion to an isothermal internal change in a *closed* system, we derive from Eqs. (8.7) and (14.13)

$$T\,\Delta S \geq \Delta U - W \qquad \text{(closed system, isothermal process)}, \quad (15.9)$$

where W denotes the work done on the system and the inequality sign applies to an irreversible process while the equality sign refers to the limiting case of a reversible process. Comparing (15.8) and (15.9), we find

$$\Delta F \leq W \qquad \text{(closed system, isothermal process)}. \quad (15.10)$$

This relation may be written more explicitly:

$$-\Delta F = -W_{\text{rev}}, \qquad -\Delta F > -W_{\text{irrev}} \quad (15.11)$$
$$\text{(closed system, isothermal process)}.$$

Here, $-W_{\text{rev}}$ and $-W_{\text{irrev}}$ denote the work done *by* the system for reversible and irreversible paths, respectively.

If we consider fixed initial and final states, then ΔF has a given value and (15.11) tells us that the work done by the closed system during a reversible isothermal change of state exceeds the work done during an irreversible isothermal change of state. Therefore, $-\Delta F$ is sometimes called the "maximum work" in isothermal processes.

We now introduce the *Gibbs function* G^α of a region:

$$G^\alpha \equiv H^\alpha - T^\alpha S^\alpha, \quad (15.12)$$

where H^α is the enthalpy of the region defined by Eq. (9.1):

$$H^\alpha \equiv U^\alpha - \sum_i \lambda_i^\alpha l_i^\alpha, \quad (15.13)$$

the symbols λ_i^α and l_i^α representing the work coefficients and work coordinates, respectively. Therefore, the quantity G^α defined by Eq. (13.15) and already used formally is indeed the Gibbs function.

The Gibbs function G of a nonhomogeneous system is given by

$$G \equiv \sum_\alpha G^\alpha \qquad \text{(heterogeneous system)}, \quad (15.14)$$

$$G \equiv \int_V \hat{G}\,dV \qquad \text{(continuous system)}. \quad (15.15)$$

Here, $\hat{G}$ denotes the local density of the Gibbs function:

$$\hat{G} = \hat{H} - T\hat{S}, \qquad (15.16)$$

$\hat{H}$ being the local enthalpy density.

Thus, the Gibbs function is an extensive property having the dimension of energy. It is also called "Gibbs free energy."

Since we have for the total enthalpy H

$$H = \sum_{\alpha} H^{\alpha} \qquad \text{(heterogeneous system)}, \qquad (15.17)$$

$$H = \int_V \hat{H}\, dV \qquad \text{(continuous system)}, \qquad (15.18)$$

it follows from Eqs. (15.5), (15.6), (15.12), and (15.14)–(15.16) that for systems of *uniform temperature* ($T^{\alpha} = T$) a simple relation holds:

$$G = H - TS \qquad (T \quad \text{uniform}) \qquad (15.19)$$

analogous to Eq. (15.7). Considering an *isothermal* change of the internal state of the system ($T =$ const), we obtain from Eq. (15.19)

$$\Delta G = \Delta H - T\,\Delta S \qquad \text{(isothermal process)}, \qquad (15.20)$$

in close analogy to Eq. (15.8).

Restricting the discussion to an isothermal internal change in a *closed system at fixed work coefficients*, we derive from Eqs. (9.11) and (14.13)

$$T\,\Delta S \geq \Delta H - W^* \qquad (15.21)$$

(closed system, isothermal process, λ_i^{α} constant).

Here, W^* is the excess work, i.e., the work done on the system minus the quasistatic work of deformation and of electric or magnetic polarization. Comparing (15.20) and (15.21) we find:

$$\Delta G \leq W^* \qquad (15.22)$$

(closed system, isothermal process, λ_i^{α} constant).

This inequality is analogous to the relation (15.10).

Combining Eqs. (13.5), (15.12), and (15.13), we obtain a relation valid for *any region* (phase α or volume element α):

$$G^{\alpha} = \sum_k n_k^{\alpha} \mu_k^{\alpha}, \qquad (15.23)$$

n_k^α and μ_k^α representing the amount of substance and the chemical potential of species k in the region α. If we take into account the definitions (3.10), (3.12), (6.2), and (6.3), we derive from Eq. (15.23)

$$\bar{G}^\alpha = \sum_k x_k^\alpha \mu_k^\alpha, \qquad \hat{G}^\alpha = \sum_k c_k^\alpha \mu_k^\alpha, \tag{15.24}$$

where $\bar{G}^\alpha$, $\hat{G}^\alpha$, x_k^α, and c_k^α denote the molar Gibbs function, the density of the Gibbs function, the mole fraction of species k, and the molarity of species k, respectively.

Equation (15.24) may be compared to Eq. (6.8), thus showing that μ_k^α is identical with the partial molar Gibbs function. This can also be concluded directly from Eqs. (6.5) and (13.17). According to the first equation in (15.24), the chemical potential of the only substance present in a one-component region coincides with the molar Gibbs function of that region.

For a *simple region* we have, on account of Eqs. (9.3), (15.12), and (15.13)

$$G^\alpha = U^\alpha + P^\alpha V^\alpha - T^\alpha S^\alpha \qquad \text{(simple region).} \tag{15.25}$$

Here, P^α and V^α denote pressure and volume, respectively. For a simple system of uniform temperature T and of uniform pressure P, we derive from Eqs. (15.5), (15.6), (15.14), (15.15), and (15.25)

$$G = U + PV - TS \qquad \text{(simple system, } T \text{ uniform, } P \text{ uniform),} \tag{15.26}$$

where V is the total volume of the system.

Equations (15.25) and (15.26) are the familiar definitions of the quantity G as introduced by Gibbs in 1875.

Let us restrict the following discussion to internal changes in simple systems of fixed temperature T and fixed pressure P, that is, to isothermal and isobaric processes in simple systems. It then follows immediately from Eq. (15.22) that

$$\Delta G \leq W^* \tag{15.27}$$
(closed simple system, isothermal and isobaric process)

or, more explicitly

$$-\Delta G = -W^*_{\text{rev}}, \qquad -\Delta G > -W^*_{\text{irrev}} \tag{15.28}$$
(closed simple system, isothermal and isobaric process),

in complete analogy to the relations (15.10) and (15.11).

When the closed system is an experimental arrangement capable of doing work other than quasistatic work and dissipated work, then the quantity $-W^*$ is said to be "useful work." The best known example of such an arrangement is a galvanic cell which can perform electric work. Thus, (15.28) tells us that, for a given isothermal and isobaric change of state in a closed system, the useful work $-W^*_{\text{rev}}$ done by the system in a reversible path exceeds the useful work $-W^*_{\text{irrev}}$ performed by the system in an irreversible path. Therefore, $-\Delta G$ is sometimes called the "maximum useful work" in isothermal and isobaric processes. We shall revert to the subject of galvanic cells in Section XXI.

In Table III, we summarize the connections between the several thermodynamic functions (internal energy U, enthalpy H, entropy S, Helmholtz function F, Gibbs function G), the work (W or W^*) done on the system, and the heat (Q) supplied to the system. For brevity, we restrict this summary to internal changes in closed simple systems of uniform pressure. Table III has been constructed by means of the relations (7.1), (8.7), (9.13), (14.13), (14.14), (15.10), and (15.27).

Let us finally consider an irreversible change which is both isothermal and isobaric and which involves a homogeneous or a heterogeneous closed simple system. We here exclude galvanic cells. Then we obviously deal with one if the following types of internal changes: (a) mixing or unmixing processes, (b) phase transitions, (c) chemical reactions. In each of these processes, we start with a homogeneous or heterogeneous system of specified chemical content (fixed masses and compositions of all the phases concerned) and we end up in a homogeneous or heterogeneous system of different chemical content. Since temperature and pressure are the same at the beginning and at the close of the process, the changes of volume accompanying the process are uniquely determined. Therefore, we may choose a path of the process where the volume changes are brought about reversibly and hence we may, without loss of generality, compute the changes of all the functions of state under the condition $W^* = 0$. Thus, we conclude from the last line in Table III

$$\Delta H = Q, \tag{15.29}$$

$$T\,\Delta S > Q, \tag{15.30}$$

$$\Delta G < 0. \tag{15.31}$$

Equation (15.29) justifies the designations "heat of mixing," "heat of vaporization," and "heat of reaction" used for the corresponding changes

TABLE III

CHANGES OF THERMODYNAMIC FUNCTIONS FOR SPECIFIED PROCESSES (INTERNAL CHANGES) IN CLOSED SIMPLE SYSTEMS OF UNIFORM PRESSURE

Kind of process	ΔU	ΔH	ΔS	ΔF	ΔG
Adiabatic	$\Delta U = W$	$\Delta H = \int V\, dP + W^*$	$\Delta S \geq 0$	—	—
Adiabatic, isobaric	$\Delta U = -P\,\Delta V + W^*$	$\Delta H = W^*$	$\Delta S \geq 0$	—	—
Isothermal	$\Delta U = Q + W$	$\Delta H = Q + \int V\, dP + W^*$	$T\,\Delta S \geq Q$	$\Delta F \leq W$	—
Isothermal, isobaric	$\Delta U = Q - P\,\Delta V + W^*$	$\Delta H = Q + W^*$	$T\,\Delta S \geq Q$	$\Delta F \leq -P\,\Delta V + W^*$	$\Delta G \leq W^*$

of enthalpy. We infer from Eqs. (15.29)–(15.31) that the Gibbs function G always decreases, while the enthalpy H and the entropy S may increase or decrease.

In particular, we derive from (15.29) and (15.30)

$$T\,\Delta S > \Delta H. \tag{15.32}$$

Thus, only three of the four conceivable sign combinations for ΔH and ΔS occur in nature: the case $\Delta H > 0$, $\Delta S < 0$ is ruled out by (15.32). We have for $\Delta H > 0$, $\Delta S > 0$: $T|\Delta S| > |\Delta H|$; and for $\Delta H < 0$, $\Delta S < 0$: $|\Delta H| > T|\Delta S|$.

These sign combinations are summarized in Table IV. The examples* refer to heterogeneous chemical reactions between pure solid and liquid phases at 25°C and 1 atm.

TABLE IV

SIGN COMBINATIONS FOR ΔH AND ΔS IN ISOTHERMAL–ISOBARIC IRREVERSIBLE PROCESSES IN CLOSED SIMPLE SYSTEMS AND EXAMPLES CONCERNING CHEMICAL REACTIONS BETWEEN PURE SOLID OR LIQUID PHASES AT 25°C AND 1 ATM

ΔH	ΔS	Example (heterogeneous chemical reaction)
+	+	Ag (solid) + HgCl (solid) → Hg (liquid) + AgCl (solid)
+	−	Impossible
−	+	Pb (solid) + 2HgCl (solid) → 2Hg (liquid) + $PbCl_2$ (solid)
−	−	Pb (solid) + 2AgCl (solid) → 2Ag (solid) + $PbCl_2$ (solid)

XVI. Fundamental Equations

We consider a single region and omit the superscript α. We rewrite the generalized Gibbs equation (13.4):

$$dS = (1/T)\,dU - (1/T)\sum_i \lambda_i\,dl_i - (1/T)\sum_k \mu_k\,dn_k. \tag{16.1}$$

We observe that the thermodynamic temperature T, the work coefficients

* Cf. Gerke (1922). Analogous examples can be found for the process of mixing the pure liquid components in a liquid two-component system (see Haase and Rehage (1955)).

λ_i, and the chemical potentials μ_k may be obtained from the entropy S and its first derivatives with respect to the internal energy U, the work coordinates l_i, and the amounts of substance n_k. An analogous statement holds for the other thermodynamic functions, too. Thus, the enthalpy (9.1)

$$H = U - \sum_i \lambda_i l_i, \tag{16.2}$$

the Helmholtz function (15.1)

$$F = U - TS, \tag{16.3}$$

and the Gibbs function (15.12)

$$G = H - TS \tag{16.4}$$

may be expressed by S, U, l_i, n_k, and the first derivatives of the function $S(U, l_i, n_k)$. The same is true for the function $U(S, l_i, n_k)$. Following Massieu (1869), we call such a function a *characteristic function*. A differential equation of the type (16.1) is said to be a *fundamental equation*. This concept is due to Gibbs (1875), who applied it to simple regions, i.e., to regions having only one work coordinate (the volume V) and one conjugate work coefficient (the negative pressure $-P$).

We quote Gibbs' own words (reprinted 1948) (only substituting our symbols for his symbols): "The distinction between equations which are, and which are not, *fundamental*, in the sense in which the word is here used, may be illustrated by comparing an equation between*

$$U, S, V, m_1, m_2, \ldots, m_N$$

with one between

$$U, T, V, m_1, m_2, \ldots, m_N.$$

As

$$T = (\partial U/\partial S)_{V,m},$$

the second equation may evidently be derived from the first. But the first equation cannot be derived from the second; for an equation between

$$U, (\partial U/\partial S)_{V,m}, V, m_1, m_2, \ldots, m_N$$

* The symbols $m_1, m_2, \ldots m_N$ denote the masses of the N species present in the region. Gibbs always uses masses instead of amounts of substance.

is equivalent to one between

$$(\partial S/\partial U)_{V,m},\ U,\ V,\ m_1,\ m_2,\ \ldots,\ m_N,$$

which is evidently not sufficient to determine the value of S in terms of the other variables."

Rearranging Eq. (16.1) and using Eqs. (16.2)–(16.4), we obtain a set of fundamental equations:

$$dU = T\,dS + \sum_i \lambda_i\,dl_i + \sum_k \mu_k\,dn_k, \tag{16.5}$$

$$dH = T\,dS - \sum_i l_i\,d\lambda_i + \sum_k \mu_k\,dn_k, \tag{16.6}$$

$$dF = -\,S\,dT + \sum_i \lambda_i\,dl_i + \sum_k \mu_k\,dn_k, \tag{16.7}$$

$$dG = -\,S\,dT - \sum_i l_i\,d\lambda_i + \sum_k \mu_k\,dn_k. \tag{16.8}$$

Each of the functions $U(S, l_i, n_k)$, $H(S, \lambda_i, n_k)$, $F(T, l_i, n_k)$, $G(T, \lambda_i, n_k)$ is a characteristic function. When applied to a simple region, the relations (16.5)–(16.8) coincide with Gibbs' fundamental equations.

The method of changing the independent variables by introducing new functions like (16.2)–(16.4) into relations like (16.1) or (16.5) is an example of a Legendre transformation.*

From Eqs. (16.5)–(16.8), we obtain

$$\begin{aligned} \lambda_i &= \left(\frac{\partial U}{\partial l_i}\right)_{S,l_j,n_k} = \left(\frac{\partial F}{\partial l_i}\right)_{T,l_j,n_k}, \\ -l_i &= \left(\frac{\partial H}{\partial \lambda_i}\right)_{S,\lambda_j,n_k} = \left(\frac{\partial G}{\partial \lambda_i}\right)_{T,\lambda_j,n_k} \end{aligned} \tag{16.9}$$

where l_j or λ_j denotes all work coordinates or work coefficients except l_i or λ_i. From the same set of fundamental equations, we find

$$\begin{aligned} \mu_k &= \left(\frac{\partial U}{\partial n_k}\right)_{S,l_i,n_j} = \left(\frac{\partial H}{\partial n_k}\right)_{S,\lambda_i,n_j} \\ &= \left(\frac{\partial F}{\partial n_k}\right)_{T,l_i,n_j} = \left(\frac{\partial G}{\partial n_k}\right)_{T,\lambda_i,n_j}, \end{aligned} \tag{16.10}$$

where n_j stands for all amounts of substance except n_k itself.

* This statement appears to be due to Ehrenfest (1911) (see also Ulich and Jost (1970)).

Applying the cross-differentiation rule to Eq. (16.8), we derive

$$\left(\frac{\partial \mu_k}{\partial T}\right)_{\lambda_i, n_k} = -\left(\frac{\partial S}{\partial n_k}\right)_{T, \lambda_i, n_j} = -S_k, \tag{16.11}$$

$$\left(\frac{\partial \mu_k}{\partial \lambda_i}\right)_{T, \lambda_j, n_k} = -\left(\frac{\partial l_i}{\partial n_k}\right)_{T, \lambda_i, n_j} = -l_{ik}. \tag{16.12}$$

Here, λ_j refers to all work coefficients except λ_i. In (16.11) and (16.12), the definition (6.5) has been used. Accordingly, S_k and l_{ik} are the partial molar entropy and the partial molar work coordinate, respectively.

For a simple region of pressure P and volume V, Eq. (16.12) reads

$$(\partial \mu_k / \partial P)_{T, n_k} = V_k, \tag{16.13}$$

where V_k denotes the partial molar volume.

Again, by means of the cross-differentiation rule, we find from Eqs. (16.7) and (16.8)

$$\left(\frac{\partial S}{\partial l_i}\right)_{T, l_j, n_k} = -\left(\frac{\partial \lambda_i}{\partial T}\right)_{l_i, n_k} \tag{16.14}$$

$$\left(\frac{\partial S}{\partial \lambda_i}\right)_{T, \lambda_j, n_k} = \left(\frac{\partial l_i}{\partial T}\right)_{\lambda_i, n_k}. \tag{16.15}$$

For simple regions, these relations reduce to

$$\left(\frac{\partial S}{\partial V}\right)_{T, n_k} = \left(\frac{\partial P}{\partial T}\right)_{V, n_k}, \tag{16.16}$$

$$\left(\frac{\partial S}{\partial P}\right)_{T, n_k} = -\left(\frac{\partial V}{\partial T}\right)_{P, n_k}, \tag{16.17}$$

the so-called Maxwell relations (Maxwell, 1885).

Combining Eqs. (16.5), (16.6), (16.14), and (16.15), we derive the formulas

$$\left(\frac{\partial U}{\partial l_i}\right)_{T, l_j, n_k} = \lambda_i - T\left(\frac{\partial \lambda_i}{\partial T}\right)_{l_i, n_k}, \tag{16.18}$$

$$\left(\frac{\partial H}{\partial \lambda_i}\right)_{T, \lambda_j, n_k} = T\left(\frac{\partial l_i}{\partial T}\right)_{\lambda_i, n_k} - l_i \tag{16.19}$$

which for simple regions simplify to

$$\left(\frac{\partial U}{\partial V}\right)_{T,n_k} = T\left(\frac{\partial P}{\partial T}\right)_{V,n_k} - P, \tag{16.20}$$

$$\left(\frac{\partial H}{\partial P}\right)_{T,n_k} = V - T\left(\frac{\partial V}{\partial T}\right)_{P,n_k}. \tag{16.21}$$

The last two equations are relevant to the evaluation of the Gay-Lussac–Joule experiment (p. 27) and of the Joule–Thomson experiment (p. 31).

Let us denote any function of state by y. We divide the relations (16.2)–(16.4) by y, take the total differentials, and compare these to Eq. (13.6). We then obtain the following set of equations which take the place of the fundamental equations (16.5)–(16.8):

$$d(U/y) = T\,d(S/y) + \sum_i \lambda_i\, d(l_i/y) + \sum_k \mu_k\, d(n_k/y), \tag{16.22}$$

$$d(H/y) = T\,d(S/y) - \sum_i (l_i/y)\, d\lambda_i + \sum_k \mu_k\, d(n_k/y), \tag{16.23}$$

$$d(F/y) = -(S/y)\,dT + \sum_i \lambda_i\, d(l_i/y) + \sum_k \mu_k\, d(n_k/y), \tag{16.24}$$

$$d(G/y) = -(S/y)\,dT - \sum_i (l_i/y)\, d\lambda_i + \sum_k \mu_k\, d(n_k/y). \tag{16.25}$$

If we set y equal to the total mass, to the total amount of substance, and to the volume of the region, we arrive at the fundamental equations for the specific quantities, the molar quantities, and the generalized densities, respectively.

We now consider an anisotropic region polarized by an electrostatic field. Then, according to Table I, the work coefficients λ_i are the stress components $\tau_1, \ldots, \tau_6$ and the components $\mathfrak{E}_1, \mathfrak{E}_2, \mathfrak{E}_3$ of the electric field strength, the conjugate work coordinates l_i being the strain components $e_1, \ldots, e_6$ multiplied by the reference volume V_0 and the components $\mathfrak{P}_1, \mathfrak{P}_2, \mathfrak{P}_3$ of the electric polarization multiplied by the actual volume $V = V_0(1 + e_1 + e_2 + e_3)$. Putting $y = V$ or $y = V_0$ and writing [see Eqs. (3.12) and (6.3)]

$$\hat{G} \equiv G/V, \qquad \hat{S} \equiv S/V, \qquad c_k \equiv n_k/V$$

or

$$\hat{G}' \equiv G/V_0, \qquad \hat{S}' \equiv S/V_0, \qquad c_k' \equiv n_k/V_0,$$

we derive from Eq. (16.25)

$$d\hat{G} = -\hat{S}\,dT - (V_0/V)\sum_{i=1}^{6} e_i\,d\tau_i - \sum_{i=1}^{3} \mathfrak{P}_i\,d\mathfrak{E}_i + \sum_k \mu_k\,dc_k \tag{16.26}$$

or

$$d\hat{G}' = -\hat{S}'\,dT - \sum_{i=1}^{6} e_i\,d\tau_i - (V/V_0)\sum_{i=1}^{3} \mathfrak{P}_i\,d\mathfrak{E}_i + \sum_k \mu_k\,dc_k', \tag{16.27}$$

respectively.

Applying the identities

$$\frac{\partial^2\hat{G}'}{\partial\tau_i\,\partial\tau_j} = \frac{\partial^2\hat{G}'}{\partial\tau_j\,\partial\tau_i} \qquad \text{and} \qquad \frac{\partial^2\hat{G}}{\partial\mathfrak{E}_i\,\partial\mathfrak{E}_j} = \frac{\partial^2\hat{G}}{\partial\mathfrak{E}_j\,\partial\mathfrak{E}_i}$$

to Eqs. (16.27) and (16.26), respectively, we find

$$\left(\frac{\partial e_i}{\partial\tau_j}\right)_{T,\mathfrak{E}_i,c_k'} = \left(\frac{\partial e_j}{\partial\tau_i}\right)_{T,\mathfrak{E}_i,c_k'} \qquad (i, j = 1, 2, \ldots, 6), \tag{16.28}$$

$$\left(\frac{\partial\mathfrak{P}_i}{\partial\mathfrak{E}_j}\right)_{T,\tau_i,c_k} = \left(\frac{\partial\mathfrak{P}_j}{\partial\mathfrak{E}_i}\right)_{T,\tau_i,c_k} \qquad (i, j = 1, 2, 3). \tag{16.29}$$

We now consider the consequences of Eqs. (16.28) and (16.29) in turn.

The derivatives occurring on the left-hand side and on the right-hand side of Eq. (16.28) are called the isothermal elastic compliance coefficients,* denoted by ζ_{ij} and ζ_{ji}, respectively. The 36 quantities of this kind which nominally exist are reduced to 21 coefficients by the relation (16.28)

$$\zeta_{ij} = \zeta_{ji} \qquad (i, j = 1, 2, \ldots, 6). \tag{16.30}$$

This symmetry of the matrix of the ζ_{ij} was already shown by Thomson (Lord Kelvin) in 1855. The physical symmetry of the crystal considered may further diminish the number of independent elastic coefficients.†

For small deformations in a nonpolarized medium, we have the generalized Hooke law (Cauchy, 1829):

$$e_i = \sum_{j=1}^{6} \zeta_{ij}\tau_j \qquad (i = 1, 2, \ldots, 6), \tag{16.31}$$

* For an isotropic region of pressure P, we have: $\tau_1 = \tau_2 = \tau_3 = -P$, $\tau_4 = \tau_5 = \tau_6 = 0$; $\zeta_{11} = \zeta_{12} = \zeta_{13} \equiv \zeta_1$, $\zeta_{21} = \zeta_{22} = \zeta_{23} \equiv \zeta_2$, $\zeta_{31} = \zeta_{32} = \zeta_{33} \equiv \zeta_3$; and the sum $\zeta_1 + \zeta_2 + \zeta_3$ becomes the third part of the isothermal compressibility.

† For a fuller discussion of the thermodynamics of elasticity, see Callen (1961).

where the ζ_{ij} only depend on temperature and composition. As a matter of fact, this equation of state is valid for many problems of practical interest in the theory of elasticity. But Eq. (16.31) has not been used in the derivation of Eq. (16.30).

The quantities occurring in Eq. (16.29) have a simple meaning if we exclude ferroelectric substances and high electric field strengths. Then experience tells us that there exists an equation of state analogous to Eq. (16.31):

$$\mathfrak{P}_i = \sum_{j=1}^{3} \gamma_{ij}\mathfrak{E}_j \qquad (i = 1, 2, 3). \tag{16.32}$$

Here the γ_{ij} depend on temperature, composition, and the stress components, but not on the electric field strength. We have by definition

$$\gamma_{ij} = \varepsilon_0(\varepsilon_{ij} - 1) \qquad (i, j = 1, 2, 3), \tag{16.33}$$

ε_0 being the permittivity of vacuum and the ε_{ij} the relative permittivities or dielectric constants.* According to Eq. (16.29), the tensor of the γ_{ij} or ε_{ij} is symmetrical:

$$\gamma_{ij} = \gamma_{ji}, \qquad \varepsilon_{ij} = \varepsilon_{ji} \qquad (i, j = 1, 2, 3). \tag{16.34}$$

Thus, the number of independent dielectric constants is reduced from nine to six. This conclusion is again due to Kelvin.

We may also derive the general relations between the thermoelastic, piezoelectric, and pyroelectric effects by making use of Eqs. (16.26) and (16.27). But details will not be given here.†

XVII. Relation of Empirical to Thermodynamic Temperature

The empirical temperature Θ has been defined in Section IV (zeroth law), while the thermodynamic temperature T was first discussed in Section XI (second law). We shall now investigate the interrelation of these two quantities.

* For an isotropic region, we have, in place of Eqs. (16.32) and (16.33), $\mathfrak{P} = \gamma\mathfrak{E} = -\gamma \operatorname{grad} \psi$, with $\gamma = \varepsilon_0(\varepsilon - 1)$ where ψ is the electric potential and ε the dielectric constant.

† A more comprehensive discussion of the thermodynamics of elastic solids in electrostatic fields is given by Li and Ting (1957).

Thermodynamic temperature T and entropy S have been simultaneously introduced by Eq. (11.1). Here and in all subsequent differential equations, T and S appear in the combinations $T\,dS$ and $S\,dT$. Thus, by a simultaneous reversal of sign in T and S, all the relations will continue to hold. We decided in Section XI that T shall be positive and thus also fixed the sign of entropy changes (cf. Section XII). But the fundamental relations still remain valid if S is replaced by aS and T by T/a, where a is a positive quantity. Hence, the definitions of T and S are incomplete until we fix the arbitrary factor a. After having established the relation between T and Θ, we shall remove this arbitrariness.

Since T is a universal function of Θ (see Sections XI and XII), we may choose some convenient experimental arrangement in which Θ is measured and compare the results to those obtained by applying the appropriate equations containing T and following from the second law.

We choose the Joule–Thomson experiment described in Section IX. We recall that the result of the measurements is a definite answer to the question of how the enthalpy H of a fixed amount of gas depends on the pressure P for a given empirical temperature Θ. We find as a first approximation

$$H(\Theta, P_{\mathrm{II}}) - H(\Theta, P_{\mathrm{I}}) = \mathrm{const}(P_{\mathrm{II}} - P_{\mathrm{I}}), \tag{17.1}$$

where P_{I} and P_{II} are two values of the pressure.

On the other hand, we know from experiments that, to the same approximation, the equation of state of a fixed amount of gas may be written

$$PV = A + BP. \tag{17.2}$$

Here, V is the volume and A and B are quantities depending only on the empirical temperature Θ.

Let us apply the fundamental equation (16.8) which contains both the entropy S and the thermodynamic temperature T. Since we treat a simple phase of fixed chemical content, Eq. (16.8) simplifies to

$$dG = -S\,dT + V\,dP, \tag{17.3}$$

G being the Gibbs function defined by (15.12):

$$G = H - TS. \tag{17.4}$$

We immediately derive from Eqs. (17.3) and (17.4)

$$(\partial G/\partial P)_T = V, \qquad G - T(\partial G/\partial T)_P = H. \tag{17.5}$$

Combining Eq. (17.2) with the last two equations or directly with (16.21) and integrating, we obtain

$$H(T, P_{\mathrm{II}}) - H(T, P_{\mathrm{I}}) = \left(A - T\frac{dA}{dT}\right)\ln\left(\frac{P_{\mathrm{II}}}{P_{\mathrm{I}}}\right) + \left(B - T\frac{dB}{dT}\right)(P_{\mathrm{II}} - P_{\mathrm{I}}). \tag{17.6}$$

The left-hand sides of Eqs. (17.1) and (17.6) have the same meaning, since T is a universal function of Θ. Therefore, we conclude

$$A - T\,dA/dT = 0, \tag{17.7}$$

or

$$A = \alpha T, \tag{17.8}$$

where α is an arbitrary constant.

Comparing Eqs. (4.2) and (17.2), we discover that A is identical to the quantity A occurring in Eq. (4.3) which defines the empirical Kelvin temperature Θ and which may be written

$$\Theta = \beta A, \tag{17.9}$$

β being a constant fixed by Eq. (4.4).

We now complete the definition of T by putting $\alpha = 1/\beta$; hence,

$$T = \Theta. \tag{17.10}$$

This convention made possible by the result (17.8) has been anticipated on several occasions.

The lower limit $T = 0$ of the thermodynamic temperature is called the *absolute zero*. We shall revert to the properties of matter extrapolated to $T = 0$ in Sections XXIII and XXIV.

In recent years, there has been a vivid discussion on states of matter referred to as states of "negative absolute temperature." Such questions arise if one considers nuclear spin systems in crystals for very short time intervals. Then it is expedient to introduce a "lattice temperature" and a "spin temperature" and the latter may become negative. Clearly, this is an extension of the concept of temperature ultimately justified by statistical mechanics. The temperature as discussed in thermodynamics is, by definition, always positive and has but one value for a given macroscopic region no matter how complex from a molecular point of view.

Nevertheless, if we allowed negative thermodynamic temperatures from the very beginning, some interesting general questions would arise. But we shall not go into details here.*

XVIII. Heat Capacity

We consider the entropy S of a region as a function of temperature T and of certain other internal state variables z_j to be specified later. Then the quantity

$$C_z \equiv T(\partial S/\partial T)_{z_j} \tag{18.1}$$

is an extensive property having the dimension of entropy and called *heat capacity* of the region. The reason for the name will appear later.

If we denote the total mass and the total amount of substance of the region by m and n, respectively, then the expressions

$$\tilde{C}_z \equiv C_z/m, \qquad \bar{C}_z \equiv C_z/n \tag{18.2}$$

are called *specific heat capacity* and *molar heat capacity*, respectively [see Eqs. (6.1) and (6.2)]. They are intensive properties having the dimensions of entropy divided by mass and of entropy divided by amount of substance, respectively.

The units to be recommended for C_z, $\tilde{C}_z$, and $\bar{C}_z$ are J K^{-1}, J K^{-1} kg^{-1}, and J K^{-1} mol^{-1}, respectively.

The most important examples relate to the case where the z_j in Eq. (18.1) are either the work coordinates l_i and the amounts of substance n_k or the work coefficients λ_i and the n_k. We then have the heat capacity at constant work coordinates

$$C_l = T\left(\frac{\partial S}{\partial T}\right)_{l_i,n_k} = \left(\frac{\partial U}{\partial T}\right)_{l_i,n_k} \tag{18.3}$$

and the heat capacity at constant work coefficients

$$C_\lambda = T\left(\frac{\partial S}{\partial T}\right)_{\lambda_i,n_k} = \left(\frac{\partial H}{\partial T}\right)_{\lambda_i,n_k}, \tag{18.4}$$

where the internal energy U and the enthalpy H have been introduced

* A clever discussion on thermodynamics and statistical mechanics at negative thermodynamic temperature is due to Ramsey (1956).

by means of the two fundamental equations (16.5) and (16.6). The best-known special cases of C_l and C_λ are the heat capacity at constant volume C_V and the heat capacity at constant pressure C_P, both relating to simple regions.

For the heat $đQ$ supplied to a closed phase during an infinitesimal internal change of state, we derive from Eqs. (10.14) and (10.14b)

$$đQ = dU - đW_{\text{diss}} - \sum_i \lambda_i \, dl_i = dH - đW_{\text{diss}} + \sum_i l_i \, d\lambda_i \quad (18.5)$$
$$\text{(closed phase).}$$

Here, $đW_{\text{diss}}$ denotes the dissipated work. Comparing this to Eqs. (18.3) and (18.4), we find

$$đQ = C_l \, dT - đW_{\text{diss}} \qquad (l_i, n_k \text{ constant}), \qquad (18.6)$$

$$đQ = C_\lambda \, dT - đW_{\text{diss}} \qquad (\lambda_i, n_k \text{ constant}). \qquad (18.7)$$

The condition $n_k = \text{const}$ requires closed phases without chemical reactions. These relations show that in the absence of dissipative effects the heat $đQ$ received by the phase during an infinitesimal increase of temperature equals the product $C_l \, dT$ or $C_\lambda \, dT$, hence the name "heat capacity" for C_l or C_λ. If, on the other hand, we consider adiabatic changes of temperature produced by electric work ($đQ = 0$, $đW_{\text{diss}} \neq 0$), then Eqs. (18.6) and (18.7) reduce to the formulas for the calorimetric determination of heat capacities (cf. Section IX).

From Eqs. (18.3) and (18.4), there follows

$$dU = C_l \, dT + \sum_i \left(\frac{\partial U}{\partial l_i}\right)_{T, n_k, l_j} dl_i \qquad (n_k \text{ constant}), \qquad (18.8)$$

$$dH = C_\lambda \, dT + \sum_i \left(\frac{\partial H}{\partial \lambda_i}\right)_{T, n_k, \lambda_j} d\lambda_i \qquad (n_k \text{ constant}). \qquad (18.9)$$

Eqs. (16.18) and (16.19) show that

$$\left(\frac{\partial U}{\partial l_i}\right)_{T, n_k, l_j} = \lambda_i - T\left(\frac{\partial \lambda_i}{\partial T}\right)_{l_i, n_k} \qquad (18.10)$$

$$\left(\frac{\partial H}{\partial \lambda_i}\right)_{T, n_k, \lambda_j} = T\left(\frac{\partial l_i}{\partial T}\right)_{\lambda_i, n_k} - l_i. \qquad (18.11)$$

Thus, we obtain from Eqs. (18.5), (18.8), and (18.9) for an adiabatic change in a phase of fixed chemical content, in the absence of dissipative effects,

$$C_l\,dT - T\sum_i \left(\frac{\partial \lambda_i}{\partial T}\right)_{l_i,n_k} dl_i = C_\lambda\,dT + T\sum_i \left(\frac{\partial l_i}{\partial T}\right)_{\lambda_i,n_k} d\lambda_i = 0 \quad (18.12)$$

(adiabatic change, $W_{\text{diss}} = 0$, n_k constant).

This formula describes effects such as adiabatic changes of volume and pressure in simple phases, adiabatic magnetization or demagnetization, and the Joule–Gough effect (adiabatic dilatation of a wire, a rod, or of other linear material).

To establish the general relation between C_l and C_λ, we start from Eqs. (16.14) and (16.15):

$$\left(\frac{\partial S}{\partial l_i}\right)_{T,l_j,n_k} = -\left(\frac{\partial \lambda_i}{\partial T}\right)_{l_i,n_k} \quad (18.13)$$

$$\left(\frac{\partial S}{\partial \lambda_i}\right)_{T,\lambda_j,n_k} = \left(\frac{\partial l_i}{\partial T}\right)_{\lambda_i,n_k}. \quad (18.14)$$

From Eqs. (18.3), (18.4), (18.13), and (18.14), we obtain

$$\begin{aligned} T\,dS &= C_l\,dT - T\sum_i \left(\frac{\partial \lambda_i}{\partial T}\right)_{l_i,n_k} dl_i \\ &= C_\lambda\,dT + T\sum_i \left(\frac{\partial l_i}{\partial T}\right)_{\lambda_i,n_k} d\lambda_i \qquad (n_k \text{ constant}) \end{aligned} \quad (18.15)$$

and thus

$$C_l - C_\lambda = T\sum_i \left(\frac{\partial \lambda_i}{\partial T}\right)_{l_i,n_k} \left(\frac{\partial l_i}{\partial T}\right)_{\lambda_i,n_k}. \quad (18.16)$$

By means of this general equation (Haase, 1969), we may compute the difference $C_l - C_\lambda$ from data on the temperature dependence of the λ_i and l_i and hence from data on the equation of state of the region considered.

We note in passing that by virtue of Eq. (18.15) the relation (18.12) may be referred to as an equation for "isentropic" changes in a region of fixed content. The epithet "isentropic" obviously means the same thing as the epithet "adiabatic without dissipative effects" if we consider a region of fixed content and thus exclude chemical reactions and exchange of matter with the surroundings.

The best-known example of Eq. (18.16) relates to a simple region (volume V, pressure P). Here, Eq. (18.16) simplifies to

$$C_P - C_V = T\left(\frac{\partial P}{\partial T}\right)_{V,n_k}\left(\frac{\partial V}{\partial T}\right)_{P,n_k} \qquad \text{(simple region).} \qquad (18.17)$$

Introducing molar quantities* such as the molar volume $\bar{V}$ and taking account of the identity [see the definitions (18.19) and (18.20)]

$$(\partial P/\partial T)_{V,n_k} = \beta/\varkappa,$$

we find

$$\bar{C}_P - \bar{C}_V = T\bar{V}\beta^2/\varkappa \qquad \text{(simple region).} \qquad (18.18)$$

Here, the expressions

$$\beta \equiv \frac{1}{V}\left(\frac{\partial V}{\partial T}\right)_{P,n_k} \qquad (18.19)$$

and

$$\varkappa \equiv -\frac{1}{V}\left(\frac{\partial V}{\partial P}\right)_{T,n_k} \qquad (18.20)$$

are the thermal expansivity and the (isothermal) compressibility.

For perfect gases (including perfect gaseous mixtures) we have on account of Eqs. (5.3), (6.2), and (18.18)–(18.20).

$$\beta = 1/T, \qquad \varkappa = 1/P, \qquad \bar{C}_P - \bar{C}_V = R \qquad \text{(perfect gas),} \qquad (18.21)$$

R denoting the gas constant.

For a heterogeneous system composed of several phases (superscript α), we define the heat capacities C_l and C_λ corresponding to the expressions (18.3) and (18.4):

$$C_l \equiv \sum_\alpha C_l^\alpha, \qquad C_\lambda \equiv \sum_\alpha C_\lambda^\alpha. \qquad (18.22)$$

From Eqs. (15.5), (15.17), (18.3), (18.4), and (18.22), we obtain for a system of *uniform temperature* T

$$C_l = \left(\frac{\partial U}{\partial T}\right)_{l_i^\alpha, n_k^\alpha} = T\left(\frac{\partial S}{\partial T}\right)_{l_i^\alpha, n_k^\alpha}, \qquad (18.23)$$

$$C_\lambda = \left(\frac{\partial H}{\partial T}\right)_{\lambda_i^\alpha, n_k^\alpha} = T\left(\frac{\partial S}{\partial T}\right)_{\lambda_i^\alpha, n_k^\alpha}, \qquad (18.24)$$

* In our notation, the specific heat capacities are $\tilde{C}_P$ and $\tilde{C}_V$, the molar heat capacities $\bar{C}_P$ and $\bar{C}_V$, and the densities of the heat capacities $\hat{C}_P$ and $\hat{C}_V$. The quantity $\hat{C}_P$, for instance, occurs in the heat conduction equation for continuous systems.

where U, H, and S are the internal energy, the enthalpy, and the entropy of the heterogeneous system. In particular, if the work coefficients λ_i^α have uniform values λ_i, then we derive from Eq. (18.24)

$$C_\lambda = \left(\frac{\partial H}{\partial T}\right)_{\lambda_i, n_k^\alpha} = T\left(\frac{\partial S}{\partial T}\right)_{\lambda_i, n_k^\alpha}. \tag{18.25}$$

The heat capacities C_l and C_λ are always finite positive quantities, except at absolute zero and at critical points. This follows from the stability conditions and is confirmed by experience.

XIX. Generalized Gibbs–Duhem Relation

When we apply Eq. (6.12) to the Gibbs function G and take account of Eq. (16.8), we obtain the following general relation between the changes of temperature T, work coefficients λ_i, and chemical potentials μ_k:

$$S\,dT + \sum_i l_i\,d\lambda_i + \sum_k n_k\,d\mu_k = 0. \tag{19.1}$$

This formula valid for any region is said to be the *generalized Gibbs–Duhem relation*. Dividing this equation by the mass, the amount of substance, or the volume of the region, we get an analogous equation where the entropy S, the work coordinates l_i, and the amounts of substance n_k are replaced by the corresponding specific quantities, molar quantities, or generalized densities, respectively. We shall restrict the discussion of Eq. (19.1) to two important examples.

The first example refers to a simple region of volume V and pressure P. We then obtain from Eq. (19.1)

$$S\,dT - V\,dP + \sum_k n_k\,d\mu_k = 0 \tag{19.2}$$

or

$$\bar{S}\,dT - \bar{V}\,dP + \sum_k x_k\,d\mu_k = 0 \tag{19.3}$$

where $\bar{S}$, $\bar{V}$, and x_k denote molar entropy, molar volume, and mole fraction of substance k, respectively. Equation (19.2) or (19.3) is called the Gibbs–Duhem relation since it is due to Gibbs (1875) and Duhem (1886).

The second example concerns an interfacial layer of volume V and area ω. Then we derive from Eq. (19.1) (see Table I)

$$S\,dT - V\,dP + \omega\,d\sigma + \sum_k n_k\,d\mu_k = 0. \tag{19.4}$$

Here, σ is the interfacial tension and P the pressure which, for curved interfaces, is to be understood as any value between the two pressures in the two adjacent bulk phases. If we denote the thickness of the surface layer by τ, then we have

$$V = \tau\omega. \tag{19.5}$$

Defining the quantities

$$S_\omega \equiv S/\omega, \qquad \Gamma_k \equiv n_k/\omega, \tag{19.6}$$

we find from Eqs. (19.4) and (19.5)

$$S_\omega\,dT - \tau\,dP + d\sigma + \sum_k \Gamma_k\,d\mu_k = 0, \tag{19.7}$$

a very useful relation (Defay *et al.*, 1966; Guggenheim, 1967).

XX. Chemical Reactions

The decomposition of the infinitesimal increase dn_k of the amount of substance n_k of species k given in Eq. (10.2) may be extended from an open phase to any open region. Thus, we have, with a slight change of notation,

$$dn_k = d_R n_k + d_a n_k, \tag{20.1}$$

$d_R n_k$ denoting the increase of n_k due to chemical reactions within the region considered and $d_a n_k$ being the increase of n_k due to exchange of matter between the region and its surroundings.

When we write down the formula for a definite chemical reaction, then the numbers appearing as coefficients in conjunction with chemical symbols are called *stoichiometric numbers* and are denoted by ν_k. The sign convention is such that ν_k is positive if the species k appears on the right-hand side and is negative if the species k appears on the left-hand side of the chemical formula. For example, in the reaction

$$N_2\ (\text{gas}) + 3H_2\ (\text{gas}) \rightleftharpoons 2NH_3\ (\text{gas}),$$

the stoichiometric numbers have the following values:

$$\nu_{N_2} = -1, \qquad \nu_{H_2} = -3, \qquad \nu_{NH_3} = 2.$$

In general, the ν_k are rational numbers.

When several chemical reactions take place simultaneously in the region considered, we assign double subscripts to the stoichiometric numbers, the first relating to the reacting species and the second referring to the particular chemical reaction. Thus, ν_{kr} is the stoichiometric number of species k in reaction r.

We define

$$đ\xi_r \equiv (d_R n_k)_r / \nu_{kr} \tag{20.2}$$

where $(d_R n_k)_r$ is the infinitesimal increase of n_k due to the chemical reaction r. The quantity ξ_r is called the *extent of reaction* of the chemical reaction r. Hence, the progress of each reaction can be described by one variable ξ_r. We write $đ\xi_r$ (inexact differential) since ξ_r is a function of state only if the system is closed and if all the reactions are independent. When we later treat a single reaction in a closed system, we will accordingly write $d\xi$ (exact differential).

To show that the definition (20.2) is consistent with the definition of the stoichiometric numbers ν_{kr}, we recall that the law of conservation of mass requires the equation

$$\sum_k M_k \nu_{kr} = 0 \tag{20.3}$$

to hold for each reaction, M_k being the molar mass of species k. Inserting this into Eq. (20.2), we obtain

$$\sum_k M_k (d_R n_k)_r = 0,$$

which expresses the fact that the total mass of the region is left unaltered by the occurrence of a chemical reaction inside the region.

We obviously have

$$d_R n_k = \sum_r (d_R n_k)_r;$$

hence, by Eqs. (20.1) and (20.2)

$$dn_k = \sum_r \nu_{kr}\, đ\xi_r + d_a n_k, \tag{20.4}$$

a relation valid for any region with an arbitrary number of chemical reactions.

If we denote the chemical potential of the reacting species k by μ_k, then the quantity

$$A_r \equiv -\sum_k \nu_{kr}\mu_k \tag{20.5}$$

is the *affinity* of reaction r. This concept is due to De Donder (1922) (see Prigogine and Defay (1954)). The affinity depends on the temperature T, the work coefficients λ_i, and the composition of the region considered.

The extent of reaction ξ_r is an extensive property having the dimension of amount of substance. The affinity A_r is an intensive property having the dimension of molar energy. The units recommended for ξ_r and A_r are therefore mol and J mol^{-1}, respectively.

Combining Eqs. (20.4) and (20.5), we obtain

$$\sum_k \mu_k\, dn_k = -\sum_r A_r\, đ\xi_r + \sum_k \mu_k\, d_a n_k. \tag{20.6}$$

This equation may be substituted into the fundamental equations (16.5)–(16.8).

When we consider a single chemical reaction, we may omit the subscript r. We thus have, according to Eqs. (20.4)–(20.6),

$$dn_k = \nu_k\, đ\xi + d_a n_k, \tag{20.7}$$

$$A = -\sum_k \nu_k \mu_k, \tag{20.8}$$

$$\sum_k \mu_k\, dn_k = -A\, đ\xi + \sum_k \mu_k\, d_a n_k. \tag{20.9}$$

We define

enthalpy of reaction $$h \equiv \sum_k \nu_k H_k, \tag{20.10}$$

entropy of reaction $$s \equiv \sum_k \nu_k S_k, \tag{20.11}$$

where H_k and S_k are the partial molar enthalpy and the partial molar entropy of species k, respectively. Denoting any partial molar quantity by Z_k, we generalize the definitions (20.10) and (20.11):

$$z \equiv \sum_k \nu_k Z_k. \tag{20.12}$$

The extensive property Z to which Z_k is connected may be, for example, the enthalpy H, the entropy S, or the Gibbs function G. It is obvious from Eqs. (20.8) and (20.12) that the affinity A might be written $-g$, since the chemical potential μ_k is the partial molar Gibbs function. Taking into account the relation (13.19)

$$\mu_k = H_k - TS_k, \tag{20.13}$$

we derive from Eqs. (20.8), (20.10), and (20.11)

$$A = Ts - h. \tag{20.14}$$

Using Eqs. (16.11), (20.8), (20.11), and (20.14), we find

$$\left(\frac{\partial A}{\partial T}\right)_{\lambda_i, n_k} = s, \tag{20.15}$$

$$T\left(\frac{\partial A}{\partial T}\right)_{\lambda_i, n_k} - A = h. \tag{20.16}$$

All these formulas hold for a single reaction in both closed and open regions.

We restrict the following discussion to a *single reaction in a closed region.* We then have, by virtue of Eqs. (20.7) and (20.9),

$$dn_k = \nu_k \, d\xi, \tag{20.17}$$

$$\sum_k \mu_k \, dn_k = -A \, d\xi. \tag{20.18}$$

For any extensive property Z, we may write, owing to Eqs. (6.10), (20.12), and (20.17)

$$dZ = \sum_k Z_k \, dn_k = z \, d\xi \qquad (T, \lambda_i \quad \text{constant}); \tag{20.19}$$

hence,

$$\left(\frac{\partial Z}{\partial \xi}\right)_{T, \lambda_i} = z, \tag{20.20}$$

or, in particular, for $Z = H$, $Z = S$, and $Z = G$, respectively,

$$\left(\frac{\partial H}{\partial \xi}\right)_{T, \lambda_i} = h, \qquad \left(\frac{\partial S}{\partial \xi}\right)_{T, \lambda_i} = s, \qquad \left(\frac{\partial G}{\partial \xi}\right)_{T, \lambda_i} = -A. \tag{20.21}$$

If we insert the last relations into Eq. (15.12)

$$G = H - TS, \tag{20.22}$$

we recover Eq. (20.14).

We now extend the discussion to a *single reaction in a closed heterogeneous system* of uniform temperature T and of uniform work coefficients λ_i. For the sake of simplicity, we impose the conditions that each reacting species occurs in but one phase of the system and that there are no phase transitions such as melting or vaporization. Then phase superscripts can be omitted in quantities such as n_k and Z_k, and the symbol dn_k always refers to a change brought about by the chemical reaction considered.

An example is the heterogeneous reaction (see Table IV)

$$\text{Ag (solid)} + \text{HgCl (solid)} \rightarrow \text{Hg (liquid)} + \text{AgCl (solid)} \qquad (25°\text{C, 1 atm}), \tag{20.23}$$

where we have a simple system composed of four pure phases all at the same temperature and pressure.

Evidently, we can define the stoichiometric numbers ν_k, the extent of reaction ξ, and the affinity A in a way analogous to the previous definitions. Thus, for the heterogeneous reaction (20.23), we have by Eqs. (20.8) and (20.17)

$$\nu_{\text{Ag}} = -1, \qquad \nu_{\text{HgCl}} = -1, \qquad \nu_{\text{Hg}} = 1, \qquad \nu_{\text{AgCl}} = 1, \tag{20.24}$$

$$d\xi = dn_k/\nu_k = -dn_{\text{Ag}} = -dn_{\text{HgCl}} = dn_{\text{Hg}} = dn_{\text{AgCl}}, \tag{20.25}$$

$$A = -\sum_k \nu_k \mu_k = \mu_{\text{Ag}} + \mu_{\text{HgCl}} - \mu_{\text{Hg}} - \mu_{\text{AgCl}}. \tag{20.26}$$

In view of the fact that the intensive properties T and λ_i have uniform values and that there is only one extent of reaction ξ and only one affinity A, we can likewise use the relations (20.10)–(20.22). The quantity Z (including G, S, and H) now refers to the whole heterogeneous system, while the partial molar quantities Z_k (including μ_k, S_k, and H_k) relate to that phase in which the species k occurs.

Let us consider a chemical reaction proceeding at fixed temperature T and at fixed work coefficients λ_i. When the extent of reaction increases by $\Delta\xi$, the corresponding increase ΔZ of any extensive property Z is given by

$$\Delta Z = \int z\, d\xi \qquad (T, \lambda_i \quad \text{constant}), \tag{20.27}$$

as follows immediately from Eq. (20.19). Applying this formula to the Gibbs function G or to the enthalpy H, for instance, we find

$$\Delta G = -\int A\, d\xi \qquad (T, \lambda_i \quad \text{constant}), \tag{20.28}$$

$$\Delta H = \int h\, d\xi \qquad (T, \lambda_i \quad \text{constant}), \tag{20.29}$$

where Eq. (20.21) has been used. The quantity z ($-A$, h, etc.) still depends on the compositions of the phases involved in the reaction. Since, in general, these compositions change with the progress of the reaction, z is a function of ξ. There are, however, exceptions, such as the heterogeneous reaction (20.23), where all phases consist of but one substance. Then z is constant and Eqs. (20.27)–(20.29) can be written

$$\Delta Z = z\,\Delta\xi, \quad \Delta G = -A\,\Delta\xi, \quad \Delta H = h\,\Delta\xi \quad (T, \lambda_i \text{ constant}). \tag{20.30}$$

If we stipulate that

$$\Delta\xi = \xi^\dagger \tag{20.31}$$

or, according to Eq. (20.17),

$$\Delta n_k = \nu_k n^\dagger \tag{20.32}$$

where

$$\xi^\dagger = n^\dagger = 1 \quad \text{mol}, \tag{20.33}$$

then we say that there occurs "one formula equivalent of reaction." We obtain from Eqs. (20.30) and (20.31)

$$\Delta Z = z\xi^\dagger, \quad \Delta G = -A\xi^\dagger, \quad \Delta H = h\xi^\dagger \quad (T, \lambda_i \text{ constant}). \tag{20.34}$$

Here, ΔZ, ΔG, and ΔH denote the increases of Z, G, and H due to the occurrence of one formula equivalent of the reaction.

The foregoing discussion shows that we have to distinguish carefully between the integral quantity of reaction ΔZ and the differential quantity of reaction z. In particular, ΔH is the integral enthalpy of reaction, while h is the differential enthalpy of reaction. The quantity ΔH, is the quantity called "heat of reaction" in Sections IX and XV. It is an extensive property, while h is an intensive property. Even in the exceptionally simple case (20.34), ΔH and h have different dimensions.*

It also should be noted that the quantity z defined by Eq. (20.12) refers to the *instantaneous* description of the reaction and that z is still meaningful for an open region, that is, for an open phase or for a volume element of a continuous system. The quantity ΔZ, however, only relates to a definite change in a closed system at fixed values of the temperature and of the work coefficients.

* The nearly universal practice of writing ΔH for quantities such as differential enthalpy of reaction, molar heat of vaporization, molar heat of mixing, etc. is therefore misleading.

The possible sign combinations for ΔG, ΔH, and ΔS and thus for $-A$, h, and s follow from our considerations in Section XV (see Table IV). But it should be stressed that the inequality

$$\int A\, d\xi > 0 \tag{20.35}$$

implied by the relations (15.31) and (20.28) can be derived directly from Eqs. (14.12) and (20.18), the only essential limitation being the uniformity of temperature. If we define ξ is such a way that $\Delta\xi > 0$, then we obtain

$$A > 0. \tag{20.36}$$

Obviously, the validity of this formula is not restricted to reactions taking place at constant temperature and constant work coefficients.

XXI. Galvanic Cells

A *galvanic cell* is a closed heterogeneous system containing one or more electrolytes and two or more electronic conductors (usually metals). For the sake of brevity, we shall only treat cells which are simple systems of uniform temperature and of uniform pressure and thus can be described by the temperature, the pressure, and the compositions of all the phases involved.

An example is the galvanic cell represented by the phase diagram

$$\text{Pt} \mid \text{Ag} \mid \text{AgCl} \mid \text{MCl (aq.)} \mid \text{HgCl} \mid \text{Hg} \mid \text{Pt} \qquad (25°\text{C, 1 atm}), \tag{21.1}$$

where all phases are solid except the two liquid phases Hg (liquid mercury) and MCl(aq.) (aqueous solution of a chloride such as HCl or KCl). The platinum wires (Pt) are called "terminals," the two-phase systems Ag + AgCl and Hg + HgCl are said to be the "electrodes." We assume local equilibrium to be established everywhere. This implies dissociation equilibrium in the solution and partition equilibrium with respect to the distribution of Cl^- at the two electrode–solution interfaces. Then the galvanic cell is called a "reversible cell," since nothing happens at zero electric current provided we do not consider the effect of the solubility of AgCl or HgCl in the solution, which is negligible within experimental accuracy. Any other kind of galvanic cell is said to be an "irreversible cell" since then, at zero electric current, irreversible processes take place inside the cell—for example, diffusion and chemical reactions. A galvanic cell at zero electric current is called an open cell.

We define the *electromotive force* (emf) of any galvanic cell (whether reversible or irreversible) by the following convention*:

The electromotive force of a galvanic cell is the difference between the electric potential of the terminal on the right and the electric potential of the similar terminal on the left, the cell being open and the local heterogeneous equilibria at the phase boundaries and the local chemical equilibria being established.

We now restrict the discussion to "chemical cells," i.e., to reversible cells of uniform temperature and uniform pressure. Here, the passage of a definite electric charge is uniquely coupled to a definite extent of a heterogeneous reaction. Thus, in the cell (21.1) one formula equivalent of the reaction (20.23)

$$\text{Ag (solid)} + \text{HgCl (solid)} \rightarrow \text{Hg (liquid)} + \text{AgCl (solid)} \qquad (25°\text{C, 1 atm}) \tag{21.2}$$

occurs when the amount of substance of the electrons entering the system at the right and leaving it at the left is $n^{\dagger} = 1$ mol. This corresponds to the passage of the charge $n^{\dagger}Le$ of positive electricity through the cell from left to right, L denoting the Avogadro constant and e the elementary charge. Since the quantity

$$\mathfrak{F} = Le \approx 96490 \quad \text{C mol}^{-1} \tag{21.3}$$

is called the *Faraday constant*, the charge

$$n^{\dagger}\mathfrak{F} \approx 96490 \quad \text{C}$$

is briefly referred to as one faraday (symbol: Far).

If we change the direction of the electric current flowing through the cell, then the chemical reaction associated with the charge transfer is reversed, too. Let E be the difference between the electric potential of the terminal on the left and the electric potential of the terminal on the right produced by an external source of electricity. If we continuously vary E in either direction until we reach the point where E is counterbalanced by the electromotive force Φ of the cell ($E = -\Phi$), then we have the limiting case of zero electric current and of a reversible chemical

* This is an extended form of the "Stockholm Convention," a recommendation issued by the International Union of Pure and Applied Chemistry (Stockholm, 1953).

change. The electric work W^* done on the cell is Eq, where q is the charge of positive electricity transferred from left to right. In the limiting case of the reversible change, we therefore obtain

$$W^*_{\text{rev}} = -q\Phi.$$

Here, W^*_{rev} has the same meaning as in Eq. (15.28) involving isothermal and isobaric processes in closed simple systems. We thus find for the increase of the Gibbs function G

$$\Delta G = W^*_{\text{rev}} = -q\Phi. \tag{21.4}$$

Now, Eq. (20.34) is always applicable since, even for cells containing phases of variable compositions, the reversible change is a fictitious chemical reaction in which all the instantaneous compositions are held constant. Equation (20.34) refers to one formula equivalent of reaction and this can be made to correspond to the passage of one faraday (1 Far) through the cell from the left to the right:

$$\Delta G = -An^{\dagger}, \qquad q = n^{\dagger}\mathfrak{F}.$$

Thus, we have

$$\mathfrak{F}\Phi = A, \tag{21.5}$$

A denoting the affinity of the "cell reaction," i.e., of the chemical reaction associated with the flow of the positive charge 1 Far from left to right at the given values of temperature and pressure.

If the cell reaction actually occurs outside the galvanic cell, it is an irreversible change proceeding in a definite direction. Let us call this the "natural direction." Combining Eq. (21.5) with the relations (15.31) and (20.34) or with the general inequality (20.36), we obtain

$$A > 0, \qquad \Phi > 0 \qquad \text{(natural direction)}.$$

The phase diagram (21.1) happens to have been written down in such a way that the cell reaction (21.2) proceeds in the natural direction, the affinity A being given explicitly by Eq. (20.26). But when we interchange the electrodes in (21.1) then the signs of both the electromotive force Φ and the affinity A are reversed ($\Phi < 0$, $A < 0$), the cell reaction now taking place in the opposite direction.

Let us consider the chemical cell

$$\text{Pt} \mid \text{Pb} \mid \text{PbSO}_4 \mid \text{H}_2\text{SO}_4\ \text{(aq.)} \mid \text{H}_2\text{(Pt)} \mid \text{Pt} \qquad (25°\text{C},\ 1\ \text{atm}), \tag{21.6}$$

the cell reaction being

$$\tfrac{1}{2}\mathrm{Pb}\ (\mathrm{solid}) + \tfrac{1}{2}\mathrm{H_2SO_4}\ (\mathrm{aq.}) \rightarrow \tfrac{1}{2}\mathrm{H_2}\ (\mathrm{gas}) + \tfrac{1}{2}\mathrm{PbSO_4}\ (\mathrm{solid}). \qquad (21.7)$$

The affinity A of this reaction can be derived from Eq. (20.8):

$$A = \tfrac{1}{2}\mu_{\mathrm{Pb}} + \tfrac{1}{2}\mu_{\mathrm{H_2SO_4}} - \tfrac{1}{2}\mu_{\mathrm{H_2}} - \tfrac{1}{2}\mu_{\mathrm{PbSO_4}}. \qquad (21.8)$$

Here, μ_k is the chemical potential of substance k.

There is a striking difference between the cells (21.1) and (21.6). In the cell (21.1), both electrodes are permeable to the same species (the anionic species Cl^-); therefore, the electrolyte does not appear in the cell reaction (21.2); consequently, the electromotive force Φ of this cell, given by Eqs. (20.26) and (21.5), is independent of the electrolyte concentration and of the nature of the cationic species. In the cell (21.6), however, the electrode on the right, H_2(Pt), is permeable to the cationic species H^+, while the electrode on the left, Pb + $PbSO_4$, is permeable to the anionic species SO_4^{2-}; thus, the electrolyte H_2SO_4 participates in the cell reaction (21.7); consequently, the electromotive force Φ of the cell (21.6), given by Eqs. (21.5) and (21.8), depends on the composition of the electrolyte solution. In order to complete the description of the cell (21.6), we have to specify the concentration of the sulfuric acid. Even the signs of Φ of A are determined by the acid concentration. It is this type of cell which allows of an experimental investigation of the thermodynamic properties of electrolyte solutions.

The electromotive force Φ may be considered to be a function of temperature T and pressure P if we assign fixed values to the compositions of all relevant phases. Then we obtain from Eqs. (20.15), (20.16), and (21.5)

$$\mathfrak{F}(\partial\Phi/\partial T)_P = s, \qquad (21.9)$$

$$\mathfrak{F}[T(\partial\Phi/\partial T)_P - \Phi] = h \qquad (21.10)$$

where s and h are the entropy of reaction and enthalpy of reaction of the cell reaction, respectively. Thus, A, s, and h may be derived from experimental determinations of the emf and its temperature dependence. The value of h may be checked by direct calorimetric measurements (see Section IX).

The relations (21.5), (21.9), and (21.10) are implicitly or explicitly contained in papers by Gibbs (1878, 1887) and by Helmholtz (1882). Equation (21.10) is often called the Gibbs–Helmholtz relation. This name is also given to Eq. (20.16).

If we express $\mathfrak{F}$ and Φ in the units C mol^{-1} and V, respectively, then, by means of Eqs. (21.5), (21.9), and (21.10), we obtain A, h, and Ts in the unit J mol^{-1}.

As an example, we mention the results (Gerke, 1922) of the emf measurements on the cell (21.1):

$$\Phi = 0.0455 \quad \text{V}, \qquad \partial\Phi/\partial T = 0.000338 \quad \text{V K}^{-1};$$

hence, by Eqs. (21.3), (21.5), (21.9), and (21.10) for $T = 298.15$ K

$$A = 4390 \text{ J mol}^{-1}, \qquad s = 32.61 \text{ J K}^{-1}\text{ mol}^{-1}, \qquad h = 5333 \text{ J mol}^{-1},$$

all these quantities relating to the reaction (21.2) (see also Table IV).

XXII. Equilibrium

According to Section II, there is equilibrium inside any system if no changes occur within the system even if the system is cut off from all external influences, except stationary external force fields possibly present. The external fields must, in general, be excluded from the conditions of enclosure, since, otherwise, phenomena such as the equilibrium distribution of matter in a gravitational or centrifugal field could not be treated. But we shall restrict the discussion to problems without external fields. Then the system, after the enclosure mentioned, is an isolated system. Thus, the relations (11.8) and (11.9) hold. Hence, the entropy S of an isolated system in equilibrium cannot increase. The equilibrium state is therefore distinguished from all neighboring nonequilibrium states by the highest value of the entropy. If a virtual first-order displacement from equilibrium into neighboring nonequilibrium states is denoted by the symbol δ (variation of first order), then we obtain

$$\delta S \leq 0 \qquad \text{(isolated system).} \tag{22.1}$$

This is the general criterion for equilibrium after Gibbs and Planck.

If all virtual displacements are bidirectional, that is, possible in both directions, then the equality sign holds in (22.1). If there are among the virtual displacements unidirectional changes, i.e., variations possible in one direction only, then the general form of (22.1) must be retained; this is essential for the derivation of the stability conditions.

For brevity, we treat the case of bidirectional variations only. Thus, we rewrite (22.1) as

$$\delta S = 0 \qquad \text{(isolated system).} \tag{22.2}$$

In this case, the entropy shows a stationary value at equilibrium, i.e., a maximum with horizontal tangent when plotted against any variable measuring the distance from equilibrium.

To illustrate the procedure of deriving equilibrium conditions, we shall restrict the further discussion to two simple examples. For more complicated problems of equilibrium (including stability) the reader should consult later chapters.

The first example concerns a single chemical reaction (extent of reaction ξ, affinity A) in a simple homogeneous system. The Gibbs equation (13.11) is now to be applied to the virtual displacements considered in Eq. (22.2). In view of the fact that the variations of both internal energy and volume vanish for an isolated system, we obtain from Eqs. (13.11), (20.18), and (22.2)

$$A\,\delta\xi = 0.$$

Since this holds for any value of $\delta\xi$, we derive

$$A = 0 \qquad \text{(chemical equilibrium).} \tag{22.3}$$

This is the condition for *homogeneous chemical equilibrium*. It can be shown that for any number of reactions in an arbitrary system (homogeneous, heterogeneous, or continuous system) the affinity of any reaction is zero at equilibrium.

The second example relates to a heterogeneous system consisting of two simple phases (superscripts $'$ and $''$), chemical reactions being excluded. Thus, there may be an exchange of matter or flow of heat between the two phases and an increase of volume of one phase at the expense of the volume of the other phase. Hence, the problem is to find the equilibrium distribution of mass, energy, and volume between the two phases of the discontinuous system. The Gibbs equation (13.11) is now to be applied to the virtual displacements in each phase and is to be combined with the general condition (22.2) and with the subsidiary conditions expressing the fact that the system is isolated. These conditions state that for the whole system the variations of entropy, internal energy, volume, and amount of substance of each component vanish at equilibrium. Hence,

$$\left(\frac{1}{T'} - \frac{1}{T''}\right)\delta U' + \left(\frac{P'}{T'} - \frac{P''}{T''}\right)\delta V' - \sum_k \left(\frac{\mu_k{'}}{T'} - \frac{\mu_k{''}}{T''}\right)\delta n_k{'} = 0,$$

T denoting the temperature, U the internal energy, P the pressure, V the volume, μ_k the chemical potential of component k, and n_k the amount of substance k. Since each of the variations $\delta U'$, $\delta V'$, $\delta n_k'$ is independent, there follows

$$T' = T'' \quad \text{(thermal equilibrium)}, \tag{22.4}$$

$$P' = P'' \quad \text{(mechanical equilibrium)}, \tag{22.5}$$

$$\mu_k' = \mu_k'' \quad \text{(distribution equilibrium)}. \tag{22.6}$$

These are the conditions for *heterogeneous equilibrium*. They can easily be extended to a multiphase system.

The condition (22.4), at first sight, seems to be trivial, since we used a similar relation in the definition of temperature. But it should be stressed that it has been shown here that Eq. (22.4) is not only valid for equilibrium across a thermally conducting wall, but also for any kind of contact between two phases.

The condition (22.5) ceases to be applicable if the interface dividing the two phases is rigid and semipermeable (i.e., permeable only to a certain number of species), as in the problems of osmotic equilibrium.

The condition (22.6) continues to hold for any kind of contact (including semipermeable membranes) provided the subscript k refers to only those species that can pass the interface between the two phases. If the subscript k relates to an ionic (or electronic) species, then μ_k is to be replaced by the "electrochemical potential"

$$\eta_k = \mu_k + z_k \mathfrak{F} \psi$$

where z_k is the charge number of species k, $\mathfrak{F}$ the Faraday constant, and ψ the electric potential.

XXIII. Nernst Heat Theorem

The "Nernst heat theorem" is concerned with the behavior of matter when extrapolated to absolute zero, that is, to the limiting value $T = 0$ of the thermodynamic temperature T.

According to Eqs. (18.4) and (18.25), the entropy S of any homogeneous or heterogeneous system at uniform temperature T' and at uniform work coefficients may be written

$$S = S_0 + \int_0^{T'} (C_\lambda/T)\, dT \tag{23.1}$$

provided there is no phase change between $T = 0$ and T'. Here, C_λ is the heat capacity at constant work coefficients and S_0 is the limiting value of the entropy at absolute zero, obtained by smooth extrapolation from the lowest temperature accessible by experiment to $T = 0$, the work coefficients and the amounts and compositions of all the phases involved being kept constant. Both quantum theory and experience show that

$$\lim_{T \to 0} C_\lambda = 0. \tag{23.2}$$

Thus, the integral in Eq. (23.1) converges and S_0 has a finite value which might be zero or nonzero. The quantity S_0 may still depend on the work coefficients and on the chemical content of the system.

Let us consider an isothermal change I → II at the temperature T' and let us denote the increase of entropy due to this change by $S_{\text{II}} - S_{\text{I}}$. We then have, in view of Eq. (23.1),

$$S_{\text{II}} - S_{\text{I}} = S_0^{\text{II}} - S_0^{\text{I}} + \int_0^{T'} \frac{C_\lambda^{\text{II}} - C_\lambda^{\text{I}}}{T}\, dT, \tag{23.3}$$

the superscripts I and II referring to the initial and final states of the system. The isothermal process I → II may be (a) a change of a work coefficient, e.g., a change of pressure or of magnetic field strength, in a single phase of fixed content, (b) a phase transition, e.g., an allotropic change or melting or evaporation, at fixed work coefficients, T' now being the equilibrium temperature, (c) a chemical reaction at fixed work coefficients, (d) a mixing or unmixing process at fixed work coefficients.

The last change involves mixtures, while all the other processes can also occur with pure phases. The quantity on the left-hand side of Eq. (23.3) can be determined from experiments. Knowing the integral in (23.3) from direct calorimetric measurements, we are thus able to compute the expression $S_0^{\text{II}} - S_0^{\text{I}}$.

Before proceeding any further, we have to discuss the concept of "internal metastability." For brevity, we shall restrict this discussion to pure phases.

Consider any one-component homogeneous system of fixed content and at a given temperature and at given values of all the work coefficients. Then, usually, there is no possibility of a change of state, e.g., of a change of volume. Phases showing this normal behavior are said to be ***internally stable***. In a few cases, however, there is indeed a change possible, at least in principle. Phases that exhibit this unusual kind of behavior are called *internally metastable*. Here, obviously, variables such as temperature and pressure do not suffice to determine the state of the system. We must introduce at least one more variable. Such a variable is called an "internal variable." If the internal variable has not reached its equilibrium value, then there is lack of "internal equilibrium" and the phase considered is internally metastable. Microscopically, the internal variable relates to some kind of randomness in the arrangement of the molecules. In an internally metastable system, there is a molecular arrangement which has been "frozen" at a higher temperature so that the arrangement at lower temperatures no longer corresponds to internal equilibrium. If, then, the metastability could be removed, there would be an internal change leading to internal equilibrium and thus to an internally stable phase.

Examples of internally metastable phases are: glasses (supercooled liquids at temperatures below the glass temperature), as well as crystals of CO, NO, N_2O, and H_2O, at very low temperatures. In the crystals mentioned, an orientational randomness is retained while the non-realizable ordered orientation corresponds to the internally stable state. In glasses, there are similar phenomena, but these are less well understood quantitatively. In the vicinity of the glass temperature, it is sometimes possible to observe the transformations glass → liquid (internally stable) or glass → crystal (internally stable).

As will be evident by now, the question of whether a phase is internally stable or metastable has nothing to do with metastability in phase diagrams. Thus, at 25°C and 1 atm, elementary carbon may occur as graphite or as diamond. But diamond is metastable with respect to graphite under the conditions mentioned. Nevertheless, both crystalline forms are internally stable, even at the lowest temperatures accessible by experiment.

We are now ready to formulate the modernized form of the *Nernst heat theorem*:

(1) If I and II refer to different values of a work coefficient in any phase of fixed content, then*

$$S_0^{\mathrm{I}} = S_0^{\mathrm{II}}. \tag{23.4}$$

* From Eqs. (16.15), (16.17), and (18.19) we conclude that, in view of Eq. (23.4), for isotropic media both the thermal expansivity and the temperature derivative of the magnetic polarization (at constant magnetic field strength, pressure, and composition) vanish at absolute zero.

(2) If the change I → II represents a phase transition or a chemical reaction taking place between pure phases all internally stable, then

$$S_0^{\mathrm{I}} = S_0^{\mathrm{II}}. \tag{23.5}$$

(3) If, for a given amount of a pure component, I denotes an internally metastable state and II denotes an internally stable state, then

$$S_0^{\mathrm{I}} > S_0^{\mathrm{II}}. \tag{23.6}$$

The relations (23.4) and (23.5) lead to the simple conclusion that, for any pure substance in internal equilibrium, S_0 is independent of the work coefficients and of the chemical state.

Nernst (1906) was the first to enunciate statements equivalent to (23.4) and (23.5). The rigorous formulation, in particular, with respect to internal metastability, the critical evaluation of careful measurements, and the interpretation in terms of the concepts of statistical mechanics are due to later authors (Fowler, Lewis, Planck, Schottky, and Simon).*

To take account of all the facts just considered, Planck (1911) proposed the convention valid for any pure substance

$$S_0 = 0 \qquad \text{(internally stable phase).} \tag{23.7}$$

We then have, on account of (23.6),

$$S_0 > 0 \qquad \text{(internally metastable phase).} \tag{23.8}$$

By this "Planck convention," the entropy of any system in any state is uniquely fixed.

An alternative name often used for the Nernst heat theorem is "third law of thermodynamics." This designation is also applied to the "theorem of unattainability of absolute zero" to be treated in the next section. Neither use can be recommended. The following main objections should be borne in mind:

(1) The zeroth, first, and second laws of thermodynamics introduce certain basic thermodynamic functions, while the "third law" does not.

(2) There are other general laws concerning the limiting behavior of matter, e.g., the formulas for the entropy of a gas at high temperatures or the limiting laws for thermodynamic functions of mixtures at infinite

* For historical details, see Haase and Jost (1956).

dilution. These relations are of comparable importance to and of a similar character as the Nernst heat theorem.

(3) The unattainability of absolute zero is by no means as fundamental as it appeared at the time of its discovery (see Section XXIV).

XXIV. Unattainability of Absolute Zero

The "theorem of unattainability of absolute zero" was first stated by Nernst in 1912. It may be reformulated in the following way (Guggenheim, 1967):

It is impossible by any procedure, no matter how idealized, to reduce the temperature of any system to absolute zero in a finite number of finite operations.

Nernst (1912, 1918) claimed to have shown that: (1) the unattainability of absolute zero follows from the first and second laws of thermodynamics, and (2) the Nernst heat theorem is derivable from the unattainability of absolute zero.

Though later authors (Einstein among them) objected to the first statement, Nernst maintained that his proof was correct. Thus, the question remained unsettled (see Simon (1951)). Modern textbooks simply ignore the problem.

The second statement, corroborated by a proof due to Lorentz (1913), relating to a special case, has been accepted by most authors. Modern textbooks usually make the sweeping statement that the Nernst heat theorem and the unattainability theorem are equivalent.

We shall now prove (Haase, 1956b, 1957) that the truth lies between Nernst's assertions and the statements of modern authors.

We may restrict the discussion of unattainability to adiabatic processes, since any change can be decomposed into an isothermal and an adiabatic change, the isothermal process contributing nothing to the lowering of temperature.

A. Notation

We consider a homogeneous or heterogeneous system of uniform temperature T and of uniform work coefficients in which there is an adiabatic change from the initial state I (initial temperature T_1) to the final state II (final temperature T_2). We use the superscripts I and II in any quantity to denote two different states (other than different temperatures) such as two pressures in a given phase, two phases of a given substance at fixed pressure, and so on. We here include mixtures and processes of mixing or unmixing. Here, H is the enthalpy, S the entropy of the system. The subscript 0 in any quantity refers to the value obtained by smooth extrapolation to $T = 0$.

B. Implications of the First and Second Laws

We first want to find out which conclusions can be drawn from the first and second laws of thermodynamics.

According to the first law (see Sections VIII and IX), the change of internal energy or of enthalpy equals zero for an adiabatic process, if the work coordinates or the work coefficients are kept constant, respectively, and if there is no excess work (see Section VII). We restrict the discussion to the enthalpy H, hence, to the case of fixed work coefficients. (The reasoning for the other case is completely analogous.) We then have, on account of Eqs. (9.14) and (18.25), for an adiabatic change I → II at fixed work coefficients and without excess work,

$$H_0^{\text{II}} - H_0^{\text{I}} + \int_0^{T_2} C_\lambda^{\text{II}}\,dT - \int_0^{T_1} C_\lambda^{\text{I}}\,dT = 0, \tag{24.1}$$

C_λ denoting the heat capacity at constant work coefficients, which is a positive quantity for $T > 0$. Obviously, this equation relates to processes such as phase changes, chemical reactions, and mixing or unmixing processes, carried out adiabatically, all work coefficients being kept constant. Thus, in the following discussion, changes of work coefficients (e.g., compression or expansion, magnetization or demagnetization) are excluded.

If the initial temperature T_1 and the final temperature T_2 coincide ($T_1 = T_2 \equiv T'$), the process I → II becomes isothermal. We may think of a phase change, a chemical reaction, or a mixing (unmixing) process, taking place at constant temperature, the work coefficients being left unaltered again. We now apply the relation (15.22), following from the first and second laws, according to which the Gibbs function cannot increase for an isothermal process I → II, carried out without excess work and at fixed work coefficients. Combining this result with Eqs. (15.19) and (18.25), we obtain

$$H_0^{\text{II}} - H_0^{\text{I}} - T'(S_0^{\text{II}} - S_0^{\text{I}}) + \int_0^{T'} (C_\lambda^{\text{II}} - C_\lambda^{\text{I}})\,dT - T'\int_0^{T'} \frac{C_\lambda^{\text{II}} - C_\lambda^{\text{I}}}{T}\,dT \leq 0. \tag{24.2}$$

Here, the equality sign or the inequality sign holds for a reversible or an irreversible path of the isothermal change I $\to$ II, respectively. The superscripts I and II have the same meaning in Eqs. (24.1) and (24.2). They may, for example, denote graphite and diamond at a given pressure.

The integrals in (24.2) converge and $S_0^{II} - S_0^{I}$ is always finite (see the beginning of Section XXIII). Thus, we find, if we proceed to the limit $T' \to 0$ in (24.2),

$$H_0^{II} - H_0^{I} \leq 0. \tag{24.3}$$

This is an important result, as we shall see immediately.

The relation (24.3) becomes an equality if the isothermal process I $\to$ II is reversible when extrapolated to $T = 0$. An example is the fusion or solidification of helium, which can be made to occur reversibly at absolute zero under the appropriate equilibrium pressure, since here the melting-point curve (pressure as a function of temperature for the solid–liquid equilibrium) continues down to $T = 0$. Thus, in passing, we note that the enthalpy of fusion (heat of fusion) of helium becomes zero at $T = 0$.* If, on the other hand, the isothermal change I $\to$ II is irreversible and remains so when extrapolated to absolute zero, then the inequality sign applies in (24.3). An example is a mixing or unmixing process at constant pressure.

Inserting (24.3) into (24.1) and remembering that $C_\lambda > 0$ for $T > 0$, we discover that the first integral in (24.1) and hence the final temperature T_2 cannot vanish, however small a value we assign to the initial temperature T_1. Therefore, absolute zero cannot be reached in a finite number of finite operations.

We have thus proved that the unattainability of absolute zero is implied by the first and second laws for processes such as phase changes, chemical reactions, mixing or unmixing processes.

Again, the second law states that the entropy never decreases for an

* A similar result is valid for any one-component, two-phase equilibrium provided the coexistence curve continues down to absolute zero. We can directly prove the vanishing of any "heat of phase transition" at $T = 0$ for such a system when we apply the equilibrium condition (22.6), which also follows from the first and second laws alone. Thus, in view of Eqs. (15.12) and (15.23) or (15.24), we have for any temperature T: $H^{I} - TS^{I} = H^{II} - TS^{II}$, the superscripts I and II referring to the two coexisting phases of the given substance. Proceeding to the limit $T \to 0$ along the equilibrium line, we obtain $H_0^{I} = H_0^{II}$. This equation coincides with the relation (24.3) if only the equality sign in (24.3) is considered.

adiabatic change [see Eqs. (11.8) and (11.9)]. Hence, we derive from Eq. (23.1)

$$S_0^{\mathrm{II}} - S_0^{\mathrm{I}} + \int_0^{T_2} \frac{C_\lambda^{\mathrm{II}}}{T} \, dT - \int_0^{T_1} \frac{C_\lambda^{\mathrm{I}}}{T} \, dT \geq 0. \tag{24.4}$$

Here, the equality sign applies when the two states can be connected by a reversible adiabatic path, while the inequality sign holds if the adiabatic process I → II can only take place irreversibly. The relation (24.4) is valid for *any* adiabatic change, in contrast to Eq. (24.1).

For processes at constant work coefficients, such as phase changes, chemical reactions, mixing or unmixing processes, the first and second laws imply the unattainability of absolute zero, as we have just seen. Thus, the first integral in (24.4) cannot vanish however small a value we assign to the second integral. Since the second integral has a negative sign, the term $S_0^{\mathrm{II}} - S_0^{\mathrm{I}}$ must be zero or negative to ensure a nonzero value of the final temperature T_2. Hence, we have

$$S_0^{\mathrm{II}} - S_0^{\mathrm{I}} \leq 0 \tag{24.5}$$

for the type of processes considered. When the adiabatic change I → II can only occur irreversibly, so that the $>$ is valid in (24.4), then in (24.5) both $<$ and $=$ are possible. If, however, the adiabatic process I → II can be made to take place reversibly, so that the equality sign applies in (24.4), then the reversible adiabatic change II → I is also possible. Thus, both (24.5) and the condition $S_0^{\mathrm{I}} - S_0^{\mathrm{II}} \leq 0$ must be fulfilled so that the equality sign holds in (24.5).

Let us apply the inequalities (24.3) and (24.5) to the irreversible process of mixing or unmixing in a system of two or more components, the work coefficients having fixed values. The superscripts I and II now relate to the pure components and to the mixture, or vice versa. In (24.5), we cannot decide which sign is valid, but in (24.3) the inequality sign holds if the isothermal change I → II in (24.2) remains irreversible when extrapolated to absolute zero. Hence, we derive the following statements implied by the first and second laws:

(a) If at absolute zero the mixture is metastable with respect to the pure components so that the process of unmixing is irreversible, then at $T = 0$ the enthalpy of mixing is positive and the entropy of mixing is positive or zero.

(b) If at absolute zero the pure components are metastable with respect to the mixture so that the process of mixing is irreversible, then at $T = 0$

the enthalpy of mixing is negative and the entropy of mixing is negative or zero.

Now, the question arises of whether the detailed statements (23.5) and (23.6) of the Nernst heat theorem (Section XXIII), relating to chemical reactions, phase transitions, and internal changes in pure phases, can also be derived from the unattainability of absolute zero and thus from the first and second laws. The answer is negative. An adiabatic change of the type treated here is an irreversible process, in principle. (Think of the transition diamond → graphite, taking place adiabatically at constant pressure. This cannot occur along the equilibrium curve!) Thus, the inequality sign holds in (24.4) and we cannot decide which sign applies in (24.5) even if we specify the states I and II.

Since the first and second laws do not give any information about the quantity $S_0^{II} - S_0^{I}$ in the case of changes of work coefficients, the statement (23.4) of the Nernst heat theorem, too, is not implied by the first and second laws.

C. Implications of the Nernst Heat Theorem

We turn to changes of work coefficients in a phase of fixed content. Examples are compression or expansion, magnetization or demagnetization. Here, the relations (24.1) and (24.2) are no longer valid. But the second law in the general form (24.4) still holds. Thus, we need a statement about $S_0^{II} - S_0^{I}$. This is provided by Eq. (23.4), which is part of the Nernst heat theorem (Section XXIII):

$$S_0^{I} = S_0^{II}, \tag{24.6}$$

the superscripts I and II relating to different values of a work coefficient. Inserting Eq. (24.6) into (24.4), we conclude that the first integral and hence T_2 cannot vanish.

Thus, we have shown that the unattainability of absolute zero follows from the Nernst heat theorem (in conjunction with the second law) for processes such as compression or expansion, magnetization or demagnetization.

As a next step, we want to investigate what can be inferred from the other statements of the Nernst heat theorem, the second law again being taken for granted.

The relations (23.5) and (23.6) of the Nernst heat theorem can be combined to give (24.5). If we insert this into the inequality (24.4) following from the second law, we again reach the conclusion that the final temperature T_2 can never be zero. Thus, the Nernst heat theorem is sufficient, but not necessary, to ensure the unattainability of absolute zero for processes in pure phases at constant work coefficients.

D. Implications of the Unattainability Theorem

Having demonstrated what follows from the first and second laws alone and from the combination of the second law with the Nernst heat theorem, we now turn to the question: What can be concluded if we take the second law and the unattainability of absolute zero for granted?

This means that we look at the inequality (24.4), supposing that the first integral cannot vanish in principle. The general condition for this is $S_0^{II} - S_0^{I} \leq 0$.

When the two states I and II can be connected by a reversible adiabatic path, as in changes of work coefficients in a given phase of fixed content, then the equality sign holds in (24.4) and this must be true for both the change I → II and the reverse process II → I. Hence, the equality $S_0^{I} = S_0^{II}$ results, in agreement with Eq. (23.4) or (24.6), which is one of the three detailed statements of the Nernst heat theorem. If I and II denote two phases of the same substance both in internal equilibrium and if there is a coexistence curve continuing down to absolute zero, then the adiabatic processes I → II and II → I can be made to occur reversibly, that is, along the equilibrium line (at variable pressure). This again leads to $S_0^{I} = S_0^{II}$, since the independence of S_0 on the pressure has already been proved. Therefore, in this special case, we arrive at Eq. (23.5), another detailed statement of the Nernst heat theorem.

When, however, the adiabatic change I → II can only take place irreversibly, then the inequality sign is valid in (24.4) and we are unable to discriminate between the two alternatives $S_0^{II} - S_0^{I} = 0$ and $S_0^{II} - S_0^{I} < 0$. This applies to the following processes involving pure phases: phase transitions without coexistence curve ending up at absolute zero, chemical reactions, and internal changes. Hence, the relations (23.5) and (23.6) of the Nernst heat theorem cannot be completely derived from the unattainability theorem.

Thus, the Nernst heat theorem is only *partly* implied by the unattainability theorem. Since, on the other hand, the Nernst heat theorem is sufficient to ensure the unattainability of absolute zero (for processes involving pure phases), the two theorems are obviously not equivalent.

E. Conclusions

We sum up the main conclusions of this section:

(1) The Nernst heat theorem is not implied by the first and second laws.

(2) The unattainability theorem can be derived from the first and second laws in certain cases.

(3) The unattainability theorem follows from the Nernst heat theorem in the remaining cases.

(4) The two theorems are not equivalent.

Nomenclature

A	affinity
C_z	heat capacity
$\tilde{C}_z$	specific heat capacity
$\bar{C}_z$	molar heat capacity
c_k	molarity of substance k
E	energy
$\mathfrak{E}$	electric field strength
e_i	strain component
F	Helmholtz function
$\mathfrak{F}$	Faraday constant
G	Gibbs function
H	enthalpy
H_k	partial molar enthalpy of substance k
$\mathfrak{H}$	magnetic field strength
h	enthalpy of reaction
$\mathfrak{J}$	magnetic polarization
L	Avogadro constant
l_i	work coordinate
M_k	molar mass of substance k
m	mass
m_k	mass of substance k
n	amount of substance
n_k	amount of substance k
P	pressure
$\mathfrak{P}$	electric polarization
Q	heat
q	(electric) charge
R	gas constant
S	entropy
S_k	partial molar entropy of substance k
s	entropy of reaction
T	thermodynamic temperature
U	internal energy
V	volume
V_0	reference volume
V_k	partial molar volume of substance k
W	work
x_k	mole fraction of substance k
y	arbitrary function of state
Z	extensive property
$\tilde{Z}$	specific quantity
$\bar{Z}$	molar quantity
$\hat{Z}$	generalized density
Z_k	partial molar quantity of substance k
z	quantity of reaction
β	thermal expansivity
Θ	empirical Kelvin temperature
$\varkappa$	compressibility
λ_i	work coefficient
μ_k	chemical potential of substance k
ν_k	stoichiometric number of substance k
ξ	extent of reaction
ϱ	density
ϱ_k	mass concentration of substance k
σ	interfacial tension
τ_i	stress component
Φ	electromotive force
χ_k	mass fraction of substance k
ω	surface area

References

Born, M. (1921). *Physik. Z.* **22**, 218, 249, 282.

Born, M. (1949). "Natural Philosophy of Cause and Chance." Oxford Univ. Press (Clarendon), London and New York.

Callen, H. B. (1961). "Thermodynamics." Wiley, New York.

Carathéodory, C. (1909). *Math. Ann.* **67**, 355.

Carathéodory, C. (1925). *Sitzber. Preuss. Akad. Wiss. Physik. Math. Kl.* p. 39.

Defay, R. (1929). *Bull. Acad. Roy. Belg.* (*Cl. Sc.*) **15**, 678.

Defay, R., Prigogine, I., Bellemans, A., and Everett, D. H. (1966). "Surface Tension and Adsorption." Longmans, Green, New York.

Ehrenfest, P. (1911). *Z. Physik. Chem.* **77**, 227.

Ehrenfest-Afanassjewa, T. (1925). *Z. Physik* **33**, 933.

Falk, G., and Jung, H. (1959). Axiomatik der Thermodynamik. *In* "Handbuch der Physik" (S. Flügge, ed.), Vol. III/2, p. 119. Springer, Berlin.

Gerke, R. H. (1922). *J. Am. Chem. Soc.* **44**, 1684.

Gibbs, J. W. (reprinted 1948). Thermodynamics. "The Collected Works of J. Willard Gibbs," Vol. I, p. 88. Yale Univ. Press, New Haven, Connecticut.

Guggenheim, E. A. (1967). "Thermodynamics." North-Holland Publ., Amsterdam.

Haase, R. (1948). *Z. Naturforsch.* **3a**, 285.

Haase, R. (1956a). "Thermodynamik der Mischphasen." Springer, Berlin.

Haase, R. (1956b). *Z. Physik. Chem.* (*Frankfurt*) **9**, 355.

Haase, R. (1957). *Z. Physik. Chem.* (*Frankfurt*) **12**, 1.

Haase, R. (1959). *Med. Grundlagenforsch.* **2**, 717.

Haase, R. (1969). "Thermodynamics of Irreversible Processes." Addison-Wesley, Reading, Massachusetts.

Haase, R., and Jost, W. (1956). *Naturwissenschaften* **43**, 481.

Haase, R., and Rehage, G. (1955). *Z. Elektrochem.* **59**, 994.

Landsberg, P. T. (1961). "Thermodynamics." Wiley (Interscience), New York.

Li, J. C. M., and Ting, T. W. (1957). *J. Chem. Phys.* **27**, 693.

Pippard, A. B. (1957). "Elements of Classical Thermodynamics." Cambridge Univ. Press, London and New York.

Planck, M. (1926). *Sitzber. Preuss. Akad. Wiss. Physik. Math. Kl.* p. 453.

Prigogine, I., and Defay, R. (1954). "Chemical Thermodynamics," translated by D. H. Everett. Longmans, Green, New York.

Ramsey, N. F. (1956). *Phys. Rev.* **103**, 20.

Simon, F. E. (1951). *Z. Naturforsch.* **6a**, 397.

Ulich, H., and Jost, W. (1970). "Kurzes Lehrbuch der Physikalischen Chemie." Steinkopff, Darmstadt.

Chapter 2A

Equilibrium, Stability, and Displacements

A. SANFELD

Introduction

The aims of Chapters 2A–2C are as follows: In Chapter 2A, we develop the physicochemical thermodynamic properties of matter in the particular cases of equilibrium states, equilibrium displacements, and equilibrium stability. The best way to treat this problem is to use the concept of chemical affinity introduced by De Donder and developed by Prigogine,

Defay, Glansdorff, and their co-workers. To do this, we base our formulation on the fundamental book of Prigogine and Defay (1967).

In Chapter 2B, we introduce the concept of irreversibility in a local way.

In Chapter 2C, we discuss surface phenomena, following the method of Bakker and Gibbs–Defay.

I. Thermodynamic Potentials

We shall briefly review some properties of the thermodynamic potentials developed exhaustively in Chapter 1.

First, as seen in Chapter 1, the classical definition of the second law is that, for all reversible changes in a closed system at a uniform temperature T,

$$dS = dQ/T \qquad \text{(reversible)}. \tag{1.1}$$

For a closed system, the state is defined by the variables T, V, and ξ (ξ is the extent of reaction, or simply the reaction coordinates (see Chapter 1, Section XX), so that

$$S = S(T, V, \xi). \tag{1.2}$$

For all irreversible changes in a closed system,

$$dS > dQ/T \qquad \text{(irreversible)}. \tag{1.3}$$

Following Clausius, we may now introduce a new quantity dQ' always positive, which represents the difference between $T\,dS$ and dQ in the course of an irreversible change. It is defined by

$$dS - (dQ/T) \equiv dQ'/T > 0 \qquad \text{(irreversible)}. \tag{1.4}$$

Equations (1.1) and (1.3) can now be combined to give

$$dS = dQ/T + dQ'/T, \tag{1.5}$$

with

$$dQ' = 0 \qquad \text{(reversible)},$$
$$dQ' > 0 \qquad \text{(irreversible)}.$$

Clausius called dQ' the *uncompensated heat*, which is always positive or zero; in classical thermodynamics, it played a purely qualitative part. It was used to delimit reversible changes for which $dQ' = 0$, and when dealing with nonequilibrium states it was sufficient to write $dQ' > 0$ without attempting an explicit calculation of its value.

The term "uncompensated heat" is not a particularly happy choice. The uncompensated heat dQ' is never the heat received by the system, but arises from irreversible changes taking place in the *interior* of the system.

The entropy of a system can vary for *two reasons and for two reasons only*: either by the transport of entropy d_eS to or from the surroundings through the boundary surface of the system, or by the *creation of entropy* d_iS inside the system. We have then [see Chapter 1, Eq. (11.2)]

$$dS = d_eS + d_iS, \tag{1.6}$$

and for a closed system

$$d_eS = dQ/T \tag{1.7}$$

$$d_iS = dQ'/T. \tag{1.8}$$

The entropy created in the system is thus equal to the Clausius uncompensated heat divided by the absolute temperature; this gives the uncompensated heat a physical significance (see Chapter 1, Section XI).

The inequality (1.4) states that the creation of entropy is always positive, that is, irreversible processes can only create entropy, they cannot destroy it.

We note that for an *isolated* system

$$dS = d_iS > 0. \tag{1.9}$$

Now, thermodynamic potentials are special state functions whose properties may be characterized, if certain variables are maintained constant, by a decrease when irreversible processes take place in the system. Thus, the thermodynamic potentials indicate the presence of irreversible phenomena in changes in which the corresponding variables are maintained constant.

Limiting our work to systems submitted only to quasistatic work (see Chapter 1, Section VII), we use the following four sets of variables:

$$\begin{matrix} S, V, n_1, \ldots, n_c; & T, V, n_1, \ldots, n_c; \\ S, p, n_1, \ldots, n_c; & T, p, n_1, \ldots, n_c; \end{matrix} \tag{1.10}$$

where $n_1, \ldots, n_c$ are the numbers of moles of components $1, \ldots, c$. The other symbols are defined in Chapter 1.

It may be interesting to note that each set of variables constains a thermal variable (S or T), a mechanical variable (p or V), and chemical variables ($n_1, \ldots, n_c$) related to the extents of reactions ($\xi_1, \ldots, \xi_c$).

The choice of a set of variables is only governed by practical factors. To each set of variables there corresponds a thermodynamic potential.

Combining the first law of thermodynamics [see Chapter 1, Eqs. (8.3) and (7.1)]

$$dU = dQ - p\,dV \tag{1.11}$$

with the second law (1.5) expressed in a modern form by De Donder (1923) (see also Duhem, 1899, 1911; Planck, 1927, 1930; Poincare, 1908), we obtain

$$dQ = dU + p\,dV = T\,dS - dQ'. \tag{1.12}$$

We now put formula (1.12) into various forms corresponding to various experimental conditions under which physicochemical changes can take place:

1. *Internal energy* U

From (1.12), we get

$$dU = T\,dS - p\,dV - dQ'. \tag{1.13}$$

Thus, for all irreversible reactions taking place *at constant* S and V,

$$dU = -dQ' < 0. \tag{1.14}$$

Thus, an irreversible change at constant entropy and volume is accompanied by a *decrease* in the internal energy. The internal energy thus plays the part of an indicator of irreversible processes for changes at constant S and V.

2. *Enthalpy* H

From the definition of the heat content or enthalpy [see Chapter 1, Eq. (9.3)]

$$H = U + pV, \tag{1.15}$$

(1.13) may be written

$$dH = T\,dS + V\,dp - dQ'. \tag{1.16}$$

For an irreversible reaction at constant S and p, we have therefore

$$dH = -dQ' < 0. \tag{1.17}$$

Thus, an irreversible change at constant entropy and *pressure* is accompanied by a decrease in the *enthalpy*. The enthalpy is the thermodynamic potential associated with the physical variables S and p.

3. *Helmholtz Free Energy F*

We define the Helmholtz free energy F by the relation [see Chapter 1, Eq. (16.3)]

$$F = U - TS, \tag{1.18}$$

and we see, from (1.13) and (1.18), that

$$dF = -S\,dT - p\,dV - dQ'. \tag{1.19}$$

For an irreversible change at constant T and V, we have therefore

$$dF = -dQ' < 0; \tag{1.20}$$

the function F is the thermodynamic potential associated with the variables T, V.

Thus, an irreversible change at constant temperature and volume is accompanied by a decrease in the Helmholtz free energy.

4. *Gibbs Free Energy G*

We define the Gibbs free energy G by the relation

$$G = U - TS + pV = H - TS. \tag{1.21}$$

Combining (1.13) with (1.21), we have

$$dG = -S\,dT + V\,dp - dQ'. \tag{1.22}$$

For an irreversible change at constant T and p, (1.22) reduces to

$$dG = -dQ' < 0. \tag{1.23}$$

Thus, an irreversible change at constant temperature and pressure is accompanied by a decrease in the Gibbs free energy.

Remark. The function $-S$ is a thermodynamic potential corresponding to the variables U, V, n_1, ..., n_c. Indeed, (1.12) gives, at constant U and V,

$$-dS = -dQ'/T < 0. \tag{1.24}$$

This conclusion is expected because a closed system where U and V are maintained constant is in fact an *isolated system*. This is the classical statement that the entropy of an isolated system increases with time.

II. Affinity

We shall consider here uniform systems in the absence of gravity, i.e., systems where pressure, temperature and composition are the same within each phase.

The diffusion equilibrium is thus reached within each phase of the system. The mechanical equilibrium excludes the barycentric motion and the viscosity, while the thermal equilibrium excludes transport of heat.

In fact, it is well known that a real system submitted to variations of temperature and pressure is not strictly uniform. The concept of a uniform system is thus an idealization of reality, but the approximation may be accepted if the variations of T and p are very slow.

If chemical reaction and phase changes take place in the same manner at each point of the system and if the exchanges of heat with the external world are slow enough to maintain a uniform temperature throughout all the system, the only irreversible processes are then general changes which can be expressed in terms of a reaction coordinate ξ. The production of entropy must be determined solely by ξ, and with De Donder (1922) we may write

$$\boxed{dQ' = A\, d\xi \geq 0.} \tag{2.1}$$

The inequality corresponds to spontaneous reaction, while the equality corresponds to the equilibrium.

De Donder introduced the function of state A called the affinity of the reaction, which does not depend upon the kind of transformation considered but depends solely on the state of the system at a particular instant.

In fact, De Donder introduced his relation (2.1) as a hypothesis, but Defay (1938) showed afterwards that, for a given value of $d\xi$, dQ' will be the same whatever may be the values of dp and dT during the change under consideration. With the variables T, p, ξ, we have

$$dU = \left(\frac{\partial U}{\partial T}\right)_{p\xi} dT + \left(\frac{\partial U}{\partial p}\right)_{T\xi} dp + \left(\frac{\partial U}{\partial \xi}\right)_{Tp} d\xi \tag{2.2}$$

and also

$$dV = \left(\frac{\partial V}{\partial T}\right)_{p\xi} dT + \left(\frac{\partial V}{\partial p}\right)_{T\xi} dp + \left(\frac{\partial V}{\partial \xi}\right)_{Tp} d\xi. \tag{2.3}$$

Let us define the following quantities (see Chapter 1, Section XVIII):

$$\left(\frac{\partial U}{\partial T}\right)_{p\xi} + p\left(\frac{\partial V}{\partial T}\right)_{p\xi} = C_{p\xi} \tag{2.4}$$

$$\left(\frac{\partial U}{\partial p}\right)_{T\xi} + p\left(\frac{\partial V}{\partial p}\right)_{T\xi} = h_{T\xi} \tag{2.5}$$

$$\left(\frac{\partial U}{\partial \xi}\right)_{Tp} + p\left(\frac{\partial V}{\partial \xi}\right)_{Tp} = -r_{Tp} \tag{2.6}$$

The quantities $C_{p\xi}$ (heat capacity of the system at constant pressure and composition), $h_{T\xi}$ (latent heat of pressure change at constant temperature and composition), and r_{Tp} (heat of reaction at constant T and p) are the thermal coefficients in the variables T, p, and ξ.

Substitution in (1.11) then leads to

$$dQ = C_{p\xi}\, dT + h_{T\xi}\, dp - r_{Tp}\, d\xi. \tag{2.7}$$

Now, we may write

$$dS = \left(\frac{\partial S}{\partial T}\right)_{p\xi} dT + \left(\frac{\partial S}{\partial p}\right)_{T\xi} dp + \left(\frac{\partial S}{\partial \xi}\right)_{Tp} d\xi. \tag{2.8}$$

Combining (2.7) and (2.8) with the Clausius equation

$$dQ' = T\, dS - dQ, \tag{2.9}$$

we obtain

$$\frac{dQ'}{dt} = \left[T\left(\frac{\partial S}{\partial T}\right)_{p\xi} - C_{p\xi}\right]\frac{dT}{dt} + \left[T\left(\frac{\partial S}{\partial p}\right)_{T\xi} - h_{T\xi}\right]\frac{dp}{dt}$$
$$+ \left[T\left(\frac{\partial S}{\partial \xi}\right)_{Tp} + r_{Tp}\right]\frac{d\xi}{dt}, \tag{2.10}$$

where dT/dt and dp/dt, which are the changes of the temperature and pressure in unit time, are quantities which can vary arbitrarily and may be given either positive or negative values. On the other hand, we consider the speed of reaction $d\xi/dt$ as a function of state (Prigogine and Defay, 1967, p. 18; De Donder, 1937, p. 936, 1938, p. 15), so that

$$d\xi/dt = \mathbf{v}(T, p, \xi). \tag{2.11}$$

Moreover, the three coefficients of dT/dt, dp/dt, and $d\xi/dt$ are functions of T, p, ξ. Equation (2.10) shows that, for a given state of the system, if the coefficients of dT/dt and dp/dt are not zero, then by assigning dT/dt and dp/dt suitable values, we may give dQ'/dt any value we wish; in particular, we can make dQ'/dt negative, which is contrary to the second law. It is thus necessary for these coefficients to be zero, that is,

$$T\left(\frac{\partial S}{\partial T}\right)_{p\xi} - C_{p\xi} = 0; \qquad T\left(\frac{\partial S}{\partial p}\right)_{T\xi} - h_{T\xi} = 0. \tag{2.12}$$

Equation (2.10) reduces then to the form (2.1) with

$$A = r_{Tp} + T(\partial S/\partial \xi)_{Tp}. \tag{2.13}$$

Because r_{Tp} and S are functions of T, p, and ξ, it results that A is also a function of T, p, and ξ, and thus it is a function of the state of the system.

Combining De Donder's inequality (2.1) with (2.11), we may write

$$dQ'/dt = A\, d\xi/dt = A \cdot \mathbf{v} \geq 0, \tag{2.14}$$

whence,

$$\begin{aligned} A &> 0, & \mathbf{v} &\geq 0 \\ A &< 0, & \mathbf{v} &\leq 0 \\ A &= 0, & \mathbf{v} &= 0, \end{aligned}$$

for, if we had $\mathbf{v} \neq 0$ with $A = 0$, we should have a chemical reaction proceeding at a finite rate in a reversible manner, which is impossible.

Conclusions

(a) The affinity is always of the same sign as the rate of reaction and thus gives the sign of the rate.

(b) When the affinity is zero, the system is in equilibrium.

The converse of this second statement is, however, not true:

$$\mathbf{v} \neq 0, \qquad \text{whence} \quad dQ' > 0 \quad \text{and} \quad A\mathbf{v} > 0 \qquad \begin{cases} \mathbf{v} > 0 & \text{gives } A > 0 \\ \mathbf{v} < 0 & \text{gives } A < 0 \end{cases}$$

$$\mathbf{v} = 0, \qquad \text{whence} \quad dQ' = 0 \quad \text{and} \quad A\mathbf{v} = 0 \qquad \begin{cases} \mathbf{v} = 0 \ \text{and} \ A = 0: & \text{true equilibrium} \\ \mathbf{v} = 0 \ \text{and} \ A \neq 0: & \text{false equilibrium.} \end{cases}$$

We describe a system as being in a state of false equilibrium when no reaction proceeds even though the affinity of the reaction is not zero. The necessary and sufficient conditions for true equilibrium in a chemical reaction is then given by

$$A(T, p, \xi) = 0. \tag{2.15}$$

In place of (2.2), we may write, in the variables T, V, ξ,

$$dU = \left(\frac{\partial U}{\partial T}\right)_{V\xi} dT + \left(\frac{\partial U}{\partial V}\right)_{T\xi} dV + \left(\frac{\partial U}{\partial \xi}\right)_{TV} d\xi. \tag{2.16}$$

By comparing (2.16) and (1.11), we may write

$$dQ = C_{V\xi}\, dT + l_{T\xi}\, dV - r_{TV}\, d\xi, \tag{2.17}$$

where

$$(\partial U/\partial T)_{V\xi} = C_{V\xi}, \tag{2.18}$$

$$(\partial U/\partial V)_{T\xi} + p = l_{T\xi}, \tag{2.19}$$

$$(\partial U/\partial \xi)_{TV} = -r_{TV}. \tag{2.20}$$

The quantity $C_{V\xi}$ is the heat capacity at constant volume and composition. $l_{T\xi}$ is the latent heat of volume change of the system, and r_{TV} is the heat of reaction at constant T and V.

The same arguments as above [see (2.12)] lead to

$$T(\partial S/\partial T)_{V\xi} - C_{V\xi} = 0; \qquad T(\partial S/\partial V)_{T\xi} - l_{T\xi} = 0. \tag{2.21}$$

and

$$A = r_{TV} + T(\partial S/\partial \xi)_{TV}. \tag{2.22}$$

It is easy to see that, *at low temperatures*, a measurement of r_{TV} gives as first approximation the value of the affinity A according to Berthelot's point of view. But, generally, A and r_{TV} are different because of the term $T(\partial S/\partial \xi)_{TV}$. If this term is important, then it is possible to obtain

$$A > 0 \qquad \text{with} \quad r_{TV} < 0 \qquad \text{(endothermic reaction).}$$

III. Affinity and Thermodynamic Potentials

Now combining (2.1) respectively with (1.13), (1.16), (1.19), and (1.22), we obtain (Prigogine and Defay, 1967, Chapter IV)

$$dU = T\,dS - p\,dV - A\,d\xi \tag{3.1}$$

$$dH = T\,dS + V\,dp - A\,d\xi \tag{3.2}$$

$$dF = -S\,dT - p\,dV - A\,d\xi \tag{3.3}$$

$$dG = -S\,dT + V\,dp - A\,d\xi. \tag{3.4}$$

For a closed system, we have then

$$U = U(S, V, \xi),\ H = H(S, p, \xi),\ F = F(T, V, \xi),\ G = G(T, p, \xi). \tag{3.5}$$

We now compare (3.1)–(3.4) with the corresponding total differentials, and so obtain the connection between (1) an extensive thermal variable S and an intensive thermal variable T, (2) an extensive mechanical variable V and an intensive mechanical variable p, and (3) an extensive chemical variable ξ and an intensive chemical variable A:

$$\left(\frac{\partial U}{\partial S}\right)_{V\xi} = T, \qquad \left(\frac{\partial U}{\partial V}\right)_{S\xi} = -p, \qquad \left(\frac{\partial U}{\partial \xi}\right)_{VS} = -A; \tag{3.6}$$

$$\left(\frac{\partial H}{\partial S}\right)_{p\xi} = T, \qquad \left(\frac{\partial H}{\partial p}\right)_{S\xi} = V, \qquad \left(\frac{\partial H}{\partial \xi}\right)_{pS} = -A; \tag{3.7}$$

$$\left(\frac{\partial F}{\partial T}\right)_{V\xi} = -S, \qquad \left(\frac{\partial F}{\partial V}\right)_{T\xi} = -p, \qquad \left(\frac{\partial F}{\partial \xi}\right)_{VT} = -A; \tag{3.8}$$

$$\left(\frac{\partial G}{\partial T}\right)_{p\xi} = -S, \qquad \left(\frac{\partial G}{\partial p}\right)_{T\xi} = V, \qquad \left(\frac{\partial G}{\partial \xi}\right)_{Tp} = -A; \tag{3.9}$$

the affinity is thus equal to the *slope* with respect to ξ of the thermodynamical potential related to the appropriate variables.

It is now easy to find the so-called Gibbs–Helmholtz equation by combining (1.18) and (3.8); we then find

$$U = F - T(\partial F/\partial T)_{V\xi}. \tag{3.10}$$

We may now derive a further set of important relations from (3.6)–(3.9). With the second derivatives, we have, for example, in the variables T, V, ξ and T, p, ξ,

$$\left(\frac{\partial S}{\partial V}\right)_{T\xi} = \left(\frac{\partial p}{\partial T}\right)_{V\xi}, \quad \left(\frac{\partial S}{\partial \xi}\right)_{TV} = \left(\frac{\partial A}{\partial T}\right)_{V\xi}, \quad \left(\frac{\partial p}{\partial \xi}\right)_{TV} = \left(\frac{\partial A}{\partial V}\right)_{T\xi}, \tag{3.11}$$

$$\left(\frac{\partial S}{\partial p}\right)_{T\xi} = -\left(\frac{\partial V}{\partial T}\right)_{p\xi}, \quad \left(\frac{\partial S}{\partial \xi}\right)_{Tp} = \left(\frac{\partial A}{\partial T}\right)_{p\xi}, \quad \left(\frac{\partial V}{\partial \xi}\right)_{Tp} = -\left(\frac{\partial A}{\partial p}\right)_{T\xi}. \tag{3.12}$$

Other groups of equations and many applications are given in the fundamental book of Prigogine and Defay (1967, p. 54).

From a general point of view, we observe that a partial derivative of a thermal variable (T or S) with respect to one of the mechanical variables (p or V) is equal to the partial derivative of the conjugate mechanical variable (V or p) with respect to the other thermal variable (S or T). Similar statements hold for the other pairs of variables (T, S) and (A, ξ) and (p, V) and (A, ξ).

In further sections, we will see some applications of Eq. (3.12).

IV. The Gibbs Chemical Potential

A. The Chemical Potential

Let us suppose that the functions of state U and S introduced by the first and second laws of thermodynamics also exist in open systems, so that we may write

$$U = U(T, V, n_1, \ldots, n_c) \equiv U(S, V, n_1, \ldots, n_c) \tag{4.1}$$

$$S = S(T, V, n_1, \ldots, n_c). \tag{4.2}$$

Now, in closed systems, we used the symbols dn_γ and dV to represent the changes, respectively, in number of moles n_γ and in volume in the time dt ($dt > 0$). However, in open systems, we have to introduce a new symbol for differentiation, δ, to avoid any confusion with the symbol d. The δn_γ are chosen quite arbitrarily and they represent virtual or true variations of any kind in the number of moles n_γ.

From (3.1), we then have

$$\delta U = \left(\frac{\partial U}{\partial S}\right)_{Vn} \delta S + \left(\frac{\partial U}{\partial V}\right)_{Sn} \delta V + \sum_{\gamma} \left(\frac{\partial U}{\partial n_{\gamma}}\right)_{SVn_{\beta}} \delta n_{\gamma}, \qquad (4.3)$$

where the subscript n means that all the n_{γ} remain constant during the derivation and the subscript n_{β} means that all n's except n_{γ} remain constant.

We may write now the following equalities:

$$\left(\frac{\partial U}{\partial S}\right)_{Vn} = \left(\frac{\partial U}{\partial S}\right)_{V\xi} \quad \text{and} \quad \left(\frac{\partial U}{\partial V}\right)_{Sn} = \left(\frac{\partial U}{\partial V}\right)_{S\xi} \qquad (4.4)$$

because the state of an open system in which all the n_{γ} remain constant changes exactly as it would in a closed system in which ξ remains constant.

Combining (4.4), (4.3), and (3.6), we obtain

$$\delta U = T\,\delta S - p\,\delta V + \sum_{\gamma} \left(\frac{\partial U}{\partial n_{\gamma}}\right)_{VSn_{\beta}} \delta n_{\gamma}. \qquad (4.5)$$

If we use H, F, and, G, we obtain

$$\delta H = T\,\delta S + V\,\delta p + \sum_{\gamma} \left(\frac{\partial H}{\partial n_{\gamma}}\right)_{pSn_{\beta}} \delta n_{\gamma} \qquad (4.6)$$

$$\delta F = -S\,\delta T - p\,\delta V + \sum_{\gamma} \left(\frac{\partial F}{\partial n_{\gamma}}\right)_{TVn_{\beta}} \delta n_{\gamma} \qquad (4.7)$$

$$\delta G = -S\,\delta T + V\,\delta p + \sum_{\gamma} \left(\frac{\partial G}{\partial n_{\gamma}}\right)_{Tpn_{\beta}} \delta n_{\gamma}. \qquad (4.8)$$

Nevertheless, we can write δH, δF, and δG in another way; i.e., from (4.5) and the definitions (1.15), (1.18) and, (1.21), we deduce that

$$\delta H = T\,\delta S + V\,\delta p + \sum_{\gamma} \left(\frac{\partial U}{\partial n_{\gamma}}\right)_{SVn_{\beta}} \delta n_{\gamma} \qquad (4.9)$$

$$\delta F = -S\,\delta T - p\,\delta V + \sum_{\gamma} \left(\frac{\partial U}{\partial n_{\gamma}}\right)_{SVn_{\beta}} \delta n_{\gamma} \qquad (4.10)$$

$$\delta G = -S\,\delta T + V\,\delta p + \sum_{\gamma} \left(\frac{\partial U}{\partial n_{\gamma}}\right)_{SVn_{\beta}} \delta n_{\gamma}. \qquad (4.11)$$

Comparing these equations with (4.6), (4.7), and (4.8), we thus have [see Chapter 1, Eqs. (16.5), (16.8)]

$$\delta U = T\,\delta S - p\,\delta V + \sum_\gamma \mu_\gamma\,\delta n_\gamma \tag{4.12}$$

$$\delta H = T\,\delta S + V\,\delta p + \sum_\gamma \mu_\gamma\,\delta n_\gamma \tag{4.13}$$

$$\delta F = -S\,\delta T - p\,\delta V + \sum_\gamma \mu_\gamma\,\delta n_\gamma \tag{4.14}$$

$$\delta G = -S\,\delta T + V\,\delta p + \sum_\gamma \mu_\gamma\,\delta n_\gamma, \tag{4.15}$$

where the symbol μ_γ, called by Gibbs (see Prigogine and Defay (1967), p. 68) the chemical potential of component γ, is the common value of the derivatives

$$\mu_\gamma \equiv \left(\frac{\partial U}{\partial n_\gamma}\right)_{SVn_\beta} = \left(\frac{\partial H}{\partial n_\gamma}\right)_{Spn_\beta} = \left(\frac{\partial F}{\partial n_\gamma}\right)_{TVn_\beta} = \left(\frac{\partial G}{\partial n_\gamma}\right)_{Tpn_\beta} \tag{4.16}$$

As the thermodynamic potentials are extensive functions, the chemical potential μ_γ is an intensive quantity; it is the partial molar quantity corresponding to a thermodynamic potential. Thus, the chemical potential describes the *local* properties of the system (for electrochemical systems, see Sanfeld (1968a)).

If the system is composed of many phases (α, δ, ...) at the same pressure and the same temperature, it is easy to show (Prigogine and Defay, 1965, p. 75) that, for example, (4.15) and (4.16) must be replaced by

$$\delta G = -S\,\delta T + V\,\delta p + \sum_\alpha \sum_\gamma \left(\frac{\partial G}{\partial n_\gamma^\alpha}\right)_{Tpn_{\beta\neq\gamma}^{\delta\neq\alpha}} \delta n_\gamma^\alpha \tag{4.17}$$

$$\mu_\gamma^\alpha = \left(\frac{\partial U}{\partial n_\gamma^\alpha}\right)_{VSn_\beta^\delta} = \left(\frac{\partial H}{\partial n_\gamma^\alpha}\right)_{Spn_\beta^\delta} = \left(\frac{\partial F}{\partial n_\gamma^\alpha}\right)_{TVn_\beta^\delta} = \left(\frac{\partial G}{\partial n_\gamma^\alpha}\right)_{Tpn_\beta^\delta} \tag{4.18}$$

B. Affinity and Thermodynamic Potentials as Functions of the Chemical Potentials

In accordance with the *principle of conservation of mass*, the stoichiometric equation for a reaction r may be written [see Chapter 1, Eq. (20.3)]

$$\sum_\alpha \sum_\gamma \nu_{\gamma r}^\alpha M_\gamma = 0, \tag{4.19}$$

where M_γ is the molecular weight of component γ. Now, from Eqs. (4.19) and (4.18) and the form of A given by (3.9), combined with [see Chapter 1, Eq. (20.2)]

$$d\xi_r = d_r n_\gamma{}^\alpha / \nu_{\gamma r}^\alpha, \tag{4.20}$$

we obtain

$$A_r = -\sum_\alpha \sum_\gamma \nu_{\gamma r}^\alpha \frac{\partial G}{\partial n_\gamma{}^\alpha} = -\sum_\alpha \sum_\gamma \nu_{\gamma r}^\alpha \mu_\gamma{}^\alpha. \tag{4.21}$$

At the true equilibrium of reaction, we have

$$A_r = 0 \tag{4.22}$$

or

$$\sum_\alpha \sum_\gamma \nu_{\gamma r}^\alpha \mu_\gamma{}^\alpha = 0 \qquad (r = 1, 2, \ldots, r). \tag{4.23}$$

Practical examples of chemical and passage equilibrium will be given in further sections.

Now, the extensive function G given by (4.17) is homogeneous of the first degree with the extensive variables $n_1{}^1, \ldots, n_\gamma{}^\alpha, \ldots, n_c{}^\delta$ as independent variables. We may then apply to G Euler's theorem for homogeneous functions, i.e.,

$$G = \sum_\alpha \sum_\gamma n_\gamma{}^\alpha \mu_\gamma{}^\alpha. \tag{4.24}$$

The chemical potentials $\mu_\gamma{}^\alpha$ of the γ constituents in the α phases of the system are thus functions of the independent variables T, p, $n_1{}^1, \ldots, n_\gamma{}^\alpha, \ldots, n_c{}^\delta$. On the one hand, they define all the thermodynamic properties of the system studied, and on the other hand they enable us to write the conditions of chemical equilibrium in the form (4.23) which is at once both simple and general.

The total volume V of the system can be written in function of the volume of each homogeneous phase V^α so that

$$V = \sum_\alpha V^\alpha(T, p, n_1{}^\alpha, \ldots, n_c{}^\alpha). \tag{4.25}$$

In the same way, if we neglect all the surface energies and entropies between the phases, then the thermodynamic potentials may be written as

$$G = \sum_\alpha G^\alpha \tag{4.26}$$

and the chemical potential may be written simply as

$$\mu_\gamma^\alpha = \left(\frac{\partial G}{\partial n_\gamma^\alpha}\right)_{Tpn_\beta^\delta} = \left(\frac{\partial G^\alpha}{\partial n_\gamma^\alpha}\right)_{Tpn_\beta^\delta} \tag{4.27}$$

Similarly, it is easy to show that

$$h_\gamma^\alpha = \left(\frac{\partial H^\alpha}{\partial n_\gamma^\alpha}\right)_{Tpn_\beta^\delta} \tag{4.28}$$

$$s_\gamma^\alpha = \left(\frac{\partial S^\alpha}{\partial n_\gamma^\alpha}\right)_{Tpn_\beta^\delta}; \tag{4.29}$$

μ_γ^α, h_γ^α, and s_γ^α are thus the partial molar quantities corresponding, respectively, to the extensive functions G^α, H^α, and S^α.

C. The Gibbs–Duhem Equation

The differentiation of (4.24) combined with the corresponding equation (4.17) gives the fundamental Gibbs–Duhem equation

$$S\,\delta T - V\,\delta p + \sum_\alpha \sum_\gamma n_\gamma^\alpha\,\delta\mu_\gamma^\alpha = 0. \tag{4.30}$$

If the system is described by the variables T, p, $\mu_1^1, \ldots, \mu_\gamma^\alpha, \ldots, \mu_c^\delta$, one of these variables is not independent. The relation (4.30) enables us to calculate one of the increments δT, δp, or $\delta\mu_\gamma^\alpha$ when the others are known.

For an isothermal, isobaric change, (4.30) reduces to

$$\sum_\alpha \sum_\gamma n_\gamma^\alpha\,\delta\mu_\gamma^\alpha = 0 \tag{4.31}$$

Applying Euler's theorem, (4.31) becomes (now dropping all the subscripts related to the number of moles)

$$\sum_\alpha \sum_\gamma n_\gamma^\alpha\left(\frac{\partial \mu_\gamma^\alpha}{\partial n_\beta^\alpha}\right)_{Tp} = 0. \tag{4.32}$$

Or, from the so-called reciprocity relation (Prigogine and Defay, 1967, p. 70)

$$\left(\frac{\partial \mu_\gamma^\alpha}{\partial n_\beta^\alpha}\right)_{Tp} = \left(\frac{\partial \mu_\beta^\alpha}{\partial n_\gamma^\alpha}\right)_{Tp} \qquad (\gamma, \beta = 1, 2, \ldots, c), \qquad (4.33$$

Eq. (4.32) may be rewritten as

$$\sum_{\alpha}\sum_{\gamma} n_\gamma^\alpha\left(\frac{\partial \mu_\beta^\alpha}{\partial n_\gamma^\alpha}\right)_{Tp} = 0. \tag{4.34}$$

Remark. From the second partial derivative of $G(T, p, n_1, \ldots, n_c)$ and from Eq. (4.15) and (3.9), we have

$$\left(\frac{\partial \mu_\gamma}{\partial T}\right)_{pn} = -\left(\frac{\partial S}{\partial n_\gamma}\right)_{Tp} = -s_\gamma \tag{4.35}$$

$$\left(\frac{\partial \mu_\gamma}{\partial p}\right)_{Tn} = \left(\frac{\partial V}{\partial n_\gamma}\right)_{Tp} = v_\gamma. \tag{4.36}$$

Differentiating $H = U + pV$ and $G = U - TS + pV$, we obtain

$$h_\gamma = u_\gamma + pv_\gamma \tag{4.37}$$

$$\mu_\gamma = u_\gamma - Ts_\gamma + pv_\gamma. \tag{4.38}$$

The specific molar entropy s_γ determines the variation of μ_γ with T, while the specific molar enthalpy determines the variation of μ_γ/T with T. Indeed, from (4.37) and (4.38), we find

$$\mu_\gamma = h_\gamma - Ts_\gamma, \tag{4.39}$$

and thus

$$\partial(\mu_\gamma/T)/\partial T = -h_\gamma/T^2. \tag{4.40}$$

D. Comparison with the Formulation of Lewis and Randall*

As we have seen previously, the affinity A is a function of the instantaneous state of the system given by the derivative of the thermodynamic potential G with respect to ξ, at constant T and p for *each* state of the matter [see Eq. (3.9)]

$$A = -(\partial G/\partial \xi)_{Tp}. \tag{4.41}$$

Now, the symbol ΔG used by Lewis and Randall means a *finite* variation of the function, e.g.,

$$\Delta G = G_{\text{final}} - G_{\text{initial}} \qquad \text{(two states)}. \tag{4.42}$$

In fact, the fundamental difference between A and ΔG is not related to

* Lewis and Randall (1923).

the existence of two states of matter, because, in $A = A(T, p, \xi)$, the extent of reaction ξ is also related to the initial state of matter:

$$\xi = (n_\gamma - n_\gamma^{\ i})/\nu_\gamma , \tag{4.43}$$

where $n_\gamma^{\ i}$ is the number of moles of the components γ when $t = 0$. The difference comes from the relation (4.24). For one phase, Eq. (4.24) reduces to

$$G = \sum_\gamma n_\gamma \mu_\gamma , \tag{4.44}$$

and thus

$$\Delta G = \left(\sum_\gamma n_\gamma \mu_\gamma\right)_{\text{final}} - \left(\sum_\gamma n_\gamma^{\ i} \mu_\gamma^{\ i}\right)_{\text{initial}} . \tag{4.45}$$

Since μ_γ does not necessarily have the same value in the initial and in the final states,

$$\Delta G \neq \sum_\gamma \mu_\gamma (n_\gamma - n_\gamma^{\ i}). \tag{4.46}$$

From (4.43) and (4.46), we obtain the inequality

$$\Delta G_\xi \neq \sum_\gamma \nu_\gamma \mu_\gamma , \tag{4.47}$$

where

$$\Delta G_\xi = \Delta G / \xi . \tag{4.48}$$

According to the Lewis approach, the chemical potentials μ_γ are implicitly considered as constant, or have a mean value so that, if $\xi = 1$,

$$\Delta G = \sum_\gamma \nu_\gamma \langle \mu_\gamma \rangle . \tag{4.49}$$

This hypothesis is quite correct if the initial composition, temperature, and pressure are only weakly modified by the chemical reactions. Strictly speaking, for a chemical reaction

$$\nu_A A + \nu_B B = \nu_C C + \nu_D D,$$

A, for a given well-defined state, is

$$A = \nu_C \mu_C + \nu_D \mu_D - \nu_A \mu_A - \nu_B \mu_B , \tag{4.50}$$

and thus does not simply represent a decrease in free energy or a difference between two states. Because the use of a mean value gives rise to some difficulties in the neighborhood of equilibrium, where G takes a minimum value, it would be better to define ΔG as a single operator of the type (4.50). For details about this problem, see the fundamental work of Prigogine and Defay (1967, Chapter V).

E. Passage from a Closed to an Open System

As we already saw, the functions $A(T, p, \xi)$ and $G(T, p, \xi)$ are well defined in closed systems, with the definition (2.1),

$$A\, d\xi = dQ',$$

the relation [Eqs. (3.9)]

$$A = -(\partial G/\partial \xi)_{Tp},$$

and the formula [Eq. (3.4)]

$$\delta G = -S\, \delta T + V\, \delta p - A\, \delta \xi.$$

For an open system, we *presumed* that the function G must also exist in the form

$$G = G(T, p, n_1, \ldots, n_c). \tag{4.51}$$

Now, when an open system undergoes a change in which all the n_γ remain constant, the change is exactly as it would be in a closed system in which ξ remains constant, so that the two first derivatives of (3.9) become

$$(\partial G/\partial T)_{pn} = -S \tag{4.52}$$

$$(\partial G/\partial p)_{Tn} = V. \tag{4.53}$$

The last derivative of (3.9) can also be written in the form

$$\left(\frac{\partial G}{\partial \xi}\right)_{Tp} = \sum_\gamma \left(\frac{\partial G}{\partial n_\gamma}\right)_{Tp} \frac{dn_\gamma}{d\xi} \tag{4.54}$$

because, in $G = G(T, p, \xi)$, ξ depends on n_γ. Now, the derivative $dn_\gamma/d\xi$ means $d_i n_\gamma/d\xi$, because, in closed systems, the change dn_γ in the number of moles of γ in a time interval dt will be only the change arising as a result of internal (subscript i) chemical reactions, and

$$d_i n_\gamma/d\xi = \nu_\gamma. \tag{4.55}$$

On the other hand, from the form (4.51), it results that in the derivative $(\partial G/\partial n_\gamma)_{Tp}$ the variations of n_γ are of any kind, e.g., result from transport across the boundaries of the system to or from the surroundings.

Combining (4.54), (4.55), and (4.16), we obtain the well-known passage equation from closed to open systems

$$(\partial G/\partial \xi)_{Tp} = \sum_\gamma \nu_\gamma \mu_\gamma. \tag{4.56}$$

Inserting Eq. (4.55) and (4.21) in Eq. (3.4) then gives *for open systems*

$$\delta G = -S\,\delta T + V\,\delta p + \sum_\gamma \mu_\gamma\,\delta n_\gamma. \tag{4.15}$$

The passage from a closed system (where the function A is well defined) to an open one (where the intensive functions μ_γ are locally defined) is thus realized without any difficulty.

V. The Phase Rule

A. General Demonstration

We will now consider the number of phases which can coexist and the relationship of the properties of each component to the equilibrium behavior of the system. First, let us remember that a phase composed of one or several components is a macroscopic homogeneous portion of a system.

Since all gases are completely miscible, there can never be more than one gas phase; however, a number of independent liquid or solid phases may be formed. The phase rule, like all basic thermodynamic relationships, is independent of assumptions pertaining to the particular nature of matter; it is not concerned with the quantities of the various phases, but only with *intensive variables*. The Gibbs phase rule only permits us to *fix arbitrarily a certain number of intensive variables when the system is in equilibrium.*

We consider a system with c components, ϕ phases at the same pressure and the same temperature, and r' distinct chemical reactions (omitting those which consist solely of the passage reactions).

In the systems considered here,* composed of c constituents and ϕ phases, the intensive variables are

$$T, p, N_1^{1}, \ldots, N_c^{1}, \ldots, N_1^{\phi}, \ldots, N_c^{\phi}, \tag{5.1}$$

where the mole fractions N_γ^α of γ in the phase α are defined by

$$N_\gamma^\alpha = n_\gamma^\alpha / \sum_\gamma n_\gamma^\alpha. \tag{5.2}$$

Nevertheless, the $(2 + c\phi)$ intensive variables are not all independent, and, at equilibrium, general relations bind these variables.

Whatever the state of the system, we have ϕ relations of the type

$$\sum_\gamma N_\gamma^\alpha = 1. \tag{5.3}$$

The equilibrium conditions are:

(1) There are $c(\phi - 1)$ conditions of transfer equilibrium of each constituent among the various phases

$$\mu_\gamma^{1} = \mu_\gamma^{2} = \cdots = \mu_\gamma^{\phi} \qquad (\gamma = 1, \ldots, c). \tag{5.4}$$

(2) There are r' conditions of equilibrium for the chemical reactions,

$$A_r = -\sum_\gamma \sum_\alpha \nu_{\gamma r} \mu_\gamma^\alpha = 0 \qquad (r = 1, \ldots, r'). \tag{5.5}$$

The number of relations among the $2 + c\phi$ intensive variables is thus

$$\phi + c(\phi - 1) + r'. \tag{5.6}$$

The variance or the number of degrees of freedom, i.e., *the number of independent intensive variables*, is thus given by (Gibbs, 1928; De Donder, 1920; Jouguet, 1921; Bowden, 1938)

$$w = 2 + (c - r') - \phi. \tag{5.7}$$

Remark. If a component is insoluble in the phase α, the supplementary insolubility condition

$$N_\gamma^\alpha = 0 \tag{5.8}$$

gives rise to a loss of a condition like (5.4). Thus, the phase rule (5.7) remains valid.

* For charged systems, see Sanfeld (1968). For anisotropic systems, see Steinchen (1970).

B. Examples

1. *One-component systems, $c = 1$ and $r' = 0$*

For a one-component system, a phase diagram is usually a plot of pressure against temperature.

a. Water Phase Diagram. The behavior of water is shown in a limited region in Fig. 1. The lines represent the equilibrium between the phases, and thus establish the fixed relationship between pressure and temperature. We consider three cases:

(*i*) The system consists of only one phase (vapor, liquid, or solid): $\phi = 1$ and $w = 2$. The system is divariant and the areas between the lines representing the equilibrium between the phases correspond to regions in which only a single phase exists and in which T and p may be varied independently.

(*ii*) The system consists of two phases (liquid and vapor, liquid and solid, or solid and vapor): $\phi = 2$ and $w = 1$. The system is monovariant. If the pressure is fixed, then the equilibrium temperature of the coexisting two phases will be a function of the chosen pressure. Equilibrium between any two phases can exist only if the value of T for a particular p falls on the line dividing the areas (see Fig. 1), e.g., liquid and gas. At each line grouping the representative points of two-phase systems, the affinity of passage of the substance from one phase to the other is zero.

(*iii*) The system consists of three phases (vapor, liquid, and solid): $\phi = 3$ and $w = 0$. We cannot fix either T or p arbitrarily and the three phases coexist only at the point of intersection of the three equilibrium

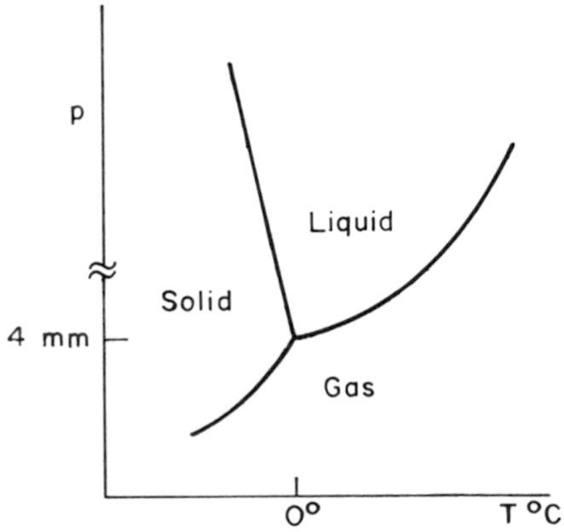

FIG. 1. Phase diagram for water in the vicinity of the triple point.

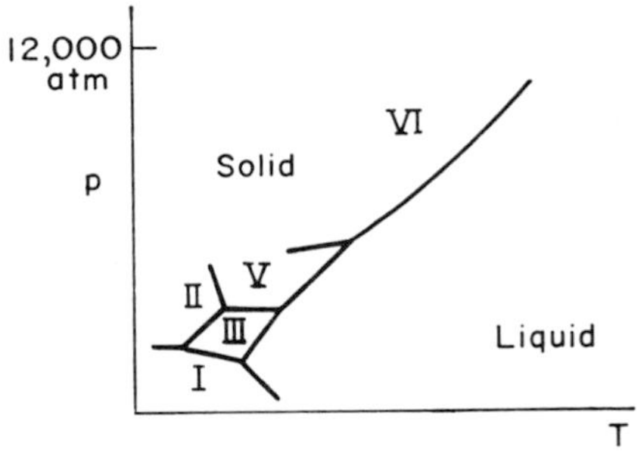

FIG. 2. Phase diagram for water at high pressure.

lines. This triple point is invariant. For water, the coexistence of the three phases occurs at a temperature of $+7.6\ 10^{-3}$ °C and a pressure of 4.6 mm of mercury.

At high pressures, a number of different crystalline forms (different phases) of ice have been observed. The phase diagram (see Fig. 2) is not complete because only limited information is available. Four triple points corresponding to the coexistence of the three forms of ice (ice II, III, and IV) are shown.

More than three independent phases cannot coexist in equilibrium in a one-component system.

b. Sulfur Phase Diagram. The four phases, clinorhombic, orthorhombic, liquid, and vapor, cannot all be in simultaneous equilibrium (Fig. 3). Any two of these are separated by a line along which they can coexist, while there are three triple points corresponding to the different forms of sulfur. Along the line *AC*, there is equilibrium between orthorombic (OR) and clinorhombic (CR); along *DA*, between orthorhombic and vapor (V); along *AB*, between clinorhombic and vapor; along *BE*, between vapor and liquid (L), etc.

The systems composed of three phases are invariant and their states are represented by isolated points or triple points (*A*, *B*, and *C*). At the equilibrium, it is impossible to find systems composed of four phases because the variance would then be less than zero.

2. *Binary Systems*

The phase diagrams, particularly for binary or ternary systems, are of great practical importance in metallurgy and physical chemistry, and much work has been done on their classifications.

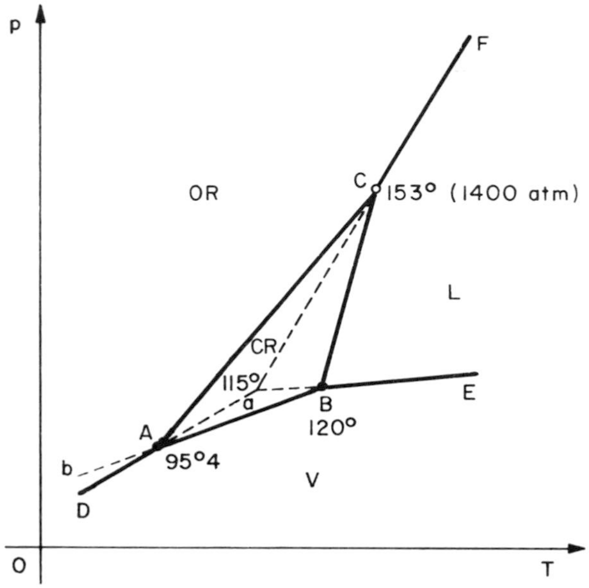

FIG. 3. Phase diagram of sulfur.

Here we have several cases:

(*i*) $\phi = 1$, $w = 3$. For example, two components in a vapor state without chemical reaction. The temperature, the pressure, and the composition can be fixed arbitrarily.

(*ii*) $\phi = 2$, $w = 2$. A mixture of alcohol and water in the presence of their vapor (without air). Two variables (the pressure and the temperature, or the pressure and the composition in one phase, or the temperature and the composition in one phase) can be fixed arbitrarily.

a. Equilibrium between Vapor and Liquid. (1) We consider first the case where the temperature has been taken as constant. The total pressure will be determined by the composition of either phase. Thus, we may represent the composition of either phase by the abscissa in a rectangular system of coordinates, and the pressures at constant temperature as ordinate (see Fig. 4).

The vaporization curve $p(N_2')$ gives the equilibrium pressure, i.e., the vapor pressure of the solution at the considered temperature T and for a given composition of solution, N_2'. For the same pressure (for example, p_D), the vapor phase (C'') is in equilibrium with the solution (C'). Nevertheless, the composition of the vapor phase N_2'' is generally

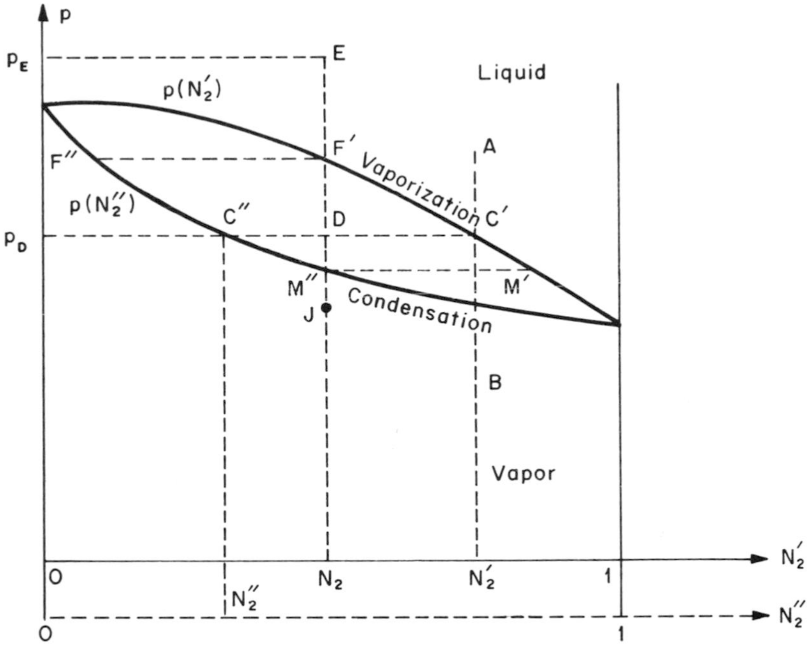

FIG. 4. Vapor–liquid phase diagram at constant temperature.

different from that of the liquid phase. The curve $p(N_2'')$ is called *condensation curve*.

The point A situated above curve $p(N_2')$ only represents a liquid phase, although the point B situated behind curve $p(N_2'')$ only represents a vapor phase.

Excluding all the false equilibria, the point D within the area enclosed by two curves represents a two-phase system consisting of the liquid C' and the vapor C'' in such proportions that the overall composition of the whole system is N_2. If n is the total number of moles in the system, it is easy to see that

$$\frac{n''}{n} = \frac{N_2' - N_2}{N_2 - N_2''} = \frac{DC'}{C''D}. \tag{5.9}$$

Thus, when D approaches the condensation curve, more and more of the system will be vaporized, and when D is close to the vaporization curve, the system will nearly all be in the liquid phase.

Fractional distillation. Let us assume that initially we have a solution in the state E. The decrease of pressure from p_E to p_D yields a partial vaporization and results in the formation of two phases C' and C''. The proportion of these two phases is given by (5.9).

The mechanism is very simple: (1) From E to F' no vaporization. (2) At F', the vaporization begins. The state of the vapor phase produced is represented by F'' ($N_1^{F''} > N_1^{F'}$). (3) In the liquid phase, the fraction of the component 1 decreases, and thus F' goes down to C' along the vaporization curve, while the state of the vapor phase sinks along the condensation curve from F'' to C''.

If the pressure decreases from p_D to $p_{M''}$, all the liquid phase becomes a vapor phase, and, below this pressure, only the vapor phase subsists with the same composition as the initial E liquid phase (point J).

Now, let us assume that we stop the vaporization or the condensation when the composition of the two obtained phases is different. This phase is then submitted to a new partial vaporization or condensation, and so on. This process is called fractional distillation.

From the experimental point of view, a one-step separation can occur on a plate of a fractionation column (Keeson, 1939; Bosnjakovic, 1935; Brown *et al.*, 1955; Ponchon, 1921; Savarit, 1922; Daniels and Alberty, 1961); the number of plates (separation steps) required to give the desired fractionation depends on the shape of the phase diagram. The effectiveness of a given fractionation column is indicated by the number of *theoretical plates* corresponding to the number of separation steps which it performs.

(2) We now consider the case where the pressure has been taken as constant. The independent variables are p and N_2'. The curve $T(N_2')$, boiling point as a function of liquid composition N_2', is called the *boiling curve* (Fig. 5).

A horizontal line cuts the boiling and condensation curves at P' and P''. These points, corresponding to the same temperature, give the composition of the liquid, N_2', and the vapor, N_2'', which are in equilibrium with one another at the temperature T. If we neglect the effect of surface tension, a bubble containing vapor and the solution represents two phases at the same pressure.

Azeotropic mixtures. Maximum- or minimum-boiling mixtures are called *azeotropes* (Fig. 6). An argument based on the Gibbs–Konovalow theorem shows that the composition of the two phases must be identical at the point of maximum or minimum.

In an azeotropic system, one phase may be transformed to the other at *constant temperature, pressure* and *composition* without affecting the

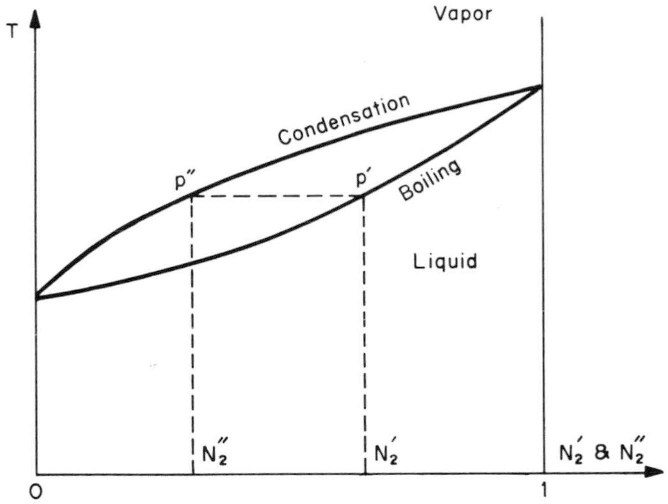

FIG. 5. Vapor–liquid phase diagram at constant pressure.

equilibrium state (Fig. 6) (see, for example, Wade and Merriman (1911)).

The azeotrope behaves in some respects like a pure substance, since it distils at constant temperature and pressure without change in composition. If, however, the pressure at which the distillation is carried out is altered, the composition is also altered (see Fig. 7), and hence the substance corresponds to a mixture, and not a pure one. Azeotropic mixtures are not uncommon.

Water and hydrogen chloride form an azeotropic mixture with a maximum boiling point of 108.6°C at 1760 mm Hg pressure, when the

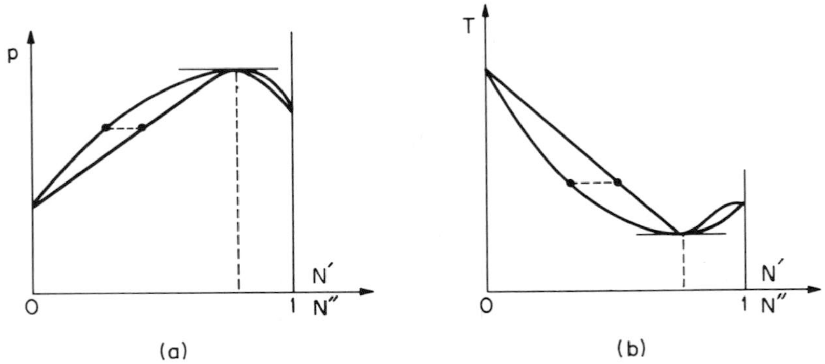

FIG. 6. Vapor–liquid equilibrium in a binary system forming an azeotrope.

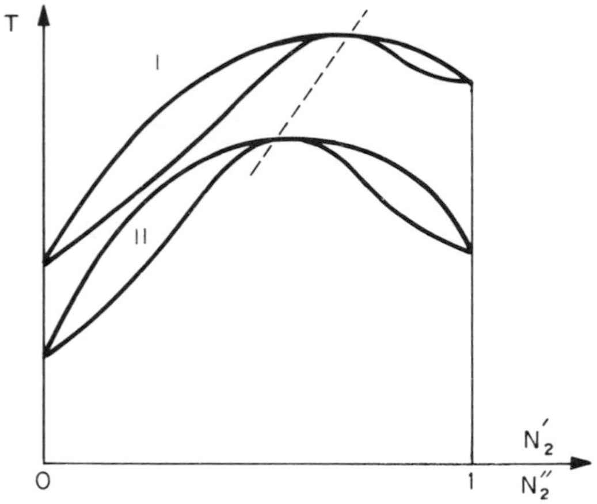

FIG. 7. Variation of the azeotropic maximum with pressure.

weight per cent of HCl is 20.222. Ethylalcohol and water form an azeotrope with a minimum boiling point of 78.15°C (95.57% ethanol). For azeotropic data on a large number of systems, see Horsley (1952) and MacDougall (1926, p. 181).

Remark. At each minimum in the isobaric curves, there corresponds a maximum in the isothermal curves, and conversely.

b. Equilibrium between Solution and Crystal. (1) *Melting-point curves. Solid solution.* The phase diagram is similar in many respects to that foı

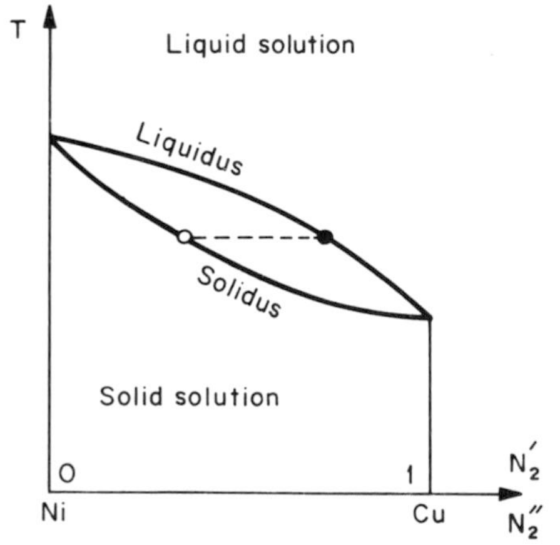

FIG. 8. Solid–liquid equilibrium (p = const) where solid solutions are formed.

the vapor pressure of binary liquids. Examples are the systems Ni + Cu (Fig. 8) and $HgBr_2 + HgI_2$ (Fig. 9). The liquidus gives the composition of the liquid phase, and the solidus gives the composition of the solid solution in equilibrium with the liquid solution. An extremum in the phase diagram might be expected if the two components form an intermediate compound. However, the extremum itself is not a proof of the existence of such a compound.

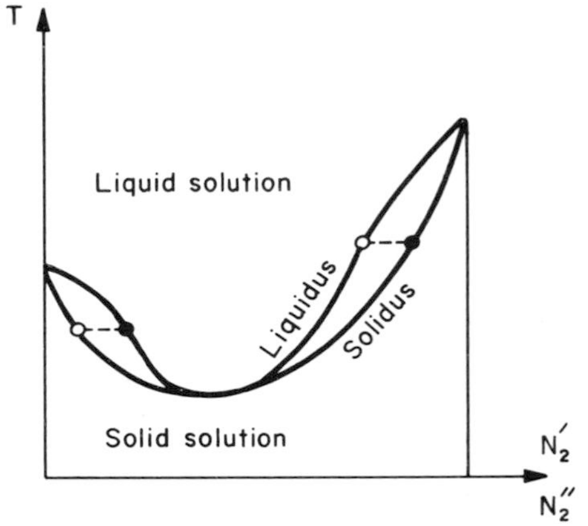

FIG. 9. Azeotropic liquid–solid phase diagram.

If the composition at which the maximum occurs is independent of the pressure, then it is probably associated with a compound. The pressure test is useful in the study of vapor–liquid mixtures.

For the solid-state problem, an X-ray examination of the crystal structure provides more reliable evidence.

(2) *Melting-point curves. Eutectics.* The two phases give a divariant system where p and N_2' are arbitrary. The phase diagram is given in Fig. 10. Here, A and B are the melting points of pure components 1 and 2, AE is the equilibrium curve for the solution (composition N_2') and the crystals of pure component 1, and BE is the equilibrium curve for the solution (composition N_2') and the crystals of pure component 2. The point of intersection of these two curves is called the eutectic point. At this point, the system has the eutectic composition at the eutectic temperature and the solution is in equilibrium with pure crystals of 1 and 2. The system is thus univariant, and the coexistence of the three phases is governed by the arbitrary choice of the pressure.

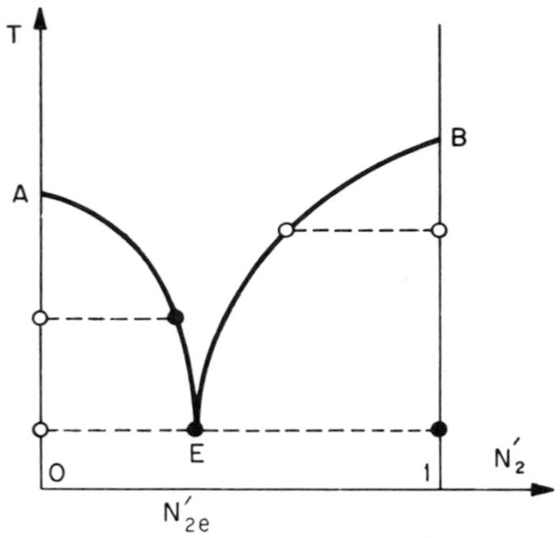

FIG. 10. Freezing-point curves of a binary system forming a eutectic (p = const).

In the presence of air, the variance is not altered, because air can be regarded as a single component, since its composition remains constant throughout (aside from the vapors derived from the solution).

(3) *False binary systems.* If the two components A and B of a binary system react together to form an addition component C so that $A + B = C$, the variance is not affected, because $c = 3$, $r' = 1$, $c - r' = 2$. Figure 11 shows the appearance of the phase diagram when the two pure substances form an addition compound.

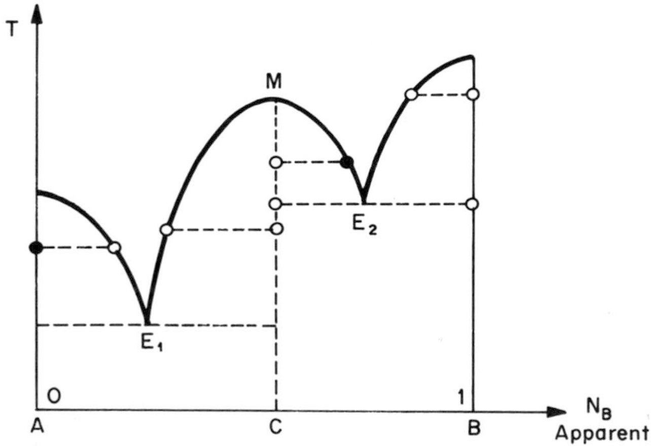

FIG. 11. Eutectic phase diagram for binary systems forming an addition compound.

As abscissa, we plot the apparent molar fraction of component B, i.e., the molar fraction given by the experimentator which ignores the existence of the addition compound.

In the absence of solid solutions, we obtain two eutectics E_1 and E_2. At E_1, there is an equilibrium: solution + crystal A + crystal C; at E_2, there is an equilibrium: solution + crystal C + crystal B.

The maximum M represents the melting point of a well-defined compound, the crystal C. At this maximum, the apparent composition remains constant when the pressure is changing.

One speaks of a congruent melting point (Prigogine and Defay, 1967, p. 377) when the solid compound melts to form a liquid phase which has the same composition as the solid. In the $NaF–MgF_2$ system (Eggers *et al.*, 1964, p. 263–265), the intermediate compound has the formula $NaMgF_3$. Two eutectics are observed, $NaF–NaMgF_3$ and $NaMgF_3–MgF_2$.

An eutectic may disappear, and is then replaced by a *transition point* r where there are three phases in equilibrium (Fig. 12).

One speaks of an incongruent melting point for the intermediate compound if, when melting occurs, the liquid has a composition different from that of the compound and a new solid phase is also formed.

According to Counts *et al.* (1953), (see also Eggers *et al.* (1964)), the compound $CaF_2–BeF_2$ is unstable above 890°C. During the transition in the melting process, we have three different phases in equilibrium, and the system is at an invariant point called *peritectic* one.

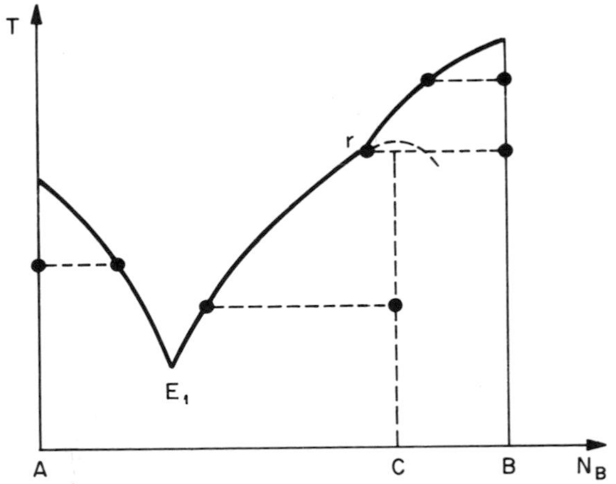

FIG. 12. Eutectic phase diagram with a transition point.

3. *Ternary Systems*

On account of the great complexity of the phenomena observable in the case of three-component systems, we shall have to be content with a brief discussion of some of the simpler cases (Vogel, 1937). The composition of a ternary mixture is best represented by a point in an equilateral triangle whose vertices 1, 2, and 3 represent the three pure components. If the side of the triangle is taken as unity, then the mole fractions N_1, N_2, and N_3 in the solution under consideration are given by the distances, measured along lines parallel to the sides of the triangle, of the point P from the sides of the triangle remote from vertices 1, 2, and 3, respectively (Figs. 13 and 14). This representation ensures automatically that $N_1 + N_2 + N_3 = 1$. The sides of the triangle represent the binary systems (1 + 2), (2 + 3), (3 + 1), and the vertices represent pure components. Now let us consider systems without chemical reactions. Different cases are possible: $\phi = 1$, $w = 4$ (case A); $\phi = 2$, $w = 3$ (case B); $\phi = 3$, $w = 2$ (case C); $\phi = 4$, $w = 1$ (case D); $\phi = 5$, $w = 0$ (case E).

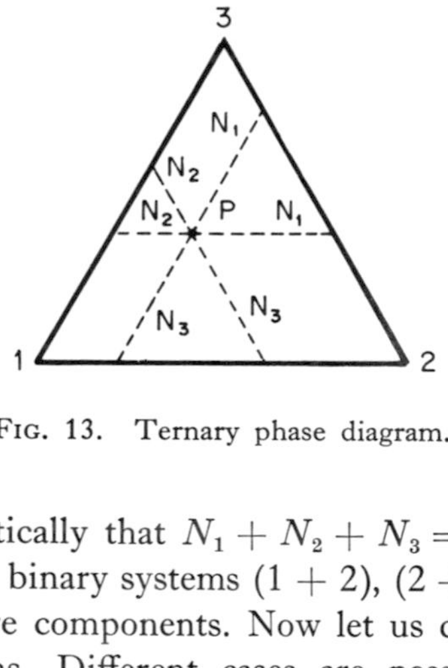

FIG. 13. Ternary phase diagram.

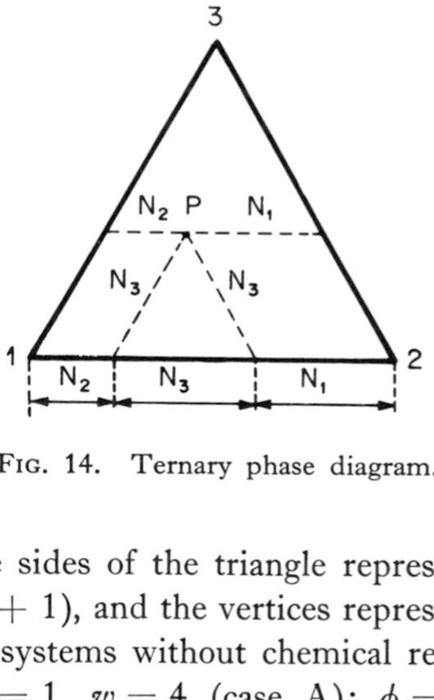

FIG. 14. Ternary phase diagram.

Example 1. *Case A.* A gas mixture with three components. The variables p, T, N_2, N_3 are given arbitrarily.

Example 2. *Case B.* Solidification of a ternary solution without solid solution. With two phases, the system is trivariant; we can, for example, consider the pressure and composition of the solution N_2', N_3' and see how the equilibrium temperature changes with these variables. If the pressure is taken as constant, then we may construct a diagram in which the equilibrium temperature between liquid and one of the solid phases is shown as a function of N_2' and N_3'. We obtain a surface in three-dimensional space.

Example 3. *Case E.* Ternary eutectic in the presence of vapor of the solution without air. An example of a classical ternary system with binary and ternary eutectics is the alloy system Bi + Pb + Sn [for details see Prigogine and Defay (1962), p. 185, MacDougall (1926)].

Remarks. (1) The diagrams become very complicated if the various components can form solid solutions in one another (Findlay, 1951).

(2) If the three components are linked by a chemical reaction, the variance is reduced by one and the behavior is simplified. Example: $CaCO_3$ (solid) $\rightleftharpoons$ CaO (solid) + CO_2 (gas). The system here is monovariant, and, at equilibrium, its state is completely specified when the temperature is fixed.

(3) To cover a complete specification of the state of a system in terms of both *intensive* and *extensive* variables, we must calculate how many variables must be fixed to *determine completely* the equilibrium state of a closed system. This problem, related to the *conditions of enclosure* (Prigogine and Defay, 1967, p. 186), is treated exhaustively in the work of Duhem (1899, vol. IV).

VI. The Equilibrium Constant

A. Introduction

In Sections II and IV, we discussed the conditions of thermodynamic equilibrium in a system subject to various constraints, and found that the conditions could be expressed by saying that chemical affinity is zero when equilibrium prevails. Now, the chemical affinity is determined by the temperature, pressure, and composition of the system, so that, if this function is a known function of the variables just mentioned, then the thermodynamic behavior of the system can be readily predicted. Because the affinity and the chemical potentials are connected by Eq. (4.21), in order to obtain specific numerical answers for the equilibrium composition in a particular system, we must know the explicit form of the equation of state

$$\mu = \mu(T, p, N_1^1, \ldots, N_c^\alpha).$$

In this section, it will be part of our work to show somewhat more explicitly how the state of equilibrium depends on the quantities of the various substances present, first, when all the substances are ideal gases, and second, when the system is a dilute solution.

From a practical point of view, if substances which can react with one another are brought together, it will be found that, in general, when all change *seems* to have ceased, the original substances are still present, although their concentrations may be extremely small.

We commonly say that the original and final substances are in chemical equilibrium.

Experimenters are especially interested in determining the extent to which a given chemical reaction will take place and in finding out how the final state of equilibrium depends, not only on the temperature and the pressure, but on the concentrations of the reacting substances.

B. The Single-Component Perfect Gas

The perfect gas is a fictitious substance, defined by certain properties which are not possessed by any actual substance, but which are supposed to be approached by every actual gas as its pressure is indefinitely diminished. We may state, then, that the perfect gas is a substance which fulfills the two following conditions:

(1) Its internal energy is a function of temperature alone (Joule's law), or, in other words,

$$\partial U(T)/\partial V = 0. \tag{6.1}$$

(2) At constant temperature, the volume V occupied by a given number of moles of gas varies in inverse proportion to the pressure (Boyle's law)

$$pV = nf(T), \tag{6.2}$$

where $f(T)$ is a function independent of the nature of the gas.

The combination of these two laws gives the well-known equation of state

$$pV = nRT, \tag{6.3}$$

where R is the gas constant ($R = 0.08205$ l atm deg^{-1} mol^{-1} or 1.987 cal mol^{-1} deg^{-1}).

Now let us assume that no chemical reaction can occur in the gas. Any transformation of a system remaining uniform is reversible, and, thus, from (1.13), (6.3), and (2.18), we have, per mole,

$$ds = c_v(T)\, d(\ln T) + R\, d(\ln v), \tag{6.4}$$

where c_v is the partial molar heat capacity at constant volume. Integration of (6.4) from the initial state $s^{\mathrm{i}}(T^{\mathrm{i}}, v^{\mathrm{i}})$ to the final state $s(T, v)$ gives (for an application, see Chapter 8, Section VIII)

$$s = s^*(T) - RT \ln p, \tag{6.5}$$

where

$$s^*(T) = s^{\mathrm{i}}(T^{\mathrm{i}}, v^{\mathrm{i}}) + \int_{T^{\mathrm{i}}}^{T} \frac{c_v(T)}{T}\, dT - R \ln v^{\mathrm{i}} + R \ln RT. \tag{6.6}$$

Combining (6.5), (6.3), and (4.38), we obtain

$$\mu = \mu^*(T) + RT \ln p. \tag{6.7}$$

For the free energy F of a perfect gas, Eq. (1.18), (6.5), and (6.1) give

$$F = n[f^*(T) + RT \ln(n/V)], \tag{6.8}$$

where

$$f^*(T) = e(T) - Ts^*(T) + RT \ln RT. \tag{6.9}$$

C. The Multicomponent Perfect Gas

We first suppose that we have several *separate* single-component perfect gases. If they are mixed together and if the free energy of the mixture is equal to the sum of the free energies that each of the gases would have if it alone were to occupy the same volume at the same temperature, then they make a multicomponent perfect gas.

Applying Eq. (6.8), we obtain

$$F = \sum_{\gamma} F_\gamma(T, V, n_\gamma) = \sum_{\gamma} n_\gamma[f_\gamma^*(T) + RT \ln(n_\gamma/V)]. \tag{6.10}$$

To find the equation of state, we must use (3.8) and (6.10):

$$p = -\left(\frac{\partial F}{\partial V}\right)_{T, n_1, \ldots, n_c} = \sum_{\gamma} n_\gamma \frac{RT}{V}, \tag{6.11}$$

and thus

$$pV = nRT, \tag{6.12}$$

where $n = \Sigma_\gamma n_\gamma$. The equation of state is thus formally identical to the corresponding equation of state of a single-component perfect gas.

From (6.10), (3.8), and (3.10), it is then easy to show that

$$S = \sum_\gamma S_\gamma(T, V, n_\gamma), \tag{6.13}$$

$$U = \sum_\gamma U_\gamma. \tag{6.14}$$

Let us *define* the partial pressure p_γ of the component γ by

$$p_\gamma = pN_\gamma. \tag{6.15}$$

This definition is true whether the gases in the mixture are perfect or not. It follows then from (6.15) and (6.11) that

$$p = \sum_\gamma p_\gamma, \qquad p_\gamma = n_\gamma RT/V = C_\gamma RT \qquad \text{(Dalton's law)}. \tag{6.16}$$

The partial pressure of γ is thus equal to the pressure which would be exerted by n_γ moles of pure γ in the same volume, and at the same temperature. It is a purely mathematical construct with no direct physical meaning.

The chemical potential of the γ component in the mixture is most conveniently computed by recalling that [see Eq. (4.16)]

$$\mu_\gamma = (\partial F/\partial n_\gamma)_{TVn_\beta}. \tag{6.17}$$

Evaluating this derivative from Eqs. (6.10) and (6.17) gives

$$\boxed{\mu_\gamma = \mu_\gamma^*(T) + RT \ln p_\gamma,} \tag{6.18}$$

where

$$\mu_\gamma^*(T) = f_\gamma^*(T) + RT + RT \ln RT. \tag{6.19}$$

The chemical potential of a single component γ in a mixture of perfect gases is equal to the chemical potential that component would have alone if it were at the same temperature and the reduced pressure p_γ.

We now evaluate the chemical affinity by applying (4.21) and (6.18). This gives

$$A = RT \ln[K_p(T)/p_1^{\nu_1}, \ldots, p_c^{\nu_c}], \tag{6.20}$$

where the function $K_p(T)$ is *defined* by

$$K_p(T) = \exp\left\{-\left[\sum_\gamma \nu_\gamma \mu_\gamma^*(T)\right] \Big/ RT\right\}. \tag{6.21}$$

At the equilibrium, $A = 0$ and (6.20) reduces to the well known law of Guldberg–Waage

$$K_p(T) = p_1^{\nu_1}, \ldots, p_c^{\nu_c}. \tag{6.22}$$

The quantity $K_p(T)$, called the equilibrium constant, is characteristic of the reaction under consideration and is a function only of the temperature.

The partial pressures used in (6.22) are equilibrium quantities.

If reactants and products (assumed to be perfect) are mixed under nonequilibrium conditions, a spontaneous change is expected until the partial pressure quotient reaches the equilibrium value. In certain circumstances, it is more convenient to express the equilibrium constant as a molar fraction or a concentration quotient.

Replacing in (6.18) p_γ respectively by pN_γ and $C_\gamma RT$ as in Eqs. (6.15) and (6.16), we get

$$\mu_\gamma = \mu_\gamma^\circ(T, p) + RT \ln N_\gamma \tag{6.23}$$

and

$$\mu_\gamma = \mu_\gamma^{\circ\circ}(T) + RT \ln C_\gamma, \tag{6.24}$$

where

$$\mu_\gamma^\circ(T, p) = \mu_\gamma^*(T) + RT \ln p \tag{6.25}$$

and

$$\mu_\gamma^{\circ\circ}(T) = \mu_\gamma^*(T) + RT \ln RT. \tag{6.26}$$

A similar treatment as above may be developed by using (4.21) and (6.23) or (6.24). We then obtain

$$\boxed{A = RT \ln[K_N(T, p)/N_1^{\nu_1}, \ldots, N_c^{\nu_c}]} \tag{6.27}$$

$$\boxed{A = RT \ln[K_c(T)/C_1^{\nu_1}, \ldots, C_c^{\nu_c}],} \tag{6.28}$$

where the functions K_N and K_c are *defined* by

$$K_N = \exp\left\{-\left[\sum_\gamma \nu_\gamma \mu_\gamma{}^\circ(T, p)\right] \Big/ RT\right\} \tag{6.29}$$

$$K_c = \exp\left\{-\left[\sum_\gamma \nu_\gamma \mu_\gamma^{\circ\circ}(T)\right] \Big/ RT\right\}. \tag{6.30}$$

At the equilibrium, (6.27) and (6.28) then reduce to

$$K_N(T, p) = N_1^{\nu_1}, \ldots, N_c^{\nu_c} \tag{6.31}$$

$$K_c(T) = C_1^{\nu_1}, \ldots, C_c^{\nu_c}. \tag{6.32}$$

The equilibrium constants may be used to evaluate the chemical affinity, or, alternatively, the latter may be used to evaluate the equilibrium constants. From (6.25), (6.26), (6.29), and (6.30), it is easy to show that the three equilibrium constants are related by the formulas

$$K_N(T, p) = p^{-\nu} K_p(T) \tag{6.33}$$

$$K_c(T) = (RT)^{-\nu} K_p(T), \tag{6.34}$$

where ν is the algebraic sum of the stoichiometric coefficients for the reaction and is given by

$$\nu = \sum_\gamma \nu_\gamma. \tag{6.35}$$

For the reaction considered as a multicomponent ideal gas

$$4HCl(g) + O_2(g) = 2H_2O(g) + 2Cl_2(g),$$

we have

$$K_p(T) = p_{H_2O}^2 p_{Cl_2}^2 / p_{HCl}^4 p_{O_2}$$

$$K_N(T, p) = N_{H_2O}^2 N_{Cl_2}^2 / N_{HCl}^4 N_{O_2}$$

$$K_c(T, p) = C_{H_2O}^2 N_{Cl_2}^2 / C_{HCl}^4 C_{O_2}$$

and the three constants are related by

$$K_N(T, p) = pK_p(T) = CK_c(T, p)$$

where $C = \sum_\gamma C_\gamma$; thus, $K_N(T, p)$ is here proportional to pressure.

D. Dependence of the Equilibrium Constant on Temperature and Pressure

1. *The Clausius–Kirchhoff Equations*

Let us first show how the thermal coefficients are interrelated. Thus, we will be able to find the dependence of the equilibrium constant on temperature and on pressure.

The internal energy U being a function of T, V, and ξ, we have the identities

$$\frac{\partial^2 U}{\partial V\,\partial T} \equiv \frac{\partial^2 U}{\partial T\,\partial V}, \quad \frac{\partial^2 U}{\partial V\,\partial \xi} \equiv \frac{\partial^2 U}{\partial \xi\,\partial V}, \quad \frac{\partial^2 U}{\partial T\,\partial \xi} \equiv \frac{\partial^2 U}{\partial \xi\,\partial T}. \tag{6.36}$$

Thus from (2.17)–(2.20), we obtain

$$\frac{\partial}{\partial T}\left(\frac{\partial U}{\partial \xi}\right)_{TV} = -\left(\frac{\partial r_{TV}}{\partial T}\right)_{V\xi} = \left(\frac{\partial C_{V\xi}}{\partial \xi}\right)_{TV} \quad \text{(Kirchhoff–Clausius)} \tag{6.37}$$

$$\left.\begin{aligned} \frac{\partial}{\partial V}\left(\frac{\partial U}{\partial \xi}\right)_{TV} &= -\left(\frac{\partial r_{TV}}{\partial V}\right)_{T\xi} = \left[\frac{\partial(l_{T\xi} - p)}{\partial \xi}\right]_{TV} \quad \text{(De Donder)} \\ \left(\frac{\partial C_{V\xi}}{\partial V}\right)_{T\xi} &= \left[\frac{\partial(l_{T\xi} - p)}{\partial T}\right]_{V\xi} \quad \text{(Clausius).} \end{aligned}\right\} \tag{6.38}$$

We may proceed in an analogous manner in the variables T, p, and ξ. From (1.15) and (2.4)–(2.6), we may write

$$C_{p\xi} = (\partial H/\partial T)_{p\xi} \tag{6.39}$$

$$V + h_{T\xi} = (\partial H/\partial p)_{T\xi} \tag{6.40}$$

$$r_{Tp} = -(\partial H/\partial \xi)_{Tp}. \tag{6.41}$$

The enthalpy H being a function of T, p, and ξ, we then have from (2.7)

$$\frac{\partial}{\partial T}\left(\frac{\partial H}{\partial \xi}\right)_{Tp} = -\left(\frac{\partial r_{Tp}}{\partial T}\right)_{p\xi} = \left(\frac{\partial C_{p\xi}}{\partial \xi}\right)_{Tp} \quad \text{(Kirchhoff–Clausius)} \tag{6.42}$$

$$\left(\frac{\partial C_{p\xi}}{\partial p}\right)_{T\xi} = \left[\frac{\partial(h_{T\xi} + V)}{\partial T}\right]_{p\xi} \quad \text{(Clausius)} \tag{6.43}$$

$$\frac{\partial}{\partial p}\left(\frac{\partial H}{\partial \xi}\right)_{Tp} = -\left(\frac{\partial r_{Tp}}{\partial p}\right)_{T\xi} = \left[\frac{\partial(h_{T\xi} + V)}{\partial \xi}\right]_{T} \quad \text{(Clausius).} \tag{6.44}$$

Now, since $C_{p\xi}$ is a function of state of the system, we have

$$\left(\frac{\partial C_{p\xi}}{\partial \xi}\right)_{Tp} = \sum_{\gamma} \left(\frac{\partial C_{p\xi}}{\partial n_{\gamma}}\right)_{Tpn_{\beta}} \frac{dn_{\gamma}}{d\xi} = \sum_{\gamma} \nu_{\gamma} c_{p\gamma}, \tag{6.45}$$

where $c_{p\gamma}$ is the partial molar heat capacity of γ at constant pressure, defined by

$$c_{p\gamma} = (\partial C_{p\xi}/\partial n_{\gamma})_{Tpn_{\beta}}. \tag{6.46}$$

Combining (6.45) and (6.42), we obtain

$$\boxed{(\partial r_{Tp}/\partial T)_{p\xi} = -\sum_{\gamma} \nu_{\gamma} c_{p\gamma} \qquad \text{(Kirchhoff)}.} \tag{6.47}$$

This equation is of importance since it enables us to calculate the heat of a reaction at any temperature provided that it is known at one temperature, and that we know the partial molar heat capacities of the components taking part in the reaction (see the example in Section VIII).

The Clausius–Kirchhoff equations are quite general because they do not suppose any particular state of matter (perfect or nonperfect gas, ideal liquid or nonideal liquid, solid).

2. *Influence of Temperature on Equilibrium Constants*

The influence of temperature on the equilibrium constant is now easy to establish. Indeed, the entropy being a function of state, we have

$$\partial^2 S/\partial T\, \partial \xi = \partial^2 S/\partial \xi\, \partial T, \tag{6.48}$$

and then, from the first of Eqs. (2.12), combined with (2.13) and (6.41), we can write

$$\boxed{\frac{\partial (A/T)}{\partial T} = -\frac{r_{Tp}}{T^2}.} \tag{6.49}$$

Thus, combining (6.49) and (6.27), we find the well-known equation of van't Hoff

$$\boxed{\left(\frac{\partial [\ln K_N(T, p)]}{\partial T}\right)_p = -\frac{r_{Tp}}{RT^2}.} \tag{6.50}$$

From (6.50) and (6.33), we also obtain

$$\boxed{\frac{d[\ln K_p(T)]}{dT} = -\frac{r_{Tp}}{RT^2},} \tag{6.51}$$

or, in another form,

$$\frac{d[R \ln K_p(T)]}{d(1/T)} = r_{Tp}. \tag{6.52}$$

If the reaction is accompanied by an absorption of heat ($r_{Tp} < 0$), the equilibrium constant increases with temperature, while, for an exothermic reaction ($r_{Tp} > 0$), it decreases.

Example. For the dissociation of water vapor, $2H_2O = 2H_2 + O_2$, the reaction is endothermic and thus is favored by a high temperature.

These equations enable us to calculate the rate at which the equilibrium constant is changing with the temperature when r_{Tp} is known; and if we know r_{Tp} as a function of the temperature, then (6.51), may be integrated. Thus, from the equilibrium constant at any temperature, we may calculate its value at any other temperature.

By integration of (6.51) between T^i (initial) and T, we get

$$\ln K_p(T) - \ln K_p(T^i) = -\int_{T^i}^{T} (r_{Tp}/RT^2)\, dT. \tag{6.53}$$

From (6.47), it is then possible to know the explicit form of $r_{Tp}(T)$ and then to calculate the integral (6.53). A detailed example is given in Section VIII.

Remarks. (1) The same equations as above may be written in the variables T, V, and ξ. To derive the relationship between the thermal coefficients $C_{V\xi}$, $l_{T\xi}$, and r_{TV} with T, V, and ξ as variables, it is most convenient to consider first the total differential dU in the variables T, V, and ξ [see Eq. (2.16)] and then to replace dV by the value given in terms of T, p, and ξ [see Eq. (2.3)].

Employing Eqs. (2.2), (2.4)–(2.6), and (2.18)–(2.20), we can write the required general relations between the thermal coefficients in the two sets of variables T, V, ξ and T, p, ξ (see Chapter 1, Section XVIII):

$$C_{p\xi} = C_{V\xi} + l_{T\xi}(\partial V/\partial T)_{p\xi} \tag{6.54}$$

$$h_{T\xi} = l_{T\xi}(\partial V/\partial p)_{T\xi} \tag{6.55}$$

$$r_{Tp} = r_{TV} - l_{T\xi}(\partial V/\partial \xi)_{Tp}. \tag{6.56}$$

For a perfect gas, (2.19) and (6.55) reduce, respectively, to

$$l_{T\xi} = p \qquad \text{(perfect gas)} \tag{6.57}$$

$$h_{T\xi} = p(\partial V/\partial p)_{T\xi} = -V \qquad \text{(perfect gas).} \tag{6.58}$$

Furthermore,

$$(\partial V/\partial T)_{p\xi} = nR/p \qquad \text{(perfect gas)} \tag{6.59}$$

$$(\partial V/\partial \xi)_{Tp} = \nu RT/p \qquad \text{(perfect gas).} \tag{6.60}$$

Equations (6.54) and (6.56) thus become

$$C_{p\xi} - C_{V\xi} = nR \qquad \text{(Mayer's formula)} \tag{6.61}$$

$$r_{Tp} - r_{TV} = -\nu RT. \tag{6.62}$$

It is then easy to show that Eqs. (6.49) and (6.51) may be written in the form

$$\left(\frac{\partial(A/T)}{\partial T}\right)_{V\xi} = -\frac{r_{TV}}{RT^2} \tag{6.63}$$

$$\frac{d[\ln K_c(T)]}{dT} = -\frac{r_{TV}}{RT^2}. \tag{6.64}$$

(2) From the experimental point of view, if r_{Tp} is independent of T, then (6.53) may be written

$$\ln K_p(T) = (r_{Tp}/RT) + \text{const.} \tag{6.65}$$

An example of a reaction for which $\ln K_p$ is a linear function of $1/T$ over a wide temperature range is given in Section VIII.

(3) Influence of pressure on the equilibrium constants. Let us write the last of Eqs. (3.12) as

$$(\partial A/\partial p)_{T\xi} = -(\partial V/\partial \xi)_{Tp} = -\Delta_{Tp}, \tag{6.66}$$

where Δ_{Tp} is the change in the volume of the system produced by the chemical reactions at constant T and p,

$$\Delta_{Tp} = \sum_{\gamma} \nu_\gamma v_\gamma. \tag{6.67}$$

The variation of the affinity of reaction with the pressure is seen to be appreciable only if the reaction is accompanied by a considerable change in volume when it takes place at constant temperature and pressure. If a reaction is accompanied by an increase in volume when the pressure is kept constant, an increase in pressure will reduce the affinity of the reaction.

Example. For the gas reaction $2H_2 + O_2 = 2H_2O$, $\Delta_{Tp} < 0$ and the affinity of the reaction increases with pressure. From the experimental point of view, according to this, it could be interesting to work at high pressure.

Combining (6.67) and (6.27), we obtain

$$\partial[\ln K_N(T, p)]/\partial p = -\Delta_{Tp}/RT. \tag{6.68}$$

An increase in pressure increases the equilibrium constant if the reaction is accompanied by a decrease in volume ($\Delta_{Tp} < 0$); and, conversely, if $\Delta_{Tp} > 0$, the equilibrium constant is decreased (Le Chatelier's principle). For a perfect gas mixture, (6.67) and (6.16) give

$$\Delta_{Tp} = \nu RT/p, \tag{6.69}$$

so that (6.68) reduces to the formula

$$p\{\partial[\ln K_N(T, p)]/\partial p\} = -\nu. \tag{6.70}$$

E. Ideal Systems

In order to discuss the behavior of ideal systems, we need to express the chemical potentials of the components in a form in which the molar fractions or the partial pressures appear explicitly.

One-Phase System

An ideal system may be defined by the equation

$$\mu_\gamma = \zeta(T, p) + RT \ln N_\gamma. \tag{6.71}$$

Mixtures of perfect gases and very dilute solutions are referred to as ideal systems. If the solution is ideal for all values of N_γ and for all γ, it is then called a perfect solution.

Pure components are always ideal systems because

$$\mu_\gamma = \mu_\gamma(T, p). \tag{6.72}$$

Now, combining (4.21) and (6.71), we obtain

$$A = RT \ln[K_N(T, p)/N_1^{\nu_1}, \ldots, N_c^{\nu_c}], \tag{6.73}$$

where $K_N(T, p)$ is defined by the equation

$$RT \ln K_N(T, p) = -\sum_\gamma \nu_\gamma \zeta_\gamma(T, p) \tag{6.74}$$

At the equilibrium,

$$A = 0 \quad \text{and} \quad K_N(T, p) = N_1^{\nu_1}, \ldots, N_c^{\nu_c} \quad \text{(Guldberg–Waage)}, \tag{6.75}$$

and we obtain again the van't Hoff expression (6.50) and Eq. (6.68) valid here for an ideal system,

$$\left\{\frac{\partial[\ln K_N(T, p)]}{\partial T}\right\}_p = -\frac{r_{Tp}}{RT^2} \quad \text{(van't Hoff)} \tag{6.76}$$

$$\left\{\frac{\partial[\ln K_N(T, p)]}{\partial p}\right\}_T = -\frac{\Delta_{Tp}}{RT}. \tag{6.77}$$

F. MULTIPHASE SYSTEMS

When each of the phases is ideal, we have following the definition [see (6.71)]

$$\mu_\gamma^\alpha = \zeta_\gamma^\alpha(T, p) + RT \ln N_\gamma^\alpha, \tag{6.78}$$

and the multiphase system is then called an ideal system.

Passage Reaction

If the reaction consists in a transfer of a component γ from one phase (′) to another (″), we have, from (4.21), (6.78), and (6.73),

$$A_\gamma = \mu_\gamma' - \mu_\gamma'' = RT \ln K_{\gamma N}(T, p) - RT \ln(N_\gamma''/N_\gamma'); \tag{6.79}$$

when the equilibrium of transfer is attained, (6.79) reduces to the well-known Nernst distribution law

$$N_\gamma''/N_\gamma' = K_{\gamma N}(T, p). \tag{6.80}$$

It is physically obvious, and also an experimental fact, that, for trivariant systems (for example, three components and two phases), the composition of γ in one phase, N_γ'', at constant temperature and pressure varies proportionally to the composition of γ in the other phase, N_γ', following (6.80). For a divariant system, both the temperature and pressure can be fixed arbitrarily. For example, we can study a pure liquid in the presence of a gaseous phase such as nitrogen. Neglecting the weak solubility of nitrogen in the liquid (subscript l), (6.80) becomes

$$N_l^{\mathrm{V}} = K_{lN}(T, p),$$

where V designates the vapor phase. The composition of the vapor of the liquid in the vapor phase is thus only a function of T and p.

For a monovariant system—for example, a pure liquid in the presence of its own vapor—the Nernst law $K_{lN}(T, p) = 1$ shows that the pressure depends only upon the temperature. This result is consistent with the phase rule.

Example. Chemical reactions between two different phases. For the thermal dissociation of calcium carbonate in air,

$$\mathrm{CaCO_3(s) = CaO(s) + CO_2(g)},$$

let us assume that each reacting component is present in one phase and that $CO_2(g)$ + air is a mixture of perfect gas. Then it is easy to see that

$$A = RT \ln[K_N(T, p)/N_{\mathrm{CO_2}}]$$

For equilibrium, $N_{\mathrm{CO_2}} = K_N(T, p)$, or $p_{\mathrm{CO_2}} = pK_N(T, p)$, which shows that the equilibrium partial pressure of CO_2 depends both upon the total pressure p and the temperature.

In the absence of air ($N_{\mathrm{CO_2}} = 1$), we find $K_N(T, p) = 1$, and the equilibrium pressure of carbon dioxide depends only upon the temperature.

Remark. If the condensed phases (liquids and solids) in a heterogeneous system are pure components and the gas is a mixture, it is more convenient, from a practical point of view, to express the equilibrium constant in terms of partial pressures.

Applying (4.21) and (6.18) to our system, we obtain

$$A = -\sum_i \nu_i\mu_i(T, p) - \sum_j \nu_j\mu_j^*(T) - RT\sum_j \nu_j \ln p_j, \qquad (6.81)$$

where the subscripts i and j refer, respectively, to the condensed pure constituents and to the gaseous components. We then may write

$$A = RT \ln(K_N/\prod_j p_j^{\nu_j}), \qquad (6.82)$$

where

$$RT \ln K_N = - \sum_i \nu_i \mu_i(T, p) - \sum_j \nu_j \mu_j^*(T). \tag{6.83}$$

As for condensed phases, the chemical potentials $\mu_i(T, p)$ are practically independent of pressure (Prigogine and Defay, 1967, p. 163); the constant K varies only with the temperature.

Example:

$$NiO(s) + CO(g) = Ni(s) + CO_2(g).$$

For equilibrium, (6.82) reduces to $K_N = p_{CO_2}/p_{CO}$ and K_N does not depend upon the total pressure.

VII. Standard Functions and Functions of Mixing

A. Standard Functions

In general, every intensive thermodynamic quantity θ in a uniform system $\theta = \theta(T, p, N_1, \ldots, N_c)$, can be split up arbitrarily into the sum of two functions—for example, a standard part $\theta^s(T, p)$ which depends only upon the temperature and the pressure, and a function of mixing $\theta^m(T, p, N_1, \ldots, N_c)$:

$$\theta = \theta^s(T, p) + \theta^m(T, p, N_1, \ldots, N_c). \tag{7.1}$$

For example, an ideal system characterized by (6.87) gives

$$\mu_\gamma{}^s = \zeta_\gamma(T, p) \tag{7.2}$$

$$\mu_\gamma{}^m = RT \ln N_\gamma. \tag{7.3}$$

The affinity of such a system is written

$$A = A^s + A^m, \tag{7.4}$$

where

$$A^s = - \sum_\gamma \nu_\gamma \zeta_\gamma(T, p) \tag{7.5}$$

$$A^m = - \sum_\gamma \nu_\gamma RT \ln N_\gamma. \tag{7.6}$$

The chemical potential being the partial molar quantity corresponding to the Gibbs free energy G, we have, from (3.9),

$$\partial \mu_\gamma / \partial T = -s_\gamma, \tag{7.7}$$

where the partial molar entropy of γ is defined by

$$s_\gamma = (\partial S / \partial n_\gamma)_{Tp}. \tag{7.8}$$

Now, combining (7.2), (7.3), and (7.7), we find

$$s_\gamma = -(\partial \zeta_\gamma / \partial T) - R \ln N_\gamma \tag{7.9}$$

and

$$s_\gamma^{\mathrm{s}} = -\partial \zeta_\gamma / \partial T \tag{7.10}$$

$$s_\gamma^{\mathrm{m}} = -R \ln N_\gamma. \tag{7.11}$$

In the same manner, the entropy of reaction can be written as

$$\left(\frac{\partial S}{\partial \xi}\right)_{Tp} = \left(\frac{\partial S}{\partial \xi}\right)_{Tp}^{\mathrm{s}} + \left(\frac{\partial S}{\partial \xi}\right)_{Tp}^{\mathrm{m}} = \sum_\gamma \nu_\gamma s_\gamma = -\sum_\gamma \nu_\gamma \frac{\partial \zeta_\gamma}{\partial T} - R \sum_\gamma \nu_\gamma \ln N_\gamma. \tag{7.12}$$

The quantity $(\partial S / \partial \xi)_{Tp}^{\mathrm{s}}$ is often called the standard entropy of reaction.

As the partial molar volume for an ideal system is only a function of T and p, we may write

$$v_\gamma = v_\gamma^{\mathrm{s}} = \partial \mu_\gamma^{\mathrm{s}} / \partial p = \partial \zeta_\gamma / \partial p \tag{7.13}$$

and

$$\Delta_{Tp} = \Delta_{Tp}^{\mathrm{s}} + \Delta_{Tp}^{\mathrm{m}} = \sum_\gamma \nu_\gamma v_\gamma^{\mathrm{s}}. \tag{7.14}$$

It is easy to show that, for ideal systems,

$$r_{Tp} = r_{Tp}^{\mathrm{s}} = -\sum_\gamma \nu_\gamma h_\gamma = -\sum_\gamma \nu_\gamma h_\gamma^{\mathrm{s}} = T^2 \sum_\gamma \nu_\gamma \, \partial(\zeta_\gamma / T) / \partial T \tag{7.15}$$

and

$$c_{p\gamma} = \partial h_\gamma / \partial T = \partial h_\gamma^{\mathrm{s}} / \partial T, \tag{7.16}$$

where h_γ is the partial molar enthalpy of γ.

For perfect solutions, h_γ^{s}, s_γ^{s} and v_γ^{s} are equal to the molar enthalpy. molar entropy, and molar volume of the pure component, respectively, On the other hand, the identity

$$\frac{\partial(\zeta_\gamma / T)}{\partial T} = \frac{1}{T} \frac{\partial \zeta_\gamma}{\partial T} - \frac{\zeta_\gamma}{T^2} \tag{7.17}$$

enables us to show that

$$r_{Tp}^{s} + T\left(\frac{\partial S}{\partial \xi}\right)_{Tp}^{s} = A^{s} \tag{7.18}$$

and

$$\frac{\partial r_{Tp}^{s}}{\partial T} = -\sum_{\gamma} \nu_{\gamma} c_{p\gamma}^{s}. \tag{7.19}$$

From (2.13) and (7.5),

$$r_{Tp}^{m} + T\left(\frac{\partial S}{\partial \xi}\right)_{Tp}^{m} = A^{m}. \tag{7.20}$$

The following equations are also easily verified

$$\left[\frac{\partial(A^{s}/T)}{\partial T}\right]_{p} = -\sum_{\gamma} \nu_{\gamma} \frac{\partial(\zeta_{\gamma}/T)}{\partial T} = \frac{r_{Tp}^{s}}{T^{2}}, \tag{7.21}$$

$$\left(\frac{\partial A^{s}}{\partial p}\right)_{T} = -\sum_{\gamma} \nu_{\gamma} \frac{\partial \zeta_{\gamma}}{\partial p} = -\sum_{\gamma} \nu_{\gamma} v_{\gamma}^{s} = -\Delta_{Tp}^{s}. \tag{7.22}$$

As, in general, every property $(\partial X/\partial \xi)_{Tp}$, where X is any extensive variable, can be added together in the same way as chemical reactions, it is immediately clear that this is particularly true for A^{s}, r_{Tp}^{s}, $(\partial S/\partial \xi)_{Tp}^{s}$, and Δ_{Tp}^{s}. More explicitly, *standard affinities, standard heats of reaction, standard expansions, and standard entropies of reaction can be added together in the same way as the chemical equations for the reactions themselves.*

Now, from (6.75) and (7.5), we see that

$$K_{N}(T, p) = \exp(A^{s}/RT). \tag{7.23}$$

This equation enables us to calculate the equilibrium constant of any reaction which can be obtained by a linear combination of known reactions.

B. Standard Function of Formation

Many tables give values of the standard affinities and heats of reaction at a temperature of 298.17°K (25°C) and a pressure of 1 atm (symbols A° and r_{Tp}°). It is much more convenient to consider standard affinities of reaction than equilibrium constants. This is because standard affinities can be added and subtracted in just the same way as stoichiometric equations, so that the standard affinity and the equilibrium constant of a reaction not included in the table is easily calculated.

A reaction involving a compound produced from its elements (these elements being taken in their normal physical state under specified conditions), is called a *formation reaction.*

It is also necessary to specify the physical condition of the compound which is formed, although this need not necessarily be the stable state under the conditions considered.

At 298.16°K and 1 atm pressure, the normal physical state of the elements are the gas state (for example, hydrogen, oxygen, fluorine), the liquid state (for example, mercury, bromine), the solid state (for example, sodium, rhombic sulfur, iodine). The standard values for the formation of an element in the stable physical state are, by definition, zero.

For example, $A^\circ_{I_2}$ is equal to zero for the formation of solid iodine, $I_2(s) \rightarrow I_2(s)$, but for the formation of gaseous atomic hydrogen, $\frac{1}{2}H_2(g) \rightarrow H(g)$, the standard affinity at 298.16°K and 1 atm is not zero, because atomic hydrogen is not stable at this temperature and this pressure.

In general, the standard affinity of formation $A_\gamma{}^\circ$, the standard heat of formation $r^\circ_{Tp\gamma}$, and the standard entropy change of formation $(\partial S^\circ/\partial \xi_\gamma)_{Tp}$ are defined as the standard affinity, standard heat, and standard entropy of the formation reaction of the component γ. Equation (7.18) then becomes

$$A_\gamma{}^\circ = r^\circ_{Tp\gamma} + 298.16\left(\frac{\partial S^\circ}{\partial \xi_\gamma}\right)_{Tp}. \tag{7.24}$$

In Table I, values of $r^\circ_{Tp\gamma}$ and $A_\gamma{}^\circ$ are given for some compounds (Prigogine and Defay, 1967, p. 99). The values of $(\partial S^\circ/\partial \xi_\gamma)$ are easily calculated from (7.24).

Unstable compounds such as O_3, NO, and NO_3 have negative standard affinities of formation at 218°K and 1 atm, while for the other inorganic compounds they are usually positive. For elements in an unstable physical state, A° is negative [S (monoclinic) and Cl(g)].

Now, from a practical point of view, the standard affinity A^s of a reaction can be evaluated from the knowledge of the equilibrium constant (7.23). This last quantity is given by measuring the mole fractions of the various components at the equilibrium. Conversely, if the standard affinity is known, we can evaluate the position of equilibrium.

A large positive standard affinity of formation means that the compound will not decompose spontaneously into its elements under the standard conditions, since the synthesis reaction is practically complete.

TABLE I

STANDARD THERMODYNAMIC FUNCTIONS OF FORMATION AT $T = 298.16°K$, $p = 1$ ATM

Substance	State	r°_{Tp} (kcal mol^{-1})	A° (kcal mol^{-1})	s° (cal mol^{-1} °K^{-1})
Ca	s	0	0	9.95
$CaCO_3$ (calcite)	s	288.450	269.780	22.2
(aragonite)	s	288.490	269.530	21.2
CaO	s	152.800	145.360	7.8
Hg	l	0	0	18.5
Hg_2	g	−27.100	—	—
Na	s	0	0	12.48
NaCl	s	98.232	91.785	17.3
H_2	g	0	0	31.211
H	g	−52.089	−48.575	27.393
F_2	g	0	0	48.6
Cl_2	g	0	0	53.286
Cl	g	−29.012	−25.192	39.457
Br_2	l	0	0	36.4
	g	−7.340	−0.751	58.639
I_2	s	0	0	27.9
	g	−14.876	−4.630	62.280
I	g	−25.482	−16.766	43.184
HI	g	−6.200	−0.310	49.314
O_2	g	0	0	49.003
O	g	−59.159	−54.994	38.469
O_3	g	−34.000	−39.060	56.8
H_2O	g	+57.798	54.635	45.106
	l	+68.317	56.690	16.716
H_2O_2	aq. ($m = 1$)	45.680	31.470	—
S (rhombic)	s	0	0	7.62
(monoclinic)	s	−0.071	−0.023	7.78
N_2	g	0	0	45.767
N	g	−85.566	−81.476	36.615
NO	g	−21.600	−20.719	50.339
NH_3	g	+11.040	3.976	46.01
CO	g	26.416	32.808	47.301
CO_2	g	94.052	94.260	51.061
CH_4	g	17.889	12.140	44.50
CH_3OH	g	48.100	38.700	56.8
	l	57.036	39.750	30.3
C_2H_5OH	g	56.240	40.300	67.4
	l	66.356	41.770	38.4
CH_3COOH	l	116.400	93.800	38.2

This does not prove, however, that the compound will not decompose to form a more stable compound.

As an example, hydrogen peroxide at 25°C and 1 atm will not decompose spontaneously to H_2 and O_2, but will completely to $H_2O(l)$ and $O_2(g)$:

$$A^\circ = -A^\circ_{H_2O_2} + A^\circ_{H_2O} + \tfrac{1}{2}A^\circ_{O_2} = 25{,}220 \quad \text{cal mol}^{-1}.$$

Other important examples are found in organic chemistry.

But the affinity only indicates the *tendency* of a reaction to proceed and says nothing about the kinetics of the reaction. In organic chemistry, we frequently find that a number of different substances can be formed from the same starting materials, so that it is usually necessary to employ a specific catalyst to accelerate the required reaction. An interesting example is given by the oxidation of acetone (Prigogine and Defay, 1967, p. 98).

Remark. In American tables, we find the mean value

$$A = -(\Delta G/\Delta\xi)_{Tp} \qquad \text{and} \qquad \Delta\xi = 1.$$

Furthermore, in place of ΔG, the symbol used usually is ΔF, so that in American tables we find the quantity

$$\Delta F = -A.$$

C. Variation of Standard Affinity with Temperature and Pressure

Let us integrate (7.21) between two temperatures T and $T_0 = 298.1°K$ at constant pressure $p_0 = 1$ atm. We find

$$\frac{A^s(T, p^\circ)}{T} - \frac{A^s(T^\circ, p^\circ)}{T^\circ} = -\int_{T^0}^{T} \frac{r^s_{Tp}}{T^2}\, dT. \tag{7.25}$$

The quantity $A^s(T^\circ, p^\circ)$ can be evaluated with the help of the standard affinity of formation and r^s_{Tp} can be calculated with the help of (7.19).

Now, to know the influence of the pressure upon the standard affinity, we must integrate (7.22) at constant temperature. The molar volumes are experimental quantities.

Finally, to calculate the affinity, we may use Eqs. (7.4)–(7.6),

$$A = A^s(T, p) - RT \sum_\gamma \nu_\gamma \ln N_\gamma . \tag{7.26}$$

VIII. Numerical Examples

Example 1. We consider a perfect gas mixture resulting from the reaction

$$2CO(g) + O_2(g) = 2CO_2(g).$$

(1) r_{Tp}^{s} at 1 atm is a function of the temperature, following Kirchhoff's law (6.47)

$$r_{Tp}^{s} = -\int \sum_{\gamma} \nu_{\gamma} c_{p\gamma}\, dT + \beta, \tag{8.1}$$

where β is the constant of integration.

Now, the molar heat capacities are usually expressed in an empirical power series of the form (Bryant, 1933; Ewell, 1940; Thacker *et al.*, 1941; Kelley, 1960; Spencer, 1945)

$$c_p = a + bT + cT^2 + \cdots$$

$$CO: \quad 6.25 + 2.091 \times 10^{-3}T - 0.459 \times 10^{-6}T^2$$

$$CO_2: \quad 6.85 + 8.533 \times 10^{-3}T - 2.475 \times 10^{-6}T^2$$

$$O_2: \quad 6.26 + 2.746 \times 10^{-3}T - 0.770 \times 10^{-6}T^2.$$

We then obtain

$$r_{Tp}^{s} = \beta + 5.06T - 5.069 \times 10^{-3}T^2 + 1.087 \times 10^{-6}T^3. \tag{8.2}$$

At 298.16°K, $r_{Tp}^{s} = \sum_{\gamma} \nu_{\gamma} r_{Tp\gamma}^{\circ}$, and, from the Table I, $r_{Tp}^{s} = 135{,}272$ cal. Putting this value in Eq. (8.2) for $T = 298.16$°K, we get the expression for r_{Tp}^{s}

$$r_{Tp}^{s} = 134{,}185 + 5.06T - 5.069 \times 10^{-3}T^2 + 1.087 \times 10^{-6}T^3. \tag{8.3}$$

(2) We now calculate the affinity as a function of the temperature by using (7.25). From Table I,

$$A^{s}(298.16°, 1 \text{ atm}) = \sum_{\gamma} \nu_{\gamma} A_{\gamma}{}^{\circ} = 122{,}904 \quad \text{cal};$$

thus,

$$A^{s}(T, 1 \text{ atm}) = \frac{122{,}904}{298}\, T - T \int_{298.16}^{T} \frac{r_{Tp}^{s}}{T^2}\, dT. \tag{8.4}$$

Putting (8.3) in the integral of (8.4), we find

$$A^s(T, 1 \text{ atm}) = 134{,}185 - 10.49T - 5.06T \ln T + 5.069 \times 10^{-3}T^2 - 0.543 \times 10^{-6}T^3. \tag{8.5}$$

(3) The equilibrium constants $K_N(T, p)$ and $K_p(T)$ can be calculated with the help of (7.23) and (6.33). The combination of (7.23), (8.5), and (6.33) gives

$$\log_{10} K_p(T) = \log_{10} K_N(T, 1 \text{ atm}) = (29.427/T) - 1.11 \log_{10} T + 5.069 \times 10^{-3}T^2 - 0.543 \times 10^{-6}T^3. \tag{8.6}$$

Applying (6.22) to our reaction, we can write

$$K_p(T) = p_{CO_2}^2 / p_{CO}^2 p_{O_2}, \tag{8.7}$$

where $K_p(T)$ is not a function of the pressure, but only of the temperature (Fig. 15).

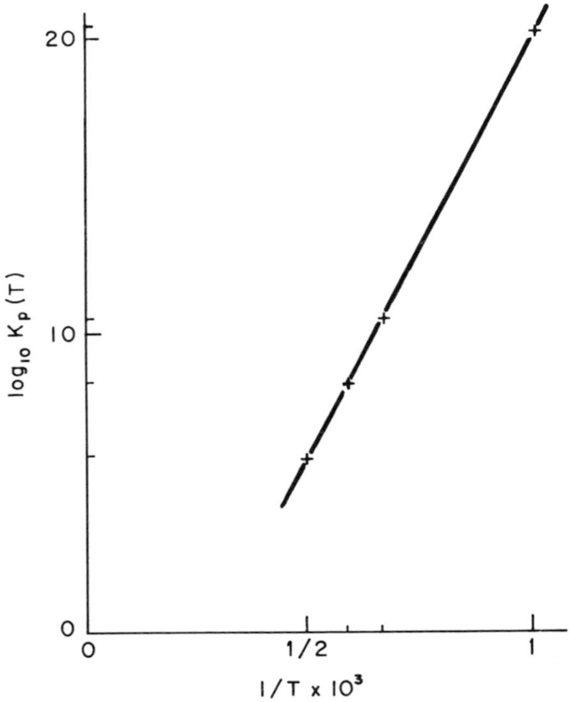

FIG. 15. Equilibrium constant as a function of temperature for the reaction 2CO + O_2 = 2CO_2.

It follows immediately from (6.15) and (8.7) that

$$K_p(T) = \frac{1}{p} \frac{N_{CO_2}^2}{N_{CO}^2 N_{O_2}}. \tag{8.8}$$

Let us now introduce the degree of dissociation α in such a way that we have at equilibrium α moles of CO, $\alpha/2$ moles of O_2, and $(1 - \alpha)$ moles of CO_2.

Equation (8.8) then becomes

$$K_p(T) = (1 - \alpha)^2(2 + \alpha)/p\alpha^3. \tag{8.9}$$

For any value of α and p, we are thus able to calculate the equilibrium temperature with the help of the Fig. 15.

For example, at $\alpha = 10^{-3}$: for $p = 1$ atm, $K_p(T) = 2 \times 10^9$ atm^{-1} and $T = 1600$°K; for $p = 0.1$ atm, $K_p(T) = 2 \times 10^{10}$ atm^{-1} and $T = 1520$°K.

Example 2. In the thermal dissociation of calcium carbonate, $CaCO_3$(s) (aragonite) = CaO(s) + CO_2(g), each constituent occurs in one phase only.

(1) r_{Tp} at 1 atm as a function of the temperature. The empirical values of the molar heat capacities are

$CaCO_3$:	26.35	(290°K–1030°K)
CO:	11.78	
CO_2:	$7 + 7.1 \times 10^{-3}T - 1.86 \times 10^{-6}T^2$.	

From Kirchhoff's law (6.47), we thus obtain

$$r_{Tp}^s = \beta + 7.57T - 3.55 \times 10^{-3}T^2 + 0.62 \times 10^{-6}T^3. \tag{8.10}$$

At 298.16°K, from Table I, we have $r_{Tp}^s = -41{,}638$ cal, and (8.10) then becomes

$$r_{Tp}^s = -43{,}600 + 7.57T - 3.55 \times 10^{-3}T^2 + 0.62 \times 10^{-6}T^3. \tag{8.11}$$

At 1200°K, $r_{Tp}^s = 38{,}500$ cal.

(2) From Table I, we have

$$A^s(298.16, 1 \text{ atm}) = \sum_\gamma \nu_\gamma A_\gamma^\circ = -29{,}910 \quad \text{cal.}$$

Now, from (7.21) and (8.11), it results

$$A^s(T, 1 \text{ atm}) = -43{,}600 - 17.4T \log_{10} T + 86.2T + 3.6 \times 10^{-3}T^2 - 0.3 \times 10^{-6}T^3. \quad (8.12)$$

At 900°K, $A^s(900°, 1 \text{ atm}) \simeq -9700$ cal.

(3) From (7.23), the equilibrium constant at 900°K and 1 atm is then equal to $K_N \simeq 3.8 \times 10^{-3}$.

(4) If the reaction is studied in the presence of air, $K_N(T, p) = N_{CO_2}$ and $p_{CO_2} = pN_{CO_2}$, and thus, at 900°K and 1 atm, $N_{CO_2} \simeq 3.8 \times 10^{-3}$, $p_{CO_2} \simeq 3.8 \times 10^{-3}$ atm.

(5) The affinity of the reaction can be calculated with the help of (7.26).

At 1 atm and 900°K and for $N_{CO_2} = 10^{-3}$,

$$A = A^s RT \ln N_{CO_2} = -9700 - RT \ln 3.8 \times 10^{-3} \simeq 2600 \text{ cal.} \quad (8.13)$$

In the absence of air, $N_{CO_2} = 1$ and $p = p_{CO_2}$, and $K_N(T, p) = 1$, which shows that the equilibrium pressure of carbon dioxide depends only on the temperature.

Example 3. Let us now consider the synthesis of ammonia, $N_2 + 3H_2 = 2NH_3$, as a mixture of perfect gases. Using Eq. (6.67), we obtain

$$\Delta_{Tp} = 2v_{NH_3} - v_{N_2} - 3v_{H_2} = -2v_{H_2}. \quad (8.14)$$

From (6.66), we see, then, that the affinity of the reaction increases with pressure.

Combining (6.66) and (8.14), we can write

$$\partial A/\partial p = 2RT/p, \quad (8.15)$$

or

$$A(T, p) - A(T, 1 \text{ atm}) = 2RT \ln p;$$

at 10^3 atm, $A(T, p) - A(T, 1 \text{ atm}) \simeq 8000$ cal. This reaction is exothermic because $r_{Tp} > 0$; $r^s_{Tp} = 22{,}800$ cal (see Table I). From (6.50), we see that the synthesis of ammonia is favored by a decrease of temperature.

Example 4. We may examine the formation of methanol in a mixture of gases

$$CO(g) + 2H_2(g) = CH_3OH(g). \quad (8.16\text{-I})$$

The heats of combustion of H_2, CO, and CH_3OH at 298.16°K and 1 atm are (see Table I)

$$H_2(g) + \tfrac{1}{2}O_2(g) = H_2O(l) \qquad r^s_{Tp(II)} = 68{,}317 \quad \text{cal} \qquad (8.16\text{-II})$$

$$CO(g) + \tfrac{1}{2}O_2(g) = CO_2(g) \qquad r^s_{Tp(III)} = 67{,}636 \quad \text{cal} \qquad (8.16\text{-III})$$

$$CH_3OH(g) + \tfrac{3}{2}O_2(g) = CO_2(g) + 2H_2O(l) \qquad r^s_{Tp(IV)} = 182{,}586 \quad \text{cal.} \qquad (8.16\text{-IV})$$

The heat of reaction of (I) is then equal to

$$r^s_{Tp(I)} = r^s_{Tp(III)} + 2r^s_{Tp(II)} - r^s_{Tp(IV)} = 21{,}684 \quad \text{cal.}$$

Using Table I, we find the same result for the reaction (I),

$$r^s_{Tp(I)} = r^\circ_{Tp\,CH_3OH} - r^\circ_{Tp\,CO} = 21{,}684 \quad \text{cal.}$$

The standard affinity $A^\circ_{(I)}$ for the reaction (I) is equal to 5892 cal. Using (7.18), we can also obtain $A^\circ_{(I)}$:

$$s^\circ_{CH_3OH(g)} = 56.8 \quad \text{cal}, \qquad s^\circ_{CO(g)} = 47.3 \quad \text{cal}, \qquad s^\circ_{H_2(g)} = 31.21 \quad \text{cal},$$

$$A^\circ_{(I)} = r^\circ_{Tp(I)} + T\sum_\gamma \nu_\gamma s_\gamma{}^\circ = 5908 \quad \text{cal.}$$

Example 5. *Entropy of a gas.* We now consider the vaporization of methanol

$$CH_3OH(l) = CH_3OH(g). \qquad (8.17)$$

We shall calculate the standard entropy of the vapor of methanol (assumed here to be perfect) $s^s_{(g)}$ (298.16, 1 atm), knowing the corresponding value of the liquid $s^s_{(l)}$ (298.16, 1 atm) and the heat of vaporization of CH_3OH at 298.16°K and 0.163 atm.

From (6.5), we find at 0.163 atm and 298.16°K

$$s^s_{(g)}(298.16°\text{K}, 1 \text{ atm}) = s^s_{(g)}(298.16°\text{K}, 0.163 \text{ atm}) + R\ln 0.163. \qquad (8.18)$$

On the other hand,

$$s^s_{(g)}(298.16°\text{K}, 0.163 \text{ atm}) = s^s_{(l)}(298.16°\text{K}, 0.163 \text{ atm}) + \Delta S_{\text{vap}}, \qquad (8.19)$$

where ΔS_{vap} is the entropy of vaporization, equal to the heat of vaporization divided by the temperature. From Table I, we take the values of

$$r^\circ_{Tp\,CH_3OH(g)} = 48{,}100 \text{ cal}, \qquad r^\circ_{Tp\,CH_3OH(l)} = 57{,}036 \text{ cal.}$$

The heat of reaction r^s_{Tp} of (8.16-I) is thus equal to -8936 cal and the heat of vaporization to 8936 cal. The entropy of vaporization is thus

$$\Delta S_{\text{vap}} = 8936/298.16 \simeq 29.98 \quad \text{cal °K}^{-1}.$$

We then obtain

$$s^s_{(g)}(298.16°\text{K}, 0.163 \text{ atm}) = s^s_{(l)}(298.16°\text{K}, 0.163 \text{ atm}) + 29.98. \quad (8.20)$$

As a first approximation, we neglect the effect of pressure upon the entropy of the liquid. We thus have, from Table I,

$$s^s_{(l)}(298.16°\text{K}, 0.163 \text{ atm}) \simeq s^s_{(l)}(298.16°\text{K}, 1 \text{ atm}) = 30.3 \text{ cal}. \quad (8.21)$$

Combining (8.18), (8.20), and (8.21), we find

$$s^s_{(l)}(298.16°\text{K}, 1 \text{ atm}) = 56.68 \quad \text{cal}.$$

It is interesting to know the influence of pressure on the entropy of liquid methanol. For methanol, the value of the dilatation coefficient is known,

$$\frac{1}{V}\left(\frac{\partial V}{\partial T}\right)_p = 0.1199 \times 10^{-3}. \quad (8.22)$$

The specific mass is about 0.8 g cm^{-3} and the molar volume is equal to 40 cm^3. It is easy now to calculate the derivative $(\partial s/\partial p)_T$ with the help of (3.12). We then obtain

$$(\partial s/\partial p)_T = -0.1199 \times 40 \times 10^{-6} = -4.8 \times 10^{-6} \quad 1 \text{ °K}^{-1}. \quad (8.23)$$

Integrating (8.23), we find

$$s^s_{(l)}(0.163 \text{ atm}) = s^s_{(l)}(1 \text{ atm}) + 4.02 \times 10^{-6} \quad 1 \text{ atm °K} \simeq 30.3 \quad \text{cal}$$

Thus, the pressure has no influence on the entropy of the liquid.

IX. Real Gases

A. Introduction

The ideal gas law may be derived directly on the basis of simple kinetic assumptions in which the translation energy of molecular motion is the main consideration.

All gases behave like ideal gases at sufficiently high temperatures and sufficiently high molar volumes. As the molar volume is decreased, however, real gases exhibit a more complicated behavior. As a gas is compressed, its properties at first deviate only slightly from those of an ideal gas. Under sufficient compression, however, every real gas undergoes a condensation to the liquid or solid state, in which condition it deviates very far indeed from ideal gas behavior. The result is that all real gases must be treated by a more realistic equation of state in such a way as to explain such phenomena as condensation, intermolecular collisions, and a variety of transport properties, such as diffusion and viscosity. The modification of the perfect gas model involves the inclusion of attractive and repulsive intermolecular forces. A detailed discussion of the intermolecular forces in gases and of the state equations is to be found in several fundamental books (Fowler and Guggenheim, 1939; Mayer and Mayer, 1940; Hirschfelder *et al.*, 1965).

Our purpose here consists in the evaluation of the thermodynamic functions from the virial equation of state, and in the introduction of the notion of fugacity.

B. The Virial Equation of State

Kammerling-Onnes (1902) suggested that, with decreasing molar volume v, the properties of every real gas can be expressed in a power series in $1/v$ of the form

$$\frac{p}{T} = \frac{R}{v}\left(1 + \frac{B(T)}{v} + \frac{C(T)}{v^2} + \cdots\right), \tag{9.1}$$

where the coefficients $B(T)$, $C(T)$, ..., are functions of the temperature. The forms of these functions depend on the types of intermolecular forces in the gas. The function $B(T)$ is called the *second virial coefficient*, $C(T)$ the *third virial coefficient*, etc. Enough terms are taken to accomodate the accuracy of the data available. The well-known equation of van der Waals may be shown to be correct only in terms to $B(T)$.

Nevertheless, for a slightly imperfect gas, we need only retain two terms of the expression, giving

$$p = \frac{nRT}{V}\left(1 + \frac{Bn}{V}\right). \tag{9.2}$$

The virial coefficients are often expressed in terms of $cm^3\,mol^{-1}$, but $B(T)$ is not the volume of the mole, because it passes through zero and becomes negative at low temperature. It has a maximum at a high temperature and then declines with rising T. The temperature corresponding to $B(T) = 0$ is called the Boyle temperature. At this temperature, the Boyle–Mariotte law $pv = \text{const}$ is observed.

C. Free Energy and Chemical Potential

We now consider a mixture of real gases defined by the variables $T, V, n_1, \ldots, n_c$.

Knowledge of $F(T, V, n_1, \ldots, n_c)$ immediately gives the state equation, because

$$p = -\partial F(T, V, n_1, \ldots, n_c)/\partial V. \tag{9.3}$$

Conversely, the knowledge of a state equation permits us to calculate the thermodynamic functions of a gas. This fact is more interesting because F is not an experimental quantity. Furthermore, observable properties of a gas are usually summarized in the form of an equation of state.

If the volume tends to infinity, the free energy F tends toward the value F^* corresponding to a perfect gas, so that, by integrating (9.3), we find

$$F(T, V, n_1, \ldots, n_c) - \lim_{V_0 \to \infty} F^*(T, V_0, n_1, \ldots, n_c) = -\lim_{V_0 \to \infty} \int_{V_0}^{V} p \, dV. \tag{9.4}$$

For a perfect gas, we may write (9.4) in the form

$$F^*(T, V, n_1, \ldots, n_c) - \lim_{V_0 \to \infty} F^*(T, V_0, n_1, \ldots, n_c) = -\lim_{V_0 \to \infty} \int_{V_0}^{V} p^* \, dV, \tag{9.5}$$

where

$$p^* = nRT/V. \tag{9.6}$$

Combining (9.4) and (9.5), we find

$$F(T, V, n_1, \ldots, n_c) - F^*(T, V, n_1, \ldots, n_c) = -\lim_{V_0 \to \infty} \int_{V_0}^{V} (p - p^*) \, dV. \tag{9.7}$$

Substituting (9.2) and (9.6), in (9.7), we have

$$F = F^* + B(RTn^2/V) \tag{9.8}$$

and, from (4.16),

$$\mu = \mu^*(T, V) + 2(BRTn/V). \tag{9.9}$$

We have, therefore, from (6.7) and (9.9), the value of the chemical potential of a real gas

$$\mu = \mu^*(T) + RT \ln(nRT/V) + 2(BRTn/V) \tag{9.10}$$

D. Fugacity

The fugacity of a pure real gas is a corrected pressure, defined in such a way that the chemical potential μ can be expressed in a classical form

$$\mu = \mu^*(T) + RT \ln \varphi. \tag{9.11}$$

This intensive quantity gives the effects of intermolecular forces on the thermodynamic properties of gases.

Comparing (9.11) and (9.10), we may write

$$\ln \varphi = \ln(nRT/V) + 2(Bn/V). \tag{9.12}$$

For a slightly imperfect gas, $2Bn/V$ is small compared with unity and (9.12) becomes, as a first approximation,

$$\varphi = \frac{nRT}{V}\left(1 + 2B\frac{n}{V}\right). \tag{9.13}$$

Comparing this relation with (9.2) yields

$$\varphi - p = n\frac{RT}{V}\frac{Bn}{V} \tag{9.14}$$

or

$$\varphi = 2p - p^*. \tag{9.15}$$

Now, in the limit of zero total pressure, the fugacity φ is identical with the pressure.

Example. At $T = 382°K$, the saturation vapor pressure of fluorobenzene is equal to 1.974 atm and the molar volume $v = 15 \times 10^3$ $cm^3\ mol^{-1}$. We then obtain $p^* = 2.085$ atm, $\varphi = 1.86$ atm.

Remark. At the Boyle temperature, $\varphi = p = p^*$.

E. Internal Energy

It is well known that in a Joule expansion a real gas tends to cool. This fact is a direct consequence of the dependence of the internal energy on the volume. Indeed, substituting (9.8) in (3.10), we find

$$U = U^* - n^2(RT^2/V)\, dB/dT, \tag{9.16}$$

where

$$U^* = F^* - T\, \partial F^*/\partial T. \tag{9.17}$$

We thus have

$$\left(\frac{\partial U}{\partial V}\right)_T = n^2 \frac{RT^2}{V^2} \frac{dB}{dT}. \tag{9.18}$$

The derivative dB/dT being positive, the interval energy increases with V. As in a Joule expansion, $dU = 0$, and, because V increases, T must decrease in order to maintain U constant.

F. Mixture of Real Gases

The state equation for a mixture of two slightly imperfect gases is

$$p = \frac{n_1 + n_2}{V} RT + \frac{RT}{V^2} [B_{11}n_1^2 + 2B_{12}n_1n_2 + B_{22}n_2^2], \tag{9.19}$$

where the coefficients B_{11}, B_{12}, and B_{22} are related to the interaction, respectively, between molecules of type 1, of types 1 and 2, and of type 2.

The free energy can thus be obtained by substituting (9.19) in (9.7):

$$F = F^* + (RT/V)[B_{11}n_1^2 + 2B_{12}n_1n_2 + B_{22}n_2^2], \tag{9.20}$$

and, from (4.16),

$$\begin{aligned} \mu_1 &= \mu_1^*(T) + RT \ln \frac{n_1RT}{V} + \frac{2RT}{V} [B_{11}n_1 + B_{12}n_2] \\ \mu_2 &= \mu_2^*(T) + RT \ln \frac{n_2RT}{V} + \frac{2RT}{V} [B_{12}n_1 + B_{22}n_2]. \end{aligned} \tag{9.21}$$

Now, in order to obtain the fugacity in terms of the virial coefficients, let us assume that for real gases (6.15) applies in the same manner, and thus

$$\Sigma p_\gamma = p \sum_\gamma N_\gamma = p. \tag{9.22}$$

Nevertheless, in this case, the partial pressure of a component is not necessarily equal to the pressure of the same component alone occupying the same volume at the same temperature.

In a real gas mixture, we may retain the simple form of (6.18) by use of the fugacity

$$\mu_\gamma = \mu_\gamma^*(T) + RT \ln \varphi_\gamma, \tag{9.23}$$

where $\mu_\gamma^*(T)$ is the same function as for the perfect gases.

Combining (9.23), (9.21), and (6.18), we obtain

$$\begin{aligned} \ln \varphi_1 &= \ln \frac{n_1 RT}{V} + 2\frac{B_{11}n_1 + B_{12}n_2}{V}, \\ \ln \varphi_2 &= \ln \frac{n_2 RT}{V} + 2\frac{B_{12}n_1 + B_{22}n_2}{V}. \end{aligned} \tag{9.24}$$

On the other hand, from (4.21) and (9.23), the affinity of a reaction is readily expressed in terms of fugacities

$$A = -\sum_\gamma \nu_\gamma \mu_\gamma^*(T) - RT \sum_\gamma \nu_\gamma \ln \varphi_\gamma. \tag{9.25}$$

We then have

$$A = RT \ln[K_p(T)/\varphi_1^{\nu_1}\varphi_2^{\nu_2} \cdots \varphi_c^{\nu_c}], \tag{9.26}$$

where

$$\ln K_p(T) = -\Big[\sum_\gamma \nu_\gamma \mu_\gamma^*(T)\Big]\Big/RT. \tag{9.27}$$

The equilibrium then gives the extension to real gases of the Guldberg–Waage law of mass action

$$K_p(T) = \varphi_1^{\nu_1} \cdots \varphi_c^{\nu_c}. \tag{9.28}$$

Instead of the fugacities used in the expression of chemical potential, it is sometimes advantageous to use a corrected mole fraction a_γ called the activity of component γ. Let us define this activity a_γ by

$$a_\gamma = \exp[(\mu_\gamma - \mu_\gamma^\circ)/RT], \tag{9.29}$$

where μ_γ° is the same function as in perfect gases; i.e., from (6.23) and (6.25),

$$\mu_\gamma^\circ(T, p) = \mu_\gamma^*(T) + RT \ln p, \tag{9.30}$$

but here p is the pressure of the real gas.

The chemical potential is then

$$\mu_\gamma = \mu_\gamma{}^\circ(T, p) + RT \ln a_\gamma. \tag{9.31}$$

The activity is thus an intensive function of $T, p, N_1, \ldots, N_c$, and contains all the effects arising from the interactions. This quantity has been introduced by Lewis and Randall (1923).

Usually, the chemical potential is expressed in a more explicit manner:

$$\mu_\gamma = \mu_\gamma{}^\circ(T, p) + RT \ln f_\gamma N_\gamma \tag{9.32}$$

where the coefficient f_γ, called the activity coefficient, is a function of $T, p, N_1, \ldots, N_c$. This coefficient may then be defined by the relation

$$f_\gamma = a_\gamma / N_\gamma. \tag{9.33}$$

In the limit of zero total pressure, the activity a_γ is identical to the mole fraction N_γ, and thus the activity coefficient is equal to unity. Accordingly, for a perfect gas mixture, all the activity coefficients are unity.

Now, it is easy to calculate the activity when one knows the fugacity. Indeed, comparing (9.32) to (9.23) and (9.30), we find

$$\varphi_\gamma = p f_\gamma N_\gamma. \tag{9.34}$$

For a pure constituent γ, $N_\gamma = 1$ and

$$\varphi_\gamma = p f_\gamma. \tag{9.35}$$

The activity coefficient of a pure gaseous component differs from unity because $\varphi \neq p$.

The chemical potential of an isolated component may then be written, following (9.30) and (9.32),

$$\mu_\gamma = \mu_\gamma{}^\circ(T, p) + RT \ln f_\gamma = \mu_\gamma{}^*(T) + RT \ln p + RT \ln f_\gamma. \tag{9.36}$$

The quantity $\mu_\gamma{}^\circ$ corresponds to the chemical potential of an isolated component in a perfect gaseous state at pressure p and temperature T. It is called the standard chemical potential $\mu_\gamma{}^s$, while the quantity $RT \ln f_\gamma N_\gamma$ is the function of mixing $\mu_\gamma{}^m$.

Let us express now the affinity in terms of activity. From (4.21) and (9.32), we find

$$A = -\sum_\gamma \nu_\gamma \mu_\gamma{}^\circ(T, p) - RT \sum_\gamma \nu_\gamma \ln f_\gamma N_\gamma \tag{9.37}$$

and thus

$$A = RT \ln[K_N(T, p)/(N_1 f_1)^{\nu_1} \cdots (N_c f_c)^{\nu_c}], \tag{9.38}$$

where

$$\ln K_N(T, p) = -\Big[\sum_\gamma \nu_\gamma \mu_\gamma^\circ(T, p)\Big]\Big/ RT. \tag{9.39}$$

The quantity $-\Sigma_\gamma \nu_\gamma \mu_\gamma^\circ(T, p)$ is called the standard affinity A^s, and thus

$$K_N(T, p) = \exp[A^s/RT]. \tag{9.40}$$

At equilibrium, the Guldberg–Waage equation becomes

$$K_N(T, p) = (N_1 f_1)^{\nu_1} \cdots (N_c f_c)^{\nu_c}. \tag{9.41}$$

As $K_N(T, p)$ is the same function as in the perfect gas, we also find [see Eq. (6.33)]

$$K_N(T, p) = p^{-\nu} K_p(T). \tag{9.42}$$

The other standard functions (superscript s) and mixing functions (superscript m) are given as follows:

$$h_\gamma = -T^2 \frac{\partial(\mu_\gamma/T)}{\partial T}, \qquad h_\gamma^{\mathrm{s}} = -T^2 \frac{\partial(\mu_\gamma^\circ/T)}{\partial T},$$

$$h_\gamma^{\mathrm{m}} = -RT^2 \frac{\partial(\ln f_\gamma)}{\partial T} \tag{9.43}$$

$$s_\gamma = -\frac{\partial \mu_\gamma}{\partial T}, \qquad s_\gamma^{\mathrm{s}} = -\frac{\partial \mu_\gamma^\circ}{\partial T},$$

$$s_\gamma^{\mathrm{m}} = -R \ln f_\gamma N_\gamma - RT \frac{\partial(\ln f_\gamma)}{\partial T} \tag{9.44}$$

$$v_\gamma = \frac{\partial \mu_\gamma}{\partial p}, \qquad v_\gamma^{\mathrm{s}} = \frac{\partial \mu_\gamma^\circ}{\partial p},$$

$$v_\gamma^{\mathrm{m}} = RT \frac{\partial(\ln f_\gamma)}{\partial p} \tag{9.45}$$

$$r_{Tp} = -\sum_\gamma \nu_\gamma h_\gamma, \qquad r_{Tp}^{\mathrm{s}} = -\sum_\gamma \nu_\gamma h_\gamma^{\mathrm{s}},$$

$$r_{Tp}^{\mathrm{m}} = -\sum_\gamma \nu_\gamma h_\gamma^{\mathrm{m}} \tag{9.46}$$

$$\left(\frac{\partial S}{\partial \xi}\right)_{Tp} = \sum_\gamma \nu_\gamma s_\gamma, \qquad \left(\frac{\partial S}{\partial \xi}\right)_{Tp}^{\mathrm{s}} = \sum_\gamma \nu_\gamma s_\gamma^{\mathrm{s}},$$

$$\left(\frac{\partial S}{\partial \xi}\right)_{Tp}^{\mathrm{m}} = \sum_\gamma \nu_\gamma s_\gamma^{\mathrm{m}}. \tag{9.47}$$

The variation of the equilibrium constant with temperature and pressure can be calculated in the same manner as for the perfect gases [see Eqs. (6.50), (6.68), and (7.21)–(7.23)]:

$$\frac{\partial[\ln K_N(T, p)]}{\partial T} = \frac{1}{R}\frac{\partial(A^s/T)}{\partial T} = -\frac{1}{R}\sum_\gamma \nu_\gamma \frac{\partial(\mu_\gamma{}^\circ/T)}{\partial T}$$
$$= -\frac{r_{Tp}^s}{RT^2} \qquad (9.48)$$

$$\frac{\partial[\ln K_N(T, p)]}{\partial p} = \frac{1}{RT}\frac{\partial A^s}{\partial p} = -\frac{1}{RT}\sum_\gamma \nu_\gamma \frac{\partial \mu_\gamma{}^\circ}{\partial p}$$
$$= \frac{-\sum_\gamma \nu_\gamma v_\gamma{}^s}{RT} = -\frac{\Delta_{Tp}}{RT}. \qquad (9.49)$$

Remark. The standard entropy of a constituent in a perfect gas mixture is equal to the entropy of the isolated constituent at the same temperature and pressure. But this is not true for real gases. Indeed, Eq. (9.44) reduces, for a real isolated gas, to

$$s_\gamma{}^m = -R \ln f_\gamma - RT\, \partial(\ln f_\gamma)/\partial T. \qquad (9.50)$$

The entropy of real isolated gases can be evaluated by calorimetric measurements. To obtain the standard entropy $s_\gamma{}^s$ (see Table I), we must subtract the above value of $s_\gamma{}^m$ from the experimental value.

X. Stability of Chemical Systems

In this section, we shall discuss the stability of equilibrium states and present some conclusions concerning the properties of thermodynamic variables in stable equilibrium systems.

A. De Donder's Method

If, in a closed system P, a reaction may take place, its state is characterized by two physical variables x and y (for example, T and p) and by a chemical variable ξ. We have

$$dn_\gamma = \nu_\gamma \, d\xi \qquad (\gamma = 1, 2, \ldots, c). \qquad (10.1)$$

Let us assume that, at time t, the system is in an equilibrium state.

The rate of reaction $\mathbf{v}_P$ is thus zero,

$$\mathbf{v}_P = 0. \tag{10.2}$$

At the same time, let us consider another system P' characterized by the same physical variables x and y as the system P but by a different chemical variable $\xi + d\xi$.

The system P' is not in general in an equilibrium state and its rate of reaction $\mathbf{v}_{P'}$ is then

$$\mathbf{v}_{P'} \neq 0. \tag{10.3}$$

In comparison with the system P, the system P' is thus a perturbed system. The perturbation consists here in the virtual displacement from ξ to $\xi + d\xi$ with constant values of x and y.

Now, we know that [see Chapter 1, Eq. (20.2)]

$$n_\gamma = n_\gamma^\circ + \nu_\gamma \xi. \tag{10.4}$$

The considered perturbation $d\xi$ is thus identical to a slight change of all the n_γ proportionally to $d\xi$,

$$\delta n_\gamma = \nu_\gamma \, d\xi. \tag{10.5}$$

At constant x and y, the system P is said to be stable with respect to the perturbations if the rate of reaction of the perturbed system P' tends to bring this system to the equilibrium state P (De Donder, 1942; Duhem, 1911, Chapter XVI).

Let us now represent the systems P and P' on the $xy\xi$ axes (see Fig. 16) and let us assume that the system P is stable. There are two cases:

(1) $\delta\xi > 0$ (system P'), $\mathbf{v}_{P'}$ is directed in the direction opposite to the ξ axis; the rate of reaction $d\xi/dt = \mathbf{v}_{P'}$ is then negative.

(2) $\delta\xi < 0$ (system P''), the rate of reaction $\mathbf{v}_{P''}$ is directed in the same direction as the ξ axis and is positive.

The criterion of stability is thus

$$\mathbf{v}_{P'} \, \delta\xi < 0. \tag{10.6}$$

There are two kinds of perturbations: those in which $\delta\xi$ can have only one sign (unilateral perturbations), and those in which $\delta\xi$ can be either positive or negative (bilateral perturbations). An example of a unilateral perturbation is the appearance of a vapor bubble in a system which was initially completely liquid, while the reverse reaction is impossible as long as there is no vapor phase.

If a stable system is perturbed, it reverts to the initial equilibrium; if it is unstable, the perturbation proceeds to a finite extent.

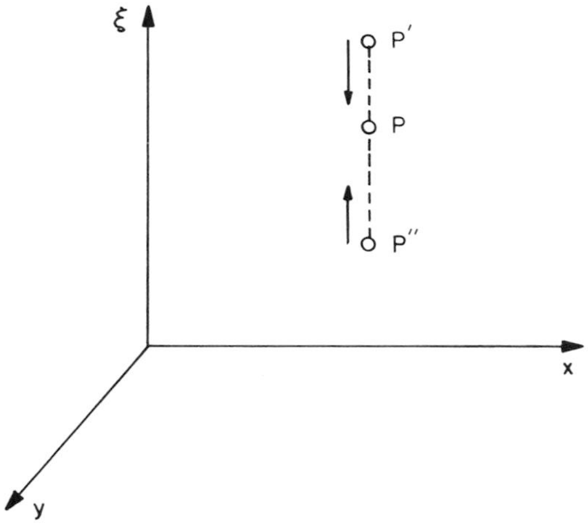

FIG. 16. Equilibrium stability of a system P.

Remark. A perturbation is not necessarily related to any external action on the system. Molecular fluctuations lead to small variations of the macroscopic quantities from their equilibrium values. There is, in fact, a relation between the probability of a fluctuation and the production of entropy which accompanies it.

B. The Prigogine–Defay Method

The method adopted by Prigogine (see Prigogine and Defay (1967), p. 205) is based upon the direct evaluation of the entropy production in the course of a perturbation and so permits a discussion of stability with respect to any kind of perturbation. The entropy production corresponding to a change from a state $P(x, y, \xi_P)$ to a state $P'(\xi_{P'})$ is given from (2.1) by

$$Q'_{PP'} = \int_{\xi_P}^{\xi_{P'}} A(\xi)\, d\xi, \tag{10.7}$$

where, for a specified process, the physical variables x and y are functions completely determined by ξ.

Developing the affinity in the form of a Taylor series, (10.7) becomes

$$Q'_{PP'} = A_P\,\Delta\xi + \frac{1}{2}\left(\frac{dA}{d\xi}\right)_P(\Delta\xi)^2 + \cdots + \frac{1}{n!}\,\frac{d^{(n-1)}A}{d\xi^{(n-1)}}\,(\Delta\xi)^n, \quad (10.8)$$

where $A_P = A(\xi_P)$ and $\Delta\xi = \xi_{P'} - \xi_P$.

For a small variation $\Delta\xi$, we have two cases:

$$(i) \qquad A_P \neq 0 \quad \text{and} \quad Q'_{PP'} = A_P\,\delta\xi; \qquad (10.9)$$

$$(ii) \qquad A_P = 0 \quad \text{and} \quad Q'_{PP'} = \tfrac{1}{2}(dA/d\xi)_P(\delta\xi)^2. \qquad (10.10)$$

The transformations characterized by Eqs. (10.8)–(10.10) are called *perturbations, where P is an equilibrium state.* They are related to the infinitesimal change $\delta\xi$ and to the conditions under which they are carried out. If, for example, x and y remain constant during the perturbations, (10.10) becomes

$$Q'_{PP'} = \tfrac{1}{2}(\partial A/\partial\xi)_{xyP}(\delta\xi)^2. \qquad (10.11)$$

Now, the state P is said to be stable with respect to the transformation $P \to P'$ if the production of entropy accompanying it is *negative*, and thus

$$\boxed{Q'_{PP'} < 0.} \qquad (10.12)$$

On the other hand, for the inverse spontaneous process, $P' \to P$, the production of entropy is positive.

C. Thermodynamic Conditions of Stability. Unilateral Perturbations

The unilateral perturbation is characterized by the fact that $\delta\xi$ can have only one sign; for example, let us take $\delta\xi > 0$. Thus, for a reaction

$$d\xi/dt \geq 0, \qquad (10.13)$$

and, from (2.1), it follows that the system will be in equilibrium provided that

$$A \leq 0. \qquad (10.14)$$

If $A_P \neq 0$ (10.12) and Eq. (10.9) yield the condition of stability

$$A_P < 0, \tag{10.15}$$

which is in fact the condition of equilibrium.

If $A_P = 0$, Eq. (10.10) gives the condition of stability

$$(dA/d\xi)_P < 0, \tag{10.16}$$

or, if x and y remain constant,

$$(\partial A/\partial \xi)_{xyP} < 0. \tag{10.17}$$

Now, combining (3.6) and (10.15), we find that, if P is a stable equilibrium state, the internal energy U increases in the perturbations,

$$(\partial U/\partial \xi)_{VSP} > 0, \tag{10.18}$$

where U is a function of ξ only (V and S are maintained constant).

On the other hand, we have from (10.10), (10.17), and (3.6)

$$(\partial U/\partial \xi)_{VSP} = 0, \qquad (\partial^2 U/\partial \xi^2)_{VSP} > 0. \tag{10.19}$$

and the internal energy U has a horizontal tangent at P; it increases during a perturbation if P is a stable equilibrium state.

With regard to the stability of phases, the initial system (unperturbed) consists of a single phase, while the final system (perturbed) contains, in addition to the original phase, a small amount of a new phase whose properties (partial molar volume, volume, composition, etc.) differ only infinitesimally from those of the original phase, or differ from them by a finite, nonzero, amount. Usually, we say that the initial phase is stable when it is stable with respect to all other phases whether infinitesimally different from them or not. In this case, the phase can never give rise spontaneously to a new phase in macroscopic amounts.

The initial phase is called a metastable phase when it is stable with respect to phases infinitesimally different from it, but there is at least one other phase with respect to which it is not stable. This means that, in the absence of nuclei, the system may remain indefinitely in equilibrium without the appearance of a new phase (supercooled liquids).

Finally, the initial phase is called an unstable phase when it is unstable with respect to the phases infinitesimally different from it. Practically, this means that the phase will disappear and give rise to one or more neighboring phases (molecular fluctuations) until we arrive at a phase which is stable with respect to adjacent phases.

Remark. Instead of (10.18), we may write the classical conditions

$$(\partial H/\partial \xi)_{pS} > 0, \tag{10.20}$$

$$(\partial F/\partial \xi)_{VT} > 0, \tag{10.21}$$

$$(\partial G/\partial \xi)_{Tp} > 0. \tag{10.22}$$

If $A_P = 0$, then we have

$$(\partial H/\partial \xi)_{pS} = 0, \qquad (\partial^2 H/\partial \xi^2)_{pS} > 0 \tag{10.23}$$

$$(\partial F/\partial \xi)_{VT} = 0, \qquad (\partial^2 F/\partial \xi^2)_{VT} > 0 \tag{10.24}$$

$$(\partial G/\partial \xi)_{Tp} = 0, \qquad (\partial^2 G/\partial \xi^2)_{Tp} > 0. \tag{10.25}$$

D. Stability with Respect to Bilateral Perturbations

We have seen earlier (Section II), that false equilibrium may be characterized by $\mathbf{v} = 0$ and $A \neq 0$. Thus, if such a system is perturbed, it does not revert to its initial situation. Only true equilibrium could be stable. The first condition of stability is thus $A_P = 0$ in a state P.

Furthermore, in the perturbed state P', the system must take a rate satisfying the inequality (10.6). We may calculate $A_{P'}$

$$A_{P'} = A_P + (\partial A/\partial \xi)_{xyP}\, \delta\xi. \tag{10.26}$$

From the stability condition $A_P = 0$, Eq. (10.26) reduces to

$$A_{P'} = (\partial A/\partial \xi)_{xyP}\, \delta\xi. \tag{10.27}$$

Now, from De Donder's inequality (2.14),

$$A_{P'}\mathbf{v}_{P'} > 0, \tag{10.28}$$

which characterizes a spontaneous process; it follows that the sign of $\mathbf{v}_{P'}$ is given by the inequality

$$(\partial A/\partial \xi)_{xyP}\, \delta\xi \mathbf{v}_{P'} > 0. \tag{10.29}$$

This relation is consistent with the criterion of stability only if

$$(\partial A/\partial \xi)_{xyP} < 0. \tag{10.30}$$

Thus, two conditions, $A_P = 0$ and (10.30), are necessary and sufficient to express the stability of a system.

Remark. At constant T and p, the two conditions are

$$A = 0, \qquad (\partial A/\partial \xi)_{Tp} = 0. \tag{10.31}$$

At constant T and V, we have

$$A = 0, \qquad (\partial A/\partial \xi)_{TV} = 0. \tag{10.32}$$

In Section III, we saw that

$$A = -\left(\frac{\partial U}{\partial \xi}\right)_{VS} = -\left(\frac{\partial H}{\partial \xi}\right)_{pS} = -\left(\frac{\partial F}{\partial \xi}\right)_{TV} = -\left(\frac{\partial G}{\partial \xi}\right)_{Tp}. \tag{10.33}$$

The conditions of stability $A = 0$ and (10.30) are:

(*i*) $\quad (\partial U/\partial \xi)_{VS} = 0, \qquad (\partial^2 U/\partial \xi^2)_{SV} > 0;$ (10.34)

when S and V are maintained constant, U is thus minimum in a stable equilibrium state (Fig. 17).

(*ii*) $\quad (\partial H/\partial \xi)_{Sp} = 0, \qquad (\partial^2 H/\partial \xi^2)_{Sp} > 0;$ (10.35)

in a stable equilibrium state, H is minimum at constant S and p.

(*iii*) $\quad (\partial F/\partial \xi)_{TV} = 0, \qquad (\partial^2 F/\partial \xi^2)_{TV} > 0;$ (10.36)

in a stable equilibrium state, F is minimum at constant T and V.

(*iv*) $\quad (\partial G/\partial \xi)_{Tp} = 0, \qquad (\partial^2 G/\partial \xi^2)_{Tp} > 0;$ (10.37)

in a stable equilibrium state, G is minimum at constant T and p.

The inequality (10.37) may be rewritten as

$$a_{Tp} < 0, \tag{10.38}$$

where

$$a_{Tp} = (\partial A/\partial \xi)_{Tp} = -(\partial^2 G/\partial \xi^2)_{Tp}. \tag{10.39}$$

Thus, from (10.33) and (4.21),

$$a_{Tp} = -\sum_{\gamma} \nu_\gamma\, \partial \mu_\gamma/\partial \xi = -\sum_{\gamma}\sum_{\beta} \nu_\gamma \nu_\beta \mu_{\gamma\beta} < 0, \tag{10.40}$$

where

$$\mu_{\gamma\beta} = \partial \mu_\gamma/\partial n_\beta. \tag{10.41}$$

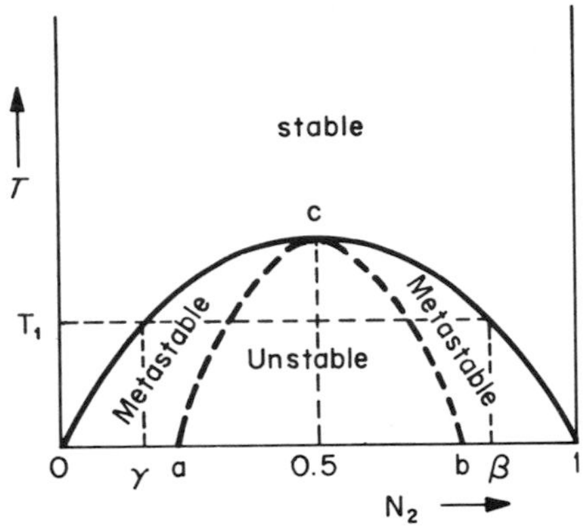

FIG. 17. Minimum internal energy for a stable equilibrium state.

Equation (10.40) can be rewritten in the form

$$a_{Tp} = \tfrac{1}{2} \sum_{\gamma} \sum_{\beta} \mu_{\gamma\beta} n_{\gamma} n_{\beta} \left(\frac{\nu_{\gamma}}{n_{\gamma}} - \frac{\nu_{\beta}}{n_{\beta}} \right)^2 < 0. \tag{10.42}$$

This condition is satisfied if $\mu_{\gamma\beta}$ is negative for all $\gamma \neq \beta$.

Prigogine and Defay (1967, Eq. (7.13)) showed that this condition is always satisfied by ideal systems. Thus, all stable states of an ideal system are stable equilibrium states at constant T and p.

E. Explicit Forms of the Stability Conditions

For a system at constant energy and volume [see Eq. (3.1)], the stability condition gives

$$\delta S \leq 0 \qquad (E, V \text{ constant}). \tag{10.43}$$

Systems which are maintained at constant energy and volume are by definition *isolated* systems. For such systems, entropy is *maximum* for stable equilibrium and

$$(\delta S)_{\text{eq}} = 0 \qquad \text{(equilibrium)} \tag{10.44}$$

$$(\delta S)_{\text{eq}} < 0 \qquad \text{(stability)} \tag{10.45}$$

$$(\delta^2 S)_{\text{eq}} < 0. \tag{10.46}$$

Let us now derive a relation for the second-order quantity $\delta^2 s$, where s

is the specific mass entropy. For a system in equilibrium, the Gibbs relation (4.12) may then be written in the classical form

$$T\,\delta s = \delta u + p\,\delta v - \sum_{\gamma} \mu_{\gamma}\,\delta N_{\gamma}, \tag{10.47}$$

where u is the specific mass internal energy, v the specific volume, N_{γ} the molar fraction, and μ_{γ} is here the chemical potential per unit mass. Let us first calculate $\delta^2 s$ using as independent variables u, v, N_{γ}; then $\delta^2 u = 0$, $\delta^2 v = 0$, and $\delta^2 N_{\gamma} = 0$, and the Gibbs formula (10.47) gives us directly

$$\delta^2 s = \delta T^{-1}\,\delta u + \delta(pT^{-1})\,\delta v - \sum_{\gamma} \delta(\mu_{\gamma} T^{-1})\,\delta N_{\gamma}. \tag{10.48}$$

Combining (10.47) and (10.48), we obtain

$$T\,\delta^2 s = -\delta T\,\delta s + \delta p\,\delta v - \sum_{\gamma} \delta\mu_{\gamma}\,\delta N_{\gamma}. \tag{10.49}$$

Let us now express the variation of the chemical potentials $\delta\mu_{\gamma}$ in the variables T, p, N_{γ}. We obtain the quadratic form [see Chapter 1 (16.11)–(16.13), (16.17), (18.3), (18.4), (18.18)–(18.20), (18.23), (18.25)],

$$\delta^2 s = -\frac{1}{T}\left[\frac{c_v}{T}(\delta T)^2 + \frac{\varrho}{\chi}(\delta v)^2 + \sum_{\gamma\beta} \mu_{\gamma\beta}\,\delta N_{\gamma}\,\delta N_{\beta}\right], \tag{10.50}$$

where

$$c_v = \left(\frac{\partial e}{\partial T}\right)_{vN_{\gamma}} = T\left(\frac{\partial s}{\partial T}\right)_{vN_{\gamma}} \tag{10.51}$$

$$c_p = T\left(\frac{\partial s}{\partial T}\right)_{pN_{\gamma}} \tag{10.52}$$

$$\chi = -\frac{1}{v}\left(\frac{\partial v}{\partial p}\right)_{TN_{\gamma}} \tag{10.53}$$

$$c_p - c_v = \frac{T}{\chi v}\left(\frac{\partial v}{\partial T}\right)^2_{pN_{\gamma}} \tag{10.54}$$

$$\mu_{\gamma\beta} = \left(\frac{\partial \mu_{\gamma}}{\partial N_{\beta}}\right)_{TpN_{\beta}} \tag{10.55}$$

Now, in the system of variables ϱu, ϱ_{γ}, it is easy to show that

$$\delta^2(\varrho s) = \varrho\,\delta^2 s, \tag{10.56}$$

ϱs being the volume entropy density; it becomes easy to calculate the quantity $\delta^2 s$.

The quadratic form has to be positive definite. From the inequality (10.46), this leads to the following stability conditions:

$$c_v > 0 \qquad \text{(thermal stability)} \tag{10.57}$$

$$\chi > 0 \qquad \text{(mechanical stability).} \tag{10.58}$$

Both the specific heat (at constant volume) and the isothermal compressibility have to be positive.

In addition, we also have

$$\sum_{\gamma\beta} \mu_{\gamma\beta}\, \delta N_\gamma\, \delta N_\beta > 0 \qquad \text{(stability with respect to diffusion).} \tag{10.59}$$

Let us consider, for example, a perturbation which consists in the appearance of a heterogeneity in the composition of a binary system which is initially uniform. The inequality (10.59) guarantees that the response of the system will restore the initial homogeneity.

The inequalities (10.57) and (10.58) ensure stability with respect to thermal and mechanical disturbances, while (10.59) ensures stability in respect to diffusion.

F. Phase Separation in Binary Mixtures

As a simple illustration of the stability condition (10.59), let us consider phase separations in binary mixtures. The stability conditions are then

$$\mu_{11} > 0, \qquad \mu_{22} > 0 \tag{10.60}$$

$$\begin{vmatrix} \mu_{11} & \mu_{21} \\ \mu_{12} & \mu_{22} \end{vmatrix} \geq 0. \tag{10.61}$$

On the other hand, from (4.33) we have

$$\mu_{12} = \mu_{21} \tag{10.62}$$

$$\begin{aligned} n_1\mu_{11} + n_2\mu_{21} &= 0, \\ n_1\mu_{12} + n_2\mu_{22} &= 0. \end{aligned} \tag{10.63}$$

Thus, the determinant in (10.61) vanishes and we have only to consider the first two inequalities. Moreover,

$$n_1^2\mu_{11} = n_2^2\mu_{22} \tag{10.64}$$

and this implies that the two inequalities in (10.60) are equivalent and that

$$\mu_{12} = \mu_{21} < 0. \tag{10.65}$$

For mixtures of perfect gases and "perfect solutions" formed by components of similar molecules, the influence of the activity coefficient f_γ may be neglected and the chemical potentials take the form [see Eq. (9.32)]

$$\mu_\gamma = \mu_\gamma{}^\circ(T, p) + RT \ln N_\gamma, \tag{10.66}$$

and it is easy to show that the inequalities (10.60) and (10.65) are verified. However, this is no longer necessarily so for regular solutions (Prigogine, 1957). Indeed, in this case,

$$\mu_1 = \mu_1{}^\circ(T, p) + RT \ln(1 - N_2) + \alpha N_2{}^2 \tag{10.67}$$

$$\mu_2 = \mu_2{}^\circ(T, p) + RT \ln N_2 + \alpha(1 - N_2)^2, \tag{10.68}$$

where α is a constant defined by

$$\alpha = N_{\mathrm{Av}} z[\varepsilon_{12} - \tfrac{1}{2}(\varepsilon_{11} + \varepsilon_{22})], \tag{10.69}$$

with N_{Av} the Avogadro number, z the number of nearest neighbors of a molecule in the considered medium, and $\varepsilon_{\gamma\beta}$ is the interaction energy between a molecule γ and a molecule β. Now,

$$\mu_{12} = \frac{\partial \mu_1}{\partial N_2} = -\frac{RT}{1 - N_2} + 2\alpha N_2. \tag{10.70}$$

If

$$2\alpha/RT > 4, \tag{10.71}$$

there exists a range of mole fractions where the stability conditions are not satisfied. We then obtain phase separations, and the phase diagram is represented schematically in Fig. 18.

There exists a critical point $N_2 = 0.5$, $T_c = (\alpha/2)R$. Above this critical point, the two components are mixable in all proportions. Below, we find two coexisting phases (for example, at $T = T_1$, we have two phases corresponding to $N_2 = \gamma$ and $N_2 = \beta$). Inside the region *acb*, the stability

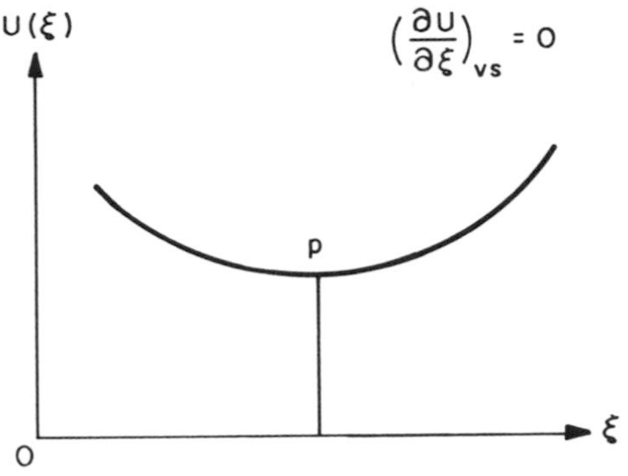

FIG. 18. Phase separation in a regular solution at constant p.

condition (10.60–10.61) is violated. This curve is called the spinodal. It separates unstable states from metastable states.

For metastable states, the stability conditions as derived in Section X, E are satisfied; however, the Gibbs free energy (if we work at constant p and T) is higher for the homogeneous mixtures than for a system formed by two coexisting phases. Metastable systems are stable with respect to small perturbations (the second-order stability conditions are satisfied, but the system is unstable with respect at least to some finite perturbations).

Remark. The limitations of the Gibbs–Duhem theory, the comparison with the kinetic theory of stability, and the important problem of the thermodynamic stability conditions for nonequilibrium states are discussed in detail by Glansdorff and Prigogine (1970).

XI. Equilibrium Displacements in Closed Systems

A. General Laws

The problems encountered when considering equilibrium displacements are similar to those met when studying stability. If the modification is due to the variables T and p as well as to ξ, and if we maintain constant the perturbed values $p + \delta p$ and $T + \delta T$, a stable system may tend to return to a *new* equilibrium state different from the initial state. This modification is called the displacement of thermodynamic equilibrium.

Let us assume, for example, that only p and T are perturbed, respectively, to $p + \delta p$ and $T + \delta T$. We thus pass from an initial system in an equilibrium state P to a perturbed system P'. Then we maintain $p + \delta p$ and $T + \delta T$ constant and the system evolves toward to a new equilibrium state P'' (Fig. 19). At each point of the line PP'', we have

$$A = 0 \tag{11.1}$$

Generally, a displacement on the surface $A = 0$ is called an equilibrium displacement. Along an equilibrium displacement,

$$\delta A = 0 \tag{11.2}$$

and thus

$$\frac{\partial A}{\partial T}\,\delta T + \frac{\partial A}{\partial p}\,\delta p + \frac{\partial A}{\partial \xi}\,\delta \xi = 0. \tag{11.3}$$

Now each derivative may be rewritten. From (2.13) and (3.12), we find

$$\partial A/\partial T = (A - r_{Tp})/T. \tag{11.4}$$

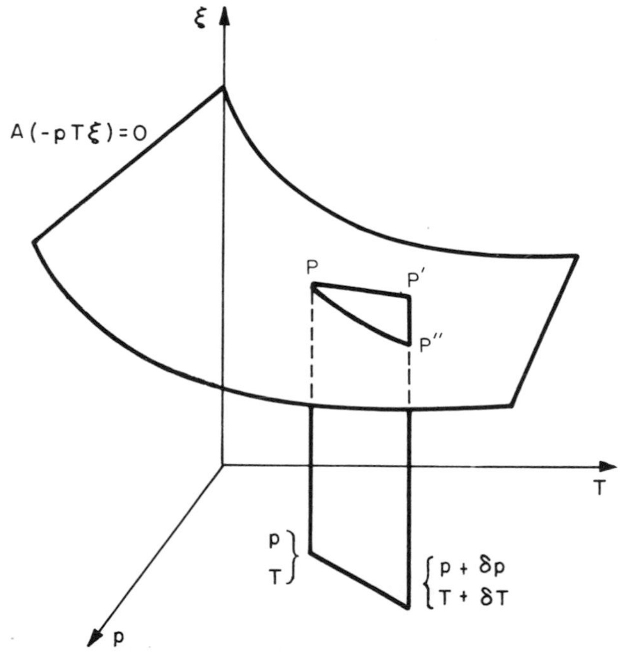

Fig. 19. Equilibrium displacement.

From (3.12),

$$\partial A/\partial p = -\sum_{\gamma} \nu_\gamma v_\gamma = -\Delta_{Tp}. \tag{11.5}$$

From (10.39),

$$\partial A/\partial \xi = a_{Tp}. \tag{11.6}$$

Combining the above six equations, we obtain the general law of the *equilibrium displacement*:

$$(r_{Tp}/T)\, \delta T + \Delta_{Tp}\, \delta p - a_{Tp}\, \delta \xi = 0. \tag{11.7}$$

1. *Isobaric Displacement*, $\delta p = 0$

Equation (11.7) then reduces to the De Donder formula (see De Donder (1925, 1927), Schottky *et al.* (1929, p. 492))

$$\left(\frac{\partial \xi}{\partial T}\right)_p = \frac{1}{a_{Tp}} \frac{r_{Tp}}{T}. \tag{11.8}$$

As the condition of stability is given by the inequality (10.38), Eq. (11.8) expresses the *van't Hoff theorem*; if a reaction is exothermic ($r_{Tp} > 0$), then $\partial \xi/\partial T < 0$, and thus an increase in temperature moves the equilibrium position of the reaction back ($\delta \xi < 0$). If the reaction is endothermic ($r_{Tp} < 0$), a rise in temperature advances the equilibrium position ($\delta \xi > 0$).

Further, De Donder's formula (11.8) makes it possibile for us to calculate the value of the derivative $(\partial \xi/\partial T)_p$.

Example. The reaction $2H_2 + O_2 = 2H_2O$ is exothermic. At ordinary temperature, we have a false equilibrium. True equilibrium exists at high temperatures and is stable at given temperature and pressure. If the temperature increases, ξ decreases, and thus water vapour dissociates partially into H_2 and O_2.

Remark. The conclusions obtained here are consistent with the equation (6.50) giving the variation of $K_N(T, p)$ with T. But, from a practical point of view, Eqs. (6.49) and (6.50) are only interesting in the case of perfect gases, where the chemical potentials are known.

2. *Isothermal Displacement*, $\delta T = 0$

From (11.7), we then have the De Donder formula

$$(\partial \xi/\partial p)_T = \Delta_{Tp}/a_{Tp}. \tag{11.9}$$

The condition of stability (10.38) and Eq. (11.9) express *Le Chatelier's theorem*: if a reaction is accompanied by an increase in volume ($\Delta_{Tp} > 0$), an increase in pressure leads to a reduction of the extent of reaction ($\delta\xi < 0$), and inversely.

3. *Isomassic Displacement*, $\delta\xi = 0$

We may write

$$\left(\frac{\partial p}{\partial T}\right)_\xi = -\frac{1}{\Delta_{Tp}}\frac{r_{Tp}}{T}. \tag{11.10}$$

Equation (11.10) is called the De Donder–Clapeyron–Clausius formula.

Let us assume that, in a chemical reaction, the composition of each constituent remains constant; then we are able to calculate the variation of the pressure with the temperature.

B. A Particular Case: $A = A(T, p)$

Let us consider a pure constituent in two phases (Fig. 20). Here, the affinity is independent of ξ because

$$A = \mu'(T, p) - \mu''(T, p). \tag{11.11}$$

Thus,

$$A = A(T, p), \tag{11.12}$$

and (11.7) reduces to the Clapeyron formula

$$\frac{\delta p}{\delta T} = -\frac{1}{\Delta_{Tp}}\frac{r_{Tp}}{T}. \tag{11.13}$$

Here, $\Delta_{Tp} = v'' - v'$ and $-r_{Tp}$ is the latent heat L.

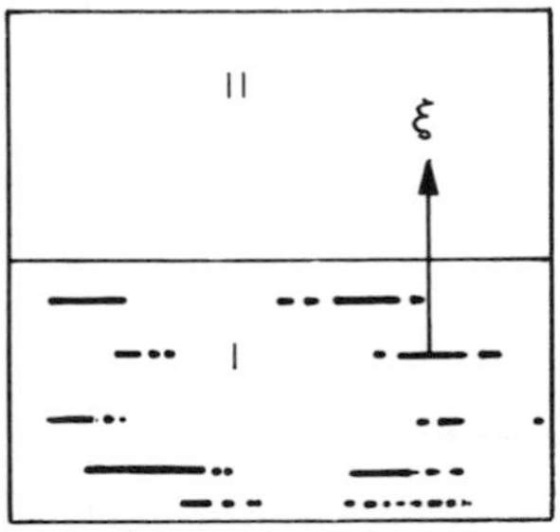

Fig. 20. Matter transfer of a pure constituent between two phases.

For example, in vaporization, v' may be neglected with repect to v'', and, for a perfect gas,

$$v'' = RT/p. \tag{11.14}$$

Combining (11.13) and (11.14), we find

$$\delta(\ln p)/\delta T = L/RT^2. \tag{11.15}$$

Integrating (11.15), we obtain

$$\ln p = -(L/RT) + \text{const.} \tag{11.16}$$

For chloroform (Fig. 21), we find for the latent heat $L_v \approx 7200$ cal mol^{-1} and for the entropy of vaporization $(\Delta S_v)_{500\text{mmHg}} \simeq 28$ cal mol^{-1} °K^{-1}.

Remark. From Eq. (2.13) at equilibrium, we may write

$$r_{Tp} = -T(\partial S/\partial \xi)_{Tp} = -T(s'' - s'). \tag{11.17}$$

Substituting (11.17) in (11.13) gives

$$\delta p/\delta T = (s'' - s')/(v'' - v'). \tag{11.18}$$

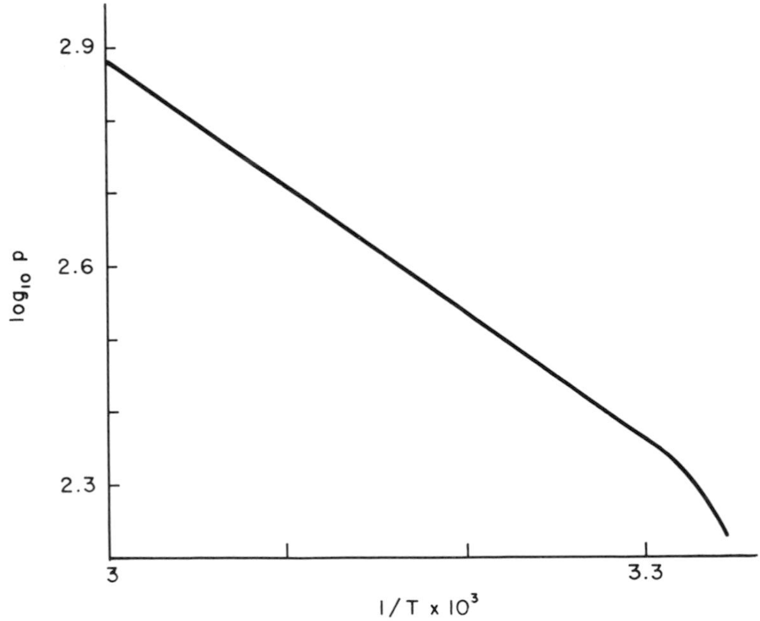

FIG. 21. Vapor pressure as a function of temperature for chloroform.

C. The Le Chatelier–Braun Theorem of Moderation

The van't Hoff and Le Chatelier theorems are characterized by the fact that in both cases the reaction occurs in such a way that it exhibits moderation of the factor perturbed.

If we increase the temperature, the reaction absorbs heat, and this tends to moderate the increase of temperature. If we increase the pressure, the reaction occurs in such a way that the volume decreases and thus tends to moderate the increase of pressure.

But these theorems of restraint or of moderation cannot be applied in the same manner when we use other variables. If we decrease the volume of a system, the reaction produced does not tend to moderate this decrease.

The problem of moderation must be studied by considering the *real* transformation produced by the system when the system is perturbed. An unambiguous answer can be deduced by using the fundamental inequality $Av > 0$.

XII. Equilibrium Displacements in Open Systems

A. Method

In open systems, the equilibrium conditions may be modified by the addition of certain constituents; for example, the addition of KOH to the mixture $HNO_3 + KOH$.

In fact, equilibrium displacement in open systems can be studied by noting the properties of equilibrium curves or equilibrium surfaces.

Furthermore, it is more convenient to use here the variables $T, p, N_1, \ldots, N_c$ or $T, p, \mu_1, \ldots, \mu_c$. Nevertheless, the fundamental problem is also related to the two equations

$$A = 0 \tag{12.1}$$

$$\delta A = 0, \tag{12.2}$$

with δA a function of the chemical potentials.

B. Two-Phase System

We consider a two-phase system and c constituents with only passage reactions from one phase to the other (Fig. 22). The equilibrium conditions (12.1) applied to the passages are

$$A_\gamma = \mu_\gamma' - \mu_\gamma'' = 0 \qquad (\gamma = 1, \ldots, c). \tag{12.3}$$

The equilibrium displacement condition (12.2) then becomes

$$\delta\mu_\gamma'' = \delta\mu_\gamma' = \delta\mu_\gamma \qquad (\gamma = 1, \ldots, c), \tag{12.4}$$

where $\delta\mu_\gamma$ is the common value of $\delta\mu_\gamma'$ and $\delta\mu_\gamma''$.

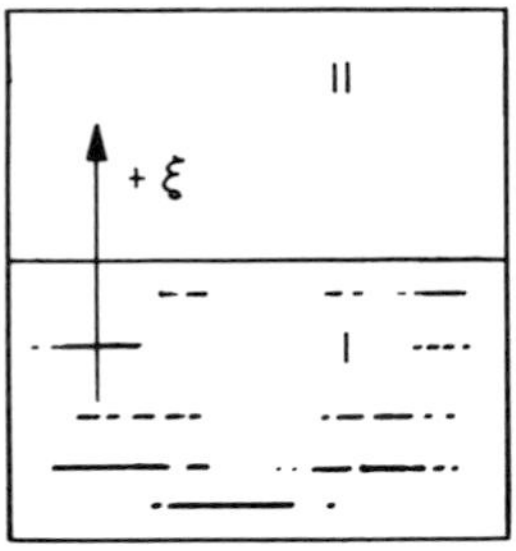

FIG. 22. Matter transfer between two phases.

Let us now write the Gibbs–Duhem equation (4.30) for each phase:

$$\begin{aligned} -s'\,\delta T + v'\,\delta p - \sum_\gamma N_\gamma'\,\delta\mu_\gamma &= 0 \\ -s''\,\delta T + v''\,\delta p - \sum_\gamma N_\gamma''\,\delta\mu_\gamma &= 0 \end{aligned} \tag{12.5}$$

where

$$s' = S'/\sum_\gamma n_\gamma', \qquad v' = V'/\sum_\gamma n_\gamma', \qquad N_\gamma' = n_\gamma'/\sum_\gamma n_\gamma'. \tag{12.6}$$

By subtraction, Eq. (12.5) becomes

$$-(s'' - s')\,\delta T + (v'' - v')\,\delta p - \sum_\gamma (N_\gamma'' - N_\gamma')\,\delta\mu_\gamma = 0. \tag{12.7}$$

Because the variance of the system is [see Eq. (5.7)]

$$w = 2 + c - \phi = c, \tag{12.8}$$

the two equations (12.5) permit us to calculate the variation of two variables of $T, p, \mu_1, \ldots, \mu_c$ when the variations of the c other variables are given.

For a pure constituent (subscript 1) in each phase, we have $N_1' = N_1'' = 1$ and (12.7) reduces to

$$\delta p/\delta T = (s'' - s')/(v'' - v'). \tag{12.9}$$

We here again find the Clapeyron equation (11.18).

C. The Gibbs–Konovalow Theorems

1. *Isothermal Equilibrium Displacement*

At constant temperature, (12.7) gives

$$(v'' - v')\,\delta p - \sum_\gamma (N_\gamma'' - N_\gamma')\,\delta\mu_\gamma = 0. \tag{12.10}$$

If, during the equilibrium displacement, a state is characterized by

$$N_\gamma' = N_\gamma'' \qquad (\gamma = 1, \ldots, c), \tag{12.11}$$

then (12.10) reduces simply to

$$\delta p = 0; \tag{12.12}$$

the pressure must pass through an extreme value (maximum, minimum, or inflection at horizontal tangent).

Equation (12.11) means that the two phases have the same composition. Such a system is called azeotropic (see Section V,B,2a). The above *Gibbs–Konovalow theorem* may then be stated in the form: If, in an isothermal equilibrium displacement of a two-phase binary system, the composition of the phases becomes the same, then the pressure must pass through an extreme value.

Example 1. Two constituents, two phases, and only passage reactions. The system is thus bivariant. If T and μ_2 are the independant variables, we thus have $\mu_1 = \mu_1(T, \mu_2)$. At constant temperature, (12.10) can be written in the form

$$(v'' - v')\frac{\delta p}{\delta \mu_2} = (N_1'' - N_1')\left(\frac{\partial \mu_1}{\partial \mu_2}\right)_T + (N_2'' - N_2'). \tag{12.13}$$

For $N_1'' = N_1'$ and $N_2'' = N_2'$, we find

$$\delta p/\delta\mu_2 = 0 \qquad \text{(horizontal tangent).} \tag{12.14}$$

From the curve $p = p(\mu_2)$ at constant temperature, we thus obtain a horizontal tangent. As μ_2 increases with N_2, the curves $p = p(N_2')$ and $p = p(N_2'')$ at T constant will give an extremum of p for the state $N_2' = N_2''$ (see Fig. 6).

Example 2. Three constituents, two phases, and only passage reactions. The system is thus trivariant. We choose T, μ_2, and μ_3 as independent variables.

If, at constant temperature, μ_2 and μ_3 vary (Fig. 23), the maximum of the surface corresponds to the state of same composition

$$N_1' = N_1'', \qquad N_2' = N_2'', \qquad N_3' = N_3''. \tag{12.15}$$

The chemical potential of a constituent as a function of its composition is given in Fig. 24.

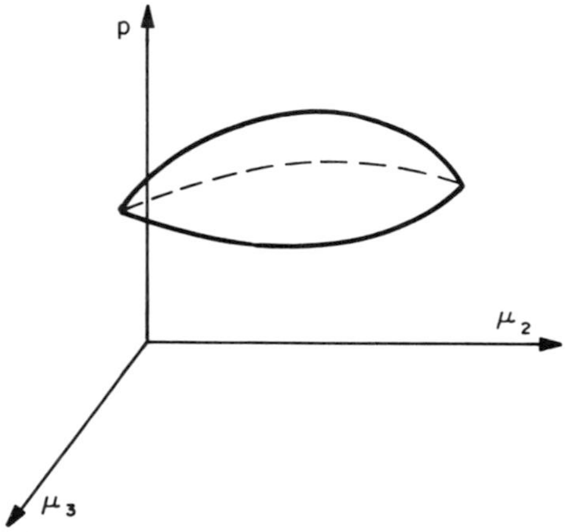

FIG. 23. Variation of the chemical potential with pressure.

2. *Isobaric Equilibrium Displacement*

In this case, (12.7) reduces to

$$(s'' - s')\,\delta T + \sum_\gamma (N_\gamma'' - N_\gamma')\,\delta\mu_\gamma = 0. \tag{12.16}$$

If the composition of the two phases is the same,

$$N_\gamma'' = N_\gamma' \qquad (\gamma = 1, \ldots, c)$$

and

$$\delta T = 0. \tag{12.17}$$

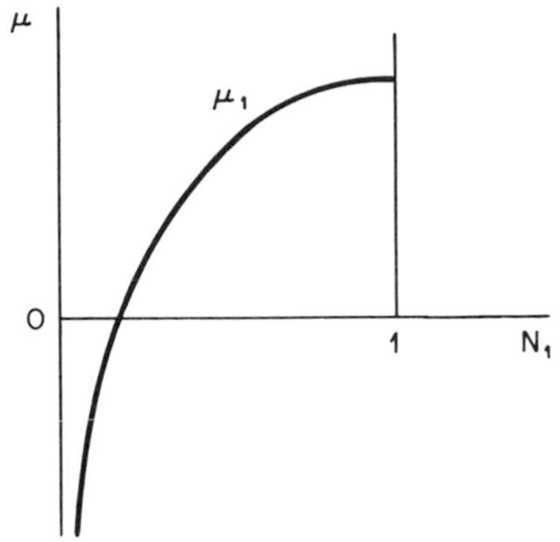

FIG. 24. Variation of the chemical potential with the composition.

This second theorem, also called *the Gibbs–Konovalow theorem*, may be expressed in the form: In an equilibrium displacement at constant pressure of a binary system, the temperature of coexistence passes through an extreme value (maximum, minimum, or inflection with a horizontal tangent) if the composition of the two phases is the same (Fig. 6).

D. The Reciprocal of the Gibbs–Konovalow Theorems

1. *Isothermal Transformation*

For a binary system, let us give $\delta\mu_2 \neq 0$ at constant temperature. The equations (12.5) can be written

$$\begin{aligned} v'\,\delta p - N_1'\,\delta\mu_1 &= N_2'\,\delta\mu_2 \\ v''\,\delta p - N_1''\,\delta\mu_1 &= N_2''\,\delta\mu_2. \end{aligned} \tag{12.18}$$

Thus,

$$\delta p = \delta\mu_2\left\{(N_1'' - N_1') \Big/ \begin{vmatrix} v' & N_1' \\ v'' & N_1'' \end{vmatrix}\right\}. \tag{12.19}$$

If the pressure passes through an extreme value, $\delta p = 0$ at this point, and, from (12.19), the phases must be of the same composition ($N_1'' = N_1'$ and $N_2'' = N_2'$).

2. *Isobaric Transformation*

Similarly, it is easy to show that, if the temperature of coexistence passes, at constant pressure, through an extremum, then the two phases must have the same composition.

XIII. Thermodynamics of Solutions

A. Introduction

A solution is a condensed phase (liquid or solid) composed of several components. Molecules of liquids and solids are strongly interacting, but the molecules of liquids are randomly distributed, while the molecules in solids are located in a regular array.

In the vicinity of the *critical temperature* and *critical pressure*, the distinction between gases and liquids vanishes. In this section, we develop the fundamental principles of the thermodynamics of solutions. In order to aid in this investigation, it is convenient to distinguish between ideal solutions and nonideal solutions. The fundamental quantity is the activity coefficient described in asymmetrical and symmetrical reference systems. This coefficient can be evaluated by different methods involving the knowledge of vapor pressures. Finally, the last subsection is devoted to a brief study of the excess functions.

B. Ideal Solutions

An ideal solution is defined in such a way that the chemical potential of each substance composing the solution has a simple functional dependence on a concrete composition variable. Under appropriate conditions, the properties of a large class of real substances may be adequately represented by the properties of ideal substances. This is often observed for dilute solutions. It is found by experiment and from molecular considerations (Prigogine, 1957) that a suitable form of the chemical potential of an ideal component γ can be written

$$\mu_\gamma = \mu_\gamma{}^\circ(T, p) + RT \ln N_\gamma, \tag{13.1}$$

where $\mu_\gamma{}^\circ(T, p)$ is some reference value of the chemical potential. If (13.1) is valid, then the solution is called a perfect solution. From

(13.1), we have

$$v_\gamma = \frac{\partial \mu_\gamma}{\partial p} = \frac{\partial \mu_\gamma^\circ(T, p)}{\partial p} \tag{13.2}$$

$$\frac{h_\gamma}{T^2} = -\frac{\partial(\mu_\gamma/T)}{\partial T} = -\frac{\partial(\mu_\gamma^\circ/T)}{\partial T}. \tag{13.3}$$

Thus, at constant T and p, the molar volumes v_γ and the molar enthalpies h_γ are constant over the whole region of ideality. On the other hand, it is easy to show that, for a binary solution,

$$v = (1 - N_2)v_1 + N_2 v_2, \tag{13.4}$$

where

$$v = V/n = \sum_\gamma N_\gamma v_\gamma. \tag{13.5}$$

As v_1 and v_2 are constant at a given temperature and pressure, the molar volume v is thus a linear function of the mole fraction.

For a perfect solution, v_1 and v_2 are the molar volumes of the pure components and the mixing process is accompanied by neither expansion nor contraction. In the same way, if the molar global enthalpy h is defined by

$$h = H/n = \sum_\gamma N_\gamma h_\gamma, \tag{13.6}$$

the mixing of two components at constant T and p is accompanied by neither an absorption nor evolution of heat.

For example (Fig. 25), the mixing of two systems by passage of component 1 from a pure phase ('') to a phase (') (solution of components 1 and 2) is characterized by the heat of reaction (see Section II)

$$r_{Tp} = h_1'' - h_1'. \tag{13.7}$$

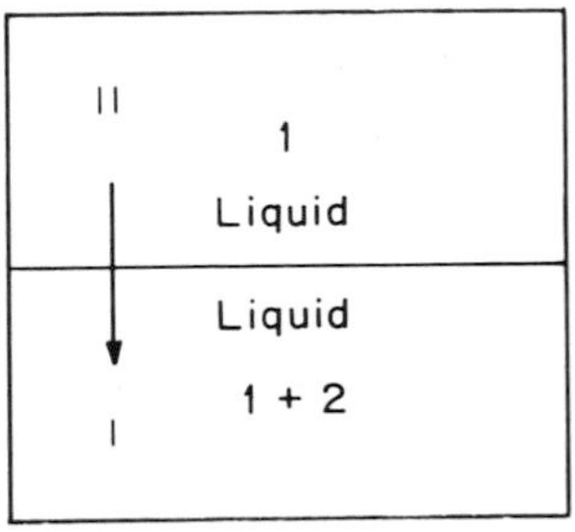

FIG. 25. Matter transfer of a pure constituent 1 ('') to a solution 1 + 2 (').

Since in a perfect solution h_1' is only a function of T and p, (13.7) reduces to $r_{Tp} = 0$. In other words, if, at constant pressure ($dp = 0$), the mixing is carried out adiabatically ($dQ = 0$), no temperature change occurs. Indeed, it is clear from (2.7) that

$$C_{p\xi}\, dT = r_{Tp}\, d\xi.$$

As $r_{Tp} = 0$, $dT = 0$.

C. Vapor Pressure of Perfect Solutions

The equilibrium condition between a solution (l) and its vapor (v) is given for two components 1 and 2 by

$$\mu_1^l = \mu_1^{\mathrm{v}}, \qquad \mu_2^l = \mu_2^{\mathrm{v}}. \tag{13.8}$$

Let us now assume that the solution is perfect and that the vapor is a mixture of perfect gases. Then, combining (13.8) with (13.1) and (6.18), we obtain

$$p_1 = k_1 N_1, \qquad p_2 = k_2 N_2, \tag{13.9}$$

where

$$\left.\begin{aligned} k_1 &= \exp\left\{\frac{\mu_1^{\circ l}(T,p) - \mu_1^{v*}(T)}{RT}\right\} \\ k_2 &= \exp\left\{\frac{\mu_2^{\circ l}(T,p) - \mu_2^{v*}(T)}{RT}\right\}. \end{aligned}\right. \tag{13.10}$$

At ordinary conditions, μ_1° and μ_2° are practically independent of pressure, and thus k_1 and k_2 are functions of T only.

The vapor pressure of each component in a perfect solution is thus proportional to its mole fraction.

For $N_1 = 1$, p is the vapor pressure p_1° of pure component γ, and thus $k_1 = p_1^\circ$. The so-called Raoult law is

$$p_1 = p_1^\circ N_1, \qquad p_2 = p_2^\circ N_2 \tag{13.11}$$

and the total pressure p of a perfect solution is a linear function of the molar fraction. Indeed,

$$p = p_1 + p_2 = p_1^\circ(1 - N_2) + p_2^\circ N_2. \tag{13.12}$$

D. Dilute Real Solutions

A large class of dilute real solutions may be considered as ideal systems. Equation (13.1) is then verified from $N_2 = 0$ to $N_2 = N_2$ (2 is the index of the solute). This limit of N_2 depends upon the nature of the system. For nonelectrolytes, the range of validity is much larger than for strong electrolytes. Equation (13.9) holds in the neighborhood of $N_2 = 0$, $N_1 = 1$. At $N_1 = 1$, $k_1 = p_1{}^\circ$, and

$$(p_1{}^\circ - p_1)/p_1{}^\circ = N_2. \tag{13.13}$$

The Raoult law (13.11) is valid only for the solvent. It is interesting to note that the relative lowering of the vapor pressure of the solvent is a function of the molar fraction of the solute; it is independent of the nature of the dissolved substance.

For the solute, (13.9) is often called Henry's law.

E. Real Solutions

In this section, we consider a real mixture of components for which the laws of Raoult and Henry are not satisfied.

This fact must be related to the form of the chemical potential. Let us arbitrarily choose a standard function $\mu_\gamma{}^\circ(T, p)$ and let us define a new quantity called activity of the component γ by the relation

$$a_\gamma = \exp\left\{\frac{\mu_\gamma(T, p, N_1, \ldots, N_c) - \mu_\gamma{}^\circ(T, p)}{RT}\right\} \tag{13.14}$$

or

$$\mu_\gamma = \mu_\gamma{}^\circ(T, p) + RT \ln a_\gamma. \tag{13.15}$$

The activity a_γ is a function of $T, p, N_1, \ldots, N_c$.

The decomposition of μ_γ into two functions $\mu_\gamma{}^\circ(T, p)$ and $a_\gamma(T, p, N_1, \ldots, N_c)$ is arbitrary; the quantity a_γ is then related to the choice of a reference system based on the limit of the ratio a_γ/N_γ for particular conditions. This ratio is called the activity coefficient f_γ. Deviations from the laws of perfect solutions may be expressed formally by introducing the activity coefficients f_γ in the expression for the chemical potential of a perfect solution. The method due to Lewis and Randall (1923) permits one formally to extend the properties of perfect solutions to actual solutions in a most elegant way. The chemical potential now takes the form

$$\mu_\gamma = \mu_\gamma^\circ(T, p) + RT \ln f_\gamma N_\gamma . \tag{13.16}$$

In this case, (13.9) must be replaced by

$$p_1 = k_1 N_1 f_1 , \qquad p_2 = k_2 N_2 f_2 , \tag{13.17}$$

where $k_1(T)$ and $k_2(T)$ are given by (13.10).

As the affinity is given by the classical relation

$$A = -\sum_\gamma \nu_\gamma \mu_\gamma , \tag{13.18}$$

we obtain from (13.16) the general expression

$$A = RT \ln[K_a(T, p)/(f_1 N_1)^{\nu_1} \cdots (f_c N_c)^{\nu_c}], \tag{13.19}$$

where

$$-\sum_\gamma \nu_\gamma \mu_\gamma^\circ(T, p) = RT \ln K_a(T, p). \tag{13.20}$$

At equilibrium, $A = 0$ and the Guldberg–Waage law may be written

$$(f_1 N_1)^{\nu_1} \cdots (f_c N_c)^{\nu_c} = K_a(T, p) \tag{13.21}$$

F. Asymmetrical Reference System

In the asymmetrical reference system, the dilute ideal solution is taken as reference system for a study of less dilute solutions. As the solution becomes more dilute, (13.16) approaches (13.1) with the same standard function. In this case, the asymmetrical property

$$\begin{aligned} N_2 &\to 0, \qquad f_1 \to 1 \\ N_1 &\to 1, \qquad f_2 \to 1 \end{aligned} \tag{13.22}$$

means that

$$f_1 \to 1 \qquad \text{for} \quad N_2 \to 0$$

while

$$f_1 \to 1 \qquad \text{for} \quad N_1 \to 1.$$

In other words, (13.17) becomes

$$p_1 = p_1^* f_1 , \qquad p_2 = p_2^* f_2 , \tag{13.23}$$

where

$$p_1^* = p_1^\circ N_1, \qquad p_2^* = k_2 N_2. \tag{13.24}$$

In Fig. 26, p_1^* and p_2^* are represented, respectively, by the Raoult line P_1O_2 and Henry line O_1Q_1. These two lines are tangential to the real experimental curves p_1 and p_2. From the curves p_1 and p_2 and from the Raoult and Henry laws, it is easy to calculate the activity coefficients. Indeed, from (13.23), we find

$$f_1 = p_1/p_1^*, \qquad f_2 = p_2/p_2^*. \tag{13.25}$$

Unfortunately, f_2 is given by the way of the tangent O_1Q_1, and this line cannot be drawn with good precision.

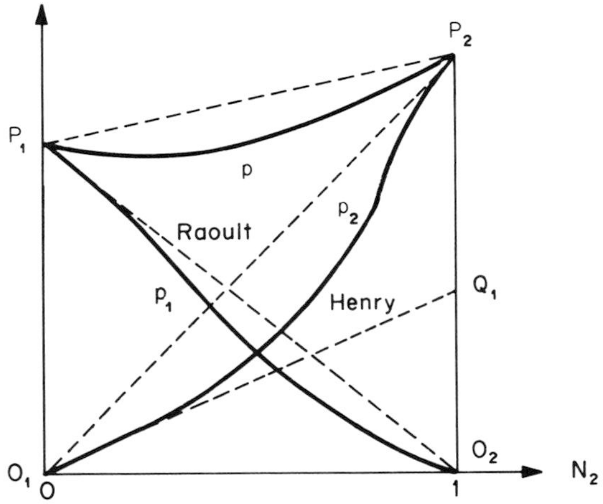

FIG. 26. The Raoult and Henry laws.

From a practical point of view, it is more convenient to use the relation between f_1 and f_2 derived from the Gibbs–Duhem equation. We shall use this method in the symmetrical reference system (see Section *H* on Boissonnas' Method).

G. Symmetrical Reference System

Let us now study the properties of a solution in the range $N_2 = 0$ (pure solvent) to $N_2 = 1$ (pure solute). In the symmetrical reference system, we define $\mu_\gamma^\circ(T, p)$ by the relation

$$\mu_\gamma^\circ(T, p) \equiv \mu_\gamma(T, p, N_\gamma = 1), \tag{13.26}$$

where $\mu_\gamma(T, p, N_\gamma = 1)$ is the chemical potential of the pure component. From (13.26) and (13.16),

$$f_\gamma = 1 \qquad \text{for} \quad N_\gamma = 1, \tag{13.27}$$

which means that the activity coefficient of each pure component is equal to the unity. Equations (13.17) give

$$\begin{aligned} N_1 &= 1, \qquad p_1^\circ = k_1, \\ N_2 &= 1, \qquad p_2^\circ = k_2, \end{aligned} \tag{13.28}$$

and thus

$$p_1 = p_1^* f_1, \qquad p_2 = p_2^* f_2, \tag{13.29}$$

where

$$p_1^* = p_1^\circ N_1, \qquad p_2^* = p_2^\circ N_2; \tag{13.30}$$

thus,

$$f_1 = p_1/p_1^*, \qquad f_2 = p_2/p_2^*. \tag{13.31}$$

The "ideal" pressures p_1^* and p_2^* are the Raoult lines P_1O_2 and O_1P_2 (Fig. 27).

From the experimental values of the total pressure p, it becomes easy to calculate f_1 and f_2. Indeed

$$p_1 = pN_1^{\mathrm{V}}, \qquad p_2 = pN_2^{\mathrm{V}}. \tag{13.32}$$

The curves p_1 and p_2 can be calculated by titrating the solution (N_1^{V} and N_2^{V} are then known). The deviations from ideality are described as negative or positive. In the first case, this means that the total pressure p in less than the ideal pressure (line P_1P_2) or that $f_\gamma < 1$ (example: chloroform + ethyl ether solutions). In the second case, $f_\gamma > 1$ (example: methylal + carbon disulphide) (Prigogine and Defay, 1967, pp. 338–339).

Remarks. (1) From a molecular point of view, we say that the vapor pressure is constant with respect to the composition if the interactions between the molecules 1–1, 1–2, and 2–2 are equal. If the attractions 1–2 < 1–1, 1–2, the vapor pressure is raised and $f_\gamma > 1$. Alternatively, if 1–2 > 1–1, 1–2, then $f_\gamma < 1$. However, entropy effects also modify the activity coefficients (Prigogine, 1957).

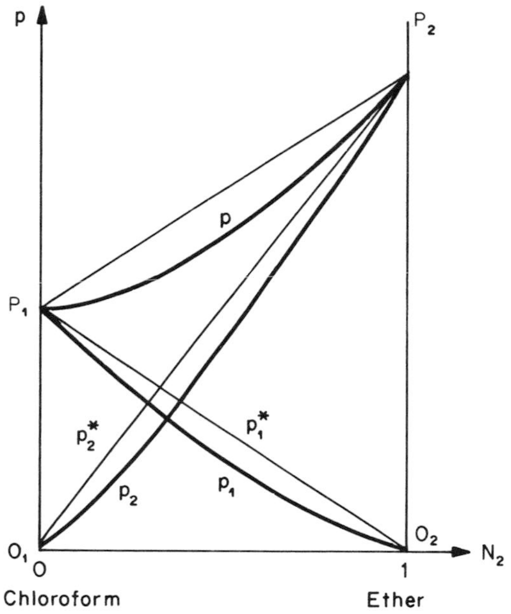

FIG. 27. Deviation from ideality for the system chloroform–ether.

(2) If deviations from ideality are large enough, the total vapor pressure p passes through a maximum or a minimum (azeotropy).

H. The Boissonnas Method

If the composition of the vapor is unknown, p_1 and p_2 can be still calculated.

From the Gibbs–Duhem relation [Eq. (4.30)] at constant T and p,

$$\sum_{\gamma} N_\gamma \, d\mu_\gamma = 0, \tag{13.33}$$

and thus, (13.16) leads to

$$N_1\left[\frac{\partial(\ln f_1)}{\partial N_2}\right]_{Tp} + N_2\left[\frac{\partial(\ln f_2)}{\partial N_2}\right]_{Tp} = 0. \tag{13.34}$$

In the symmetrical reference system, (13.34) reduces to the well-known Duhem–Margules relation

$$N_1\left[\frac{\partial(\ln p_1)}{\partial N_2}\right]_{T} + N_2\left[\frac{\partial(\ln p_2)}{\partial N_2}\right]_{T} = 0. \tag{13.35}$$

This equation is valid only for an ideal vapor phase. Equation (13.35) can be transformed into

$$\frac{\partial p_2}{\partial N_2} = \frac{1}{1 - (p_1/p_2)(N_2/N_1)} \frac{\partial p}{\partial N_2}. \tag{13.36}$$

For a very dilute solution, $p_1 = p_1{}^\circ N_1$, and thus

$$dp_1 = p_1{}^\circ \, dN_1 \qquad \text{and} \qquad dp_2 = dp + p_1{}^\circ \, dN_2. \tag{13.37}$$

Boissonnas (1939) suggested that the composition range be divided into a number of equal intervals ΔN_2 and that Eqs. (13.36) and (13.37) be used in the approximate forms

$$\Delta p_2 = \frac{\Delta p}{1 - (p_1/p_2)(N_2/N_1)} \tag{13.38}$$

with

$$\Delta p_2 = \Delta p + p_1{}^\circ \, \Delta N_2 \qquad (\text{for} \quad N_2 = 0). \tag{13.39}$$

From the curves p_1 and p_2, it is easy to calculate f_1 and f_2.

Remark. For an azeotrope, the two phases have the same composition at the extremum and $p_1/N_1 = p_2/N_2$. The relation (13.36) then becomes wrong in the neighborhood of the extremum. Practically, we apply the method of Boissonnas first from $N_2 = 0$ to a point neighboring the extremum and then from $N_1 = 0$ until the vicinity of the azeotrope. If, at the temperature considered, the vapor pressure p_2 (solute) is small in comparison with p_1 (solvent), we may write

$$p = p_1. \tag{13.40}$$

From the experimental value of p, it becomes easy to calculate f_1, because

$$f_1 = p_1/p_1{}^\circ N_1. \tag{13.41}$$

On the other hand, integrating Eq. (13.34), we find the value of f_2

$$\ln f_2 = -\int_{N_2=1}^{N_2} (N_1/N_2) \, d(\ln f_1). \tag{13.42}$$

An interesting example of this case (butyl sebacate at 20°C dissolved in methanol) was studied by Colmant (1954).

Analytical forms of activity coefficients are given in several publications (Prigogine and Defay, 1967, p. 339).

1. *Fugacity in Solution*

The fugacity φ_γ of a component γ is defined in a solution by the relation

$$\mu_\gamma = \mu_\gamma^*(T) + RT \ln \varphi_\gamma, \tag{13.43}$$

where $\mu_\gamma^*(T)$ is the standard function of a perfect gas.

When a solution (l) is in equilibrium with its vapor (v), we have

$$\begin{aligned} \mu_\gamma^l &= \mu_\gamma^*(T) + RT \ln \varphi_\gamma^l \\ \mu_\gamma^{\mathrm{v}} &= \mu_\gamma^*(T) + RT \ln \varphi_\gamma^{\mathrm{v}}, \end{aligned} \tag{13.44}$$

where φ_γ^l is the fugacity in the solution and $\varphi_\gamma^{\mathrm{v}}$ that in the vapor. At equilibrium,

$$\varphi_\gamma^l = \varphi_\gamma^{\mathrm{v}}. \tag{13.45}$$

This means that, for two phases in equilibrium, each component has the same fugacity in the two phases.

Now, combining (13.43) and (13.18), we find the affinity of a reaction in solution

$$A = RT \ln[K_\varphi(T)/\varphi_1^{\nu_1} \cdots \varphi_c^{\nu_c}]. \tag{13.46}$$

Nevertheless, as vapor pressures of numerous dissolved substances are small, fugacities are not frequently used.

2. *Molar Concentration and Molality*

We know that the molar concentration of a solute C_s is given by

$$C_s = n_s/V = N_s/(N_1 v_1 + \sum_s N_s v_s). \tag{13.47}$$

If the solution is very dilute, $N_1 \simeq 1$, and

$$C_s \simeq N_s/v_1^\circ, \tag{13.48}$$

where v_1° is the specific molar volume of the pure solvent; the chemical potential given by (13.16) then reduces to

$$\mu_s = \mu_s^{\circ\circ}(T, p) + RT \ln f_s^c C_s \tag{13.49}$$

where

$$\mu_s^{\circ\circ}(T, p) = \mu_s^\circ(T, p) + RT \ln v_1^\circ(T, p). \tag{13.50}$$

For a reaction between dissolved substances, the affinity is given by

$$A = RT \ln[K^\circ(T, p)/(f_2^c C_2)^{\nu_2} \cdots (f_c^c C_c)^{\nu_c}], \tag{13.51}$$

where

$$RT \ln K^\circ(T, p) = -\sum_s \nu_s \mu_s^{\circ\circ}(T, p). \tag{13.52}$$

It follows that

$$K^\circ(T, p) = K(T, p)(v_1^\circ)^{-\nu}, \tag{13.53}$$

and, at equilibrium,

$$(f_2^c C_2)^{\nu_2} \cdots (f_c^c C_c)^{\nu_c} = K^\circ(T, p). \tag{13.54}$$

On the other hand, the molality of a dissolved substance m_s is defined by

$$m_s = 1000 n_s / n_1 M_1 = 1000 N_s / N_1 M_1. \tag{13.55}$$

For a dilute solution, $N_1 \simeq 1$ and

$$m_s \simeq 1000 N_s / M_1. \tag{13.56}$$

The chemical potential given by (13.16) then reduces to

$$\mu_s = \mu_s^\dagger(T, p) + RT \ln f_s^m m_s \tag{13.57}$$

where

$$\mu_s^\dagger(T, p) = \mu_s^\circ + RT \ln(M_1/1000). \tag{13.58}$$

The affinity of a reaction between dissolved substances is given by

$$A = RT \ln[K^\dagger(T, p)/(f_2^m m_2)^{\nu_2} \cdots (f_c^m m_c)^{\nu_c}] \tag{13.59}$$

and

$$K^\dagger(T, p) = K(T, p)(M_1/1000)^{-\nu}. \tag{13.60}$$

At equilibrium, we have

$$(f_2^m m_2)^{\nu_2} \cdots (f_c^m m_c)^{\nu_c} = K^\dagger(T, p). \tag{13.61}$$

Remarks. (1) In the three systems, the molar fractions, the concentrations and molalities, and the activity coefficients (f_s, f_s^c, f_s^m) have the same values in a dilute solution, but the three activities ($f_s N_s$, $f_s^c C_s$, $f_s^m m_s$) are different.

(2) The standard affinities usually found in the literature are quantities related to molalities in dilute solutions.

I. Excess Functions

The main problem with which we shall be concerned in this section is the effect of mixing two or more substances. In other words, we shall compare the properties of the mixture to those of the pure substances. In order to separate the effects of mixing from the effects of changes in temperature or pressure, we shall always compare the mixture to the pure components taken at the same pressure and temperature. The total volume V° of separate components with specific volumes $v_\gamma{}^\circ$ is given by

$$V^\circ = \sum_\gamma n_\gamma v_\gamma{}^\circ. \tag{13.62}$$

The change of the volume due to mixing V_{m}, is then

$$V_{\mathrm{m}} = V - V^\circ = \sum_\gamma n_\gamma (v_\gamma - v_\gamma{}^\circ), \tag{13.63}$$

where V is the real volume of the solution.

Similarly, we have the enthalpy of mixing, the entropy of mixing, and the Gibbs free energy of mixing:

$$H_{\mathrm{m}} = \sum_\gamma n_\gamma (h_\gamma - h_\gamma{}^\circ) \tag{13.64}$$

$$S_{\mathrm{m}} = \sum_\gamma n_\gamma (s_\gamma - s_\gamma{}^\circ) \tag{13.65}$$

$$G_{\mathrm{m}} = \sum_\gamma n_\gamma (\mu_\gamma - \mu_\gamma{}^\circ), \tag{13.66}$$

where the index $^\circ$ refers to the pure components in the same physical state as into the solution.

Other important quantities of the same kind are the molar functions

$$v_{\mathrm{m}} = \sum_\gamma N_\gamma (v_\gamma - v_\gamma{}^\circ) \tag{13.67}$$

$$h_{\mathrm{m}} = \sum_\gamma N_\gamma (h_\gamma - h_\gamma{}^\circ) \tag{13.68}$$

$$s_{\mathrm{m}} = \sum_\gamma N_\gamma (s_\gamma - s_\gamma{}^\circ) \tag{13.69}$$

$$g_{\mathrm{m}} = \sum_\gamma N_\gamma (\mu_\gamma - \mu_\gamma{}^\circ). \tag{13.70}$$

The enthalpy of mixing H_{m} is the heat absorbed by the system at constant T and p because

$$\int dQ = \int dH = H - H_0 = H_m. \tag{13.71}$$

Now, a solution will be described as perfect if the Gibbs free energy of mixing $g_m{}^p$ takes the following simple form

$$g_m{}^p = RT \sum_\gamma N_\gamma \ln N_\gamma. \tag{13.72}$$

The corresponding heat of mixing $h_m{}^p$ and volume of mixing $v_m{}^p$ are zero. Alternatively, these properties may be used to test the validity of the laws of perfect solutions for a given mixture. For the entropy of mixing in a perfect solution, we find the positive expression

$$s_m{}^p = -R \sum_\gamma N_\gamma \ln N_\gamma > 0. \tag{13.73}$$

Thus, in a perfect solution, all thermodynamic functions of mixing except those containing the entropy are zero.

The entropy of mixing (13.73) has a simple meaning. To obtain its interpretation, we must make use of the Boltzmann formula relating the number of accessible configurations of the system to the entropy. This problem is developed by Prigogine (1957).

From the experimental point of view, the discrepancies between real and perfect systems are better illustrated by the excess functions. The difference between the thermodynamic functions of mixing and the value corresponding to a perfect solution (superscript p) at the same T and p and composition will be called the thermodynamic excess function (denoted by subscript e). Thus,

$$v_e = v_m, \quad h_e = h_m, \quad s_e = s_m - s_m{}^p, \quad g_e = g_m - g_m{}^p. \tag{13.74}$$

The Gibbs free energy of mixing may be represented by the following expressions [cf. (13.16) and (13.70)]

$$g_m = RT \sum_\gamma N_\gamma \ln f_\gamma N_\gamma, \tag{13.75}$$

where f_γ is defined in the symmetrical reference system. On the other hand, since

$$\frac{\partial \mu_\gamma}{\partial p} = v_\gamma, \qquad \frac{\partial \mu_\gamma}{\partial T} = -s_\gamma, \qquad \frac{\partial(\mu_\gamma/T)}{\partial T} = -\frac{h_\gamma}{T^2}, \tag{13.76}$$

we obtain

$$\begin{aligned} v_{\mathrm{m}} &= RT \sum_{\gamma} N_{\gamma}\, \partial(\ln f_{\gamma})/\partial p \\ h_{\mathrm{m}} &= -RT^2 \sum_{\gamma} N_{\gamma}\, \partial(\ln f_{\gamma})/\partial T \\ s_{\mathrm{m}} &= -R \sum_{\gamma} N_{\gamma} \ln f_{\gamma} N_{\gamma} - RT \sum_{\gamma} n_{\gamma}\, \partial(\ln f_{\gamma})/\partial T; \end{aligned} \tag{13.77}$$

the excess thermodynamic quantities then become

$$v_{\mathrm{e}} = RT \sum_{\gamma} N_{\gamma}\, \partial(\ln f_{\gamma})/\partial p \tag{13.78}$$

$$h_{\mathrm{e}} = -RT^2 \sum_{\gamma} N_{\gamma}\, \partial(\ln f_{\gamma})/\partial T \tag{13.79}$$

$$s_{\mathrm{e}} = -R \sum_{\gamma} N_{\gamma} \ln f_{\gamma} - RT \sum_{\gamma} N_{\gamma}\, \partial(\ln f_{\gamma})/\partial T \tag{13.80}$$

where the excess entropy is defined by

$$Ts_{\mathrm{e}} = h_{\mathrm{e}} - g_{\mathrm{e}}. \tag{13.81}$$

The excess volume v_{e} is found by measuring the mixing volume, h_{e} by the mixing heat, and g_{e} by vapor-pressure curves. The relation (13.81) then gives the value of the excess entropy. The orders of magnitude of the excess free energy, excess enthalpy, and excess entropy for mixtures of saturated hydrocarbons are shown in Figs. 28–31 for four systems studied by different authors.*

We observe that the contribution of the excess enthalpy and of the excess entropy (multiplied by T) to the excess free energy are of comparable magnitude.

In the case n-heptane + n-hexadecane (Fig. 28), g_{e} is negative (activity coefficients smaller than 1) and we have negative deviations from Raoult's law, since, when the activity coefficients are smaller than one, the partial vapor pressures are smaller than those of perfect solutions. However, for the other systems, we have positive deviations from Raoult's law and the partial vapor pressures are higher than those of a perfect

* For the system n-heptane + n-hexadecane at 20°C, see Brönsted and Koefeld (1946), Van der Waals and Hermans (1949). For the systems h-hexane + cyclohexane at 20°C and n-octane + tetraethylmethane at 50°C, see Prigogine and Mathot (1950a,b). For the system 2,2,4-trimethylpentane + hexadecane at 24.9°C, see Van der Waals (1950).

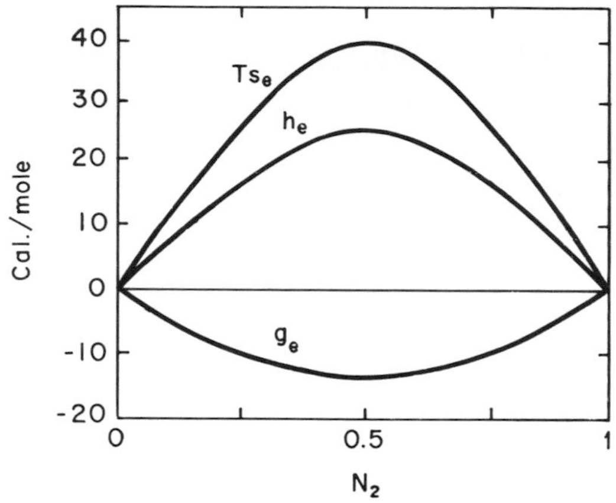

FIG. 28. Excess functions for the system (1) *n*-heptane + (2) *n*-hexadecane at 20°C.

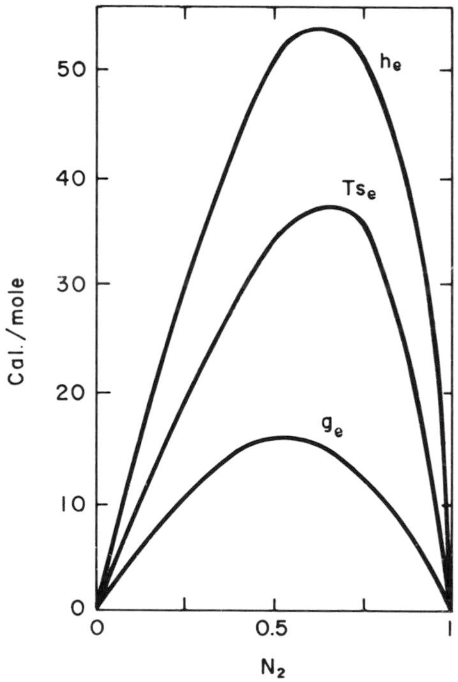

FIG. 29. Excess functions for the system (1) *n*-hexane + (2) cyclohexane at 20°C.

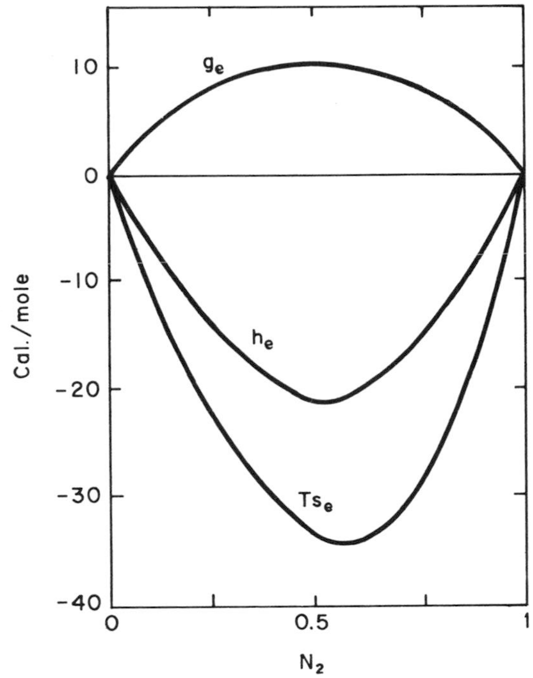

FIG. 30. Excess functions for the system 2-2-4-trimethylpentane (1) + *n*-hexadecane (2) at 24.9°C.

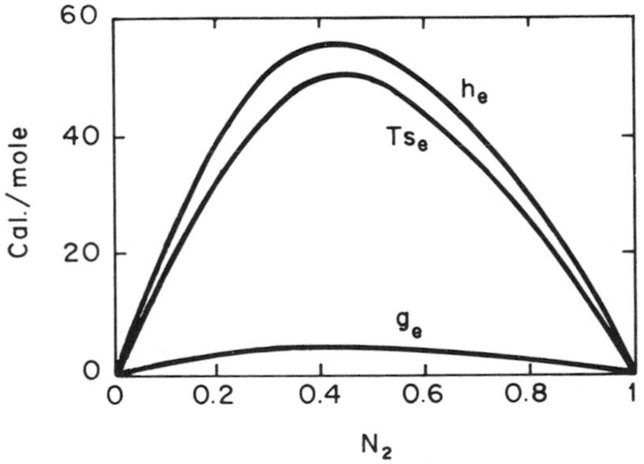

FIG. 31. Excess functions for the system (1) *n*-octane + (2) tetraethylmethane at 50°C.

solution. Generally, the excess functions have a simple parabolic form, but for some systems (like systems containing alcohol or water), they may be more complicated. The discrepancies to ideality are due to energy and entropy effects [difference of interactions (h_e) 1–1, 2–2, 1–2; entropy effects (s_e) related to the differences in molecular size].

In the system 2,2,4-trimethylpentane + hexadecane (Fig. 30), the entropy effect partially compensates the energy effect in such a way that $g_e \simeq 0$ (perfect solutions at 24.9°C). In the system *n*-octane + tetraethylmethane (Fig. 31), the excess entropy is negative, and this means that the mixing entropy is smaller than its value corresponding to a perfect solution. The orientation disorder is larger in the pure substance than in the mixture.

An interesting case is the system water + triethylamine (Fig. 32) (Haase and Rehage, 1955). The excess entropy s_e is negative and larger in absolute value than s_m^p; the mixing real entropy s_m is then negative. Nevertheless, the mixing exists because of the effect of the mixing en-

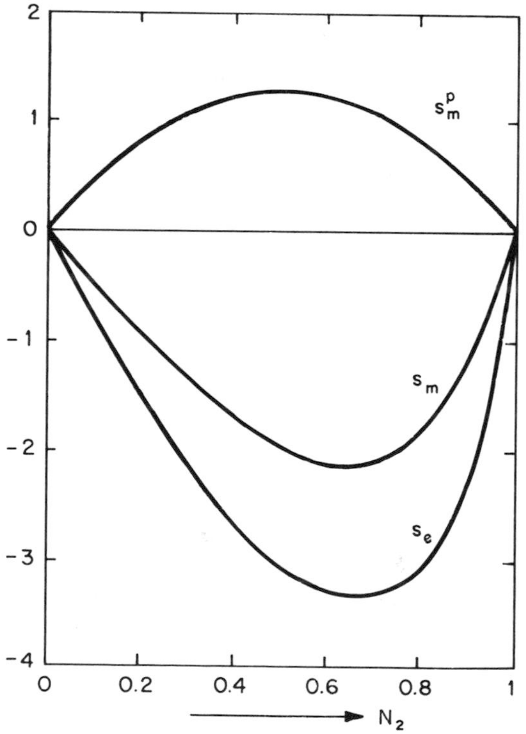

FIG. 32. Excess entropy for the system triethylamine + water.

thalpy h_m. The condition of stability of miscibility is given by $g_m < 0$, and thus $h_m - Ts_m < 0$, which means that, if s_m is negative, h_m is also negative and large in absolute value. The mixture is exothermic.

Now, it is interesting to consider two limit cases for which the discrepancies are only related, on one hand, to the energetic effects, and on the other hand, to the entropy effects. First, from the above relations, it is easy to show that

$$\partial G_e/\partial T = -S_e, \tag{13.82}$$

where

$$G_e = RT \sum_{\gamma} n_{\gamma} \ln f_{\gamma} \tag{13.83}$$

$$\partial G_e/\partial n_{\gamma} = RT \ln f_{\gamma} \tag{13.84}$$

$$H_e = -RT^2 \sum_{\gamma} n_{\gamma}\, \partial(\ln f_{\gamma})/\partial T \tag{13.85}$$

$$\partial H_e/\partial n_{\gamma} = -RT^2\, \partial(\ln f_{\gamma})/\partial T. \tag{13.86}$$

1. *Regular Solutions*

The term "regular solutions" was introduced by Hildebrand* to describe mixtures whose excess entropy was found experimentally to be zero. Deviations from ideality arise entirely from the energetic term:

$$|\, h_e \,| \gg T\,|\, s_e \,|, \qquad g_e \simeq h_e, \qquad s_e \simeq 0. \tag{13.87}$$

For a binary system, we obtain

$$\partial S_e/\partial n_1 = 0, \qquad \partial S_e/\partial n_2 = 0 \tag{13.88}$$

and

$$\partial^2 G_e/\partial T\, \partial n_1 = 0, \qquad \partial^2 G_e/\partial T\, \partial n_2 = 0, \tag{13.89}$$

$$\partial(RT \ln f_1)/\partial T = 0, \qquad \partial(RT \ln f_2)/\partial T = 0, \tag{13.90}$$

so that

$$\ln f_1 \propto 1/T, \qquad \ln f_2 \propto 1/T. \tag{13.91}$$

The activity coefficients for regular solutions are thus inversely proportional to the absolute temperature.

* See Fowler and Guggenheim (1939), Hildebrand (1929), Guggenheim (1935), Prigogine (1957).

2. *Athermal Solutions*

Here, the deviations from ideality arise entirely from the entropy term, the mixing heat being zero (this is the origin of the term athermal):

$$T\,|\,s_e\,| \gg |\,h_e\,|, \qquad g_e \simeq -Ts_e, \quad h_e \simeq 0. \tag{13.92}$$

For a binary system,

$$\partial H_e/\partial n_1 = 0, \qquad \partial H_e/\partial n_2 = 0, \tag{13.93}$$

so that

$$\partial(\ln f_1)/\partial T = 0, \qquad \partial(\ln f_2)/\partial T = 0. \tag{13.94}$$

The activity coefficients of athermal solutions are independent of temperature.

XIV. Osmotic Pressure

Measurements of osmotic pressure are frequently used to determine activity coefficients and solution molecular weights. The measurements are particularly useful in the determination of the properties of polymer solutions. The system under consideration consists of a liquid solution of components (phase $'$) separated from pure liquid (phase $''$) by a nondeformable heat-conducting membrane permeable to component 1 alone (Fig. 33). The temperature is assumed to be uniform in the system.

The affinity of passage of component 1 from the phase $''$ to the phase $'$ is

$$A = \mu_1'' - \mu_1' = \mu_1^\circ(T, p'') - \mu_1^\circ(T, p') - RT \ln f_1' N_1', \tag{14.1}$$

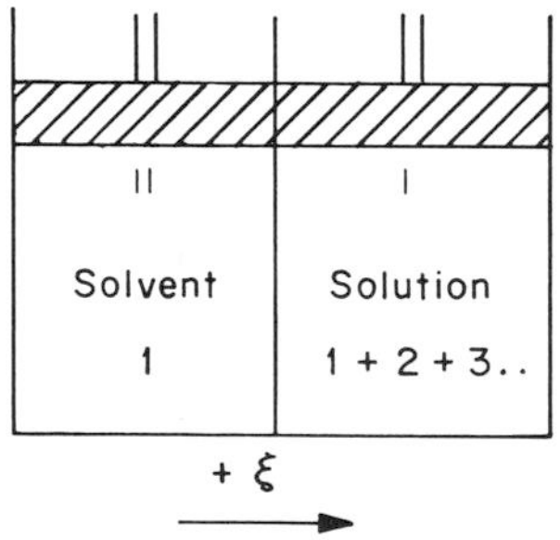

FIG. 33. Osmotic phenomena.

where μ_1° is the same function in the two phases. Let us assume now that $p'' = p'$. Equation (14.1) reduces to

$$(A)_{p''=p'} = -RT \ln f_1' N_1'. \tag{14.2}$$

This affinity cannot be equal to zero unless $N_1' = 1$, which means that the solvent is pure in each phase. In other words, when substances are dissolved in the solution, we have $N_1' < 1$ and the affinity $(A)_{p''=p'}$ is positive; the solvent tends to pass from the pure solvent ($''$) to the solution ($'$) and increases in this solution phase. Hence, equilibrium cannot be established unless $p'' \neq p'$. The difference

$$\Pi = p' - p'' \tag{14.3}$$

is called osmotic pressure.

At equilibrium, (14.1) reduces to

$$\mu_1^\circ(T, p'') - \mu_1^\circ(T, p') = RT \ln f_1' N_1'. \tag{14.4}$$

For a displacement along an equilibrium line at T constant, we have

$$\frac{\partial \mu_1^\circ}{\partial p''} \, \delta p'' - \frac{\partial \mu_1^\circ}{\partial p'} \, \delta p' = RT \, \delta(\ln f_1' N_1'). \tag{14.5}$$

with

$$\frac{\partial \mu_1^\circ}{\partial p''} = v_1^{\mathrm{s}}(T, p''), \qquad \frac{\partial \mu_1^\circ}{\partial p'} = v_1^{\mathrm{s}}(T, p'), \tag{14.6}$$

where the quantity v_1^{s} is the standard molar volume.

Neglecting the compressibility of component 1, (14.5) and (14.3) give

$$\delta \Pi = -(RT/v_1^\circ) \, \delta(\ln f_1' N_1'), \tag{14.7}$$

where v_1° is the molar volume of the solvent extrapolated to zero pressure. Integrating (14.7) from $N_1' = 1$ ($\Pi = 0$) to $N_1' = N_1'$, we obtain

$$\Pi = -(RT/v_1^\circ) \ln f_1' N_1'. \tag{14.8}$$

For a sucrose solution of 0.88 mol l^{-1}, the experimentally observed (Eucken, 1934) osmotic pressure is about 27 atm.

Instead of characterizing deviations from ideality of the solvent by its activity coefficient f_1, it is often advantageous to introduce the osmotic coefficient Γ of Bjerrum and Guggenheim defined by the equality

$$\Gamma \ln N_1 = \ln f_1 N_1. \tag{14.9}$$

Comparing this with (14.8), we see that

$$\Pi = -\Gamma(RT/v_1^\circ) \ln N_1'. \tag{14.10}$$

For very dilute solutions, (14.10) reduces to

$$\Pi = \Gamma RT \sum_s C_s'. \tag{14.11}$$

For an ideal dilute solution ($C_s' \leq 10^{-6}$ mol l^{-1}), we find the classical van't Hoff equation

$$\Pi = RT \sum_s C_s', \tag{14.12}$$

which shows that the osmotic pressure is independent of the nature of the solvent.

For a weak electrolyte $AB \rightleftharpoons A^- + B^+$, Eq. (14.11) may be rewritten

$$\Pi = \Gamma RT(1 + \alpha)C_{\mathrm{AB}}^\circ, \tag{14.13}$$

where α is the dissociation degree and C_{AB}° the total concentration of AB. The coefficient Γ may be calculated if we know the experimental values of Π and α (known, for instance, by electrical conductivity measurements).

Remark. Comparing (14.2) to (14.8), it follows that the affinity for $p'' = p'$ is given by

$$(A)_{p'=p''} = v_1^\circ \Pi. \tag{14.14}$$

XV. Equilibrium Curves between Two Phases

A. Generalization of the Nernst Distribution Law

We know that the affinity A_γ of transfer of component γ from one phase ($'$) to the other ($''$) is given by

$$A_\gamma = \mu_\gamma' - \mu_\gamma'' = \mu_\gamma'^\circ(T, p) - \mu_\gamma''^\circ(T, p) + RT \ln(f_\gamma' N_\gamma' / f_\gamma'' N_\gamma''). \tag{15.1}$$

At equilibrium,

$$f_\gamma'' N_\gamma'' / f_\gamma' N_\gamma' = K_\gamma(T, p), \tag{15.2}$$

where we have put

$$RT \ln K_\gamma(T, p) = \mu_\gamma'(T, p) - \mu_\gamma''(T, p). \tag{15.3}$$

For ideal phases, (15.2) reduces to the classical Nernst distribution law (see 6.80)

$$N_\gamma''/N_\gamma' = K_{\gamma N}(T, p). \tag{15.4}$$

The quantity K_γ is called the distribution or partition coefficient. At T and p constant, the equilibrium constant K_γ is independent of mole fractions.

Remark. The distribution law of matter in an electric field was established a few years ago by Sanfeld and co-workers (1968c).

B. Van Laar Relation

For a general displacement along an equilibrium line, (15.2) becomes

$$\delta\left[\ln \frac{f_\gamma'' N_\gamma''}{f_\gamma' N_\gamma'}\right] = \delta[\ln K_\gamma(T, p)] = \frac{\partial(\ln K_\gamma)\,\delta p}{\partial p} + \frac{\partial(\ln K_\gamma)\,\delta T}{\partial T}. \tag{15.5}$$

We thus find from (9.48) and (9.49) that

$$\delta\left[\ln \frac{f_\gamma'' N_\gamma''}{f_\gamma' N_\gamma'}\right] = -\frac{r^{\text{s}}_{Tp}(\gamma)}{RT^2}\,\delta T - \frac{\Delta^{\text{s}}_{Tp}(\gamma)}{RT}\,\delta p \tag{15.6}$$

This is the well-known van Laar equation, which we now apply to different cases:

1. *Binary Systems Forming a Eutectic*

Let us consider first the equilibrium between the solution (′) and one pure solid phase (″) and the transfer of component 1 from the solution to the solid phase (″):

$$N_1'' = 1, \qquad N_2'' = 0, \qquad f_1'' = 1.$$

Applying the van Laar relation (15.6) for a displacement at constant p, we obtain

$$\delta(\ln f_1' N_1') = \frac{L_{f_1}}{R}\,\frac{\delta T}{T^2}, \tag{15.7}$$

where

$$L_{f_1} = r^{\text{s}}_{Tp}(1) \tag{15.8}$$

is the latent heat of fusion at the pressure considered.

For a pure system, $N_1' = 1$ and $N_1'' = 1$, the variance is equal to unity and we called T_{10} the fusion temperature of component 1 at the pressure considered.

Integrating (15.7) from a point $N_1' = 1$ ($f_1' = 1$, $T = T_{10}$) to the required value of N_1', we obtain the equilibrium curve of a solution with crystals 1:

$$\ln f_1' N_1' = \frac{L_{f_1}}{R}\left[\frac{1}{T_{10}} - \frac{1}{T}\right]. \tag{15.9}$$

If we assume the solution phase to be perfect, this simplifies to

$$\ln N_1' = \frac{L_{f_1}}{R}\left[\frac{1}{T_{10}} - \frac{1}{T}\right], \tag{15.10}$$

where $T < T_{10}$.

This approximate equation is that of the line of coexistence at a given pressure given in Fig. 34.

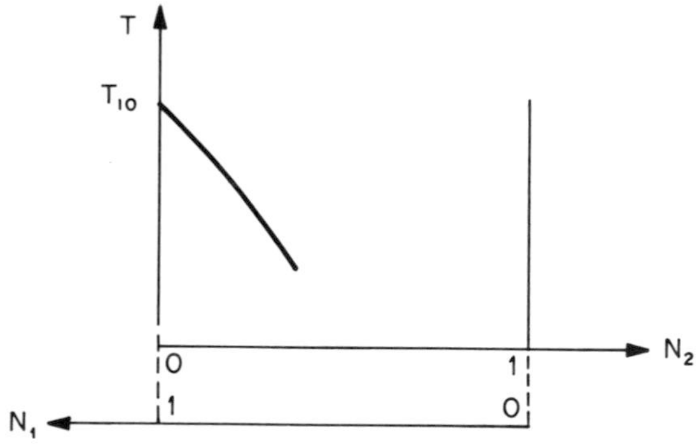

FIG. 34. Shape of freezing-point curve.

We define the depression θ of the freezing point of the solution by

$$\theta = T_{10} - T.$$

For very dilute solutions, $\theta \ll T$ and $1/T \simeq \theta/T_{10}^2$; thus, (15.9) reduces to

$$\frac{L_{f_1}\theta}{RT_{10}^2} \simeq \Gamma \sum_s N_s', \tag{15.11}$$

where the osmotic coefficient Γ is defined by (14.9).

The lowering of the freezing point is then given by

$$\theta = \Gamma \frac{RT_{10}^2}{L_{f_1}} \sum_s N_s'. \tag{15.12}$$

If a solution is both very dilute and ideal, $\Gamma = 1$, and we have, in terms of molalities,

$$\theta = \frac{RT_{10}^2}{L_{f_1}} \frac{M_1}{1000} \sum_s m_s' = \theta_c \sum_s m_s'; \tag{15.13}$$

the quantity θ_c is called the cryoscopic or freezing-point constant. It depends only on the nature of the solvent. For water, $\theta_c = 1.86$, while for benzene, $\theta_c = 5.08$.

For a substance 2, the coexistence line BE (Fig. 35) is represented by the equation

$$\ln f_2' N_2' = \frac{L_{f_2}}{R} \left[\frac{1}{T_{20}} - \frac{1}{T} \right]. \tag{15.14}$$

At the eutectic point, crystals of both 1 and 2 and the solution coexist in equilibrium and the two equations (15.9) and (15.14) are simultaneously satisfied.

If the solution is perfect, N_{2e}' and T_e (values at the eutectic point) can be calculated from

$$\begin{aligned} \ln N_{2e}' &= \frac{L_{f_2}}{R} \left[\frac{1}{T_{20}} - \frac{1}{T_e} \right] \\ \ln(1 - N_{2e}') &= \frac{L_{f_1}}{R} \left[\frac{1}{T_{10}} - \frac{1}{T_e} \right]. \end{aligned} \tag{15.15}$$

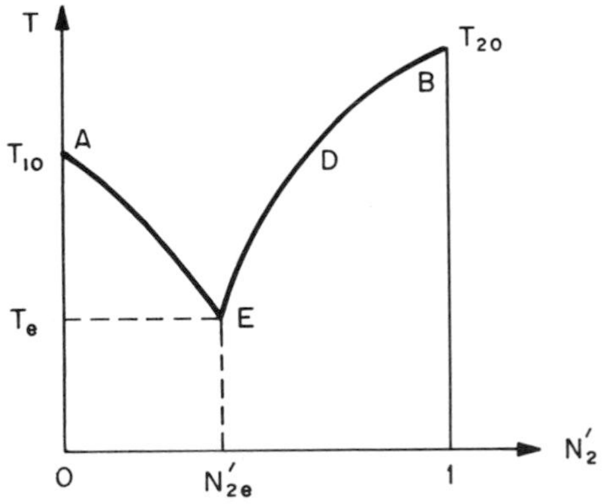

FIG. 35. Crystallization or freezing-point curves at constant pressure, with eutectic point.

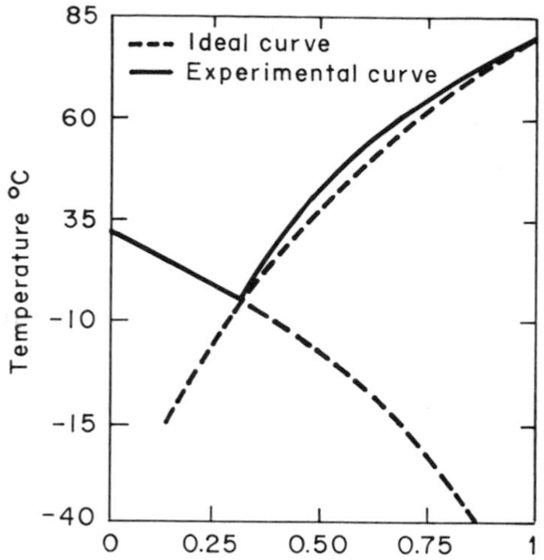

FIG. 36. Freezing point curves for $o + p$-chloronitrobenzene.

An example (Hollman, 1900; Kohman, 1925; Timmermans, 1936, p. 54) is given in Fig. 36.

Remarks. (1) In the case of saturated salt solutions (excluding the formation of mixed crystals)—for example, $H_2O + NaCl$—the line AE is called the freezing curve, while the line EB is called the solubility curve of e.g., NaCl. A better definition is "crystallization" curves.

The solubility is obtained by solving Eq. (15.14); it increases with the temperature (line ED).

The heat of fusion of ice in the presence of aqueous solutions of NaCl and $CaCl_2 \cdot 6H_2O$ has been calculated by Defay and Sanfeld (1959).

(2) The equilibrium between a crystalline substance i and its ions in solution is governed by the equation

$$\mu_i^{s}(T, p) = \nu_+\mu_+ + \nu_-\mu_-, \tag{15.16}$$

where μ_i^{s} is the chemical potential of the solid. For example, for the system $NaCl = Na^+ + Cl^-$, the equilibrium condition is

$$\mu_{NaCl}^{s} = \mu_{Na^+} + \mu_{Cl^-} = \mu_{Na^+}^{\circ}(T, p) + \mu_{Cl^-}^{\circ}(T, p) + RT \ln f_{Na^+}^{c} C_{Na^+} \times f_{Cl^-}^{c} C_{Cl^-}) \tag{15.17}$$

or

$$f_{Na^+}^{c} C_{Na^+} f_{Cl^-}^{c} C_{Cl^-} = \exp\left\{\frac{\mu_{NaCl}^{s} - \mu_{Na^+}^{\circ} - \mu_{Cl^-}^{\circ}}{RT}\right\}. \tag{15.18}$$

At a given temperature and pressure, the product of activities is thus a constant called the solubility product K_S. For very dilute solutions, (15.18) reduces to

$$C_{Na^+}C_{Cl^-} = K_S(T, p). \tag{15.19}$$

Since the equilibrium Nernst tension of a cell depends on ion concentrations, it is clear that the solubility product K_S can be determined by electromotive-force measurements (Petré *et al.*, 1969).

2. *Mixed Crystals*

Let us now consider the same components in a liquid and a solid phase. At constant pressure, N_2'' and N_2' can be calculated with the aid of the van Laar equations (15.6):

$$\begin{aligned} \delta\left[\ln \frac{f_1''N_1''}{f_1'N_1'}\right] &= -\frac{r^{s}_{Tp}(1)}{RT^2}\,\delta T \\ \delta\left[\ln \frac{f_2''N_2''}{f_2'N_2'}\right] &= -\frac{r^{s}_{Tp}(2)}{RT^2}\,\delta T. \end{aligned} \tag{15.20}$$

Details of calculations are given by Prigogine and Defay (1967, Chapter XXIII). For the system Cu + Ni, the theoretical and experimental curves (Seltz, 1934) are in good agreement, showing that this system is nearly ideal (Fig. 37).

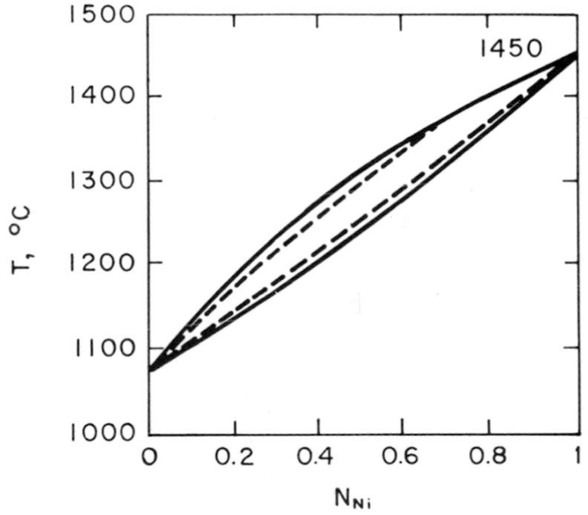

FIG. 37. Phase diagram of system Cu + Ni (———) observed; (– – – –) ideal.

On the other hand, treating the two phases as regular solutions, Scatchard and Hamer (1935) have calculated the liquidus and solidus curves. The results for the systems Ag + Pd and Au + Pt (Scatchard and Hamer, 1935; Doerinckel, 1907; Grigorjew, 1929) are shown in Figs. 38 and 39.

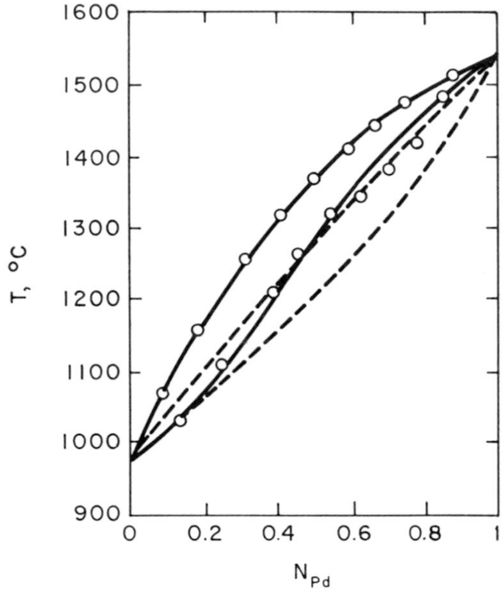

FIG. 38. Freezing-point diagram for the system Ag + Pd.

Remarks. (1) The systems Mn–Cu, Fe–V, Ni–Pd form azeotropes (see Section V).

(2) Transition from mixed crystals to addition compounds and eutectics with the phenomenon of miscibility gaps has been observed for many systems (Timmermans, 1936, p. 76).

3. *Boiling-Point Law*

When a solute species 2 may be regarded as nonvolatile, the vapor phase ($''$) contains only the component 1 (Fig. 40). We then have $N_1'' = 1$ and $f_1'' = 1$, and, at p constant, (15.6) reduces to

$$\delta(\ln f_1'N_1') = \frac{r^{\mathrm{s}}_{Tp}(1)}{RT^2}\,\delta T, \tag{15.21}$$

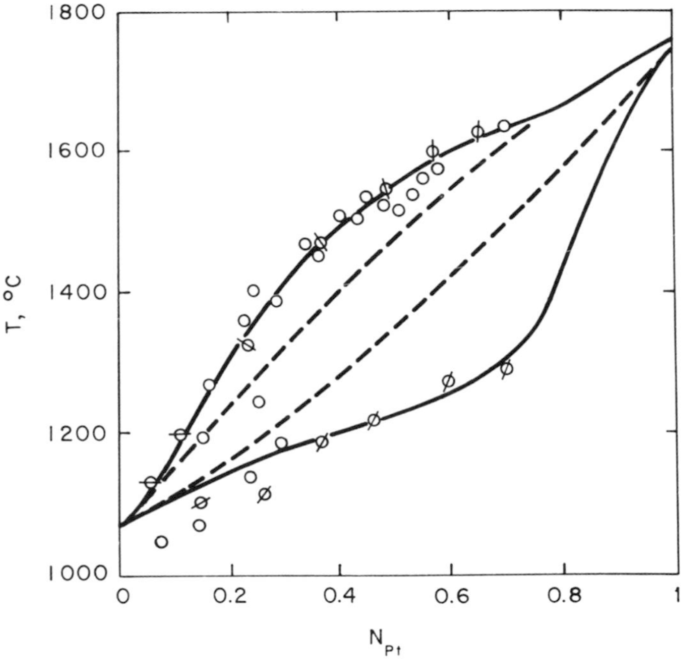

FIG. 39. Freezing-point diagram for the system Au + Pt. (———) calculated; (- - - -) ideal; (○) measured (Doerinckel); (∅) measured (Grigorjew).

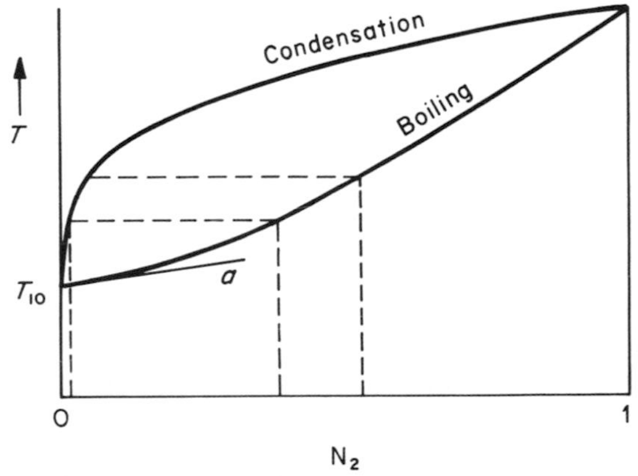

FIG. 40. Boiling-point and condensation curves for solution of nonvolatile solute.

where $r^{s}_{Tp}(T)$ has a negative value (to permit vaporization, the system must absorb an amount of heat), and thus

$$r^{s}_{Tp}(1) = -L_v(1), \tag{15.22}$$

where $L_v(1)$ is the heat of vaporization of pure component 1, boiling at the considered temperature.

Integrating (15.21), we find the equation of the boiling curve

$$\ln f_1'N_1' = \frac{L_v(1)}{R}\left[\frac{1}{T} - \frac{1}{T_{10}}\right]. \tag{15.23}$$

For a perfect solution,

$$\ln N_1' = \frac{L_v(1)}{R}\left[\frac{1}{T} - \frac{1}{T_{10}}\right], \tag{15.24}$$

and thus $T > T_{10}$, which means an elevation of boiling temperature resulting from the addition of a nonvolatile solute.

Equation (15.23) may be rewritten in molalities (see Eqs. 15.11), (15.12), and (15.13)

$$\theta = \Gamma\theta_e \sum_s m_s', \tag{15.25}$$

where the boiling point or ebullioscopic constant θ_e is defined in dilute solutions by

$$\theta_e = \frac{RT_{10}^2}{L_v(1)} \frac{M_1}{1000}. \tag{15.26}$$

In a perfect binary solution, the elevation of boiling point is proportional to the molality of the dissolved substance. The coefficient θ_e is only a function of the solvent. For water, $\theta_e = 0.51$°C; for benzene, $\theta_e = 2.53$°C.

4. *Boiling Curves for Immiscible Liquids*

For immiscible liquids (for example, hydrocarbon + water), we observe a eutectic-point vapor–liquid (Fig. 41). The only difference between this system and the liquid–solid system is that here the condensed phase is the liquid state. Two immiscible liquid components boil at the vapor–liquid eutectic-point.

Example. The boiling point of cooking fat is about 170°C. An addition of water decreases this temperature to under 100°C. Neglecting the small effect of surface tension, the pressure inside the bubble is identical to

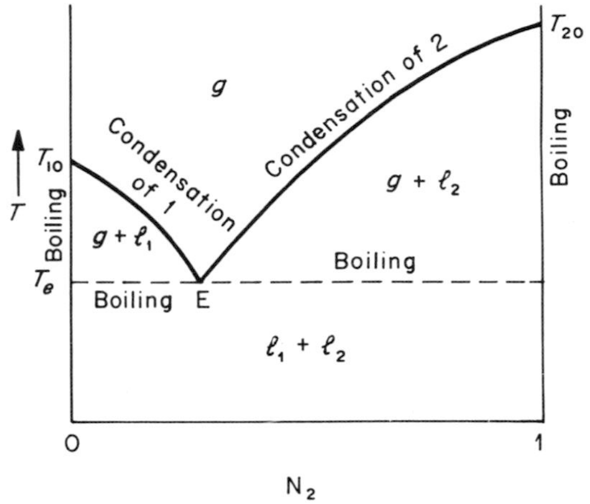

FIG. 41. Evaporation of immiscible liquids at constant pressure.

that in the liquid. If the bubble forms at the boundary of two liquids (1 and 2), the pressure in the bubble is equal to the pressure p in the liquid, but $p = p_1 + p_2$, where p_1 and p_2 are the partial vapor pressures of components 1 and 2. At atmospheric pressure, $p = p_a$ and the boiling point occurs at a temperature lower than that which should give $p_1 = p_a$ and $p_2 = p_a$. This phenomenon has an industrial application: steam distillation. Let us now calculate the boiling temperature of two liquids in contact. The evaporation of immiscible liquids at constant temperature is given in Fig. 42.

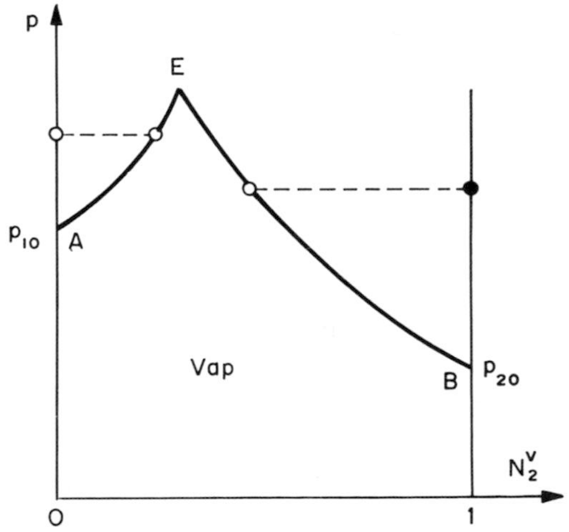

FIG. 42. Evaporation of immissible liquids at constant temperature.

Applying the van Laar relation at T constant, we obtain for the vapor phase ($''$)

$$\delta(\ln N_1'') = -\frac{\Delta^{s}_{Tp}(1)}{RT}\,\delta p, \tag{15.27}$$

where $\Delta^{s}_{Tp}(1)$ is the dilatation related to the transfer of component 1 from the liquid to the vapor. As a first approximation, let us assume that

$$\Delta^{s}_{Tp}(1) \simeq RT/p. \tag{15.28}$$

Putting (15.28) in (15.27) and integrating from $N_1'' = 1$ to N_1'', we get

$$\ln N_1'' = \ln(p_{10}/p) \tag{15.29}$$

curve AE: $\quad N_1'' = p_{10}/p$

curve BE: $\quad N_2'' = p_{20}/p.$

At the eutectic point, we have

$$N''_{1e} = p_{10}/p_e, \qquad N''_{2e} = p_{20}/p_e; \tag{15.30}$$

thus,

$$N''_{1e}/N''_{2e} = p_{10}/p_{20} \tag{15.31}$$

$$p_e = p_{10} + p_{20}. \tag{15.32}$$

The ratio of the molar fractions of the two components in the eutectic vapor phase is equal to the ratio of the vapor pressures of the pure substances. The eutectic vapor–liquid pressure at a given temperature is equal to the sum of the vapor pressures of the pure substances at this temperature (Gay-Lussac's law).

Remark. Vapor–liquid equilibria of partially miscible systems can also be found (Prigogine and Defay, 1967, p. 356).

References

Boissonnas, C. G. (1939), *Helv. Chim. Acta* **22**, 541.

Bosnjakovic, F. (1935). "Technische Thermodynamik." Steinkopff, Dresden.

Bowden, S. T. (1938). "The Phase, Rule and Phase Reactions." Macmillan, London.

Brönsted, J., and Koefeld, J. (1946). *Kgl. Danske Videnskab. Selskab Mat. Fys. Medd.* **22**, pt. 7.

Brown, G. G., *et al.* (1955). "Unit Operations." Wiley, New York.

Bryant, W. M. D. (1933). *Ind. Eng. Chem.* **25**, 820.

Clausius, R. (1850). "Theorie Mecanique de la Chaleur," Vol. I, p. 54. Cited in G. Poggendorff (1850). *Annalen der Physik und Chemie* **79**.

Colmant, P. (1954). *Bull. Soc. Chim. Belges* **63**, 13.

Counts, W. E., Roy, R., Osborn, E. F., (1953). *J. Am. Ceram. Soc.* **36**, 42.

Daniels, F., and Alberty, R. A. (1961). "Physical Chemistry." Wiley, New York.

De Donder, T. (1920). "Leçons de Thermodynamique et de Chimie Physique," p. 125. Gauthier-Villars, Paris.

De Donder, T. (1922). *Bull. Acad. Roy. Belg.* (*Cl. Sc.*) **7**, 1917, 205.

De Donder, T. (1923). "Leçons de Thermodynamique et de Chimie Physique" (F. H. van den Dungen and G. van Lerberghe, eds.), Paris.

De Donder, T. (1925). *Compt. Rend.* **180**, 1334.

De Donder, T. (1927). *l'Affinité, (Gauthier-Villars, Paris).*

De Donder, T. (1937). *Bull. Acad. Roy. Belg.* (*Cl. Sc.*) **23**, 936.

De Donder, T. (1938). *Bull. Acad. Roy. Belg.* (*Cl. Sc.*) **24**, 15.

De Donder, T. (1942). *Bull. Acad. Roy. Belg.* (*Cl. Sc.*) **28**, 496.

Defay, R. (1938). *Bull. Acad. Roy. Belg.* (*Cl. Sc.*) **24**, 347.

Defay, R., and Sanfeld, A. (1959). *Bull. Soc. Chim. Belges* **68**, 295.

Doerinckel, F. (1907). *Z. Anorg. Chem.* **54**, 345.

Duhem, P. (1899). "Traité Élémentaire de Mécanique Chimique," Vol. IV. Paris.

Duhem, P. (1911). "Traité d'Énergétique," Vol. 2. Gauthier-Villars, Paris.

Eggers, D. F., Gregory, N. W., Halsey, G. D., and Rabinovitch, B. S. (1964). "Physical Chemistry," pp. 263–265. Wiley, New York.

Eucken, A. (1934). "Grundriss der Physikalischen Chemie," 4th ed. Akademische Verlagsgesellschaft-M.B.H. Leipzig.

Ewell, R. H. (1940). *Ind. Eng. Chem.* **32**, 147.

Findlay, A., Campbell, A. N., and Smith, N. O. (1951). "Phase Rule and Its Application." Dover, New York.

Fowler, R. H., and Guggenheim, E. A. (1939). "Statistical Thermodynamics." Cambridge Univ. Press, London and New York.

Fowler, R. H., and Guggenheim, E. A. (1956). "Statistical Thermodynamics," 2nd ed. Cambridge Univ. Press, London and New York.

Gibbs, J. W. (1928). Collected work. "Thermodynamics," Vol. 1, p. 96. Longmans, Green, New York.

Glansdorff, P., and Prigogine, I. (1970). "Entropy, Stability, and Structure." Wiley, New York. To be published.

Grigorjew, A. T. (1929). *Z. Anorg. Allgem. Chem.* **178**, 97.

Guggenheim, E. A. (1935). *Proc. Roy. Soc.* **A148**, 304.

Haase, R., and Rehage, G. (1955). *Z. Elektrochem.* **59**, 994.

Hildebrand, J. H. (1929). *J. Am. Chem. Soc.* **51**, 69.

Hirschfelder, J. O., Curtiss, C. F., and Bird, R. B. (1965). "Molecular Theory of Gases and Liquids." Wiley, New York.

Hollman, A. F. (1900). *Rec. Trav. Chim.* **19**, 101.

Horsley, L. H. (1952). Azeotropic data. *Advan. Chem. Ser.* American Chemical Society.

Jouguet, E. (1921). *J. École Polytech.* (*Paris*) [2], **21**, 62.

Kamerlingh-Onnes, H. (1902). *Proc. Sci. Amsterdam* **4**, 125

Keeson, W. H. (1939). "Thermodynamische Theorie van het Rectificatieproces." Conten's, Amsterdam.

KELLEY, K. K. (1960). High temperature heat content capacity and entropy data for the elements and inorganic compounds. *U. S. Bur. Mines Bull.* **584**.

KOHMAN, G. T. (1925). *J. Phys. Chem.* **25**, 1048.

LEWIS, G. N., and RANDALL, M. (1923). "Thermodynamics." Mc Graw-Hill, New York.

MAC DOUGALL, F. H. (1926). "Thermodynamics and Chemistry." Wiley, New York.

MAYER, J. E., and MAYER, M. G. (1940). "Statistical Mechanics." Wiley, New York.

PÉTRÉ, G., STEINCHEN, A., and SANFELD, A. (1969). "Travaux Pratiques de Chimie Physique," Univ. of Brussels Press, Brussels.

PLANCK, M. (1927). "Thermodynamik," 3rd ed., English transl. London.

PLANCK, M. (1930). "Thermodynamik," 9th ed., Hirzel, Berlin and Leipzig.

POINCARE, H. (1908). "Thermodynamique," 2nd ed. Gauthier-Villars, Paris.

PONCHON, M. (1921). *Tech. Moderne* **13**, 20 and 55.

PRIGOGINE, I. (1957). "The Molecular Theory of Solutions." North-Holland Publ., Amsterdam.

PRIGOGINE, I., and DEFAY, R. (1962). "Chemical Thermodynamics," translated by D. H. Everett. Longmans, Green, New York.

PRIGOGINE, I., and DEFAY, R. (1965). "Chemical Thermodynamics," translated by D. H. Everett. Longmans, Green, New York.

PRIGOGINE, I., and DEFAY, R. (1967). "Chemical Thermodynamics," translated by D. H. Everett. Longmans, Green, New York.

PRIGOGINE, I., and MATHOT, V. (1950a). *J. Chem. Phys.* **18**, 765.

PRIGOGINE, I., and MATHOT, V. (1950b). *Bull. Soc. Chim. Belges* **52**, 111.

SANFELD, A. (1968). *In* "Introduction to Thermodynamics of Charged and Polarized Layers" (I. Prigogine, ed.), Monograph n° 10. Wiley, New York.

SAVARIT, R. (1922). *Arts et Métiers* **64**, 142, 178.

SCATCHARD, G., and HAMER, W. J. (1935). *J. Am. Chem. Soc.* **57**, 1810.

SCHOTTKY, W., ULICH, H., and WAGNER, C. (1929). "Thermodynamik," p. 492. Springer Berlin.

SELTZ, H. (1934). *J. Am. Chem. Soc.* **56**, 307.

SPENCER, H. M. (1945). Hydrocarbons and related compounds. *J. Am. Chem. Soc.* **67**, 1859, and references cited therein.

STEINCHEN, A. (1970). Thesis, Free Univ. of Brussels, Brussels.

THACKER, C. M., FOLKINS, H. O., and MILLER, E. L. (1941). *Ind. Eng. Chem.* **33**, 584.

TIMMERMANS, J. (1936). "Les Solutions Concentrées. Masson, Paris.

VAN DER WAALS, J. H. (1950). Thesis, Groningen.

VAN DER WAALS, J., and HERMANS, J. J. (1949). *Rec. Trav. Chim.* **68**, 181.

VOGEL, D. (1937). Die Heterogenen Gleichgewichte. Masing, "Handbuch des Metallphysik," Vol. II. Leipzig.

WADE, J., and MERRIMAN, R. W. (1911). *J. Chem. Soc.* **99**, 1004.

Chapter 2B

Irreversible Processes

A. SANFELD

I. Introduction

The thermodynamic properties of matter can be derived from two axioms (the two laws). However, the results of classical thermodynamics soon become too restrictive for the description of most real phenomena. Its main limitation lies in the fact that it can only describe equilibrium properties. In the classical reasoning, one has to suppose that the system goes from the initial to the final state through an infinite series of equilibrium states; it has to follow a "reversible" path.

One may, of course, prove that such a path exists, but among the possible paths connecting the initial and the final states, the actual one followed by the system is certainly not the reversible one, which is a mere fiction. The overwhelming majority of phenomena at the macroscopic scale are irreversible.

The true promoter of irreversible thermodynamics is De Donder (see De Donder and Van Rysselberghe (1936)). His main idea was that one could be able to go further than the usual statement of the second law, which is essentially an inequality (the entropy production can never be negative), and give an explicit quantitative evaluation of the entropy production.

In the meantime, Onsager (1931) established his celebrated "reciprocal relations" connecting the coefficients which occur in the linear phenomenological laws that describe irreversible processes.

These reciprocal relations reflect, on the macroscopic level, the time-reversal invariance of the microscopic equation of motion.

In 1945, Casimir reformulated the reciprocal relations so that they would be valid for a larger class of irreversible phenomena than had been previously considered by Onsager.

Meixner (1943) and Prigogine (1947) set up a consistent phenomenological theory of irreversible processes, incorporating both Onsager's reciprocity theorem and the explicit calculation, for a certain number of physical situations, of the so-called entropy source strength (which is, in fact, the uncompensated heat of Clausius).

In the last twenty years, the theory of the thermodynamics of irreversible processes has grown very rapidly. Stationary states, nonlinear thermodynamics, and order and dissipation were studied in great detail by Glansdorff and Prigogine (1970) and Prigogine (1967). New aspects arise when an electromagnetic or an electric field acts on a material system; this problem was developed by Prigogine *et al.* (1953), de Groot and Mazur (1962), Defay (1954), and Sanfeld (1968). Let us also mention an interesting approach to electrochemical kinetics made by Van Rysselberghe (1963).

In this short survey, we shall give a brief summary of nonequilibrium thermodynamics close to equilibrium.

II. Conservation of Energy in Open Systems

In the absence of an external field and of macroscopic kinetic energy, the principle of conservation of energy is (see Chapter 1)

$$dU = dQ - p\,dV. \tag{2.1}$$

When considering an open system, formula (2.1) must be replaced by

$$dU = d\phi - p\,dV, \tag{2.2}$$

where $d\phi$ is the flow of energy or, more exactly, the enthalpy received by the system during time dt, due to heat transfer and exchange of matter. From Eq. (2.2), we can deduce the change of total enthalpy

$$dH = d\phi + V\,dp. \tag{2.3}$$

Let us note that, for open systems, the quantity $p\,dV$ is not necessarily a real work.

It is generally stated that energy or enthalpy is completely defined, apart from an arbitrary additive constant. Hence, instead of H, we may also write

$$H' = H + \alpha, \tag{2.4}$$

where α is the additive constant. However, for a homogeneous system, the extensive variable H is proportional to the mass m of the system and, of course, the same is true for H'. Thus, α is also proportional to m and is no longer a constant for an open system, and we can write

$$h_m' = h_m + \beta, \tag{2.5}$$

where

$$h_m = H/m, \qquad h_m' = H'/m \tag{2.6}$$

where β is now a "real" constant.

III. Equality of Exchange Flow of Energy

We consider two systems I and II separated by a real or conceptual surface Ω (Fig. 1). The resultant flow $d^{\mathrm{I}}\phi$ received by system I during the time interval dt is equal to the energy flow through Ω, $d_{\mathrm{i}}{}^{\mathrm{I}}\phi$, plus the energy flow $d_{\mathrm{e}}{}^{\mathrm{I}}\phi$ from the exterior. If p is uniform, (2.2) reads

$$d_{\mathrm{i}}{}^{\mathrm{I}}\phi + d_{\mathrm{e}}{}^{\mathrm{I}}\phi = dU^{\mathrm{I}} + p\,dV^{\mathrm{I}}. \tag{3.1}$$

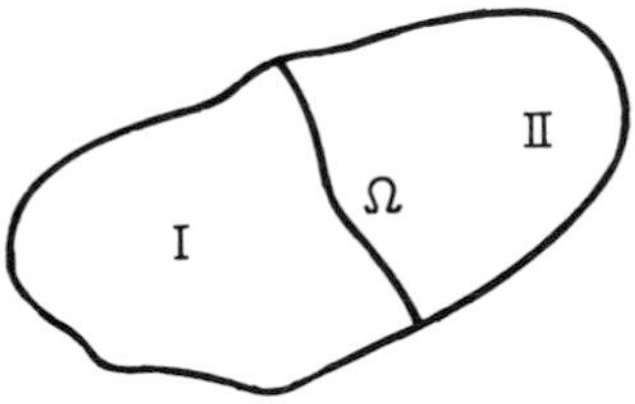

FIG. 1.

Similarly, we obtain for system II

$$d_i^{II}\phi + d_e^{II}\phi = dU^{II} + p\,dV^{II}. \tag{3.2}$$

For the whole system (I + II), we have

$$d_e^{I}\phi + d_e^{II}\phi = d(U^{I} + U^{II}) + p(dV^{I} + dV^{II}). \tag{3.3}$$

Examining (3.1)–(3.3), we find that

$$d_i^{II}\phi = -d_i^{I}\phi. \tag{3.4}$$

Relation (3.4) says that the energy flow received by the system I from the system II is equal and of opposite sign to that received by the system II from the system I.

If the two systems I and II are separated by a porous membrane and if the pressure in these two systems is, respectively, p^{I} and p^{II}, Eq. (3.3) becomes

$$d_e^{I}\phi + d_e^{II}\phi = d(U^{I} + U^{II}) + p^{I}\,dV^{I} + p^{II}\,dV^{II}. \tag{3.5}$$

For such a system, the relation (3.4) can be demonstrated in the same way as before. From equation (2.3), we may write the energy flow in the form

$$d\phi = \frac{\partial H}{\partial T}\,dT + \left(\frac{\partial H}{\partial p} - V\right)dp + \sum_{\gamma} \frac{\partial H}{\partial n_\gamma}\,dn_\gamma. \tag{3.6}$$

Let us now split dn_γ into two parts: $d_i n_\gamma$, due to the chemical reactions in the system, and $d_e n_\gamma$ coming from the exterior. We then may separate $d\phi$ into two flows, respectively called the thermal flow (th) and the convective-diffusive flow (c.d.), defined by

$$d\phi_{\text{th}} = \frac{\partial H}{\partial T}\,dT + \left(\frac{\partial H}{\partial p} - V\right)dp + \sum_{\gamma} \frac{\partial H}{\partial n_\gamma}\,d_i n_\gamma \tag{3.7}$$

$$d\phi_{\text{c.d.}} = \sum_{\gamma} h_\gamma\,d_e n_\gamma. \tag{3.8}$$

Applying such a decomposition to the systems I and II, it can easily be shown that the thermal flows and the convective-diffusive flows are not always conservative through the surface Ω.

IV. Entropy Production in Open Systems

The second law of thermodynamics asserts that, inside any system (open or closed), entropy can be created but never destroyed, i.e.,

$$d_iS \geq 0. \tag{4.1}$$

For an open system, the entropy balance is

$$dS = d_eS + d_iS, \tag{4.2}$$

where d_eS is the entropy received by the system and d_iS the entropy created inside the system. In closed systems, d_eS is equal to dQ/T. In open systems, we have to take into account that, when matter enters the system, it transports its entropy. If the system is nonuniform in temperature, we may divide it in small volumes each having a well-defined temperature, so that

$$S = \int_v s_v \, \delta V, \tag{4.3}$$

where s_v is the entropy per unit volume in each volume element. More explicitely, s_v is the entropy per unit volume of a uniform system with the same temperature and concentrations as those existing in the volume element of the nonuniform system. Since condition (4.1) has to be realized inside any system, it also prevails in each volume element of the nonuniform system. If the only irreversible process taking place in the system is a chemical reaction, the entropy production per unit time and unit volume, σ_{chem}, is given by

$$\sigma_{\text{chem}} = AV_v/T \geq 0, \tag{4.4}$$

where V_v is the rate of the reaction per unit volume. In a uniform system, $\sigma_s = (1/V)\, d_iS/dt$, while, in a nonuniform system, [see Eq. (4.3)], $\sigma_s = d_is_v/dt$ (σ_s is the entropy source).

If there are several chemical reactions ($\varrho = 1, 2, \ldots, c$) in the volume element, (4.4) reads

$$\sigma_{\text{chem}} = \sum_{\varrho} (A_\varrho V_{v\varrho}/T) \geq 0, \tag{4.5}$$

where the sum is extended to all the reactions. As the summation over all the $A_\varrho V_{v\varrho}$ has only to be positive, one of them, for example, $A_1 V_{v_1}$, might be negative if the summation over the others is large enough to realize the condition (4.5). The reaction 1 is said then to be coupled with the others.

Similarly, we can calculate the entropy production related to the diffusion and to the heat flow. Coupling may also occur between diffusion and thermal flow and also among the diffusion flows of the different components among themselves. The viscosity also gives rise to an entropy production in each volume element (Rayleigh's dissipation function).

Let us now consider a system consisting of two homogeneous and closed parts I and II, brought to the uniform temperatures T^{I} and T^{II} and contiguous to one another along the surface Ω (Fig. 2). We have

$$dS = dS^{\mathrm{I}} + dS^{\mathrm{II}} = \frac{d^{\mathrm{I}}Q}{T^{\mathrm{I}}} + \frac{d^{\mathrm{II}}Q}{T^{\mathrm{II}}}. \tag{4.6}$$

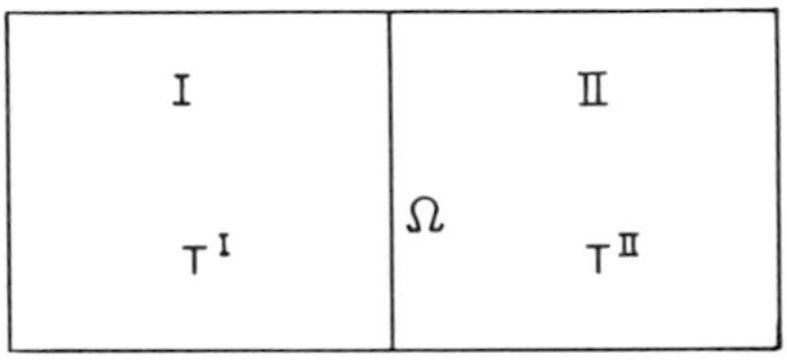

FIG. 2.

Each part receives an amount of heat given by

$$d^{\mathrm{I}}Q = d_{\mathrm{i}}{}^{\mathrm{I}}Q + d_{\mathrm{e}}{}^{\mathrm{I}}Q \tag{4.7}$$

and

$$d^{\mathrm{II}}Q = d_{\mathrm{i}}^{\mathrm{II}}Q + d_{\mathrm{e}}^{\mathrm{II}}Q, \tag{4.8}$$

where $d_{\mathrm{i}}{}^{\mathrm{I}}Q$ is the heat received by I from II and $d_{\mathrm{e}}{}^{\mathrm{I}}Q$ the heat received by I from the exterior of the whole system. Since

$$d_{\mathrm{i}}^{\mathrm{II}}Q = -d_{\mathrm{i}}{}^{\mathrm{I}}Q, \tag{4.9}$$

we have, combining (4.6)–(4.9),

$$dS = \frac{d_{\mathrm{e}}{}^{\mathrm{I}}Q}{T^{\mathrm{I}}} + \frac{d_{\mathrm{e}}^{\mathrm{II}}Q}{T^{\mathrm{II}}} + d_{\mathrm{i}}^{\mathrm{II}}Q\left[\frac{1}{T^{\mathrm{II}}} - \frac{1}{T^{\mathrm{I}}}\right] = d_{\mathrm{e}}S + d_{\mathrm{i}}S. \tag{4.10}$$

For the whole system, we thus have, according to the second law,

$$d_{\mathrm{i}}S = d_{\mathrm{i}}^{\mathrm{II}}Q\left[\frac{1}{T^{\mathrm{II}}} - \frac{1}{T^{\mathrm{I}}}\right] \geq 0, \tag{4.11}$$

which means that the heat flows from I to II when $T^{\mathrm{I}} > T^{\mathrm{II}}$. The entropy production vanishes when $T^{\mathrm{I}} = T^{\mathrm{II}}$, i.e., at the thermal equilibrium.

In a continuous system, (4.1) reads

$$\sigma_{\mathrm{th}} = \mathbf{W} \cdot \mathrm{grad}(1/T), \tag{4.12}$$

where $\mathbf{W}$ is the amount of heat which crosses, per unit time, a unit surface perpendicular to the lines of heat current. Equation (4.12) may be written in the form

$$\sigma_{\mathrm{th}} = -\left(\frac{1}{T^2}\,\mathrm{grad}\,T\right) \cdot \mathbf{W} \tag{4.13}$$

or

$$T\sigma_{\mathrm{th}} = \left(-\frac{\mathrm{grad}\,T}{T}\right) \cdot \mathbf{W}. \tag{4.14}$$

By analogy with the chemical reaction, we may call the term $-(\mathrm{grad}\,T/T)$ the thermal affinity.

As already seen, the central concept in irreversible thermodynamics is entropy production. Therefore, the entropy balance plays a fundamental role. For a uniform open system, we have already stated the fundamental Gibbs formula

$$dS = \frac{1}{T}\,dU + \frac{p}{T}\,dV - \sum_{\gamma} \frac{\mu_\gamma}{T}\,dn_\gamma. \tag{4.15}$$

After splitting dn_γ into $d_{\mathrm{e}}n_\gamma$ and $d_{\mathrm{i}}n_\gamma$, with

$$d_{\mathrm{i}}n_\gamma = \nu_\gamma\,d\xi \tag{4.16}$$

and, since

$$A = -\sum_{\gamma} \nu_\gamma \mu_\gamma, \tag{4.17}$$

Eq. (4.15) becomes

$$dS = \frac{dU}{T} + \frac{p}{T}\,dV - \sum_{\gamma} \frac{\mu_\gamma}{T}\,d_{\mathrm{e}}n_\gamma + \frac{A\,d\xi}{T}. \tag{4.18}$$

Using the energy balance of the open system [Eq. (2.2)], we get

$$dS = \underbrace{\frac{d\phi}{T} - \sum_{\gamma} \frac{\mu_\gamma}{T}\,d_{\mathrm{e}}n_\gamma}_{d_{\mathrm{e}}S} + \underbrace{\frac{A\,d\xi}{T}}_{d_{\mathrm{i}}S}. \tag{4.19}$$

We see that the entropy received by the open system is related to the energy flow and to the amount of matter which enters the system. Let us now define

$$dJ_{\text{th}} = d\phi - \sum_{\gamma} h_{\gamma}\, d_{\text{e}}n_{\gamma}; \tag{4.20}$$

With Eq. (4.20), (4.19) reads

$$dS = \frac{dJ_{\text{th}}}{T} + \sum s_{\gamma}\, d_{\text{e}}n_{\gamma} + \frac{A\, d\xi}{T}. \tag{4.21}$$

In the second term of the r.h.s. of (4.21), the entropy brought by the matter entering the system appears explicitly.

Let us now consider a system, closed as a whole, consisting of two parts I and II with pressure and temperature respectively p^{I}, T^{I} and p^{II}, T^{II} (Fig. 3). Since $d_{\text{i}}^{\text{II}}\phi = -d_{\text{i}}{}^{\text{I}}\phi$ and $d_{\text{e}}^{\text{II}}n_{\gamma} = -d_{\text{e}}{}^{\text{I}}n_{\gamma}$, where the

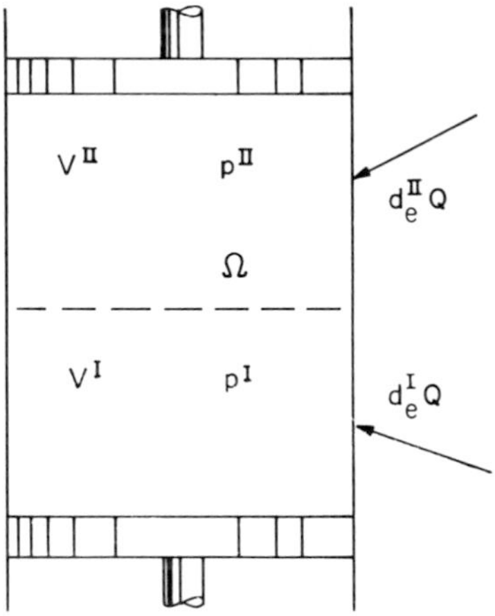

Fig. 3.

index e relates to external quantities for the considered phase, the variation of entropy of the whole system is given by

$$dS = \frac{d_{\text{e}}{}^{\text{I}}Q}{T^{\text{I}}} + \frac{d_{\text{e}}^{\text{II}}Q}{T^{\text{II}}} + \left(\frac{1}{T^{\text{II}}} - \frac{1}{T^{\text{I}}}\right) d_{\text{i}}^{\text{II}}\phi - \sum_{\gamma} \left(\frac{\mu_{\gamma}^{\text{II}}}{T^{\text{II}}} - \frac{\mu_{\gamma}{}^{\text{I}}}{T^{\text{I}}}\right) d_{\text{e}}n_{\gamma}^{\text{II}}$$
$$+ \frac{A^{\text{I}}\, d\xi^{\text{I}}}{T^{\text{I}}} + \frac{A^{\text{II}}\, d\xi^{\text{II}}}{T^{\text{II}}}, \tag{4.22}$$

where

$$d_{\text{e}}S = \frac{d_{\text{e}}{}^{\text{I}}Q}{T^{\text{I}}} + \frac{d_{\text{e}}^{\text{II}}Q}{T^{\text{II}}} \tag{4.23}$$

and

$$d_i S = \left(\frac{1}{T^{\rm II}} - \frac{1}{T^{\rm I}}\right) d_i^{\rm II}\phi + \sum_\gamma \left(\frac{\mu_\gamma{}^{\rm I}}{T^{\rm I}} - \frac{\mu_\gamma^{\rm II}}{T^{\rm II}}\right) d_e n_\gamma^{\rm II} + \frac{A^{\rm I}\, d\xi^{\rm I}}{T^{\rm I}} + \frac{A^{\rm II}\, d\xi^{\rm II}}{T^{\rm II}} \geq 0. \tag{4.24}$$

The two last terms of (4.24) are the classical terms due to the chemical reactions, while the two first terms of the r.h.s. of (4.24) are due to the energy flow and to the transfer of matter from I to II. Comparing the first term of the r.h.s. of (4.24) to the term which gives the entropy creation due to the heat flow between the two closed systems in (4.10), we see that the flow received $d_i^{\rm II}\phi$ replaces in (4.24) the heat received $d_i^{\rm II}Q$ in the closed system (4.11). The term $\Sigma_\gamma[(\mu_\gamma{}^{\rm I}/T^{\rm I}) - (\mu_\gamma^{\rm II}/T^{\rm II})]$ generalizes the concept of diffusion affinity to a system where the temperature is different in the two regions I and II.

V. Continuous Systems

When the system is inhomogeneous, we define at each point of the system intensive quantities θ_v related to the corresponding extensive quantities Θ by the relation

$$\theta_v = \delta\Theta/\delta V. \tag{5.1}$$

The densities θ_v of Θ defined by (5.1) are related to the quantities per unit mass by the equation

$$\theta_v = \frac{\delta\Theta}{\delta V} = \frac{\delta\Theta}{\delta m}\,\frac{\delta m}{\delta V} = \theta_m \varrho, \tag{5.2}$$

where $\varrho = \delta m/\delta V$ is the density, in mass per unit volume.

Let us now define the barycentric velocity of the fluid contained in a volume element δV of the system by the relation

$$\boldsymbol{\omega}\varrho = \sum_\gamma \varrho_\gamma \boldsymbol{\omega}_\gamma, \tag{5.3}$$

where ϱ_γ is the mass of the γ component in the unit volume and $\boldsymbol{\omega}_\gamma$ the velocity of the component γ. If the velocities of the different components are equal to the barycentric velocity, we say that there is no diffusion. When the velocities $\boldsymbol{\omega}_\gamma$ are different from $\boldsymbol{\omega}$, we define the diffusion velocity of the γ component by the relation

$$\boldsymbol{\Delta}_\gamma = \boldsymbol{\omega}_\gamma - \boldsymbol{\omega}; \tag{5.4}$$

since $\varrho = \Sigma_\gamma\, \varrho_\gamma$, (5.3) and (5.4) give the relation between the diffusion velocities

$$\sum_\gamma \varrho_\gamma \boldsymbol{\Delta}_\gamma = 0. \tag{5.5}$$

Let us now define the operator d/dt as the variation with time along the barycentric movement, which is related to the operator $\partial/\partial t$, the local variation with time, through

$$\frac{d}{dt} = \frac{\partial}{\partial t} + \boldsymbol{\omega} \cdot \text{grad}. \tag{5.6}$$

We may write the local variation with time of each quantity θ_v defined by (5.1) in the form of a local balance, as was first done by Prigogine (1947). We get then

$$\partial\theta_v/\partial t = -\text{div}\, \Phi[\Theta] + \sigma[\Theta], \tag{5.7}$$

where $\Phi[\Theta]$ is the flow per unit surface associated with Θ and $\sigma[\Theta]$ is the source of Θ. For example, the mass balance reads

$$\partial\varrho/\partial t = -\text{div}\, \varrho\boldsymbol{\omega}; \tag{5.8}$$

since the total mass is conservative, the source term is zero; this equation is identical to the continuity equation in hydrodynamics. As the total mass of the γ component is not conservative, because of the chemical reactions which may occur in the system, the local balance for ϱ_γ has the form

$$\begin{aligned}\partial\varrho_\gamma/\partial t &= -\text{div}\, \varrho_\gamma\boldsymbol{\omega}_\gamma + \sum_\varrho \nu_{\gamma\varrho} M_\gamma V_{v\varrho} = -\text{div}\, \varrho_\gamma\boldsymbol{\omega} - \text{div}\, \varrho_\gamma\boldsymbol{\Delta}_\gamma \\ &\quad + \sum_\varrho \nu_{\gamma\varrho} M_\gamma V_{v\varrho},\end{aligned} \tag{5.9}$$

where $V_{v\varrho}$ is the rate of the ϱth reaction per unit volume, $\nu_{\gamma\varrho}$ the stoichiometric coefficient of γ in the ϱth, reaction and M_γ the molar mass of γ. We can also calculate a balance equation for the mass fraction

$$N_\gamma = \varrho_\gamma/\varrho. \tag{5.10}$$

Using (5.6) and (5.8)–(5.10), we obtain the balance equation for N_γ

$$\varrho\, dN_\gamma/dt = -\text{div}\, \varrho_\gamma \mathbf{\Delta}_\gamma + \sum_\varrho \nu_{\gamma\varrho} M_\gamma V_{v\varrho}. \tag{5.11}$$

Similarly, we may relate the local variation with time of any quantity θ_v per unit volume, to the variation along the barycentric motion of the quantity $\theta_m = \theta_v/\varrho$ per unit mass. Combining (5.6) and (5.8), it is easy to see that

$$\varrho\, d\theta_m/dt = \partial\theta_v/\partial t + \text{div}\, \boldsymbol{\omega}\theta_v. \tag{5.12}$$

In the same way, the equation of motion of the fluid can be regarded as the balance of the local velocity of the center of mass; indeed, for a non-viscous fluid, it has the form

$$\varrho\, d\boldsymbol{\omega}/dt = \sum_\varrho \varrho_\gamma \mathbf{F}_\gamma - \text{grad}\, p, \tag{5.13}$$

where $\mathbf{F}_\gamma$ is the external force per unit mass acting on component γ. From the motion equation (5.13), it is easy to derive the balance of the kinetic energy per unit mass $l_m = \omega^2/2$:

$$\frac{dl_m}{dt} = \mathbf{F} \cdot \boldsymbol{\omega} - \frac{\boldsymbol{\omega} \cdot \text{grad}\, p}{\varrho}. \tag{5.14}$$

If the volume element δV moves in an external field, its potential balance is given by

$$dO_m/dt = -\mathbf{F} \cdot \boldsymbol{\omega}, \tag{5.15}$$

where O_m is the potential energy of the unit mass and $\mathbf{F} \cdot \boldsymbol{\omega}$ the work performed by the force $\mathbf{F}$.

If there is no diffusion, one may consider the mass element as a closed system and write its internal energy balance in the simple form

$$\frac{du_m}{dt} = -\frac{1}{\varrho}\, \text{div}\, \mathbf{W} - p\frac{dv}{dt}, \tag{5.16}$$

where u_m is the internal energy per unit mass, $v = 1/\varrho$ is the volume of the unit mass, and $\mathbf{W}$ the heat flow per unit surface. The heat flow is related to the quantity of heat dq/dt received by the mass element along its motion by the equation

$$dq/dt = -(1/\varrho)\, \text{div}\, \mathbf{W}. \tag{5.17}$$

The first term of the r.h.s. of (5.16) is thus the quantity of heat received by the mass element, while the second one is the mechanical work received by the mass element. The balance of the total energy $\theta_m = l_m + u_m + O_m$, thus reads

$$\frac{d\theta_m}{dt} = -\frac{1}{\varrho}\,\mathrm{div}\,\mathbf{W} - p\,\frac{d(1/\varrho)}{dt} - \frac{\boldsymbol{\omega}\cdot\mathrm{grad}\,p}{\varrho} \tag{5.18}$$

or

$$\frac{\partial\theta_v}{\partial t} = -\mathrm{div}(\boldsymbol{\omega}\theta_v + \mathbf{W} + p\boldsymbol{\omega}), \tag{5.19}$$

a relation which shows that there is no source of total energy. The total energy is conservative. When there is diffusion in the system, the balance of internal energy takes the form

$$\frac{du_m}{dt} = -\frac{1}{\varrho}\,\mathrm{div}\,\mathbf{W} - p\,\frac{dv}{dt} + \frac{1}{\varrho}\sum_\gamma \varrho_\gamma \mathbf{F}_\gamma\cdot\boldsymbol{\Delta}_\gamma, \tag{5.20}$$

where $\mathbf{F}_\gamma$ is the force acting on the γ component. The total energy balance, however, remains the same as for the system without diffusion.

We already saw that the variation of entropy of a homogeneous system was given by the Gibbs equation

$$T\,dS = dU + p\,dV - \sum_\gamma \mu_\gamma\,dm_\gamma. \tag{5.21}$$

Reducing this expression to unit mass, we get

$$T\,ds_m = du_m + p\,dv - \sum_\gamma \mu_\gamma\,dN_\gamma, \tag{5.22}$$

where the chemical potential μ_γ is locally defined by

$$\mu_\gamma = \left(\frac{\partial s_m}{\partial N_\gamma}\right)_{u_m v N_\beta}. \tag{5.23}$$

The basic assumption of irreversible thermodynamics is that Eq. (5.22) still holds in a certain range out of equilibrium, and that one may write it along the barycentric motion of the mass element as

$$T\frac{ds_m}{dt} = \frac{du_m}{dt} + p\,\frac{dv}{dt} - \sum_\gamma \mu_\gamma\frac{dN_\gamma}{dt}. \tag{5.24}$$

This assumption can be verified by statistical mechanics over a relatively wide range of nonequilibrium situations. This means that even in a situation globally out of equilibrium, a local equilibrium is still realized, corresponding to the Maxwellian distribution of the velocities.

Using the transformation (5.7), the mass balance of (5.11), and the internal energy balance (5.16), we may write (5.24) in the local form of a balance, i.e.,

$$\frac{\partial s_v}{\partial t} = -\operatorname{div}\left(\boldsymbol{\omega} s_v + \frac{\mathbf{W}}{T} - \sum_\gamma \varrho_\gamma \boldsymbol{\Delta}_\gamma \frac{\mu_\gamma}{T}\right) + \sum_\varrho \frac{A_\varrho V_{v\varrho}}{T} + \mathbf{W} \cdot \operatorname{grad} \frac{1}{T}$$
$$+ \frac{1}{T} \sum_\gamma \varrho_\gamma \boldsymbol{\Delta}_\gamma \cdot \left(\mathbf{F}_\gamma - T \operatorname{grad} \frac{\mu_\gamma}{T}\right), \tag{5.25}$$

where the flow of entropy J_s is given by

$$J_s = \boldsymbol{\omega} s_v + \frac{\mathbf{W}}{T} - \sum_\gamma \varrho_\gamma \boldsymbol{\Delta}_\gamma \frac{\mu_\gamma}{T}, \tag{5.26}$$

which means that the volume element may receive entropy from the exterior (1) because of the matter which enters the system with the velocity $\boldsymbol{\omega}$, (2) because of the heat flow, and (3) because of the entropy brought by the diffusion of each component with the velocity $\boldsymbol{\Delta}_\gamma$. Now, the entropy source σ_s, which is the key term of irreversible thermodynamics, thus reads

$$\sigma_s = \frac{1}{T} \sum_\varrho A_\varrho V_{v\varrho} + \mathbf{W} \cdot \operatorname{grad} \frac{1}{T} + \frac{1}{T} \sum_\gamma \varrho_\gamma \boldsymbol{\Delta}_\gamma \cdot \left(\mathbf{F}_\gamma - T \operatorname{grad} \frac{\mu_\gamma}{T}\right). \tag{5.27}$$

Equation (5.27) may be written

$$T\sigma_s = \sum_\varrho A_\varrho V_{v\varrho} - \mathbf{W} \cdot \frac{\operatorname{grad} T}{T} + \sum_\gamma \varrho_\gamma \boldsymbol{\Delta}_\gamma \cdot \left(\mathbf{F}_\gamma - T \operatorname{grad} \frac{\mu_\gamma}{T}\right), \tag{5.28}$$

where $-(\operatorname{grad} T)/T$, A_ϱ, and $[\mathbf{F}_\gamma - T \operatorname{grad}(\mu_\gamma/T)]$ are generalized forces and $\mathbf{W}$, $V_{v\varrho}$, and $\varrho_\gamma \boldsymbol{\Delta}_\gamma$ are the flows conjugate to these forces. The entropy production is thus a bilinear expression in the forces and the flows caused by them. The term $(1/T) \sum_\varrho A_\varrho V_{v\varrho}$ is the creation of entropy caused by the chemical reactions, $(\mathbf{W}/T)(\operatorname{grad} T)/T$ is the entropy creation due to the heat or to the energy flow, and $(1/T) \sum_\gamma \varrho_\gamma \boldsymbol{\Delta}_\gamma [\mathbf{F}_\gamma - T \times \operatorname{grad}(\mu_\gamma/T)]$ is the entropy created by the diffusion flows. The second law of thermodynamics requires that

$$\sigma_s \geq 0, \tag{5.29}$$

but it does not require that each term of the sum be positive. There may be coupling of irreversible processes in such a way that, while certain terms of the sum may be negative, the corresponding phenomena may yet occur, being coupled with one another. Let us denote the affinities or generalized forces by X_i and the flows by J_i. Equation (5.28) then takes the simple form

$$T\sigma_s = \sum_i J_i X_i \geq 0. \tag{5.30}$$

In the neighborhood of the thermodynamic equilibrium, where the forces X_i are small, we may expand the flows in a Taylor series around zero, as we know that flows and forces are zero at the equilibrium, and keep only the linear terms. We then postulate a linear relation between the flows and forces

$$J_i = \sum_{k=1}^{n} L_{ik} X_k, \tag{5.31}$$

where the sum is taken over all (n) irreversible processes occuring in the system.

Onsager's reciprocity theorem asserts that, in the range of validity of the linear approximation, the relationship between forces and fluxes is not arbitrary. The phenomenological coefficients L_{ik} must satisfy the symmetry condition

$$L_{ik} = L_{ki}. \tag{5.32}$$

The general proof of Onsager's rule is based upon the principle of microscopic reversibility and upon the treatment by statistical mechanics of fluctuations and of their decay (Prigogine, 1967). We cannot know *a priori* if all coefficients L_{ik} really exist. The existence of an effective coupling between two irreversible processes can only be established from experiments. If it is shown experimentally that a coupling exists between a given force X_k and a flow J_i, Onsager's relations require that the converse coupling also exists. Another criterion permits one to reduce the number of effective couplings: it is a consequence of the Curie symmetry principle, which states that a macroscopic cause never has more symmetry elements than the effect it produces. For instance, a chemical affinity (which is a scalar) can never produce a heat flow (which is a vector); the corresponding coupling coefficient thus vanishes. Near equilibrium, we state, then, the following theorem: "A coupling is possible only between phenomena having the same tensorial symmetry."

However, the symmetry requirements on the coupling of irreversible processes are no longer valid in the so-called nonlinear region [out of the range of validity of (5.31)], in which symmetry-breaking processes are possible (Glansdorff and Prigogine, 1970). We shall restrict our development to the linear region.

In accord with the experimental discoveries of Fourier, Fick, Ohm, and others, each flow is proportional to its conjugate force, the "straight" coefficient being L_{ii}. All the straight coefficients L_{ii} appear on the diagonal of the matrix of forces of the set of equations (5.31). However, the equations also state that the flow J_i may be driven by forces X_k if the "coupling coefficients" or "cross coefficients" L_{ik} $(i \neq k)$ differ from zero.

The linear law holds only for sufficiently slow processes occurring when the system is not too distant from a state of equilibrium.

The range of phenomena covered by the phenomenological equations is very wide and many processes of physical, chemical, biochemical, and physiological interest are included (Katchalski and Curran, 1965).

To illustrate equations (5.30) and (5.31), let us consider the case of two simultaneous irreversible processes, for which two phenomenological relations may be written down:

$$J_1 = L_{11}X_1 + L_{12}X_2, \qquad J_2 = L_{21}X_1 + L_{22}X_2. \tag{5.33}$$

If the two irreversible processes represent thermal conductivity and diffusion, the coefficient L_{ik} is connected with thermodiffusion, that is, with the appearance of a concentration gradient in an initially homogeneous mixture under the influence of a temperature gradient.

The corresponding entropy production (5.30) then becomes

$$\sigma_s = T^{-1}[L_{11}X_1^2 + (L_{12} + L_{21})X_1X_2 + L_{22}X_2^2] > 0, \tag{5.34}$$

and thus

$$L_{11} > 0, \qquad L_{22} > 0, \tag{5.35}$$

$$(L_{12} + L_{21})^2 < 4L_{11}L_{22}. \tag{5.36}$$

Hence, the straight coefficients are positive, but the cross-coefficients L_{12} and L_{21} may be positive or negative, their magnitude being limited only by (5.36). This is in agreement with the empirical observation that coefficients like thermal conductivity or electrical conductivity are always positive while, for example, the thermodiffusion coefficient has no definite sign.

VI. Transport and Chemical Equations

In the linear region, we may write, from (5.27), two entropy productions defined by

$$\sigma_{\text{transp}} = \mathbf{W} \cdot \text{grad} \frac{1}{T} + \frac{1}{T} \sum_{\gamma} \varrho_\gamma \mathbf{\Delta}_\gamma \cdot \left(\mathbf{F}_\gamma - T \,\text{grad} \frac{\mu_\gamma}{T} \right) \geq 0 \quad (6.1)$$

$$\sigma_{\text{chem}} = \frac{1}{T} \sum_{\varrho} A_\varrho V_{v\varrho} \geq 0. \quad (6.2)$$

We note in (6.1) that the force factor corresponding to the diffusion flow $\varrho_\gamma \mathbf{\Delta}_\gamma$ contains the quantity $\text{grad}(\mu_\gamma/T)$ which, being such that

$$\text{grad}(\mu_\gamma/T) = (1/T)\,\text{grad}\,\mu_\gamma - (\mu_\gamma/T^2)\,\text{grad}\,T, \quad (6.3)$$

is undefined to the extent of a term proportional to grad T on account of the undetermined constant contained in μ_γ. If, however, we subtract from $\mathbf{W}$ the sum $\sum_\gamma \varrho_\gamma \mathbf{\Delta}_\gamma h_\gamma$ and add the corresponding entropy production term to that due to the diffusion flows, we have

$$\mathbf{J}_{\text{th}} = \mathbf{W} - \sum_{\gamma} \varrho_\gamma \mathbf{\Delta}_\gamma h_\gamma, \quad (6.4)$$

and, in place of (6.1),

$$\sigma_{\text{transp}} = \mathbf{J}_{\text{th}} \cdot \text{grad} \frac{1}{T} + \sum_{\gamma} \varrho_\gamma \mathbf{\Delta}_\gamma \cdot \left[\frac{\mathbf{F}_\gamma}{T} + h_\gamma \,\text{grad}\left(\frac{1}{T}\right) - \text{grad} \frac{\mu_\gamma}{T} \right] \geq 0. \quad (6.5)$$

The new flux $\mathbf{J}_{\text{th}}$ is called the reduced heat flow. If we write

$$(\text{grad}\,\mu_\gamma)_T = v_\gamma \,\text{grad}\, p + \sum_{\beta=2}^{c} \left(\frac{\partial \mu_\gamma}{\partial N_\beta} \right)_{Tp} \text{grad}\, N_\beta, \quad (6.6)$$

the force factor corresponding to $\varrho_\gamma \mathbf{\Delta}_\gamma$ in (6.5) becomes

$$\mathbf{X}_\gamma = (1/T)[\mathbf{F}_\gamma - (\text{grad}\,\mu_\gamma)_T], \quad (6.7)$$

which is free of any undetermined term proportional to grad T, and (6.5) can be rewritten as

$$\sigma_{\text{transp}} = \left(\mathbf{J}_{\text{th}}\, \text{grad} \frac{1}{T} \right) + \frac{1}{T} \sum_{\gamma} \varrho_\gamma \mathbf{\Delta}_\gamma [\mathbf{F}_\gamma - (\text{grad}\,\mu_\gamma)_T] \geq 0. \quad (6.8)$$

We call the diffusion affinity of constituent γ the vectorial quantity

$$\mathbf{A}_\gamma{}^{\text{d}} = \mathbf{F}_\gamma - (\text{grad}\,\mu_\gamma)_T. \quad (6.9)$$

Instead of (5.30), it is easy to show (Prigogine, 1967) that we may introduce a new set of affinities X_i' which are linear combinations of the old ones and choose a new set of rates J_i' in such a way that the entropy production is invariant

$$\sum_i J_i X_i = \sum_i J_i' X_i'. \tag{6.10}$$

The entropy production due to ϱ chemical reactions may be written as

$$\sigma_{\text{chem}} = \sum_\varrho X_\varrho J_\varrho \qquad \text{with} \qquad J_\varrho = (V_v)_\varrho; \quad X_\varrho = A_\varrho/T. \tag{6.11}$$

For example, let us consider the well-known synthesis of HI

$$H_2 + I_2 = 2HI.$$

The affinity A and the Guldberg–Waage constant K_c are related by the relation [see Chapter IIA, Eq. (6.28)]

$$A = RT \ln[K_c(T)/C_{H_2}^{-1} C_{I_2}^{-1} C_{HI}^2]. \tag{6.12}$$

The usual kinetic expression for the reaction rate is given by

$$V = \vec{V} - \overleftarrow{V} = \vec{k} C_{I_2} C_{H_2} - \overleftarrow{k} C_{HI}^2 = \vec{k} C_{H_2} C_{I_2} [1 - (\overleftarrow{k}/\vec{k})(C_{HI}^2/C_{I_2} C_{H_2})], \tag{6.13}$$

where the coefficients $\vec{k}$ and $\overleftarrow{k}$ are, respectively, the rate constants of the forward and the backward reactions. The ratio of the kinetic constants $\vec{k}/\overleftarrow{k}$ is equal to the equilibrium constant K_c

$$K_c(T) = \vec{k}/\overleftarrow{k}, \tag{6.14}$$

so that (6.13) can be written as

$$V = -\vec{k} C_{H_2} C_{I_2} [1 - \exp(-A/RT)]. \tag{6.15}$$

For reactions close to equilibrium,

$$|A/RT| \ll 1, \tag{6.16}$$

and (6.15) reduces to

$$V = \vec{k} C_{H_2} C_{I_2} A/RT. \tag{6.17}$$

The rate V is thus proportional to the affinity A.

VII. Stationary States

Stationary states, i.e., states in which the state parameters are independent of time, play an important role in the applications of nonequilibrium thermodynamics, especially in biology. Stationary nonequilibrium states have the important property that, under certain conditions, they are characterized by a minimum of the entropy production, compatible with the external constraints imposed on the system. This property is valid only if the phenomenological coefficients are supposed to be constants. Since, in real systems, this is not true in general, it means that overall gradients of the thermodynamic parameter over the complete system have to be small enough so that the assumption of constancy of the phenomenological coefficients holds approximately.

The Glansdorff–Prigogine (1970) theorem of the minimum of the entropy production may be described as follows:

In every closed or open system satisfying the Onsager relations, where the macroscopic velocity is zero and where the local values of $T, \mu_1, \ldots, \mu_c$ at the boundary surface are fixed, the stationary state ($\partial T/\partial t = 0$, $\partial \mu_\gamma/\partial t = 0$ at each point) corresponds to a minimum of the function $P \equiv \int_V \sigma_s \, \delta V$, the entropy production of the overall system. During the evolution of the system from its initial state to its stationary state, the entropy production P per unit time decreases constantly:

$$\frac{\partial P}{\partial t} = \int_V \frac{\partial \sigma_s}{\partial t} \, \delta V = \int_V \sum_i \left(J_i \frac{\partial X_i}{\partial t} + X_i \frac{\partial J_i}{\partial t} \right) \delta V < 0, \qquad (7.1)$$

At the stationary state

$$\partial P/\partial t = 0 \qquad \text{(stationary state)}. \qquad (7.2)$$

If the Onsager conditions are not fulfilled, a theorem (Glansdorff and Prigogine, 1954) of a more general character can still be derived:

$$\int_V \sum_i J_i \frac{\partial X_i}{\partial t} \, \delta V < 0. \qquad (7.3)$$

This may be used for the study of the evolution of the system and also of the existence of cyclic evolutions around a stationary state (Prigogine and Balescu, 1955, 1956).

VIII. Diffusion in Systems at Uniform Temperature

Let us first write the Gibbs equation [see Chapter IIA, Eq. (4.8)] in terms of quantities reduced to the unit volume:

$$\delta g_v = -s_v \, \delta T + \delta p + \sum_\gamma \mu_\gamma \, \delta C_\gamma. \tag{8.1}$$

Integrating at constant T, p, and μ_γ, with g_v varying from 0 to g_v and C_γ from 0 to C_γ, we have

$$g_v = \sum_\gamma \mu_\gamma C_\gamma. \tag{8.2}$$

Differentiating this expression again and comparing the result with (8.1), we obtain

$$\delta p = s_v \, \delta T + \sum_\gamma C_\gamma \, \delta\mu_\gamma,$$

which is the Gibbs–Duhem equation in local form. At constant temperature, we may then write

$$\operatorname{grad} p = \sum_\gamma C_\gamma \operatorname{grad} \mu_\gamma. \tag{8.3}$$

Now, we are interested in the problem of diffusion velocities in unaccelerated nonviscous fluids. In this case, (5.13) reduces to

$$\operatorname{grad} p = \sum_\gamma \varrho_\gamma \mathbf{F}_\gamma = \sum_\gamma C_\gamma \mathbf{F}_\gamma{}^*, \tag{8.4}$$

where

$$\mathbf{F}_\gamma{}^* = M_\gamma \mathbf{F}_\gamma.$$

It follows from (8.4) and (8.3) that

$$\sum_\gamma C_\gamma \operatorname{grad} \mu_\gamma = \sum_\gamma C_\gamma \mathbf{F}_\gamma{}^*. \tag{8.5}$$

Let us consider a two-component system. The entropy production due to diffusion σ_{diff} is given by [see (6.8) and (6.9)]

$$\sigma_{\text{diff}} = \frac{1}{T} (C_1 \mathbf{\Delta}_1 \cdot \mathbf{A}_1^{*\text{d}} + C_2 \mathbf{\Delta}_2 \cdot \mathbf{A}_2^{*\text{d}}) \geq 0, \tag{8.6}$$

where

$$\mathbf{A}_\gamma^{*\text{d}} = \mathbf{F}_\gamma{}^* - (\operatorname{grad} \mu_\gamma)_T. \tag{8.7}$$

We have, from (8.5) and (8.7),

$$\sum_{\gamma} C_\gamma \mathbf{A}_\gamma^{*\mathrm{d}} = 0. \tag{8.8}$$

For a two-component system,

$$\mathbf{A}_2^{*\mathrm{d}} = -(C_1/C_2)\mathbf{A}_1^{*\mathrm{d}}, \tag{8.9}$$

and, from (5.5),

$$C_2 M_2 \mathbf{\Delta}_2 = -C_1 M_1 \mathbf{\Delta}_1. \tag{8.10}$$

It follows that

$$\sigma_{\mathrm{diff}} = \frac{1}{T}\left(1 + \frac{C_1 M_1}{C_2 M_2}\right) C_1 \mathbf{\Delta}_1 \cdot \mathbf{A}_1^{*\mathrm{d}} \geq 0. \tag{8.11}$$

This shows that, in a binary system, there is actually only one diffusion flux which, in the range of small affinities $\mathbf{A}_1^{*\mathrm{d}}$, will be expected to follow the proportionality relation

$$\left(1 + \frac{C_1 M_1}{C_2 M_2}\right) C_1 \mathbf{\Delta}_1 = L_{11} \mathbf{A}_1^{*\mathrm{d}}. \tag{8.12}$$

If there is no appreciable external force and if the solution is such that $C_1 M_1$ is much smaller than $C_2 M_2$, we have

$$C_1 \mathbf{\Delta}_1 = -L_{11}(\mathrm{grad}\ \mu_1)_T. \tag{8.13}$$

For perfect solutions [see Chapter 2A, Eq. (13.49), where $f_s^c = 0$],

$$C_1 \mathbf{\Delta}_1 = -L_{11} RT (1/C_1)\ \mathrm{grad}\ C_1. \tag{8.14}$$

Constituent 1 being a solute at low concentration in a solvent 2 whose diffusion velocity $\mathbf{\Delta}_2$ is practically the same as the barycentric velocity $\boldsymbol{\omega}$, formula (8.14) be identified with Fick's law

$$C_1 \mathbf{\Delta}_1 = -D_1\ \mathrm{grad}\ C_1, \tag{8.15}$$

in which the diffusion coefficient D_1 is a constant over the range of concentration C_1 corresponding to perfect behavior. It follows that we should have

$$L_{11} = (C_1/RT) D_1. \tag{8.16}$$

For further details and different aspects of the phenomenological theory of diffusion, the reader should consult the book of the Groot (1951).

IX. Diffusion in Systems at Nonuniform Temperature

We shall first consider heat conduction as occurring alone in a system of uniform composition with negligible external forces **F**. The entropy production [see (6.5)] is

$$\sigma_{\text{th}} = \mathbf{J}_{\text{th}} \cdot \text{grad}(1/T). \tag{9.1}$$

Let us consider the thermal flow J_{th} as taking place along the x axis. We then have

$$\sigma_{\text{th}} = -J_{\text{th}} \frac{1}{T^2} \frac{\partial T}{\partial x}. \tag{9.2}$$

It has been customary to take J_{th} as flux and $-(1/T_2)\, \partial T/\partial x$ as force and write

$$J_{\text{th}} = -L_{\text{th}} \frac{1}{T^2} \frac{\partial T}{\partial x}. \tag{9.3}$$

However, compatibility with Fourier's law

$$J_{\text{th}} = -l\, \partial T/\partial x \tag{9.4}$$

in which the coefficient of heat conductivity l is constant or at least insensitive to T, requires that

$$L_{\text{th}} = lT^2. \tag{9.5}$$

On the other hand, taking Fourier's law into account in (9.2), we obtain

$$\sigma_{\text{th}} = l \frac{1}{T^2} \left(\frac{\partial T}{\partial x} \right)^2. \tag{9.6}$$

Now using l in place of L_{th} as phenomenological coefficient, we are led to take as thermodynamic force $-(1/T)\, \partial T/\partial x$ instead of $-(1/T^2)\, \partial T/\partial x$ and as thermodynamic flux J_{th}/T instead of J_{th}. The advantage of using constant or at least nearly constant Onsager coefficients is of particular importance in the study of simultaneous irreversible processes. We shall now examine the effects resulting from simultaneous heat conduction and diffusion.

First, let us define a molecular velocity **u** by

$$\mathbf{u}C = \sum_{\gamma} C_{\gamma} \boldsymbol{\omega}_{\gamma}. \tag{9.7}$$

The diffusion velocities then become

$$\mathbf{\Delta}_\gamma{}^u = \boldsymbol{\omega}_\gamma - \boldsymbol{\omega}, \tag{9.8}$$

which gives

$$\sum_\gamma C_\gamma \mathbf{\Delta}_\gamma{}^u = 0. \tag{9.9}$$

We may then write the transport entropy production as

$$\sigma_{\text{transp}} = \mathbf{J}_{\text{th}} \cdot \operatorname{grad} \frac{1}{T} + \left(\frac{\mathbf{A}_1^{*\text{d}}}{T} \cdot C_1 \mathbf{\Delta}_1{}^u\right) + \left(\frac{\mathbf{A}_2^{*\text{d}}}{T} \cdot C_2 \mathbf{\Delta}_2{}^u\right). \tag{9.10}$$

Taking Eqs. (9.9) and (8.9) into account, (9.10) becomes

$$\sigma_{\text{transp}} = \mathbf{J}_{\text{th}} \cdot \operatorname{grad} \frac{1}{T} + \left[\frac{1}{T}\left(1 + \frac{C_2}{C_1}\right)\mathbf{A}_2^{*\text{d}} \cdot C_2 \mathbf{\Delta}_2\right]. \tag{9.11}$$

The phenomenological relations then give, with $L_{12} = L_{21}$,

$$\begin{aligned} C_2 \mathbf{\Delta}_2{}^u &= \frac{L_{11}}{T} \frac{C}{C_1} \mathbf{A}_2^{*\text{d}} + L_{12} \operatorname{grad} \frac{1}{T} \\ \mathbf{J}_{\text{th}} &= \frac{L_{12}}{T} \frac{C}{C_1} \mathbf{A}_2^{*\text{d}} + L_{22} \operatorname{grad} \frac{1}{T}, \end{aligned} \tag{9.12}$$

where

$$C = C_1 + C_2. \tag{9.13}$$

On the other hand, at uniform pressure, the diffusion affinity reduces to [see Chapter 2A, Eq. (13.15)]

$$\mathbf{A}_2^{*\text{d}} = -\left(\frac{\partial \mu_2}{\partial N_2}\right)_{Tp} \operatorname{grad} N_2 = -\frac{RT}{a_2} \frac{\partial a_2}{\partial N_2} \operatorname{grad} N_2, \tag{9.14}$$

and (9.12) may be written as follows:

$$\begin{aligned} C_2 \mathbf{\Delta}_2{}^u &= -DC \operatorname{grad} N_2 - \frac{L_{12}}{T^2} \operatorname{grad} T \\ \mathbf{J}_{\text{th}} &= -\frac{L_{12}}{L_{11}} DC \operatorname{grad} N_2 - \frac{L_{22}}{T^2} \operatorname{grad} T, \end{aligned} \tag{9.15}$$

where

$$D = \frac{L_{11} R}{C_1 a_2} \frac{\partial a_2}{\partial N_2}. \tag{9.16}$$

The existence of mutual influence terms in these linear equations is demonstrated by the observation of two effects: the Dufour effect and thermal diffusion, or the Soret effect (Chanu, 1958).

If grad $T = 0$, the first equation of (9.15) gives the diffusion flux at uniform temperature, and it is easy to see that D is the diffusion coefficient. Moreover, if grad $N_2 \neq 0$, there is an initial heat flow:

$$J_{\text{th}} = -(L_{12}/L_{11})DC \text{ grad } N_2 \tag{9.17}$$

and a temperature gradient is set up. This is the Dufour effect. Actually, it has only been observed in the mixing by diffusion into each other of two gases initially at the same temperature. The effect has not yet been observed in liquids.

The Soret effect is observed in the following conditions: grad $N_2 = 0$ and grad $T \neq 0$. Equations (9.15) reduce to

$$\mathbf{J}_{\text{th}} = -l \text{ grad } T \tag{9.18}$$

$$C_2 \mathbf{\Delta}_2{}^u = -(L_{12}/T_2) \text{ grad } T, \tag{9.19}$$

where the thermal conductibility coefficient l of the system is given by

$$l = L_{22}/T^2. \tag{9.20}$$

Thus, if a system originally at uniform composition is heated at one extremity and cooled at the other, the diffusing species 2 will tend to concentrate itself toward the cool end of the container if $L_{12} > 0$ or, on the contrary, toward the warm end if $L_{12} < 0$. The Soret effect has been observed and studied rather extensively in liquids, where it appears to be of the same order of magnitude as in gases.

X. Electrokinetic Effects

By means of Onsager's reciprocity relation, we shall now study an example of interference between irreversible processes, i.e., the connection between the various electrokinetic effects which have been examined by Mazur and Overbeek (1951).

Consider a system consisting of two vessels I and II, which communicate by means of a porous wall or a capillary. The temperature and the concentrations are supposed to be uniform throughout the entire system and both phases differ only with respect to pressure and electrical potentials. The entropy production due to the transfer of the constituents from vessel I to vessel II is given by

$$d_iS = (1/T) \sum_\gamma \tilde{A}_\gamma \, d\xi_\gamma = -(1/T) \sum_\gamma \tilde{A}_\gamma \, dn_\gamma{}^{\mathrm{I}}, \tag{10.1}$$

where $\tilde{A}_\gamma$ is the electrochemical affinity given by (Sanfeld, 1968),

$$\tilde{A}_\gamma = \tilde{\mu}_\gamma{}^{\mathrm{I}} - \tilde{\mu}_\gamma^{\mathrm{II}} = (\mu_\gamma{}^{\mathrm{I}} - \mu_\gamma^{\mathrm{II}}) + z_\gamma(\varphi^{\mathrm{I}} - \varphi^{\mathrm{II}}), \tag{10.2}$$

and where the electrochemical potential $\tilde{\mu}_\gamma$ is defined by

$$\tilde{\mu}_\gamma = \mu_\gamma + z_\gamma \varphi, \tag{10.3}$$

with z_γ the charge of γ and φ the electrical potential. Thus,

$$\tilde{A}_\gamma = \Delta\mu_\gamma + z_\gamma \, \Delta\varphi, \tag{10.4}$$

where Δ denotes the difference in the value of a given variable between vessel I and vessel II.

In our case, formula (10.1) can be written in the form

$$\frac{d_iS}{dt} = -\frac{1}{T} \sum_\gamma v_\gamma \frac{dn_\gamma{}^{\mathrm{I}}}{dt} \Delta p - \frac{1}{T} \sum_\gamma z_\gamma \frac{dn_\gamma{}^{\mathrm{I}}}{dt} \Delta\varphi. \tag{10.5}$$

We introduce the fluxes

$$J_{\mathrm{m}} = -\sum_\gamma v_\gamma \, dn_\gamma{}^{\mathrm{I}}/dt; \qquad I = -\sum_\gamma z_\gamma \, dn_\gamma{}^{\mathrm{I}}/dt, \tag{10.6}$$

where I is the electrical current due to a transfer of charges from I to II, and J_{m} is the resultant flow of matter (J_{m} might also be called a resultant flow of volume).

Combining (10.5) and (10.6), we find

$$\frac{d_iS}{dt} = J_{\mathrm{m}} \frac{\Delta p}{T} + I \frac{\Delta\varphi}{T}, \tag{10.7}$$

and the phenomenological equations are given by

$$I = L_{11} \frac{\Delta\varphi}{T} + L_{12} \frac{\Delta p}{T}, \qquad J_{\mathrm{m}} = L_{21} \frac{\Delta\varphi}{T} + L_{22} \frac{\Delta p}{T}, \tag{10.8}$$

with the Onsager relation $L_{12} = L_{21}$.

We have here two irreversible effects, transport of matter under the influence of a pressure difference, and electrical current due to the difference of electrical potential. Moreover, we have a cross effect related by the coefficients $L_{12} = L_{21}$, which is due to the interference of the two irreversible processes. We obtain four effects which can be studied experimentally:

(1) At zero electrical current, we obtain *the streaming potential*

$$(\Delta\varphi/\Delta p)_{I=0} = -L_{12}/L_{11}. \tag{10.9}$$

(2) At uniform pressure, we have *the electroosmotic effect*

$$(J_{\mathrm{m}}/I)_{\Delta p=0} = L_{21}/L_{11}. \tag{10.10}$$

(3) When the flux J_{m} is zero, (10.8) gives *the electroosmotic pressure*

$$(\Delta p/\Delta\varphi)_{J_{\mathrm{m}}=0} = -L_{21}/L_{22}. \tag{10.11}$$

(4) At uniform electrical potential, the *streaming current* appears

$$(I/J_{\mathrm{m}})_{\Delta\varphi=0} = L_{12}/L_{22}. \tag{10.12}$$

On application of the Onsager relation $L_{12} = L_{21}$, these four effects give the connection between the osmotic effect and the streaming effect

$$(\Delta\varphi/\Delta p)_{I=0} = (J_{\mathrm{m}}/I)_{\Delta p=0} \tag{10.13}$$

$$(\Delta p/\Delta\varphi)_{J_{\mathrm{m}}=0} = -(I/J_{\mathrm{m}})_{\Delta\varphi=0}. \tag{10.14}$$

Relation (10.13), known as Saxen's relation, had already been established by applying kinetic considerations. However, such kinetic considerations are only possible if some simplified model of the diaphragm separating the two phases is adopted—for example, if the diaphragm is identified with a capillary of uniform section. The importance of the thermodynamic demonstration resides in the fact that it holds whatever the nature of the diaphragm or the porous wall (Katchalski and Curran, 1965). This example is instructive in that it shows clearly what kind of information can be obtained from the thermodynamics of irreversible processes.

Although such methods are insufficient for the explicit calculation of the thermodynamic coefficients, they make it possible to establish a connection between effects which at first sight appear to be quite independent. Other examples ("Knudsen effect," "fountain effect," "thermoosmotic effect," "thermomechanical effect") are given in different books (Van Rysselberghe, 1963; Prigogine, 1967).

XI. Entropy Production due to Viscosity

In a viscous fluid, the equation of motion of the system is

$$\varrho\, d\boldsymbol{\omega}_i/dt = \varrho \mathbf{F}_i + \sum_j \partial T_{ij}/\partial x_j \qquad (i, j = 1, 2, 3), \tag{11.1}$$

where the quantities T_{ij} are the Cartesian components of the stress tensor T of the medium. In hydrodynamics, the tensor T is considered as symmetrical,

$$T_{ij} = T_{ji}, \tag{11.2}$$

and the Lamb pressure p is defined by

$$p = -(T_{11} + T_{22} + T_{33})/3. \tag{11.3}$$

The viscosity stress tensor T'_{ij} is defined by

$$T_{ij} = -p\, \delta_{ij} + T'_{ij}, \tag{11.4}$$

and the viscosity coefficients η and ζ are given by

$$T'_{ij} = \eta \left[\frac{\partial \omega_i}{\partial x_j} + \frac{\partial \omega_j}{\partial x_i} - \frac{2}{3}\, \delta_{ij} \frac{\partial \omega_l}{\partial x_l} \right] + \zeta\, \delta_{ij} \frac{\partial \omega_l}{\partial x_l}. \tag{11.5}$$

When the fluid is incompressible, the equation of motion of the incompressible viscous fluid then becomes

$$\varrho \frac{d\omega_i}{dt} = \varrho F_i - \frac{\partial p}{\partial x_i} + \eta\, \nabla^2 \omega_i \tag{11.6}$$

with ∇ the classical nabla symbol. This is called the Navier–Stokes equation (Landau and Lifshitz, 1963). It is then easy to show that the entropy balance becomes

$$\varrho \frac{ds_m}{dt} = -\frac{1}{T} \operatorname{div} \mathbf{W} + \frac{1}{T} \sum_\gamma \varrho_\gamma \mathbf{F}_\gamma \cdot \boldsymbol{\Delta}_\gamma - \frac{\varrho}{T} \sum_\gamma \mu_\gamma \frac{dN_\gamma}{dt}$$
$$+ \frac{1}{T} \sum_i \sum_j T'_{ij} \frac{\partial \omega_j}{\partial x_i}. \tag{11.7}$$

We see that the entropy production due to viscosity or internal friction σ_{visc} is

$$\sigma_{\text{visc}} = \frac{1}{T} \sum_i \sum_j T'_{ij} \frac{\partial \omega_j}{\partial x_i} > 0. \tag{11.8}$$

Formula (11.8) shows that the T'_{ij} are the fluxes corresponding to the forces. There are no explicit couplings between these and the transport phenomena, nor between them and the scalar chemical reactions, the viscosity tensor being of rank two. The requirement that σ_{visc} must be positive makes the viscosity coefficient η a positive definite quantity.

Remark. Other interesting irreversible processes in electrochemistry, thermoelectricity and thermoelectrochemistry are developed in great detail by Van Rysselberghe (1963) and by Harman and Honig (1967). For problems related to biophysics, see the excellent book of Katchalski and Curran (1965).

References

Casimir, H. B. G. (1945). *Rev. Mod. Phys.* **17**, 343.

Chanu, J. (1958). *C. R. H. Acad., Ser. C* **246**, 67; *J. Chim. Phy. Physicochim. Biol.* **55**, 733–43.

De Donder, Th., and Van Rysselberghe, P. (1936). "Affinity." Stanford Univ. Press, Stanford, California.

Defay, R., and Mazur, P. (1954). Bull. Sci. Chim. Belg. **63**, 573.

de Groot, S. R. (1951). "Thermodynamics of Irreversible Processes." North-Holland Publ., Amsterdam.

de Groot, S. R., and Mazur, P. (1962). "Non-Equilibrium Thermodynamics." North-Holland Publ., Amsterdam.

Glansdorff, P., and Prigogine, I. (1954). *Physica* **20**, 773.

Glansdorff, P., and Prigogine, I. (1970). "Entropy, Stability, and Structure." Wiley, New York.

Harman, T. C., and Honig, J. M. (1967). "Thermoelectric and Thermomagnetic Effects and Applications." McGraw-Hill, New York.

Katchalski, A., and Curran, P. F. (1965). "Non-Equilibrium Thermodynamics in Biophysics." Harvard Univ. Press, Cambridge, Massachusetts.

Landau, L. D., and Lifshitz, E. M. (1963). "Fluid Mechanics." Pergamon, New York.

Mazur, P., and Overbeek, J. T. G. (1951). *Rec. Trav. Chim.* **70**, 83.

Meixner, J. (1943). *Ann. Phys.* **43**, 244.

Onsager, L. (1931). *Phys. Rev.* **37**, 405; **38**, 2265.

Prigogine, I., Mazur, P., and Defay, R. (1953). *J. Chim. Phys. Physicochim. Biol.* **50**, 146.

Prigogine, I. (1947). "Etude Thermodynamique des Phénomènes Irréversibles" Desoer, Liege.

Prigogine, I. (1967). "Introduction to Thermodynamics of Irreversible Processes," 3rd ed. Wiley (Interscience), New York.

Prigogine, I., and Balescu, R. (1955). *Bull. Acad. Roy. Belg.* (*Cl. Sc.*) **41**, 917.

Prigogine, I., and Balescu, R. (1956). *Bull. Acad. Roy. Belg.* (*Cl. Sc.*) **42**, 256.

Sanfeld, A. (1968). "Introduction to the Thermodynamics of Charged and Polarized Layers" (I. Prigogine, ed.), Monograph n° 10. Wiley, London.

Van Rysselberghe, P. (1963). "Thermodynamics of Irreversible Processes." Hermann, Paris.

Chapter 2C

Thermodynamics of Surfaces

A. SANFELD

I. Mechanical Properties of an Interface

A. Surface Tension

From a macroscopic point of view, the boundary between two fluid phases is regarded as a sharply defined mathematical surface. It is clear from either molecular or microscopic considerations that this assumption cannot rigourously apply. The interface between two phases is in reality a thin region of about 10^{-6} cm thickness whose physical properties vary sharply from the bulk properties of one phase to the bulk properties of the other phase.

A classical thermodynamic study for such systems is too complicated, and, in fact, it is a matter of common observation that a fluid behaves as though it consisted of two homogeneous fluids separated by a stretched isotropic surface or membrane of infinitesimal thickness. Any local deformation of this membrane without variation of the area requires no mechanical work. The surface thus has no rigidity and is called the surface of tension. Let us divide the surface of tension into two regions I and II by a curve l (Fig. 1). If, across an element δl, of l, the region I

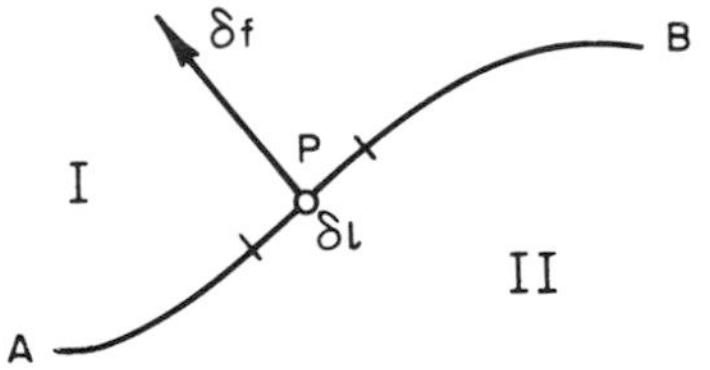

FIG. 1. Definition of surface tension at a point P on a line AB in the surface.

exerts a force $\gamma\, \delta l = \delta f$ upon region II at a point P, then γ is called the surface (or interfacial) tension at this point.

The force δf is tangential to the surface of tension, perpendicular to δl, and independant of its orientation. The existence of a surface tension is, in fact, related to the thermodynamic instability of two phases in contact with one another, and thus to the free energy of contact. The contractile behavior of the transition layer minimizes this free energy and gives rise to the macroscopic experimental quantity. The unit often used in the measurement of γ is dyn cm^{-1}.

B. Mechanical Equilibrium

In a gravity field g (z axis), we have

$$dp^{\alpha}/dz = -\varrho^{\alpha} g; \qquad dp^{\beta}/dz = -\varrho^{\beta} g, \tag{1.1}$$

where p^{α} and p^{β}, ϱ^{α} and ϱ^{β} are, respectively, the hydrostatic pressures and the densities in the two phases α and β (Fig. 2).

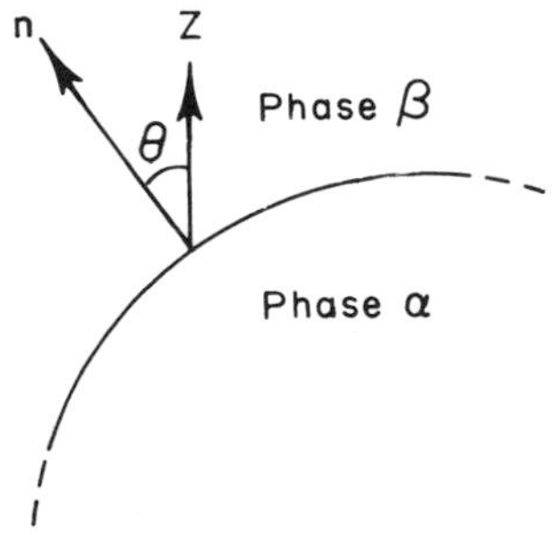

FIG. 2. Two phases α and β separated by a surface whose normal **n** is directed from α to β.

It is easy to show that the condition of local mechanical equilibrium of the surface of tension is given by

$$p^{\alpha} - p^{\beta} = \gamma(K_1 + K_2) + g\Gamma \cos\theta \tag{1.2}$$

$$d\gamma/dz = g\Gamma, \tag{1.3}$$

where Γ is the surface density, K_1 and K_2 the two principal curvatures of the surface, and θ the angle between the z axis and the perpendicular to the surface directed from α to β.

Neglecting the gravity effect on γ ($\Gamma \simeq 10^{-7}$ g cm^{-2} and $d\gamma/dz = 10^{-4}$ dyn cm^{-2}), (1.2) reduces to the classical Laplace formula

$$p^\alpha - p^\beta = \gamma(K_1 + K_2), \tag{1.4}$$

with γ uniform on the surface.

Introducing the mean curvature defined by

$$\frac{1}{r_\mathrm{m}} = \frac{1}{2}\left(\frac{1}{r_1} + \frac{1}{r_2}\right), \tag{1.5}$$

where r_1 and r_2 are the principal radii of curvature, (1.4) becomes

$$p^\alpha - p^\beta = 2\gamma/r_\mathrm{m}. \tag{1.6}$$

For a spherical surface of radius of curvature r, we then have

$$p^\alpha - p^\beta = 2\gamma/r. \tag{1.7}$$

The fundamental equation (1.4) shows that, because of its surface tension, a curved surface maintains mechanical equilibrium between two fluids at different pressure p^α and p^β.

Generally, because of the effect of gravity, $p^\alpha - p^\beta$, and thus $K_1 + K_2$, varies along the surface. In particular, $p^\alpha = p^\beta$ if $r = \infty$. Hence, a plane surface can exist only if the pressures of the fluids on the two sides are equal.

C. Application of the Laplace Formula

Consider a pure liquid forming a drop in thermodynamic equilibrium with its vapor. For mechanical equilibrium,

$$p^l - p^\mathrm{v} = 2\gamma/r. \tag{1.8}$$

For physicochemical equilibrium

$$\mu^l = \mu^\mathrm{v}, \tag{1.9}$$

where the superscripts l and v refer, respectively, to the liquid and vapor phases. For an equilibrium displacement, we obtain

$$\delta p^l - \delta p^\mathrm{v} = \delta(2\gamma/r) \tag{1.10}$$

$$\delta\mu^l = \delta\mu^\mathrm{v}. \tag{1.11}$$

The Gibbs–Duhem equation [Chapter 2A, Eq. (4.30)] may be applied to each coexisting phase, so that at constant temperature (1.11) gives

$$-v^l\,\delta p^l + v^{\mathrm{v}}\,\delta p^{\mathrm{v}} = 0 \qquad (T \quad \text{constant}). \tag{1.12}$$

Assuming the vapor to be a perfect gas and the liquid to be incompressible, (1.10) and (1.12) give, after integration, the well-known Kelvin equation

$$\ln(p^{\mathrm{v}}/p^0) = (2\gamma/r)\,v^l/RT, \tag{1.13}$$

where p^0 is the saturation vapor pressure of the macroscopic liquid phase ($r \to \infty$). This equation shows that the vapor pressure increases as the droplet decreases in size. An example for water at 18° [($\gamma = 73$ dyn cm^{-1})] is given in Table I.

TABLE I

r (cm)	p^{v}/p^0	$p^l - p^{\mathrm{v}}$ (bar)
∞	1	0
10^{-4}	1.001	1.46
10^{-5}	1.011	14.6
10^{-6}	1.115	146

The equilibrium of a drop initially satisfying (1.13) in an infinite volume of vapor of pressure p^{v} is unstable. If, by a small fluctuation, a little liquid evaporates, the droplet decreases in size and its vapor pressure exceeds that of the surrounding atmosphere: it will therefore continue to evaporate. Conversely, if a little vapor condenses, the vapor pressure of the droplet falls below that of the surroundings and further condensation will occur. This phenomenon leads to an explanation of the existence of supersaturated vapor (Dufour and Defay, 1963); this is a vapor which, in the absence of liquid, can exist at a given pressure p^{v} greater than the saturation pressure p^0 corresponding to the given temperature T.

D. Equilibrium of a Line of Contact

The Laplace equation (partial derivatives of the second degree) determines the local form of the surface tension. Now let us consider the

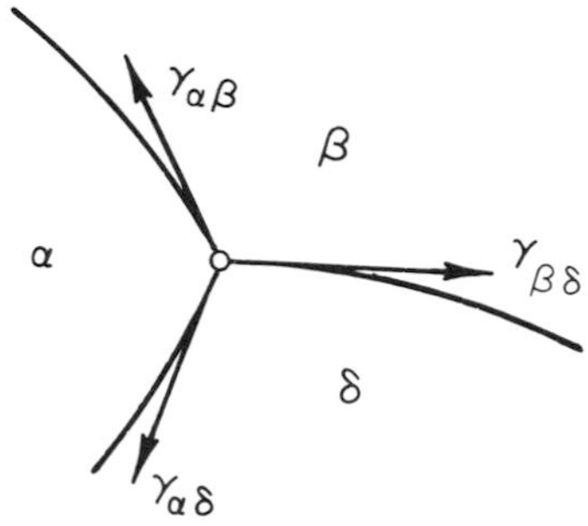

FIG. 3. Equilibrium at a line of contact: Neumann's triangle.

equilibrium of a line of contact formed by the interaction of three phases α, β, and δ (Fig. 3). As the total force acting on an element of line through P of length δl is zero, we find

$$\mathbf{1}_{\alpha\beta}\gamma_{\alpha\beta} + \mathbf{1}_{\alpha\delta}\gamma_{\alpha\delta} + \mathbf{1}_{\beta\delta}\gamma_{\beta\delta} + \mathbf{n}K\eta = 0, \tag{1.14}$$

where $\gamma_{\alpha\beta}$, $\gamma_{\alpha\delta}$, and $\gamma_{\beta\delta}$ are the tensions related to the three surfaces, and the last term corresponds to a line tension η ($\mathbf{n}$ is the normal and K the curvature of the line of contact). In fact, the value of this line tension is quite negligible ($\simeq 10^{-6}$ dyn). Equation (1.14) then reduces to the Neumann equation

$$\mathbf{1}_{\alpha\beta}\gamma_{\alpha\beta} + \mathbf{1}_{\alpha\delta}\gamma_{\alpha\delta} + \mathbf{1}_{\beta\delta}\gamma_{\beta\delta} = 0. \tag{1.15}$$

This equation may be applied to a line of contact formed by two liquid phases α and β and a solid surface s (Fig. 4). We obtain the Young equation

$$\cos\theta\gamma_{\alpha\beta} = \gamma_{\beta s} - \gamma_{\alpha s}, \tag{1.16}$$

where θ is the contact angle of the surface $\alpha\beta$ with the solid and $\gamma_{\alpha s}$

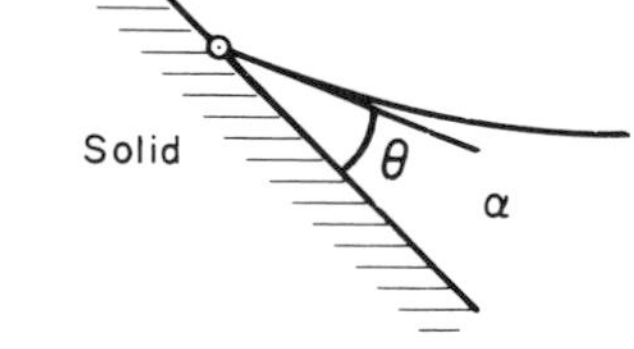

FIG. 4. Surface tension of a solid; the contact angle.

$\gamma_{\beta s}$ the solid–fluid tensions. If $|\gamma_{\beta s} - \gamma_{\alpha s}| > \gamma_{\alpha\beta}$, no equilibrium position of the line of contact can be found, and one of the two phases α or β covers the whole area of the solid: the liquid is said to wet the solid perfectly.

E. Mechanical Work of External Forces

Let us consider a two-phase system α and β, with volumes V^α and V^β separated by a surface Ω of area A. In the absence of any external force fields, the work received by the system is (Fig. 5)

$$\delta\tau = -p^\alpha\,\delta V^\alpha - p^\beta\,\delta V^\beta + \gamma\,\delta A. \tag{1.17}$$

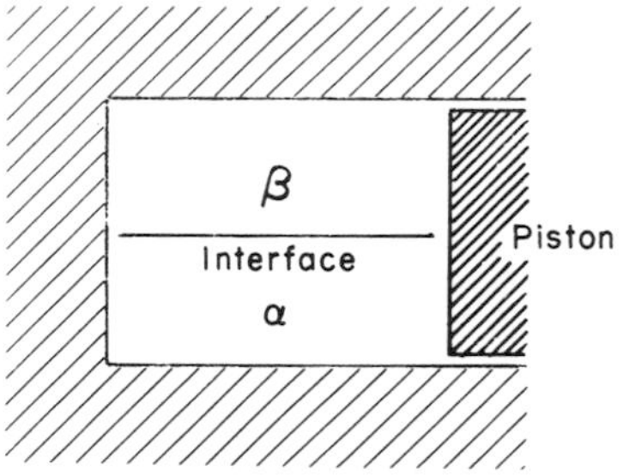

Fig. 5. Mechanical work done by a capillary system: piston subjected to surface tension.

The piston is subjected to the pressure p^α and p^β of the two fluids α and β with $p^\alpha = p^\beta$ and to the surface tension γ of the surface Ω. In Fig. 6, the piston is not subjected directly to the surface tension. The system comprises a spherical drop of liquid of volume V^α separated from the vapor by the surface Ω. The whole system is contained in a cylinder of total volume $V = V^\alpha + V^\beta$. We have

$$\delta\tau = -p^\beta\,\delta V \tag{1.18}$$

$$\delta V = \delta V^\alpha + \delta V^\beta. \tag{1.19}$$

On the other hand, for a spherical surface $\alpha\beta$,

$$\delta V^\alpha = 2r\,\delta A, \tag{1.20}$$

and thus, from (1.18)–(1.20), we again find Eq. (1.17).

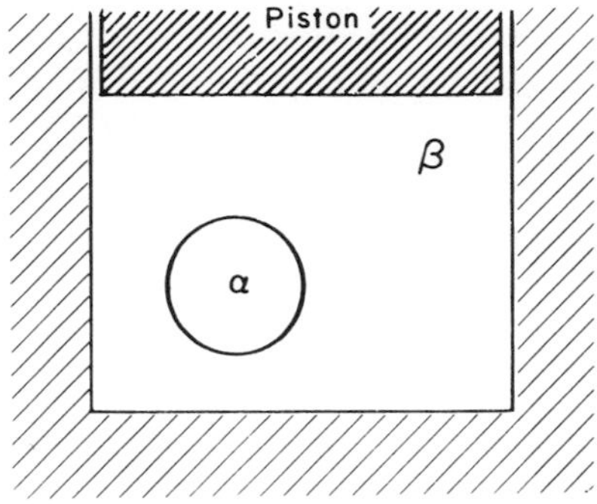

FIG. 6. Mechanical work done by a capillary system: piston not subjected to surface tension.

Remark. For a multiphase system, (1.17) must be rewritten in the form

$$\delta\tau = -\sum_{\alpha} p^{\alpha}\,\delta V^{\alpha} + \sum_{\alpha\beta} \gamma^{\alpha\beta}\,\delta A^{\alpha\beta}. \tag{1.21}$$

F. Application of the Mechanical Work Relation

At constant temperature, we know that the free energy of a closed system is equal to the work received by the system (see Chapter 2A, Section III). For a two-phase system,

$$(\delta F)_{Tn} = -p^{\alpha}\,\delta V^{\alpha} - p^{\beta}\,\delta V^{\beta} + \gamma\,\delta A. \tag{1.22}$$

Let us now consider a supersaturated vapor of pressure p^{v} in an infinite volume. The free energy of formation of a droplet phase of size r is equal to

$$\Delta F = -(4\pi r^3/3)(p^l - p^{\mathrm{v}}) + 4\pi r^2\gamma, \tag{1.23}$$

where p^l is the pressure in the droplet. From Eqs. (1.23) and (1.7), we thus obtain

$$\Delta F = 4\pi r^2\gamma/3 = \tfrac{1}{3}\gamma A, \tag{1.24}$$

where A is the area of the drop. The presence of a "germ" enables the condensation of this vapor. The germ is a droplet whose radius r can be calculated by Kelvin's equation. For water at 18°C, the free energy of formation of a germ ΔF_{g} (homogeneous nucleation) is given in Table II (Dufour and Defay, 1963).

TABLE II

p^v/p^0	r (cm)	ΔF_g (erg)	$\Delta F_g/kT$
1.011	10^{-5}	3×10^{-8}	0.75×10^6
1.115	10^{-6}	3×10^{-10}	0.75×10^4

As we know, the probability of obtaining a germ is given by the expression $\exp[-\Delta F_g/kT]$; we observe that the formation of a germ is not probable. A droplet lying on a solid surface may constitute a germ, but the free energy of formation is lower. For a spherical cap of radius r, we obtain (Fig. 7):

$$\begin{aligned}&\text{Volume of the liquid:} && \tfrac{2}{3}\pi r^3(1 - \tfrac{3}{2}\cos\theta + \tfrac{1}{2}\cos^3\theta) = v \\ &\text{Surface liquid–vapor:} && 2\pi r^2(1 - \cos\theta) = A \\ &\text{Surface liquid–solid:} && \pi r^2 \sin^2\theta = A' \\ &\text{Laplace's equation:} && p^l - p^v = 2\gamma/r \\ &\text{Young's equation:} && \gamma^{vs} - \gamma^{ls} = \gamma\cos\theta\end{aligned} \tag{1.25}$$

$$\Delta F = -(p^l - p^v)v + \gamma A + (\gamma^{ls} - \gamma^{vs})A'. \tag{1.26}$$

By combining these different equations, we obtain the relation

$$\Delta F = \tfrac{4}{3}\pi r^2\gamma(\tfrac{1}{2} - \tfrac{3}{8}\cos\theta - \tfrac{1}{8}\cos^3\theta). \tag{1.27}$$

A comparison between (1.27) and (1.24) for the same radius r leads to the conclusion that the condensation of vapor takes place preferentially on solid surfaces or on dust (heterogeneous nucleation).

Remark. If $\theta = \pi$, we find the value corresponding to homogeneous nucleation. If $\theta = 30°$, the free energy of formation is equal to one tenth of the corresponding energy of homogeneous nucleation.

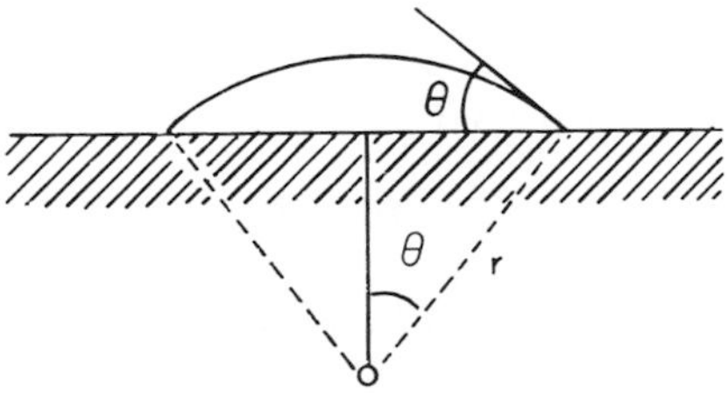

FIG. 7. Drop laid on a solid surface.

II. Pressure Tensor in Surface Layers

A. Introduction

From the molecular point of view, there exists, between two contiguous homogeneous phases in equilibrium, a zone of finite thickness with which the density, composition, and pressure tensor vary rapidly from the values characteristic of one phase to those characteristic of the other. A microscopic formulation of the surface tension may be introduced with the aid of the pressure tensor (Bakker, 1928; Kirkwood and Buff, 1949; Ono and Kondo, 1960, p. 134). In order to clarify these ideas, we shall first recall the force acting on a volume element of a fluid. We have (a) external local forces: the force $\mathbf{X}$ and the torque $\mathbf{C}$ per unit volume; and (b) the internal forces expressed by the pressure tensor in rectangular coordinates:

$$\mathsf{P} = \begin{pmatrix} P_{xx} & P_{xy} & P_{xz} \\ P_{yx} & P_{yy} & P_{yz} \\ P_{zx} & P_{zy} & P_{zz} \end{pmatrix}. \tag{2.1}$$

The motion equation of a fluid element is then given by

$$\int \delta V \varrho \, d\boldsymbol{\omega}/dt = \int \delta V \mathbf{X} - \int \delta \mathbf{A} \cdot \mathsf{P}, \tag{2.2}$$

where $\boldsymbol{\omega}$ is the velocity, $d\boldsymbol{\omega}/dt$ the acceleration, and $-\delta\mathbf{A} \cdot \mathsf{P}$ is the force acting on the surface element δA by the adjacent fluid ($\delta\mathbf{A}$ is directed along the external normal).

Application of Green's formula gives the well-known Euler equation

$$\varrho \, d\boldsymbol{\omega}/dt = \mathbf{X} - \boldsymbol{\nabla} \cdot \mathsf{P}, \tag{2.3}$$

where

$$\boldsymbol{\nabla} \cdot \mathsf{P} = \mathbf{1}_x\left(\frac{\partial}{\partial x} P_{xx} + \frac{\partial}{\partial y} P_{yx} + \frac{\partial}{\partial z} P_{zx}\right) + \mathbf{1}_y(\cdots) + \mathbf{1}_z(\cdots). \tag{2.4}$$

Now, we assume that the external local torque $\mathbf{C}$ is zero, and thus that the tensor P is symmetrical. The mechanical equilibrium condition of a fluid is given by [see Eq. (2.3)]

$$\boldsymbol{\nabla} \cdot \mathsf{P} = \mathbf{X}. \tag{2.5}$$

We shall restrict our development to systems for which $\mathbf{X} = 0$, and thus (2.5) reduces to

$$\boldsymbol{\nabla} \cdot \mathsf{P} = 0. \tag{2.6}$$

The influence of an electric field on the pressure tensor has been studied in great detail by Defay and Sanfeld (1967) (see also Sanfeld (1968)).

B. Pressure Tensor in a Plane Layer

Let us first consider rectangular axes x, y, and z, where the z axis is normal to the interface. The system is homogeneous in the directions x and y and the pressure tensor is

$$\mathsf{P} = \mathsf{P}(z) = \begin{pmatrix} p_T(z) & 0 & 0 \\ 0 & p_T(z) & 0 \\ 0 & 0 & p_N(z) \end{pmatrix}, \tag{2.7}$$

where the normal pressure p_N acts on a face parallel to the interface and the tangential pressure p_T acts on a face perpendicular to the interface.

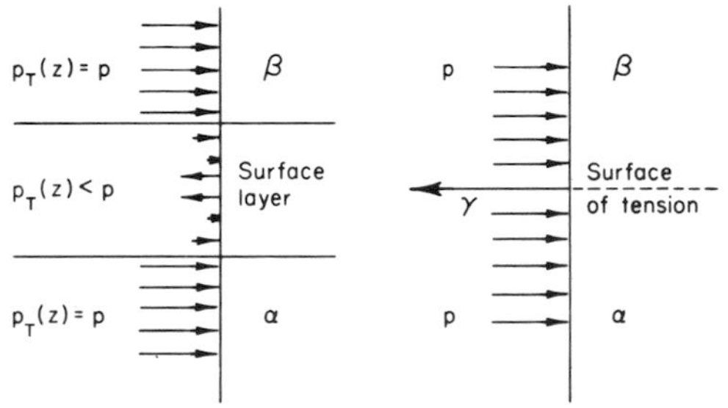

Fig. 8. Real system and equivalent model with tension surface.

Outside the surface layer, in the isotropic regions α and β of the two fluids, $p_N = p_T$. The local mechanical equilibrium condition (2.6) here reduces to

$$dp_N/dz = 0. \tag{2.8}$$

The normal pressure p_N is thus independent of z, so that, in the transition region, the normal pressure is equal to the scalar hydrostatic pressure p of the homogeneous phases. In contrast, there is no such condition for p_T. The tangential pressure may even take on negative values (Fig. 8).

The pressure deficit $\{p - p_T\}$ in the surface layer manifests itself macroscopically as a tension exerted by the fluid on the walls of the container.

The equivalence of the two models shown in Fig. 8 yields the expression

$$\int p_T(z)\,dz = \int p\,dz - \gamma \tag{2.9}$$

or

$$\gamma = \int \{p - p_T(z)\}\,dz, \tag{2.10}$$

where the region of integration is, in fact, the surface layer.

A second condition of equivalence can be expressed by the equality of the momentum forces. This determines the height z_0 of the surface of tension

$$\int z p_T(z)\,dz = \int z p\,dz - \gamma z_0; \tag{2.11}$$

thus,

$$z_0 = (1/\gamma) \int z\{p - p_T(z)\}\,dz = \int z\{p - p_T(z)\}\,dz \Big/ \int \{p - p_T(z)\}\,dz. \tag{2.12}$$

We note here that the value of γ is defined independently of the position of the surface of tension. Hence, for any transformation for which the interface remains plane, the height z at which γ is applied has no practical importance. However, this is not the case for curved surfaces.

C. Pressure Tensor in a Spherical Layer

In spherical coordinates, the pressure tensor is a function of the radius of the spherical surface involving the point under consideration:

$$\mathsf{P} = \mathsf{P}(r) = \begin{pmatrix} p_T(r) & 0 & 0 \\ 0 & p_T(r) & 0 \\ 0 & 0 & p_N(r) \end{pmatrix}. \tag{2.13}$$

In the bulk of the homogeneous phases, P is isotropic and independent of r; in phases α and β, we have

$$\mathsf{P} = p^\alpha \mathsf{I} \qquad \text{and} \qquad \mathsf{P} = p^\beta \mathsf{I},$$

where I is the unit tensor. In the transition region, p_T and p_N vary with r, and here the mechanical equilibrium condition (2.6) gives (Bakker, 1928)

$$-\partial p_N/\partial r = 2(p_N - p_T)/r. \tag{2.14}$$

Equation (2.14) may be rewritten

$$\partial(r^2 p_N)/\partial r = 2 p_T r. \tag{2.15}$$

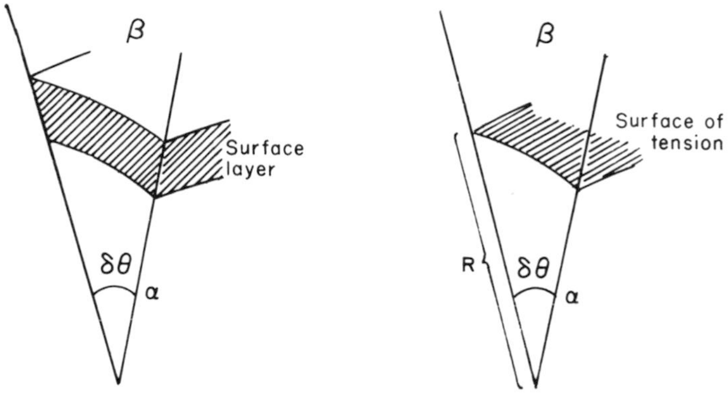

FIG. 9. Section of a system; real and equivalent models with tension surface for spherical interface.

We shall now compare our system with the model surface of tension (Fig. 9). The mechanical forces and momentum equilibrium may be expressed by

$$\delta\theta \int p_{\mathrm{T}} r \, dr = \delta\theta \int^{R} p^{\alpha} r \, dr + \delta\theta \int_{R} p^{\beta} r \, dr - \delta\theta\gamma R \tag{2.16}$$

$$\delta\theta \int p_{\mathrm{T}} r^2 \, dr = \delta\theta \int^{R} p^{\alpha} r^2 \, dr + \delta\theta \int_{R} p^{\beta} r^2 \, dr - \delta\theta\gamma R^2, \tag{2.17}$$

where R determines the position of the surface of tension and where the integrals are limited to the thickness of the layer. Let us define the quantity $p^{\alpha\beta}$ by the following relations:

$$\begin{aligned} p^{\alpha\beta} &= p^{\alpha} \quad \text{if} \quad r < R \\ p^{\alpha\beta} &= p^{\beta} \quad \text{if} \quad r > R. \end{aligned} \tag{2.18}$$

Thus, we obtain the two fundamental equations

$$\gamma = (1/R) \int \{p^{\alpha\beta} - p_{\mathrm{T}}(r)\} r \, dr \tag{2.19}$$

$$R = \int \{(p^{\alpha\beta} - p_{\mathrm{T}}(r)\} r^2 \, dr / \int \{p^{\alpha\beta} - p_{\mathrm{T}}(r)\} r \, dr. \tag{2.20}$$

The knowledge of R is thus necessary to calculate γ. It is easy to verify that these equations are consistent with the Laplace relation

$$\gamma = (p^{\alpha} - p^{\beta}) R/2. \tag{2.21}$$

Remark. For $R \to \infty$, it is easy to show that (2.19) and (2.20) reduce to (2.10) and (2.12).

III. Gibbs's Surface Model

A. The Gibbs Dividing Surface

Let us now consider a point P in a transition layer separating two homogeneous phases α and β (Fig. 10). In the vicinity of P, we find other points $P', P'', P''', \ldots$ equivalent to P (the intensive properties are identical at each of these points). Thus, we obtain a surface of uniform properties which includes $P, P', P'', \ldots$. Through each point of a line perpendicular to the layer, there exists a surface of uniform properties. All these surfaces are parallel with one another and are characterized by the coordinate λ measured along a normal to the layer.

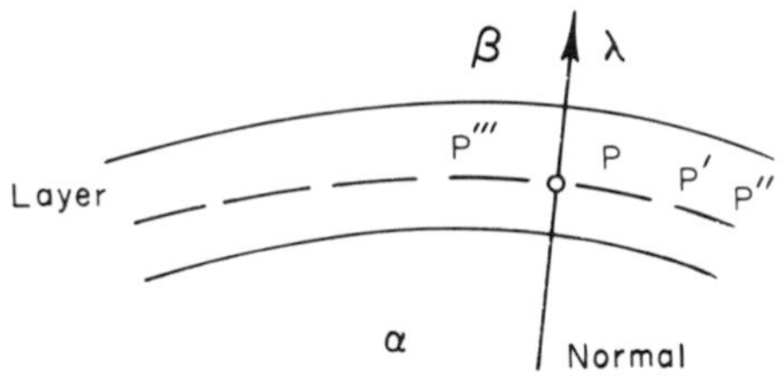

Fig. 10. Transition layer separating two homogeneous phases α and β.

In order to describe the macroscopic properties of the layer, Gibbs replaces the heterogeneous transition regions by a geometrical dividing surface which coincides with one of the surfaces of uniform properties. The dividing surface is thus in the interfacial region between the homogeneous phases α and β. In this model, the Gibbs surface is a two-dimensional phase without thickness but with well-defined physico-chemical properties (Defay *et al.*, 1966). We restrict our study to a two-phase c-components equilibrium system without external field.

B. Surface Quantities

We consider a closed surface generated by a moving normal to the dividing surface s (Fig. 11). The closed surface is of such an extent that it includes portions of the homogeneous bulk phases α and β. The Helmholtz free energy of matter enclosed by the closed surface can be adequately described by relations of the type

$$F = F^{\alpha} + F^{\beta} + F^{a}, \tag{3.1}$$

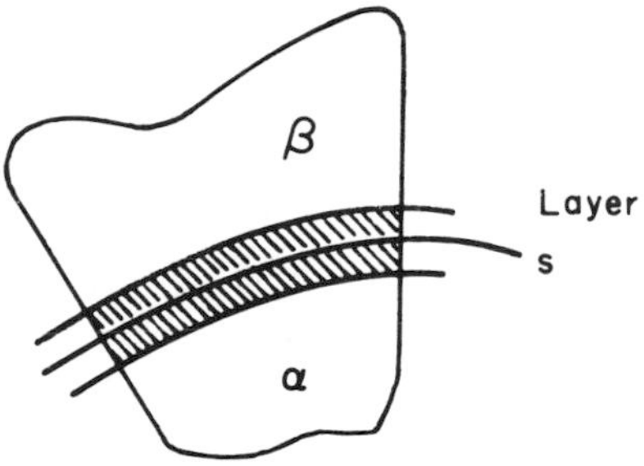

FIG. 11. Dividing surface s in a surface layer.

where F^a is the surface free energy, F^α the energy of region α, and F^β the energy of region β. If V^α and V^β are the volumes of phases α and β, f_v^α and f_v^β the free energy densities, f_s^a the surface free-energy density, and A the area, we may write

$$F^\alpha = V^\alpha f_v^\alpha, \qquad F^\beta = V^\beta f_v^\beta, \qquad F^a = A f_s^a. \tag{3.2}$$

In general, V^α, V^β, A, and f_s^a depend on the particular choice of the surface s of coordinate λ. Another dividing surface of coordinate $\lambda + d\lambda$ gives the variations

$$dV^\alpha = A\, d\lambda; \qquad dV^\beta = -A\, d\lambda; \qquad dA = (K_1 + K_2)A\, d\lambda, \tag{3.3}$$

where $K_1 + K_2$ is the mean curvature of the surface s, assumed to have the same value at each point of the surface. Since the total free energy F is independent of the choice of the dividing surface, we obtain

$$f_v^\alpha - f_v^\beta + (K_1 + K_2) f_s^a + df_s^a/d\lambda = 0. \tag{3.4}$$

A similar expression may be written for the entropies:

$$s_v^\alpha - s_v^\beta + (K_1 + K_2) s_s^a + ds_s^a/d\lambda = 0. \tag{3.5}$$

However, if n_γ is the total number of molecules of species γ in the system and C_γ^α and C_γ^β the concentrations in the two phases α and β, we find

$$n_\gamma = n_\gamma^\alpha + n_\gamma^\beta + n_\gamma^a \tag{3.6}$$

$$n_\gamma^\alpha = V^\alpha C_\gamma^\alpha, \qquad n_\gamma^\beta = V^\beta C_\gamma^\beta, \qquad n_\gamma^a = A\Gamma_\gamma, \tag{3.7}$$

where n_γ^a is the number of molecules adsorbed on the dividing surface and Γ_γ is the specific adsorption. Similarly, we obtain

$$C_\gamma^\alpha - C_\gamma^\beta + (K_1 + K_2)\Gamma_\gamma + d\Gamma_\gamma/d\lambda = 0. \tag{3.8}$$

C. Free Energy and Surface Work

For infinitesimal variations $\delta T, \ldots, \delta n_\gamma, \ldots$ and an external work $\delta\tau$, the variation of free energy is given by [see Chapter 2A, Eq. (4.14)]

$$\delta F = -S\,\delta T + \sum \mu_\gamma\,\delta n_\gamma + \delta\tau, \tag{3.9}$$

where $\delta\tau$ corresponds to displacements of the boundaries of the system and of the interface. In the case of the particular dividing surface s, we have

$$\delta\tau = -p^\alpha\,\delta V^\alpha - p^\beta\,\delta V^\beta + \tilde{\gamma}\,\delta A + A(C_1{}^*\,\delta K_1 + C_2{}^*\,\delta K_2), \tag{3.10}$$

where $\tilde{\gamma}$ is a tension which opposes the expansion of the layer, $C_1{}^*$ and $C_2{}^*$ are the rigidity coefficients which oppose a variation of the curvature. Substituting (3.10) in (3.9),

$$\delta F = -S\,\delta T + \sum_\gamma \mu_\gamma\,\delta n_\gamma - p^\alpha\,\delta V^\alpha - p^\beta\,\delta V^\beta + \tilde{\gamma}\,\delta A + A(C_1{}^*\,\delta K_1 + C_2{}^*\,\delta K_2). \tag{3.11}$$

Let us now consider a virtual variation which corresponds to the choice of a new surface s' (s and s' are parallel); we may write

$$\delta V^\alpha = A\,\delta\lambda, \qquad \delta V^\beta = -A\,\delta\lambda, \qquad \delta A = (K_1 + K_2)A\,\delta\lambda, \tag{3.12}$$

$$\delta K_1 = -K_1{}^2\,\delta\lambda, \qquad \delta K_2 = -K_2{}^2\,\delta\lambda. \tag{3.13}$$

The real system being unmodified, $\delta\tau = 0$; thus, we obtain the relation

$$p^\alpha - p^\beta = \tilde{\gamma}(K_1 + K_2) - C_1{}^*K_1{}^2 - C_2{}^*K_2{}^2, \tag{3.14}$$

which must be satisfied whatever the dividing surface.

If K_1 and K_2 are constant on the surface s and if the intensive variables $T, p^\alpha, p^\beta, \tilde{\gamma}, \ldots, \mu_\gamma, \ldots$ are given, it is easy to show that (Euler's theorem)

$$F = \sum_\gamma \mu_\gamma n_\gamma - p^\alpha V^\alpha - p^\beta V^\beta + \tilde{\gamma} A. \tag{3.15}$$

The invariance of F in comparison with the dividing surface then gives

$$p^\alpha - p^\beta = \gamma(K_1 + K_2) + d\tilde{\gamma}/d\lambda. \tag{3.16}$$

A comparison with the expression for F obtained in the previous section yields the following relations:

$$F^a = \sum_\gamma \mu_\gamma n_\gamma{}^a + \tilde{\gamma} A \tag{3.17}$$

$$f_s{}^a = \sum_\gamma \mu_\gamma \Gamma_\gamma + \tilde{\gamma}. \tag{3.18}$$

D. Primary Role of the Dividing Surface

If $K_1 = K_2$, then $C_1{}^* = C_2{}^*$; this is the special case of a plane interface. Since the thickness of the transition layer is always small with respect to the principle curvature radii, $C_1{}^*$ and $C_2{}^*$ are approximately equal to C^*, the coefficient corresponding to a plane interface. Equations (3.11) and (3.14) then become

$$dF = -S\,dT + \sum_\gamma \mu_\gamma\,dn_\gamma - p^\alpha\,dV^\alpha - p^\beta\,dV^\beta + \tilde{\gamma}\,dA + AC^*\,d(K_1 + K_2) \tag{3.19}$$

and

$$p^\alpha - p^\beta = \tilde{\gamma}(K_1 + K_2) - C^*(K_1{}^2 + K_2{}^2). \tag{3.20}$$

In comparison with the Laplace formula (1.4) for to the surface tension, it appears that the particular unrigid dividing surface ($C^* = 0$) coincides with the surface of tension. If λ_0 is the coordinate of this particular dividing surface, then $\gamma = \tilde{\gamma}(\lambda_0)$, and the variations of free energy will be

$$dF = -S\,dT + \sum_\gamma \mu_\gamma\,dn_\gamma - p^\alpha\,dV^\alpha - p^\beta\,dV^\beta + \gamma\,dA. \tag{3.21}$$

From (3.16) and (3.20), we obtain

$$d\tilde{\gamma}/d\lambda = -C^*(K_1{}^2 + K_2{}^2) = p^\alpha - p^\beta - \tilde{\gamma}(K_1 + K_2); \tag{3.22}$$

thus, on the surface of tension

$$d\tilde{\gamma}/d\lambda = 0, \qquad d^2\tilde{\gamma}/d\lambda^2 = \tilde{\gamma}(K_1{}^2 + K_2{}^2) > 0. \tag{3.23}$$

It follows that, for the surface of tension, $\tilde{\gamma}(\lambda)$ is minimum. This remarkable property was used by Gibbs (1928) in his fundamental work. On differentiation of (3.22), we see that

$$dC^*/d\lambda = -\tilde{\gamma} + C^*[(K_1 - K_2)^2(K_1 + K_2)/(K_1{}^2 + K_2{}^2)]. \tag{3.24}$$

For a plane or spherical interface, Eq. (3.24) reduces to

$$dC^*/d\lambda = -\tilde{\gamma}. \tag{3.25}$$

The ratio C^*/γ is more or less equal to 10^{-7} cm and we deduce therefore that a variation of λ of an amount of a few angströms is sufficient to modify the sign of C^*.

E. Thermodynamic Relations

We shall adopt here the surface of tension as the dividing surface. As we saw,

$$dF = -S\,dT + \sum_\gamma \mu_\gamma\,dn_\gamma - p^\alpha\,dV^\alpha - p^\beta\,dV^\beta + \gamma\,dA; \tag{3.26}$$

hence,

$$\gamma = (\partial F/\partial A)_{TV^\alpha V^\beta n}. \tag{3.27}$$

We can divide each of the extensive quantities F, S, and n_γ into three parts:

$$\begin{aligned} F &= F^\alpha + F^\beta + F^a \\ S &= S^\alpha + S^\beta + S^a \\ n_\gamma &= n_\gamma^\alpha + n_\gamma^\beta + n_\gamma^a. \end{aligned} \tag{3.28}$$

For the two volume phases

$$\begin{aligned} dF^\alpha &= -S^\alpha\,dT - p^\alpha\,dV^\alpha - \sum_\gamma \mu_\gamma\,dn_\gamma^\alpha \\ dF^\beta &= -S^\beta\,dT - p^\beta\,dV^\beta - \sum_\gamma \mu_\gamma\,dn_\gamma^\beta, \end{aligned} \tag{3.29}$$

and thus

$$dF^a = -S^a\,dT + \gamma\,dA + \sum_\gamma \mu_\gamma\,dn_\gamma^a. \tag{3.30}$$

The interface may thus be considered as a two-dimensional phase of area A and free energy F^a, with

$$S^a = -\left(\frac{\partial F^a}{\partial T}\right)_{An^a}, \qquad \mu_\gamma = \left(\frac{\partial F^a}{\partial n_\gamma{}^a}\right)_{TAn_\beta{}^a}, \qquad \gamma = \left(\frac{\partial F^a}{\partial A}\right)_{Tn^a}. \tag{3.31}$$

Furthermore, from (3.30), we have

$$\frac{\partial F^a}{\partial n_\gamma{}^\alpha} = \frac{\partial F^a}{\partial n_\gamma{}^\beta} = \frac{\partial F^a}{\partial V^\alpha} = \frac{\partial F^a}{\partial V^\beta} = 0. \tag{3.32}$$

This property is only true for extensive variables.

Indeed, the surface phase is nonautonomous; it depends directly on the intensive properties of the neighboring phases; for example, the dependence between F^a and $n_\gamma{}^a$ is given by

$$\mu_\gamma = \left(\frac{\partial F^a}{\partial n_\gamma{}^a}\right)_{Tn_\beta{}^a A}. \tag{3.33}$$

Let us now consider the specific surface free energy f^a defined by

$$f^a = F^a/A. \tag{3.34}$$

By differentiation, we obtain

$$df^a = \frac{dF^a}{A} - \frac{F^a}{A^2}\,dA. \tag{3.35}$$

From (3.35) and (3.30), we have

$$df^a = \frac{1}{A}\left[-S^a\,dT + \sum_\gamma \mu_\gamma\,dn_\gamma{}^a + \gamma\,dA\right] - \frac{F^a}{A^2}\,\delta A. \tag{3.36}$$

The variation of $n_\gamma{}^a$ may be expressed in terms of specific adsorption Γ_γ:

$$dn_\gamma{}^a = A\,dF_\gamma + \Gamma_\gamma\,dA. \tag{3.37}$$

If $dn_\gamma{}^a$ in Eq. (3.36) is replaced by the value given by (3.37), we obtain

$$df^a = s^a\,dT + \sum_\gamma \mu_\gamma\,d\Gamma_\gamma + (dA/A)(\gamma + \Sigma\,\mu_\gamma\Gamma_\gamma - f^a), \tag{3.38}$$

where the specific entropy s^a is given by the ratio S^a/A. Because f^a is not function of A,

$$df^a = -s^a\,dT + \sum_\gamma \mu_\gamma\,d\Gamma_\gamma \tag{3.39}$$

$$f^a = \gamma + \sum_\gamma \mu_\gamma\Gamma_\gamma, \tag{3.40}$$

with

$$s^a = -(\partial f^a/\partial T)_\Gamma \tag{3.41}$$

$$\mu_\gamma = (\partial f^a/\partial\Gamma_\gamma)_{T\Gamma_\beta}. \tag{3.42}$$

Then, from the differential of (3.40) and (3.39), there follows the well-known Gibbs equation

$$d\gamma = -s^a\, dT - \sum_{\gamma} \Gamma_\gamma\, d\mu_\gamma. \tag{3.43}$$

Thus,

$$s^a = -(\partial\gamma/\partial T)_\mu \tag{3.44}$$

$$\Gamma_\gamma = -(\partial\gamma/\partial\mu_\gamma)_{T\mu_\beta} \tag{3.45}$$

where the subscript μ_β indicates that the derivative is taken keeping all the μ's except μ_γ constant. This can, however, be achieved only if the system has at least $c + 1$ degrees of freedom $T, \mu_1, \ldots, \mu_c$.

For a plane surface, the highest possible variance corresponds to systems where there are no chemical reactions among the components and where the surface has only one phase on it. For a plane surface, it is impossible to vary separately all the $c + 1$ variables $T, \mu_1, \ldots, \mu_c$. It follows that the Gibbs equation (3.43) does not enable the adsorption Γ_γ to be determined on a plane surface. A determination of the adsorption is theoretically possible on a curved surface, for, in this case, the variance is $c + 1$ (for the variance of a capillary system, see Section IV).

F. Plane Surface and Relative Adsorptions

For a plane surface, the variables $T, \mu_1, \ldots, \mu_c$ cannot be varied independently. We shall therefore examine the way in which μ_1 varies as a function of $T, \mu_2, \ldots, \mu_c$.

The Gibbs–Duhem equations for the two bulk phases may be written

$$\begin{aligned} dp^\alpha &= s_v^\alpha\, dT + \sum_\gamma C_\gamma^\alpha\, d\mu_\gamma \\ dp^\beta &= s_v^\beta\, dT + \sum_\gamma C_\gamma^\beta\, d\mu_\gamma, \end{aligned} \tag{3.46}$$

where s_v^α and s_v^β are the entropy densities of phases α and β.

Moreover, for a plane surface, the mechanical equilibrium gives

$$p^\alpha = p^\beta, \tag{3.47}$$

so that we obtain for the variation of μ_1

$$d\mu_1 = -\frac{s_v^\alpha - s_v^\beta}{C_1^\alpha - C_1^\beta}\, dT - \sum_{\gamma\geq 2} \frac{C_\gamma^\alpha - C_\gamma^\beta}{C_1^\alpha - C_1^\beta}\, d\mu_\gamma. \tag{3.48}$$

We may now replace $d\mu_1$ in the Gibbs equation (3.43) to give

$$d\gamma = -(s^a)_1\, dT - \sum_{\gamma \geq 2} \Gamma_{\gamma 1}\, d\mu_\gamma, \tag{3.49}$$

where the quantities $(s^a)_1$ and $\Gamma_{\gamma 1}$ are, respectively, the relative surface entropy and the relative adsorption defined by

$$\begin{aligned} (s^a)_1 &= s^a - \Gamma_1 \frac{s_v^\alpha - s_v^\beta}{C_1^\alpha - C_1^\beta} \\ \Gamma_{\gamma 1} &= \Gamma_\gamma - \Gamma_1 \frac{C_\gamma^\alpha - C_\gamma^\beta}{C_1^\alpha - C_1^\beta}. \end{aligned} \tag{3.50}$$

Written in this form, the Gibbs equation is very much more useful, since the variables $T, \mu_2, \ldots, \mu_c$ can be completely independent. It is therefore possible to determine the relative surface entropy and relative adsorption by determining experimentally the differential coefficients

$$(s^a)_1 = -\left(\frac{\partial \gamma}{\partial T}\right)_{\mu_1 \ldots \mu_c} \tag{3.51}$$

$$\Gamma_{\gamma 1} = -\left(\frac{\partial \gamma}{\partial \mu_\gamma}\right)_{T\mu_{\beta \neq 1}}. \tag{3.52}$$

Example. The liquid phase consists of a saturated solution of nitrogen (subscript 2) in water (subscript 1) and the gaseous phase of nitrogen and water vapor. At moderately low pressures, the fugacity of nitrogen may be equated to the partial pressure p_2, so that, at constant temperature,

$$d\mu_2 = RT\, dp_2/p_2. \tag{3.53}$$

The variation of the surface tension will be

$$d\gamma = -\Gamma_{21}(RT/p_2)\, dp_2 \tag{3.54}$$

and

$$\Gamma_{21} = -(p_2/RT)\, d\gamma/dp_2. \tag{3.55}$$

Suppose that we find that at 17°C the surface tension falls by 0.1 dyn cm^{-1} when the pressure is increased from 1 to 2 atm. Then, taking the mean pressure p_2 as 1.5 atm, we have (Defay *et al.*, 1966, Chapter VII)

$$\Gamma_{21} \simeq 0.62 \times 10^{-11} \quad \text{mol cm}^{-2}.$$

Other examples can be found in the book by Defay and co-workers (1966, Chapter VII).

G. Influence of Curvature on the Surface Tension

We restrict our discussion to spherical surfaces and to one-component systems. At constant temperature,

$$dp^\alpha = C^\alpha \, d\mu, \qquad dp^\beta = C^\beta \, d\mu, \qquad d\gamma = -\Gamma \, d\mu. \tag{3.56}$$

For the surface of tension, the differential form may be written

$$dp^\alpha - dp^\beta = (2/r) \, d\gamma + 2\gamma \, d(1/r). \tag{3.57}$$

From (3.56) and (3.57), we get

$$\Gamma = -\frac{1}{2}(C^\alpha - C^\beta)\frac{d\gamma}{d(\gamma/r)} = -\frac{1}{2}(C^\alpha - C^\beta)\frac{d(\ln \gamma)/d(1/r)}{1 + (1/r)\, d(\ln \gamma)/d(1/r)}. \tag{3.58}$$

To determine Γ, we must measure the variation of γ with the curvature of the surface. From an experimental point of view, we are limited to radii $\geq 10^{-3}$ cm, and, in this range, it is impossible to observe a variation of γ.

By integration of (3.58) for small values of $1/r$, we obtain

$$\ln\left(\frac{\gamma}{\gamma_\mathrm{p}}\right) \simeq -\frac{2\Gamma}{C^\alpha - C^\beta}\left(\frac{1}{r}\right), \tag{3.59}$$

where γ_p is the surface tension of corresponding plane surface.

Example. Let us consider a drop of radius equal to 10^{-3} cm.

If $C^\alpha - C^\beta = 10^{22}$ molecules cm^{-3} and $\Gamma = 10^{15}$ molecules cm^{-2}, we find a relative variation for γ of about 0.01%.

H. Surface Activity

Every dissolved substance which modifies the surface tension of the solvent is called a surface-active agent. These substances tend to accumulate in the surface layer. They are made up primarily of a polar or an ionic soluble group and of an insoluble part (for example, an aliphatic chain). The Gibbs equation gives the relation between the relative adsorption of the surface-active agent (index s) and the slope of the γ–C_s curve,

$$\partial\gamma/\partial C_s = -\Gamma_{s1}kT/C_s. \tag{3.60}$$

The surface activity of a substance is defined as the initial value $-(\partial\gamma/\partial C_s)_{C_s\to 0}$ of the slope of the curve against C_s in the bulk phase (Fig. 12).

Traube showed that the surface activity is directly related to the length of the molecule. For different normal alcohols, he found that the surface activity was about three times higher than the previous value each time the chain was increased by one —CH_2— group. Detergents are surface-active agents at water–solid interfaces. They lower the surface tension of the interface and allow the wetting of the solid.

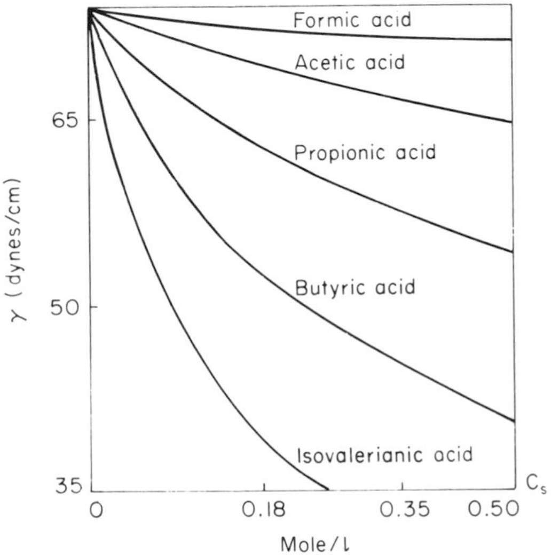

FIG. 12. Surface tension as a function of the concentration of the dissolved substance.

Remark. It is well known that many surface-active agents are capable of profoundly affecting the state of dispersion of the particles of a disperse phase (solid, liquid, or gaseous) in a liquid medium. If the tendency for spontaneous mutual adhesion to occur among the disperse particles in a system is diminished by the addition of a surface-active agent, we can say that it has a "deflocculating" action; if, on the other hand, the tendency toward mutual adhesion is increased, then the surface-active agent can be conveniently referred to as a "flocculating agent." This problem is important in colloid chemistry (Frens, 1968).

IV. Phase Rule

Consider a system of c independent components and φ bulk phases separated by s surfaces, each comprising one or several surface phases (for a treatment of the phase rule in cases where there may be chemical reactions, see (Defay and Prigogine, 1951)). We denote by $\psi \geq s$ the total number of surface phases.

The physicochemical state of the system considered is defined by the intensive variables: T (uniform); $C_1, \ldots, C_c$ (in each volume phase); $\Gamma_1, \ldots, \Gamma_c$ (in each surface phase); and $r^1, \ldots, r^\psi$ (curvature of each surface phase). The system is defined by $1 + c(\varphi + \psi) + \psi$ intensive variables. If we wish the system to be in equilibrium, we can no longer arbitrarily fix all these variables, for, in equilibrium, they are limited by the following conditions:

(a) On a given surface, the different surface phases have the same surface tension.

(b) The Laplace equation must be satisfied for each surface phase.

(c) The chemical potential of each component has the same value for each bulk or surface phase.

The total number of conditions is thus $(\psi - s) + \psi + c(\varphi + \psi - 1)$, and the number of remaining independent variables at equilibrium, a number which we call the variance of the system (see Chapter 2A, Section V) is therefore

$$w = 1 + c - (\psi - s) \tag{4.1}$$

if we suppose, as above, that the only reactions within the system are transfer reactions.

Example. For a two-phase (liquid–gas), one-component system, the variance is $w = 2$, which means that the vapor pressure of a liquid drop depends on the temperature and on its radius. If the whole surfaces are plane, we have $\varphi - 1$ relations

$$p^1 = \cdots p^\alpha = \cdots p^\varphi, \tag{4.2}$$

and the radii are infinite. It is easy to show (Defay *et al.*, 1966, Chapter VI) that, in this case,

$$w = 2 + c - \varphi - (\psi - s) \qquad \text{(plane interfaces).} \tag{4.3}$$

Finally, if each surface comprises only a single surface phase, $\psi = s$, and we again find the classical phase rule (see Chapter 2A, Section V)

$$w = 2 + c - \varphi. \tag{4.4}$$

V. Influence of Temperature on Surface Tension

We shall limit the study of the dependence on the temperature of the surface tension to single molecules (rare gases or quasispherical molecules). The surface tension decreases quite linearily with an increase in temperature from the triple point to the critical point. At the critical point, $\gamma = 0$, $s^a = 0$, and $\Gamma = 0$; thus, from the Gibbs equation, it follows that

$$d\gamma/dT = 0 \qquad \text{(critical point);} \tag{5.1}$$

the following experimental result is often used

$$\gamma \propto (T_c - T)^n, \tag{5.2}$$

where T_c is the critical temperature and n is a coefficient with $1.20 \leq n \leq 1.25$.

In the triple point–boiling point region, where $0.5 \leq T/T_c \leq 0.7$, Eq. (5.2) may be replaced by

$$\gamma \propto (aT_c - T), \tag{5.3}$$

with $a \simeq 0.93$. Experimental values of a are summarized (Fuks, 1967) in Table III.

TABLE III

	A	Kr	Xe	CH_4	N_2	O_2	CO
a	0.927	0.928	0.942	0.928	0.939	0.920	0.938

Other empirical equations are found in the literature:

1. *Eötvös Equation*

$$\gamma(v^l)^{2/3} = b(T_c - T), \tag{5.4}$$

where v^l is the molar volume of the liquid and where the coefficient b

is quite independent of the nature of the substance considered (for nonassociated substance) (see Table IV). Nevertheless, Eq. (5.4) is not valid in the neighborhood of the critical point [see Eq. (5.1)].

TABLE IV

	A	Kr	Xe	CCl_4	C_6H_6	C_6H_{12}	C_5H_{10}
b (ergs °K^{-1})	1.86	1.86	1.86	2.05	2.05	2.05	2.05

2. *The Parachor*

The concept of the parachor rests upon the equation of Kleeman (1910), Mac Leod (1923), and Sugden (1930), who found

$$\gamma^{1/4}[(v^l)^{-1} - (v^{\mathrm{v}})^{-1}] = P, \tag{5.5}$$

where v^{v} is the molar volume in the vapor phase and P is a constant characteristic of each substance. The experimental study of the critical region shows that

$$(v^l)^{-1} - (v^{\mathrm{v}})^{-1} \propto (T_c - T)^{\beta}, \tag{5.6}$$

with $0.3 \leq \beta \leq 0.35$. In the critical region, we then obtain the relation

$$\gamma \propto (T_c - T)^{4\beta} \propto (T_c - T)^{1.3 \pm 0.1}, \tag{5.7}$$

which is in a good agreement with (5.2).

The Eötvös formula breaks down in the immediate neighborhood of the critical point, and was modified by Nakayama, who replaced the factor $v^{2/3}$ by $y^{2/3}$,

$$\gamma y^{2/3} = b(T - T_c), \tag{5.8}$$

where $y^{-1} = (v^l)^{-1} - (v^{\mathrm{v}})^{-1}$. Away from T_c, $y \simeq v^l$, and (5.8) reduces to (5.4). From (5.8) and (5.7), we again find (5.2) with $n = 1 + \frac{2}{3}\beta \simeq 1.22 \pm 0.02$. Buff and Lovett (1968) recently obtained $n = 1.27 \pm 0.02$.

VI. Properties of Monolayers

A. Introduction

The study of the properties of monolayers spread out on a liquid has been the subject of numerous publications since the beginning of the present century (Davies and Rideal, 1961, Chapter 5; Gaines, 1966). Authors have, in particular, displayed great interest in the state equations deriving the various plausible formulations of surface pressure as well as of the chemical potential of the constituents in the layer. Monolayers are formed from molecules like higher fatty acids or polymers consisting of a hydrophobic and a hydrophilic part. These molecules, insoluble in water, spread out as a film with its end, the COOH group, wetted by water (hydrophilic), and its long hydrocarbon chain (hydrophobic) tending to leave the substrate. The same phenomenon arises at the oil–water interface. Interesting properties of monolayers can be investigated by pressure–area, surface-viscosity, and surface electrical-potential measurements (Guastalla, 1947; Davies and Rideal, 1961).

B. Surface Pressure

The surface pressure of a monolayer is the lowering of surface tension due to the monolayer. The molecules contained in the monolayer may be regarded as exerting a two-dimentional osmotic pressure; there is a repulsion in the plane of the surface which is measured on a floating barrier acting as a semipermeable membrane permeable to water only (Guastalla, 1947). It is this pressure opposing the contractile tension of the clean interface that is called the surface pressure Π

$$\Pi = \gamma_0 - \gamma, \qquad (6.1)$$

where γ_0 is the surface tension of the clean surface. The variation of Π with the area available to the surface-active material is represented by a Π–Ω curve, where Ω is the area per molecule.

Some measurements of this type are shown in Fig. 13 for a homologous series of fatty acids. Qualitatively, the results are the same as for the isotherms of an imperfect gas condensing to a liquid. However, instead of changing temperature, which is difficult on a water surface, as the number of carbon atoms increases, the hydrophilic forces gain over the hydrophobic part of the molecule, and so the film has a great tendency

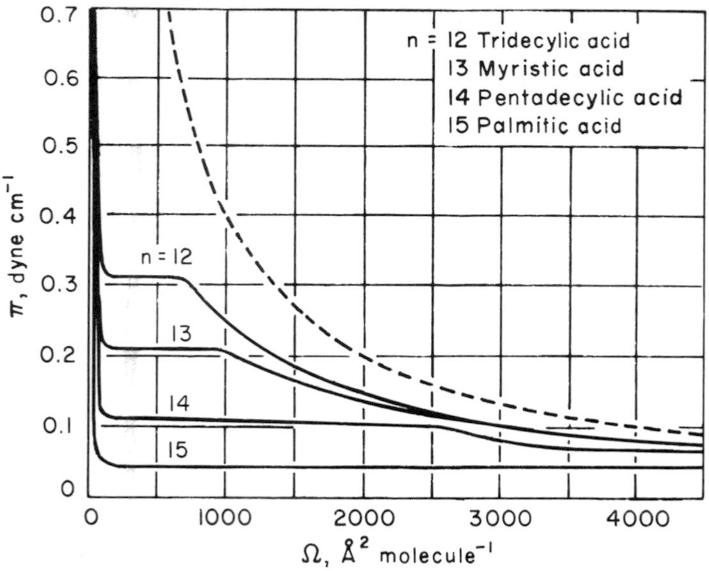

FIG. 13. Pressure–area phase diagram for a fatty-acid film spread on a water surface.

to condense. The dashed curve in Fig. 13 is the ideal gas law in two dimensions,

$$\Pi = kT/a. \tag{6.2}$$

If the film is dilute enough, that is, if the area per molecule *a* is large enough, the ideal behavior is approached. At higher densities, two-dimensional condensation may take place. As the phase on the surface is further compressed, it is bound to break out of the two-dimensional film that is only one molecule thick. Ultimately, the surface film becomes thick enough so that it is properly treated as an ordinary three-dimensional phase. If molecules of the insoluble component are so tightly packed that further compression in the form of a film becomes impossible, then they pile up to form a crystal or a floating lens which grows steadily as the area of the surface is progressively decreased.

C. Surface Viscosity

A monolayer is resistant to shear stress in the plane of the surface just as, in bulk, a liquid is retarded in its flow by viscous forces. The viscosity of the monolayer may indeed be measured in two dimensions by flow through a canal in a surface or by its drag on a ring in the surface, corresponding to the Ostwald and Couette instruments for the study of bulk viscosities. The relation between surface viscosity η_s and bulk viscosity η is given by

$$\eta_s = \eta d, \tag{6.3}$$

where d is the thickness of the surface phase (about 10^{-7} cm for many monolayers). From measurements of viscosities, it appears that hydrocarbon chains are strongly oriented.

Remark. A thermodynamic analysis of the surface potential was performed by Koenig twenty years ago (Koenig, 1951), on the base of the Volta effect. The air–water and oil–water surface potentials of monolayers in connection with the electrical double layer were developed in a great detail by Davies and Rideal (1961).

D. Gibbs's Surface Model

Let us first consider a solution of electrolyte in contact with a vapor phase. On the plane interface, a monolayer is spread out. We call $\gamma(1, \ldots, c)$ the constituents of the system, including the vapor and the liquid phases. We suppose that our system consists of laminae $\alpha = 1, \ldots, \infty$ each parallel to the interface, and containing constituents able to pass through it.

Only lamina β contains the monolayer, i.e., the insoluble nondissociated molecule (subscript 3) together with the insoluble ion of the dissociate molecule (subscript 2). For such an electrochemical system, Defay and colleagues (1966, Eq. (21.28)) have extended Gibbs's formulation outside thermodynamic equilibrium, in the molar form

$$A\,d\gamma = -\tilde{S}\,dT + \sum_{\alpha} V^{\alpha}\,dp_e - \sum_{\alpha}\sum_{\gamma=3}^{c} n_{\gamma}^{\alpha}\,d\tilde{\mu}_{\gamma}^{\alpha} - n_3\,d\tilde{\mu}_3^{\beta} - n_2\,d\tilde{\mu}_2^{\beta}, \tag{6.4}$$

where $\tilde{\mu}_{\gamma}^{\alpha}$ is the electrochemical potential of γ in phase α, equal to the sum $\mu_{\gamma}^{\alpha} + z_{\gamma}\varphi^{\alpha}$, where z_{γ} is the charge of γ and φ^{α} the electrical potential in phase α; A is the area of the plane interface; p_e is the external pressure (in the absence of gravity, p_e is equal to the pressure in the bulk phase); $\tilde{S}$ is the total entropy of the system; and V^{α} is the volume of lamina α.

Let us now apply the Gibbs surface model to Eq. (6.4) for an insoluble monolayer R–H spread on a surface β of a dilute aqueous solution of hydrochloric acid. We assume that thermodynamic equilibrium is achieved. We first place the dividing surface under the monolayer, and call V' and V'' the two volumes (the liquid and the vapor phase) delimited by this surface of area A; we get

$$\Gamma_{\mathrm{R-H}} = \frac{n^{\beta}_{\mathrm{R-H}}}{A}, \quad \Gamma_{\mathrm{R^-}} = \frac{n^{\beta}_{\mathrm{R^-}}}{A}, \quad \Gamma_{j\neq 1,2} = \frac{n_j - V'C_j' - V''C_j''}{A}. \tag{6.5}$$

At constant temperature and pressure, we may write

$$(d\gamma)_{T,p_e} = -\Gamma_{\mathrm{H_2O}}\, d\mu'_{\mathrm{H_2O}} - \Gamma_{\mathrm{H^+}}\, d\tilde{\mu}'_{\mathrm{H^+}} - \Gamma_{\mathrm{OH^-}}\, d\tilde{\mu}'_{\mathrm{OH^-}} - \Gamma_{\mathrm{Cl^-}}\, d\tilde{\mu}'_{\mathrm{Cl^-}} - \Gamma_{\mathrm{R-H}}\, d\mu^{\beta}_{\mathrm{R-H}} - \Gamma_{\mathrm{R^-}}\, d\tilde{\mu}^{\beta}_{\mathrm{R^-}}, \tag{6.6}$$

where the μ' are the chemical potentials in the bulk of the homogeneous liquid phase. The equilibrium condition may be written

$$\mu^{\beta}_{\mathrm{R-H}} = \tilde{\mu}^{\beta}_{\mathrm{R^-}} + \tilde{\mu}'_{\mathrm{H^+}}, \qquad \mu'_{\mathrm{H_2O}} = \tilde{\mu}'_{\mathrm{H^+}} + \tilde{\mu}'_{\mathrm{OH^-}}. \tag{6.7}$$

If the surface layer is neutral as a whole,

$$\Gamma_{\mathrm{H^+}} = \Gamma_{\mathrm{OH^-}} + \Gamma_{\mathrm{Cl^-}} + \Gamma_{\mathrm{R^-}}. \tag{6.8}$$

Combining (6.6), (6.7), and (6.8), we get

$$d\gamma = -\Gamma^{\mathrm{t}}_{\mathrm{H_2O}}\, d\mu'_{\mathrm{H_2O}} - 2\Gamma_{\mathrm{Cl^-}}\, d\mu'_{\mathrm{H^+Cl^-}} - \Gamma^{\mathrm{t}}_{\mathrm{R-H}}\, d\mu^{\beta}_{\mathrm{R-H}}, \tag{6.9}$$

where the total adsorptions are defined by

$$\Gamma^{\mathrm{t}}_{\mathrm{H_2O}} = \Gamma_{\mathrm{H_2O}} + \Gamma_{\mathrm{OH^-}}, \qquad \Gamma^{\mathrm{t}}_{\mathrm{R-H}} = \Gamma_{\mathrm{RH}} + \Gamma_{\mathrm{R^-}} \tag{6.10}$$

and

$$d\tilde{\mu}'_{\mathrm{H^+}} + d\tilde{\mu}'_{\mathrm{Cl^-}} = 2\, d\mu'_{\mathrm{H^+Cl^-}}. \tag{6.11}$$

On the other hand, we know that, at T and p_e constant, the chemical potentials $\mu'_{\mathrm{H_2O}}$ and $\mu'_{\mathrm{H^+Cl^-}}$ appearing in Eq. (6.9) are not independent variables; actually, in the bulk phase, we may write

$$C'_{\mathrm{H_2O}}\, d\mu'_{\mathrm{H_2O}} + C'_{\mathrm{H^+}}\, d\tilde{\mu}'_{\mathrm{H^+}} + C'_{\mathrm{OH^-}}\, d\tilde{\mu}'_{\mathrm{OH^-}} + C'_{\mathrm{Cl^-}}\, d\tilde{\mu}'_{\mathrm{Cl^-}} = 0 \tag{6.12}$$

and

$$C'_{\mathrm{H^+}} = C'_{\mathrm{OH^-}} + C'_{\mathrm{Cl^-}}. \tag{6.13}$$

Now, combining Eqs. (6.7) and (6.11)–(6.13), we find that

$$C'^{\mathrm{t}}_{\mathrm{H_2O}}\, d\mu'_{\mathrm{H_2O}} + 2C'_{\mathrm{Cl^-}}\, d\mu'_{\mathrm{H^+Cl^-}} = 0, \tag{6.14}$$

where

$$C'^{t}_{H_2O} = C'_{H_2O} + C'_{OH^-}. \tag{6.15}$$

Thus, from (6.14), in a dilute solution of HCl,

$$C'^{t}_{H_2O} \gg C'_{Cl^-} \qquad \text{and} \qquad d\mu'_{H_2O} \ll d\mu'_{H^+Cl^-}.$$

If Γ_{Cl^-} is not too small compared with $\Gamma^{t}_{H_2O}$, (6.9) then becomes

$$d\gamma = -2\Gamma_{Cl^-}\, d\mu'_{H^+Cl^-} - \Gamma^{t}_{R-H}\, d\mu^{\beta}_{R-H} \tag{6.16}$$

If the spread-out film is in a gaseous state, there are many water molecules in β; $\Gamma^{t}_{H_2O}$ is thus large for the reference surface, and $\Gamma_{Cl^-} \ll \Gamma^{t}_{H_2O}$ because the diffuse layer is not important. The term $d\mu'_{H_2O}$ may no longer be neglected.

In this case, it seems to be more convenient to use the relative adsorption in such a way that the Gibbs dividing surface yields $\Gamma^{t}_{H_2O} = 0$. The values of V' and V'' are then different from their values in the above cases, but the formulation remains valid and we may write, in terms of the relative adsorption Γ_{γ,H_2O}, the following equation:

$$d\gamma = -2\Gamma_{Cl^-,H_2O}\, d\mu'_{H^+Cl^-} - \Gamma^{t}_{R-H,H_2O}\, d\mu^{\beta}_{R-H}. \tag{6.17}$$

Measuring γ for given values of Γ^{t}_{R-H,H_2O} at constant temperature, pressure, and $\mu'_{H^+Cl^-}$, we then obtain the law

$$\gamma = \gamma(\Gamma^{t}_{R-H,H_2O}). \tag{6.18}$$

From the curve (6.18), we have

$$\frac{\partial\gamma}{\partial\Gamma^{t}_{R-H,H_2O}} = \frac{\partial\gamma}{\partial\mu^{\beta}_{R-H}}\,\frac{\partial\mu^{\beta}_{R-H}}{\partial\Gamma^{t}_{R-H,H_2O}} = -\Gamma^{t}_{R-H,H_2O}\,\frac{\partial\mu^{\beta}_{R-H}}{\partial\Gamma^{t}_{R-H,H_2O}}. \tag{6.19}$$

It is then possible to find the law $\mu^{\beta}_{R-H} = \mu^{\beta}_{R-H}(\Gamma^{t}_{R-H,H_2O})$ for a given value of $\mu'_{H^+Cl^-}$.

For example, if (6.18) is a linear law of the form

$$\gamma = \gamma^{i} - \theta_a \Gamma^{t}_{R-H,H_2O}, \tag{6.20}$$

where θ_a is a function of T, p_e, and $\mu'_{H^+Cl^-}$ We obtain

$$\partial\gamma/\partial\Gamma^{t}_{R-H,H_2O} = -\theta_a, \tag{6.21}$$

and, from (6.19),

$$(\theta_a/\Gamma^{\mathrm{t}}_{\mathrm{R-H,H_2O}})\, d\Gamma^{\mathrm{t}}_{\mathrm{R-H,H_2O}} = d\mu^{\beta}_{\mathrm{R-H}}. \tag{6.22}$$

Now integrating (6.22)

$$\mu^{\beta}_{\mathrm{R-H}} = \mu^{\beta i}_{\mathrm{R-H}} + \theta_a \ln \Gamma^{\mathrm{t}}_{\mathrm{R-H,H_2O}} \tag{6.23}$$

For different values of $\mu_{\mathrm{H^+Cl^-}}$, it is now possible to know the variation of θ_a with $\mu_{\mathrm{H^+Cl^-}}$.

Remark. It is possible that $\mu^{\beta i}_{\mathrm{R-H}}$ is also a function of $\mu_{\mathrm{H^+Cl^-}}$.

E. Various Explicit Formulations of the Chemical Potential in the Layer and in the Gibbs Dividing Surface

In absence of fields, for an ideal uncharged monolayer defined in the manner of Defay and Prigogine (see Defay *et al.* (1966), Chapter XII, § 5), we may write for each γ the expression:

$$\mu_\gamma{}^a = \zeta_\gamma{}^a(T, p) + RT \ln N_\gamma{}^c - \gamma\omega_\gamma, \tag{6.24}$$

where superscript c refers to the monolayer model and a to Gibbs' model. We also have

$$A = \sum_\gamma n_\gamma{}^c \omega_\gamma, \tag{6.25}$$

where ω_γ is the partial molar area of γ.

When the surface is saturated by only one component,

$$\omega_\gamma = A/n_\gamma{}^c = 1/\Gamma_\gamma^{c\infty}, \tag{6.26}$$

where $\Gamma_\gamma^{c\infty}$ is the saturation adsorption *in the layer* and not the Gibbs adsorption related to a division surface.

On the other hand, ω_γ may be a function of γ, i.e., the value of ω_γ may be slightly different in a saturated than in a nonsaturated layer. In the molecular models, it is always assumed that ω_γ is a constant quantity. For a nonideal monolayer in the absence of field, Eq. (6.24) must be replaced by (Defay *et al.* (1966), Chapter XII, § 5)

$$\mu_\gamma{}^c = \left(\frac{\partial F^c}{\partial n_\gamma{}^c}\right)_{Tp\gamma n_\beta{}^c} + pv_\gamma{}^c - \gamma\omega_\gamma, \tag{6.27}$$

where $v_\gamma{}^c$ is the molar volume of γ in the layer. The term $pv_\gamma{}^c$ is small compared to $\gamma\omega_\gamma$. Equation (6.27) is valid for the solvent and also for the solutes. In the case of regular solutions, many authors give the explicit expressions of F^c and $\mu_\gamma{}^c$ (see, for example, Defay *et al.* (1966) Chapter XII, § 8). With the Gibbs model, we may write [see (Defay *et al.*, 1966) (12.13) and (12.14)]

$$\mu_\gamma{}^a = \left(\frac{\partial F^a}{\partial n_\gamma{}^a}\right)_{TpAn_\beta{}^a} \tag{6.28}$$

$$\mu_\gamma{}^a = \left(\frac{\partial F^a}{\partial n_\gamma{}^a}\right)_{Tp\gamma n_\beta{}^a} - \gamma\omega_\gamma. \tag{6.29}$$

For nonideal uncharged systems in the absence of field, the use of the equation (Arcuri, 1966)

$$\mu_\gamma{}^a = \xi_\gamma{}^a(T, p) + RT \ln f_\gamma{}^a N_\gamma{}^a - \gamma\omega_\gamma \tag{6.30}$$

defines the activity coefficient $f_\gamma{}^a$.

Remark. Many authors have shown that the monolayer model is very coarse. Indeed, Defay *et al.* (1966, Chapter XII, § 10) proved that this model is inconsistent with the Gibbs formula. It seems better, following Defay *et al.* (1966) and Ono and Kondo (1960, p. 159), to use the multilayer model (infinite laminae) for which Eq. (6.24) may be written for each lamina with varying from lamina to lamina.

VII. Multilayer Model and Interfacial Orientation

A. Introduction

We know that, if a surface layer is many molecules thick, its composition may vary with position within the layer; this circumstance makes it physically reasonable to use the multilayer model developed by Defay and colleagues (1966), a model one could call intermediate between the continuous and the discontinuous models. The system is divided into uniform regions called phases; the nonuniform regions, such as the capillary layer, are subdivided into a number of laminae each sufficiently thin be considered as homogeneous.

Our purpose here is to develop capillary theory based on the multilayer model with a view to deriving an explicit formulation of interfacial orientation.

Excluding all microscopic fluctuation effects, this work deals only with systems for which orientation equilibrium occurs after the establishment of the diffusion equilibrium. All transport of matter from one region to another may be treated as a transfer of one or several components from one phase to another.

The only entropy production sources are, on the one hand, the chemical reactions and the transport from one phase to another, and, on the other hand, the orientation of every component which occurs in the laminae. Thus, in addition to the classical chemical and transport affinities, we shall have orientation affinities.

In Defay's work (1966), the orientation of the components is not treated as an independent variable. This means that the molecules, while moving from one phase to another, are supposed to be always in instantaneous orientation equilibrium with the dipole structure of the successive laminae.

If, on the contrary, the orientation equilibrium is reached after the diffusion equilibrium, orientation variables independent of diffusion variables should clearly appear in the thermodynamic formalism. This will enable us to show how the Gibbs formula may easily be extended to these systems and, for illustrative purposes, we shall discuss a very simple example of surface orientation.

Remark. The extension of our theory to electrocapillary systems has also been made (Sanfeld, 1968, Chapter 14).

B. Thermodynamics of a Closed Capillary System

We adopt the multilayer model in which each lamina, within the surface layer is considered as a homogeneous phase of infinitesimal thickness. Let us now consider a system at uniform temperature and in mechanical equilibrium, unable to exchange molecules with the surroundings. In particular, for a system containing only one plane interface, the work $d\tau$ done on the system by its surroundings is given by (1.17):

$$d\tau = -p_e \, dV + \gamma \, dA, \tag{7.1}$$

where p_e is the external (uniform) pressure acting on the system, and $V = \sum_\alpha V^\alpha$ is the total volume equal to the sum of the volumes of the individual phases (bulk phases I and II and laminae). The equations, derived from the first and second laws of thermodynamics, have the form [see Chapter 1 and Chapter 2A, Eq. (1.11)]

$$dU = dQ - \sum_{\alpha} p_{\mathrm{e}}\, dV^{\alpha} + \gamma\, dA \tag{7.2}$$

and

$$dS = d_{\mathrm{e}}S + d_{\mathrm{i}}S, \tag{7.3}$$

with $d_{\mathrm{i}}S \geq 0$. Because of the way in which the system has been defined (no thermal flux, no hydrodynamic motion), the only possible sources of entropy production are the chemical reactions, the diffusion of molecules from one part of the system to another, and the orientation variations in each lamina within the surface layer. The variations in orientation are due to strong interactions between molecules or atoms belonging to the same lamina or to neighboring laminae.

We now define the degree of advancement introduced by De Donder (1922) [see Eq. (2.1) Chapter 2A] for all possible reactions (matter transport, i.e., passage of one or more components from one phase to another, chemical, and orientation reactions). We have

$$n_{\gamma}^{\alpha} - n_{\gamma}^{\alpha 0} = {}^{\alpha-1}\xi_{\gamma}^{\ \alpha} - {}^{\alpha}\xi_{\gamma}^{\alpha+1} + \sum_{r} \nu_{\gamma r}^{\alpha} \xi_{r} \tag{7.4}$$

$$\langle m_{x_i\gamma}^{\alpha} \rangle - \langle m_{x_i\gamma}^{\alpha} \rangle^{0} = \xi_{x_i\gamma O}^{\alpha} \tag{7.5}$$

$$\langle m_{x_i\gamma}^{\alpha} \rangle - \langle m_{x_i\gamma}^{\alpha 2} \rangle^{0} = \xi_{x_i\gamma Q}^{\alpha}, \tag{7.6}$$

where ${}^{\alpha-1}\xi_{\gamma}^{\ \alpha}$ and ${}^{\alpha}\xi_{\gamma}^{\alpha+1}$ are, respectively, the degrees of advancement of the transport reactions of constituent γ from phase $\alpha - 1$ to phase α and from α to $\alpha + 1$. The index r refers here to the chemical reactions, the superscript 0 to the time $t = 0$ (origin of ξ), and $\langle m_{x_i\gamma}^{\alpha} \rangle$ and $\langle m_{x_i\gamma}^{\alpha 2} \rangle$ are the orientation variables, i.e., the mean projection and the mean-square projection on the axes x_i ($i = 1, 2, 3$) of the dipole moment per mole of γ in the phase α. Let us remark that the distribution function of the orientation may be described by the six projections $\langle m_{x_i} \rangle$ and $\langle m_{x_i}^2 \rangle$ only if we assume the classical approximation of a Gaussian distribution. The degrees of orientation $\xi_{x_i\gamma O}^{\alpha}$ and $\xi_{x_i\gamma Q}^{\alpha}$ are defined by (7.5) and (7.6).

If the molecular orientation varies during the crossing from one lamina to another, then variables ${}^{\alpha-1}\xi_{\gamma}^{\ \alpha}$, $\xi_{x_i\gamma O}^{\alpha}$, and $\xi_{x_i\gamma Q}^{\alpha}$ vary together during the crossing.

Usually, in capillary theory, orientations and diffusions are treated as independent variables, i.e., the orientation and diffusion occur simultaneously. Here, we consider the case of independent variables. An extension of De Donder's equation (2.1) Chapter 2A for all the possible reactions leads to

$$T\,d_iS = \sum_{\varrho} A_\varrho\,d\xi_\varrho + \sum_{\alpha\gamma i} A^\alpha_{x_i\gamma O}\,d\xi^\alpha_{x_i\gamma O} + \sum_{\alpha\gamma i} A^\alpha_{x_i\gamma Q}\,d\xi^\alpha_{x_i\gamma Q}, \tag{7.7}$$

where the summation symbol $\Sigma_{\alpha\gamma i}$ represents the triple summation $\Sigma_\alpha \Sigma_\gamma \Sigma^3_{i=1}$. The coefficient A_ϱ is the chemical affinity of the reaction ϱ (i.e., passage and chemical reactions). The orientation affinities $A_{x_i\gamma O}$ and $A_{x_i\gamma Q}$ are related to dipole moments $\langle m^\alpha_{x_i\gamma}\rangle$ and $\langle m^{\alpha 2}_{x_i\gamma}\rangle$.

From Eqs. (7.2), (7.3), (7.7) and from Chapter 2A Eqs. (1.7) and (1.18), we obtain

$$dF = -S\,dT - \sum_\alpha p_e\,dV^\alpha + \gamma\,dA - \sum_\varrho A_\varrho\,d\xi_\varrho - \sum_{\alpha\gamma i} A^\alpha_{x_i\gamma O}\,d\xi^\alpha_{x_i\gamma O}$$
$$- \sum_{\alpha\gamma i} A^\alpha_{x_i\gamma Q}\,d\xi^\alpha_{x\gamma_i Q}. \tag{7.8}$$

The free energy F can thus be expressed in terms of variables T, V^α, A, ξ_ϱ, $\xi^\alpha_{x_i\gamma O}$, and $\xi^\alpha_{x_i\gamma Q}$, where $\alpha = 1, 2, \ldots;\ \gamma = 1, 2, \ldots, c$; and $i = 1$, 2, 3. The derivatives of F with respect to one variable, all others remaining constant, have the form

$$\frac{\partial F}{\partial T} = -S, \qquad \frac{\partial F}{\partial V^\alpha} = -p_e, \qquad \frac{\partial F}{\partial A} = \gamma, \qquad \frac{\partial F}{\partial \xi_\varrho} = -A_\varrho \tag{7.9}$$

and

$$\frac{\partial F}{\partial \xi^\alpha_{x_i\gamma O}} = -A^\alpha_{x_i\gamma O}, \qquad \frac{\partial F}{\partial \xi_{x_i\gamma Q}} = -A^\alpha_{x_i\gamma Q}. \tag{7.10}$$

C. Thermodynamics of an Open Capillary System

The previous discussion leads to the conclusion that F is a function of the variables which determine the physicochemical state of the phases, the mode of repartition of the components among the phases, and their orientation within each phase,

$$F = F(T, V^\alpha, A, n_\gamma{}^\alpha, \langle m^\alpha_{x_i\gamma}\rangle, \langle m^{\alpha 2}_{x_i\gamma}\rangle), \tag{7.11}$$

where the symbol $n_\gamma{}^\alpha$ represents $n^{1\ldots\alpha}_{1\ldots\gamma}$, the number of moles of each component in each phase of the system. The same will be true for an open system. Let us remark now that the orientation variables are intensive. If the system is subjected to a transformation in which all $n_\gamma{}^\alpha$, $\langle m^\alpha_{x_i\gamma}\rangle$, and $\langle m_{x_i\gamma}^{\alpha 2}\rangle$ remain constant, the free energy will vary in exactly the same way as it would in a closed system where all ξ are constant. The three first equations (7.9) can thus be written

$$S = -\left(\frac{\partial F}{\partial T}\right)_x$$

where subscript $x = V^\alpha n_\gamma{}^\alpha A\langle m_{x_i\gamma}^\alpha\rangle\langle m_{x_i\gamma}^{\alpha 2}\rangle$

$$p_e = -\left(\frac{\partial F}{\partial V^\alpha}\right)_y \tag{7.12}$$

where subscript $y = Tn_\gamma{}^\alpha A\langle m_{x_i\gamma}^\alpha\rangle\langle m_{x_i\gamma}^{\alpha 2}\rangle$

$$\gamma = \left(\frac{\partial F}{\partial A}\right)_z$$

where subscript $z = TV^\alpha n_\gamma{}^\alpha\langle m_{x_i\gamma}^\alpha\rangle\langle m_{x_i\gamma}^{\alpha 2}\rangle$. On the other hand, if the system is subjected to a transformation in which only $\langle m_{x_i\gamma}^\alpha\rangle$ varies, the free energy will vary in exactly the same way as it would in a closed system where all ξ are constant except $\xi_{x_i\gamma O}^\alpha$. From (7.5), (7.6), and (7.10), we obtain

$$\begin{aligned} \frac{\partial F}{\partial\langle m_{x_i\gamma}^\alpha\rangle} &= \frac{\partial F}{\partial \xi_{x_i\gamma O}^\alpha} = -A_{x_i\gamma O}^\alpha \\ \frac{\partial F}{\partial\langle m_{x_i\gamma}^{\alpha 2}\rangle} &= \frac{\partial F}{\partial \xi_{x_i\gamma Q}^\alpha} = -A_{x_i\gamma Q}^\alpha. \end{aligned} \tag{7.13}$$

Furthermore, we define the quantity

$$\mu_\gamma{}^\alpha = \left(\frac{\partial F}{\partial n_\gamma{}^\alpha}\right)_{z'} \tag{7.14}$$

where subscript $z' = TV^\alpha A n_\beta{}^\delta\langle m_{x_i\gamma}^\alpha\rangle\langle m_{x_i\gamma}^{\alpha 2}\rangle$, as the chemical potential of component γ in the phase α for a state where the mean orientations have given values. When these mean orientations take their equilibrium values, the chemical potential reduces to the classical equation (4.18). A derivative in which only $n_\gamma{}^\alpha$ varies means that component γ added to phase α takes the preexistent orientation in this phase. From (7.4), (7.9), and (7.14), we have

$$A_\varrho = -\sum_{\alpha\gamma}\frac{\partial F}{\partial n_\gamma{}^\alpha}\,\frac{\partial n_\gamma{}^\alpha}{\partial \xi_\varrho} = -\sum \nu_\gamma{}^\alpha\mu_\gamma{}^\alpha. \tag{7.15}$$

Equation (7.8) can thus be written

$$dF = -S\,dT - \sum_{\alpha} p_e\,dV^\alpha + \gamma\,dA + \sum_{\alpha\gamma} \mu_\gamma{}^\alpha\,dn_\gamma{}^\alpha$$
$$- \sum_{\alpha\gamma i} A^\alpha_{x_i\gamma O}\,d\langle m^\alpha_{x_i\gamma O}\rangle - \sum_{\alpha\gamma i} A^\alpha_{x_i\gamma Q}\,d\langle m^{\alpha 2}_{x_i\gamma Q}\rangle. \qquad (7.16)$$

The Function F is a homogeneous function of first degree in the variables V^α, A, $n_\gamma{}^\alpha$, and thus, from Euler's equation

$$F = -\sum_{\alpha} p_e V^\alpha + \gamma A + \sum_{\alpha\gamma} n_\gamma{}^\alpha \mu_\gamma{}^\alpha \qquad (7.17)$$

where the chemical potentials $\mu_\gamma{}^\alpha$ depend on the orientations [see Eq. (7.14)] and where the summation over α includes all the bulk of both phases I and II and all the surface layers.

By differentiation of this relation, and subtracting (7.16), we obtain

$$A\,d\gamma = -S\,dT + V\,dp_e - \sum_{\alpha\gamma} n_\gamma{}^\alpha\,d\mu_\gamma{}^\alpha$$
$$- \sum_{\alpha\gamma i} A^\alpha_{x_i\gamma O}\,d\langle m^\alpha_{x_i\gamma}\rangle - \sum_{\alpha\gamma i} A^\alpha_{x_i\gamma Q}\,d\langle m^{\alpha 2}_{x_i\gamma}\rangle. \qquad (7.18)$$

The above proof does not assume the existence of equilibrium with respect to the distribution of components among the surface layers, but does assume mechanical and thermal equilibrium.

But if orientation phenomena are much slower than diffusion, we can reach a partial equilibrium state, and, for each component γ,

$$\mu_\gamma{}^{\mathrm{I}} = \mu_\gamma{}^\alpha = \mu_\gamma^{\mathrm{II}}, \qquad (7.19)$$

although the orientation affinities are different from zero. The superscript to μ_γ may then be dropped.

From (7.18) and (7.19), we find, in the Gibbs model,

$$d\gamma = -s^a\,dT - \sum_{\gamma} \Gamma_\gamma\,d\mu_\gamma - \sum_{\alpha\gamma i} A^{*\alpha}_{x_i\gamma O}\,d\langle m^\alpha_{x_i\gamma}\rangle - \sum_{\alpha\gamma i} A^{*\alpha}_{x_i\gamma Q}\,d\langle m^{\alpha 2}_{x_i\gamma}\rangle, \qquad (7.20)$$

where $A^{*\alpha}_{x_i\gamma} = A^\alpha_{x_i\gamma}/A$. This is an extension of the Gibbs equation to chemical systems where orientation reaches equilibrium a long time after the diffuse equilibrium. At the true equilibrium (diffusion and orientation), $A^{*\alpha}_{x_i\gamma O} = 0$ and $A^{*\alpha}_{x_i\gamma Q} = 0$ and (7.20) reduces to the classical equation (3.43).

Let us now suppose that we maintain a constant temperature, pressure, and composition of the bulk phase, i.e., $d\mu_\gamma = 0$. The evolution of the surface from a state of partial equilibrium (i.e., from a state where the orientation is not in equilibrium) would be given, with the aid of (7.20),

$$d\gamma = -\sum_{\alpha\gamma i} A^{*\alpha}_{x_i\gamma O}\, d\langle m^{\alpha}_{x_i\gamma}\rangle - \sum_{\alpha\gamma i} A^{*\alpha}_{x_i\gamma Q}\, d\langle m^{\alpha 2}_{x_i\gamma}\rangle. \tag{7.21}$$

This formula implies that each μ_γ is constant during the transformation, i.e., that diffusion occurs quickly enough to ensure continually that the equality (7.19) holds, by balancing the influence of the change in orientation of the local μ_γ.

D. Examples

We consider a system in a real equilibrium state. A short perturbation (friction or motion laying down the molecules) is applied so as to avoid diffusion (in the bulk phase, the temperature, the pressure, and the composition are constant and we suppose all the μ_γ are uniform in the medium). Nevertheless, this perturbation is able to reverse the molecular orientation in certain laminae. When the perturbation cancels out, the system returns to equilibrium in agreement with (7.21). This case may be related to the viscosity flow of monomolecular solutions.

From the experimental point of view (Defay and Pétré, 1970), it is well known that the rate of adsorption of sebacic acid, azelaic acid, and diols at the interface air/aqueous solution is not only diffusion-controlled; a barrier of potential energy between the substrate and the surface phase, related to the orientation of the adsorbed molecule, has to be taken into account in considering the rate of evolution to the equilibrium state.

Now we suppose that only one component of the upper lamina orients itself at the interface. Equation (7.21) may then be rewritten

$$d\gamma = -\left[\sum_i A^{*}_{x_i O}\, d\langle m_{x_i}\rangle + \sum_i A^{*}_{x_i Q}\, d\langle m^2_{x_i}\rangle\right]. \tag{7.22}$$

Let us suppose that initially (out of equilibrium) one half of the undeformable or rigid dipoles are directed vertically upward and the other half vertically downward, while, at equilibrium, all dipoles turn vertically downward. If the x_1 axis is perpendicular to the surface of the layer, the rigid moments $\langle m_{x_1}\rangle$ and $\langle m^2_{x_1}\rangle$ may be written as $m\langle\cos\beta_1\rangle$ and $m^2\langle\cos \times \beta_1\rangle$, where m is the arithmetic value of the dipole moment and $\cos\beta_1$

the direction cosine of a unit vector on the axes of a dipole in comparison with the vertical axis x_1. Furthermore, let us assume that $A_Q{}^*$ varies only slowly with $\langle m\rangle$ in such a way that contribution of the second integral $\int_1^1 A_Q{}^*\, d\langle m^2\rangle$ can be neglected. Integrating (7.22) from $\langle\cos\beta_1\rangle = 0$ to $\langle\cos\beta_1\rangle = -1$ and $\langle\cos^2\beta_1\rangle = 1$ to $\langle\cos^2\beta_1\rangle = 1$, one then has

$$\langle A_0{}^*\rangle = \Delta\gamma/m, \tag{7.23}$$

where $\langle A_0{}^*\rangle$ is the mean value of the surface orientation affinity in the integration and $\Delta\gamma$ the variation of the surface or interfacial tension due to the change of orientation.

If, by way of a perturbation, all the rigid dipoles are initially directed vertically upward and if, during equilibrium they turn downward, then, after integration, Eq. (7.22) becomes

$$\langle A_0{}^*\rangle = \Delta\gamma/2m. \tag{7.24}$$

A measurement of the surface tension for the two extreme positions of the orientations can give us a value of the mean affinity of orientation, as long as the dipoles are rigid and their moments are known.

Remarks. (1) If we assume that only one of variables $\xi^{\alpha}_{x_i\gamma O}$ varies, all others being kept constant in Eq. (7.8), and that, moreover, $\langle m_{x_1}\rangle$ varies from 0 to $-\langle m_{x_1}\rangle$ ($\langle\cos\beta\rangle$ varies from 0 to -1), we find

$$\Delta F = A\langle A^*_{x_{10}}\rangle\langle m_{x_1}\rangle. \tag{7.25}$$

Putting (7.25) in (7.23), we get

$$\langle\Delta F\rangle = A\,\Delta\gamma. \tag{7.26}$$

(2) It is easy to compare now the mean affinity $\langle A_0{}^*\rangle/\Gamma$ with $RT/\langle m\rangle$, where Γ is the number of moles per cm^2. At 15°C, $RT \simeq 2.3 \times 10^{10}$ ergs mol^{-1}. Let us choose, for example, a binary liquid whose surface is covered by 10^{-10} mol cm^{-2} of the surfactants. We suppose now that, if one half of the surface molecules of the surfactants are turning, the experimental value of $\Delta\gamma$ is 5 dyn cm^{-1}. Since the variation of the mean affinity per mole is only due to the surfactant in the upper lamina (we exclude the variation of the other component), it is easy to see from Eq. (7.23) that $\langle A_0{}^*\rangle/\Gamma$ is of order of magnitude $2.2RT/\langle m\rangle$.

(3) The influence of orientation terms on the Lippmann electrocapillary equation has recently been studied (Sanfeld, 1968).

VIII. Capillary Condensation

A. The Bubble

In order to interpretate the condensation of a vapor in the pores of a solid and thus the capillary condensation, we shall first treat the problem of the bubble.

Let us combine Eqs. (1.10) and (1.12) in such a way that

$$\delta\left(\frac{2\gamma}{r}\right) = \frac{v^l - v^v}{v^l}\,\delta p^v. \tag{8.1}$$

Then we neglect the molar volume of the liquid v^l compared with v^v and assume that the vapor behaves as a perfect gas. Equation (8.1) then becomes

$$\delta\left(\frac{2\gamma}{r}\right) = -\frac{RT}{v^l}\,\frac{\delta p^v}{p^v}. \tag{8.2}$$

On integrating (8.2) assuming v^l is constant, we obtain

$$\ln\frac{p^v}{p^0} = -\frac{2\gamma}{r}\,\frac{v^l}{RT}, \tag{8.3}$$

where p^0 is the normal vapor pressure (vapor pressure corresponding to a plane surface). Thus, the larger the curvature of the bubble, the smaller is the vapor pressure p^v. This equation explains the low vapor pressure exhibited by a liquid held in a porous solid whose walls are wetted by the liquid. When the pores are very small, then the liquid is separated from its vapor by a concave meniscus of small radius of curvature. Equation (8.3) also explains the superheating of liquids above the normal boiling point. Thus, for a bubble of vapor to form in water subjected to a pressure of 1 atm, it is necessary, because of the Laplace equation (1.7), for the pressure in the interior of the bubble to be greater than 1 atm. However, while the vapor pressure of a plane surface of water at 100°C is 1 atm, that of a concave surface will be less. Consequently, no bubbles can exist at 100°C and it is necessary to heat the water to above 100°C to achieve boiling. Powders or other impurities which favor the formation of large bubbles diminish the degree of superheating.

B. Capillary Condensation

The Kelvin equation (8.3) provides a ready interpretation of the condensation of a vapor in the pores of a solid. We consider the idealized problem in which all the pores are supposed to be cylinders of the same radius r. We suppose that the pores are partially filled with a liquid in contact with its own vapor (Fig. 14) and assume furthermore that the walls of the pores are completely wetted by the liquid. So long as the menisci are away from the mouths of the pores, all the menisci will have hemispherical surfaces of radius r. The vapor pressure of the liquid in the pores is given by (8.3); this value, denoted by p_r^{v}, is less than the normal vapor pressure p^0. Consequently, liquid can exist in a porous medium in equilibrium with unsaturated vapor. If the vapor pressure is increased slightly, condensation will occur in all pores in which the

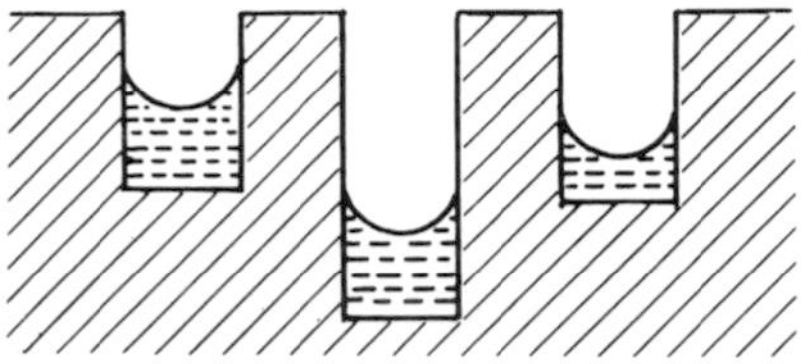

FIG. 14. Condensation in capillary pores of uniform size.

meniscus has not yet reached the mouth of the pore. In pores where the meniscus has reached the mouth of the pore, further condensation would result in an increase in radius of curvature of the surface (Fig. 15); condensation in this pore therefore ceases when the radius of curvature reaches the equilibrium value corresponding to the pressure p^{v} ($>p_r^{\text{v}}$) which is being maintained in the vapor. Thus, condensation will proceed in the partially filled pores, and be halted in the filled pores, until a point is reached at which all the pores are similarly filled and the liquid in them has everywhere the radius of curvature corresponding to p^{v}. Further increase in p^{v} results in condensation in all the pores and the flattening of the menisci, which become plane when $p^{\text{v}} = p^0$. The vapor

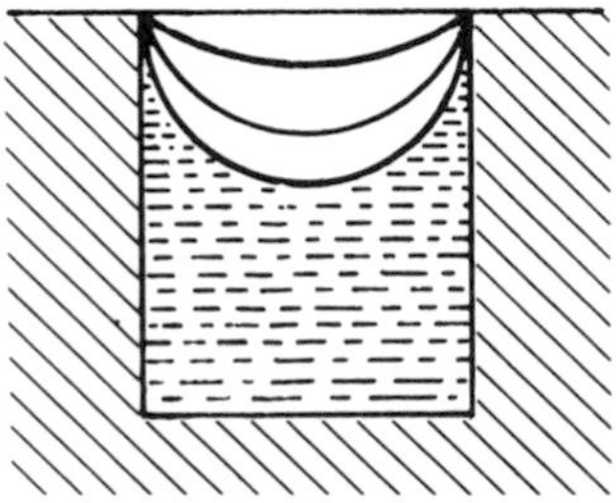

FIG. 15. Variation of curvature of meniscus in a nearly filled pore approaching saturated vapor pressure.

is now saturated and any further increase in p^v is immediately offset by condensation of bulk liquid.

Thus, when an evacuated porous solid, in which all the pores are of equal size, is exposed to a vapor whose pressure is steadily increased, the following phenomena are to be expected (cf. Fig. 16). First, from O to A, vapor will be adsorbed by the whole solid surface. The shape of this curve may be explained in general terms by the theory of Brunauer *et al.* (1938) (see also Brunauer (1944)) in which it is supposed that the vapor forms, in succession, several adsorbed layers on the solid.

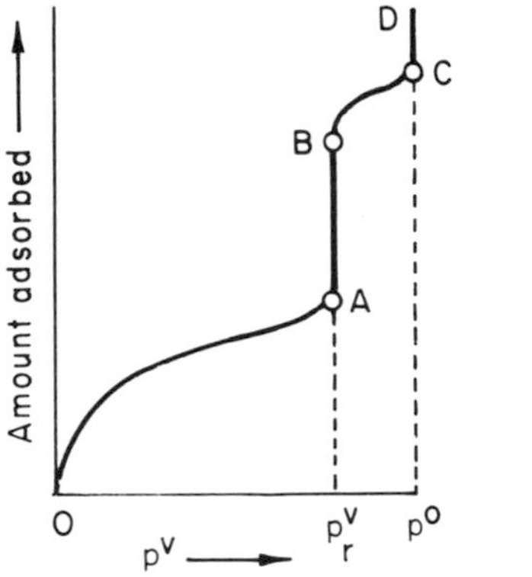

FIG. 16. Isotherm for adsorption of vapor by an ideal porous body.

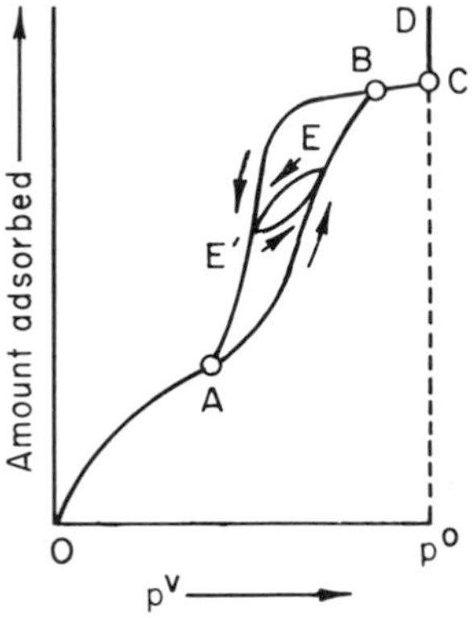

FIG. 17. Isotherm for adsorption of vapor by real porous body.

At A, capillary condensation commences, and the amount of liquid adsorbed increases at constant $p^v = p_r^v$ up to the point at which all the pores are just filled (B). Between B and C, further condensation causes the menisci to flatten until, at C, they are plane, the vapor is saturated, and condensation of bulk liquid can take place.

Real solids clearly do not have pores of the same size. Small pores will fill first and the largest will not begin to fill until the menisci in the smallest pores have already begun to flatten. It is for this reason that, for real solids (Fig. 17), the ideal vertical section AB becomes the oblique section AEB.

Furthermore, real solids are not always perfectly wetted, and their wettability may depend on various circumstances, such as the presence of an adsorbed layer of inert gas other than the vapor being studied. This problem has been developed in an exhaustive way by many authors (de Boer, 1953; Defay *et al.*, 1966).

IX. Surface Tensions of Crystals

The surface tension, regarded as a mechanical force, poses some very complicated problems in the case of solids. In particular, its direct measurement is impossible except in special circumstances in which the molecules of the solid have a certain mobility, as, for example, in metals at temperatures very close to their melting points.

But, generally, if a rectangular crystalline sheet is stretched by a force applied to two of its sides, it is clear that the tensor in the sheet is anisotropic.

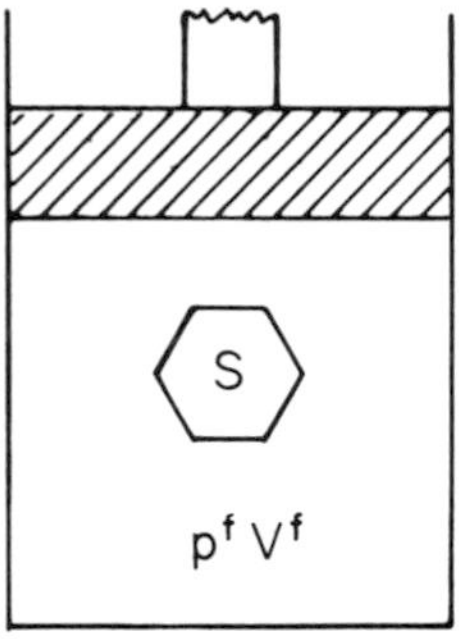

FIG. 18. Equilibrium of a crystal immersed in a fluid (superscript f) (liquid or vapor).

Nevertheless, the growth of a crystal face is possible by deposition of fluid molecules on its faces. The external mechanical work received by the system is

$$d\tau = -p^{\mathrm{f}}\, dV^{\mathrm{f}} - p^{s}\, dV^{s} + \sum_{\beta} \gamma^{\beta}\, dA^{\beta}, \tag{9.1}$$

where the quantities $\gamma^1, \gamma^2, \ldots, \gamma^\beta$ are the surface tensions, respectively, of phases $1, 2, \ldots, \beta$, and the superscript f indicates fluid. As the work $d\tau$ is equal to the work done by the piston on the fluid (Fig. 18)

$$d\tau = -p^{\mathrm{f}}\, dV = -p^{\mathrm{f}}(dV^{\mathrm{f}} + dV^{s}) = -p^{\mathrm{f}}\, dV^{\mathrm{f}} - p^{s}\, dV^{s} + (p^{s} - p^{\mathrm{f}})\, dV^{s}, \tag{9.2}$$

where we suppose the existence of a pressure p^s within the crystal far from the surface regions. For some internal point O, we draw a line h^β normal to each face β (Fig. 19). The change in volume is, to the first order in small quantities, equal to

$$dV^{s} = \sum_{\beta} A^{\beta}\, dh^{\beta}. \tag{9.3}$$

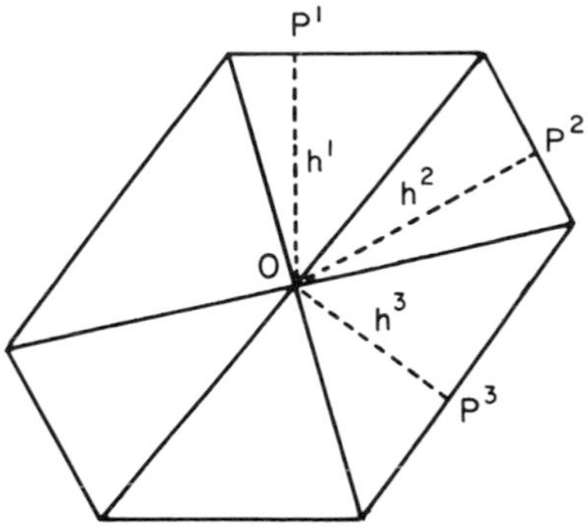

FIG. 19. Geometrical variables defining the size and shape of a crystal.

Furthermore,

$$V^s = \sum_\beta \tfrac{1}{3} A^\beta h^\beta; \tag{9.4}$$

thus

$$dV^s = \tfrac{1}{3} \sum_\beta A^\beta \, dh^\beta + \tfrac{1}{3} \sum_\beta h^\beta \, dA^\beta. \tag{9.5}$$

From (9.2), (9.3), and (9.5), we obtain

$$d\tau = -p^{\mathrm{f}} \, dV^{\mathrm{f}} - p^s \, dV^s + \sum_\beta (p^s - p^{\mathrm{f}}) \tfrac{1}{2} h^\beta \, dA^\beta. \tag{9.6}$$

By comparison of (9.6) and (9.1), we may assume that

$$\gamma^\beta = \tfrac{1}{2}(p^s - p^{\mathrm{f}}) h^\beta. \tag{9.7}$$

Thus, one possible equilibrium form of a crystal is that for which

$$\frac{\gamma^1}{h^1} = \frac{\gamma^2}{h^2} = \cdots \frac{\gamma^\beta}{h^\beta}. \tag{9.8}$$

This form, in which the distance of each face from O is proportional to the surface tension of that face, is called the Wulff form, and the set of relationship (9.8) are called Gibbs–Wulff relations.

The concept of an isotropic pressure p^s in the interior of a crystal is self consistent provided that those faces for which the tensions are larger, are the most remote from the point O. In every crystal which is in the equilibrium form, there exists a point O such that the Wulff relations are satisfied.

Let us remark that (9.7) has a form like Laplace's equation. Moreover, if, in the demonstration of the Kelvin relations, (1.13), we use (9.7) in place of (1.8), we find the vapor pressure of a Wulff crystal

$$\ln \frac{p^{\mathrm{f}}}{p^{0}} = \frac{v^{1}}{RT} \frac{2\gamma^{\beta}}{h^{\beta}}. \tag{9.9}$$

Small crystals have a larger vapor pressure than big ones. In the same vapor, big crystals grow while the small ones vanish. In the same way, in a liquid phase, small crystals are more soluble and disappear, to the benefit of big ones.

Remark. The surface tension of crystals and the thermodynamics of deformed elastic bodies have been studied in a more exhaustive way by many authors (see, for example, Gibbs (1928), Rice (1936), Herring (1953), Ghez (1968), Defay *et al.* (1966)).

References

Arcuri, C. (1966). *Koninkl. Vlaam. Acad. Wetenschap. Letters Schone Kunsten Belg. Colloq. Greuslaagrerschynselen Vloeistoffilmen*, 1966, p. 189.

Bakker, G. B. (1928). Kapillarität und Oberflachenspannung. *In* "Handbuch der Experimentalphysik," Vol. VI. Akad. Verlagsges., Leipzig.

Brunauer, S. (1944). "The Adsorption of Gases and Vapours." Oxford Univ. Press, London and New York.

Brunauer, S., Emmet, P. H., and Teller, E. (1938). *J. Am. Chem. Soc.* **60**, 309.

Buff, F. P., and Lovett, R. A. (1968). *In* "Simple Dense Fluids" (H. L. Frisch and Z. E. Salsburg, eds.). Academic Press, New York.

Davies, J. T., and Rideal, E. K. (1961). "Interfacial Phenomena." Academic Press, New York.

de Boer, J. H. (1953). "The Dynamical Character of Adsorption." Oxford Univ. Press (Clarendon), London and New York.

De Donder, Th. (1922). *Bull. Cl. Sci. Acad. Roy. Belg.* **7**, 1917, 205.

Defay, R., and Pétré, G. (1970). "Dynamic and Surface Tension, Surface and Colloid Science (A Series of Collective Volumes)" (E. Matijevic, ed.).

Defay, R., and Prigogine, I. (1951). "Tension Superficielle et Adsorption," Desoer, Liege.

Defay, R., and Sanfeld, A. (1967). *Electrochim. Acta* **12**, 913.

Defay, R., Prigogine, I., Bellemans, A., and Everett, D. H. (1966). "Surface Tension and Adsorption." Longmans, Green, New York.

Dufour, L., and Defay, R. (1963). "Thermodynamics of Clouds." Academic Press, New York.

Frens, G. (1968). The reversibility of irreversible colloids. Thesis, van't Hoff Laborat. Utrecht.

FUKS, S. (1967). Thesis, Free Univ. of Brussels, Brussels.
GAINES, JR., G. L. (1966). "Insoluble Monolayers at Liquid-Gas Interfaces." Wiley, New York.
GHEZ, R. (1968). Equilibre mécanique et de forme des petits cristaux. Thesis, Ecole Polytech. Univ. Lausanne, Lausanne.
GIBBS, J. W. (1928). "Collected Works," 2 Vols., p. 184 et seq., Longmans, Green, New York.
GUASTALLA, J. (1947). *Mem. Serv. Chim. Etat (Paris)* **33**, 265.
HERRING, C. (1953). "Structure and Properties of Solid Surfaces" (R. Gomer and C. S. Smith, eds.). Univ. of Chicago Press, Chicago, Illinois.
KIRKWOOD, J. G., and BUFF, F. P. (1949). *J. Chem. Phys.* **17**, 338.
KLEEMAN, R. D. (1910). *Phil. Mag.* **19**, 783.
KOENIG, F. O. (1951). *Compt. Rend. C. I. T. C. E. Berne*, p. 299.
MCLEOD, D. B. (1923). *Trans. Far. Soc.* **19**, 38.
ONO, S., and KONDO, S. (1960). Molecular theory of surface tension of liquids. *In* "Handbuch der Physik" (S. Flügge, ed.), Vol. X, Springer, Berlin.
RICE, J. (1936). *In* "Commentary on the Scientific Writings of J. W. Gibbs" (F. G. Donnan and A. Haas, eds.). Yale Univ. Press, New Haven, Connecticut.
SANFELD, A. (1968). "Introduction to Thermodynamics of Charged and Polarized Layers" (I. Prigogine, ed.), Monograph n° 10. Wiley, London.
SUGDEN, S. (1930). "The Parachor and Valency." G. Routbadge & Sons, London.

Chapter 3

Thermodynamic Properties of Gases, Liquids, and Solids

R. HAASE

I. Introduction

Let us consider a simple phase, that is to say, a homogeneous system which is isotropic and not subject to influence by electromagnetic fields or by surface effects. Examples of simple phases are gases, liquids, and isotropic crystals under ordinary conditions. We shall treat both pure phases and mixtures.

Usually, the most appropriate choice of independent variables is the following set: thermodynamic temperature T, pressure P, and amounts $n_1, n_2, \ldots, n_N$ of all the substances $1, 2, \ldots, N$ present in the phase considered. Sometimes—for example, for gases—it may be more expedient to take the volume V (instead of the pressure P) as one of the independent variables.

The characteristic function for the independent variables $T, P, n_1, n_2, \ldots, n_N$ is the Gibbs function G. The corresponding fundamental equation is (see p. 62)

$$dG = -S\,dT + V\,dP + \sum_{k=1}^{N} \mu_k\,dn_k. \tag{1.1}$$

Here, S denotes the entropy and μ_k the chemical potential of the substance k in the phase considered. The chemical potential coincides with the partial molar Gibbs function.

If the function $G(T, P, n_1, n_2, \ldots, n_N)$ is known, all thermodynamic properties of the phase in question, whether gaseous, liquid, or solid, may be readily derived. This can be seen by inspection of Eq. (1.1) and by considering some other fundamental thermodynamic relations, for example, the formula for the enthalpy H,

$$H = G + TS, \tag{1.2}$$

which follows directly from the definition of the Gibbs function G.

Since there is the further relation (see p. 56)

$$G = \sum_{k=1}^{N} n_k \mu_k, \tag{1.3}$$

we may also start from expressions giving the chemical potentials μ_k as functions of temperature, pressure, and composition.

The most useful composition variable for general considerations is the mole fraction x_k of substance k, defined by

$$x_k \equiv n_k/n, \tag{1.4}$$

where

$$n \equiv \sum_{k=1}^{N} n_k. \tag{1.5}$$

From (1.4) and (1.5), we derive

$$\sum_{k=1}^{N} x_k = 1. \tag{1.6}$$

We may, thus, arbitrarily choose $x_2, x_3, \ldots, x_N$ to be the independent composition variables.

Combining Eqs. (1.3) and (1.4), we find

$$\bar{G} = \sum_{k=1}^{N} x_k \mu_k, \tag{1.7}$$

where $\bar{G}$ $(= G/n)$ is the molar Gibbs function of the phase.

The chemical potentials μ_k have the simple property of being constant throughout a heterogeneous system in equilibrium (see p. 86). Thus, it is often advantageous to start from the functions $\mu_k(T, P, x_2, \ldots, x_N)$. Therefore, we recall the differential equation (see p. 63)

$$d\mu_k = -S_k \, dT + V_k \, dP + D\mu_k \tag{1.8}$$

and the relation (see p. 49)

$$\mu_k = H_k - TS_k. \tag{1.9}$$

Here, S_k denotes the partial molar entropy of substance k, V_k the partial molar volume of substance k, $D\mu_k$ an infinitesimal change of μ_k due to a change in composition at constant temperature and pressure, and H_k the partial molar enthalpy of substance k.

For a phase consisting of a single substance, the chemical potential of this substance coincides with the molar Gibbs function $\bar{G}$ of the pure phase, as follows from Eq. (1.7). Similarly, the partial molar volume, the partial molar enthalpy, and the partial molar entropy are now identical with the molar volume $\bar{V}$, the molar enthalpy $\bar{H}$, and the molar entropy $\bar{S}$, respectively. Thus, there results from Eqs. (1.8) and (1.9)

$$\bar{V} = (\partial\bar{G}/\partial P)_T, \quad \bar{S} = -(\partial\bar{G}/\partial T)_P, \quad \bar{H} = \bar{G} - T(\partial\bar{G}/\partial T)_P. \tag{1.10}$$

Hence, all the thermodynamic properties of a pure phase can be derived from the molar characteristic function $\bar{G}(T, P)$.

When comparing the thermodynamic functions of a mixture with those of the pure substances in the same state of aggregation at the given values of temperature and pressure, it is convenient to proceed from a partial molar quantity Z_k (for example, μ_k, V_k, H_k, S_k) of the substance k in the mixture to the corresponding quantity $Z_k^\bullet$ valid for the pure substance k. The $Z_k^\bullet$ is identical with the molar quantity $\bar{Z}$ of the pure phase in question, but the symbol $Z_k^\bullet$, recently recommended,* stresses the fact that it is the particular transition, mixture → pure substance k, in which we are interested.

Another problem that frequently arises in mixtures is the derivation of a partial molar quantity Z_k from the molar quantity $\bar{Z}$ ($= Z/n$, Z being any extensive quantity) which is known as a function of temperature, pressure, and composition. It is now expedient to choose $x_2, x_3, \ldots, x_N$ as independent composition variables if Z_1 is the desired quantity, to choose $x_1, x_3, \ldots, x_N$ as independent variables if Z_2 is the desired quantity, and so on. We denote the set of independent mole fractions by x_j and the conjugate partial molar quantity by Z_i. Then, according to Chapter 1, p. 18, Eqs. (6.18) and (6.19), we obtain

$$Z_i = \bar{Z} - \sum_j x_j (\partial \bar{Z}/\partial x_j)_{T,P,x_k}, \tag{1.11}$$

where the subscript x_k indicates that all independent mole fractions are to be held constant except x_j. Thus, we have, for a mixture of two substances,

$$\begin{aligned} Z_1 &= \bar{Z} - x_2(\partial \bar{Z}/\partial x_2)_{T,P}, \\ Z_2 &= \bar{Z} - x_1(\partial \bar{Z}/\partial x_1)_{T,P} = \bar{Z} + (1 - x_2)(\partial \bar{Z}/\partial x_2)_{T,P}. \end{aligned} \tag{1.12}$$

Here, the identity $x_1 + x_2 = 1$ has been used.

For the coexistence of two or more phases such as vapor–liquid equilibrium, liquid–liquid equilibrium, solid–liquid equilibrium, solid–solid equilibrium and so on, the reader should consult Chapter 2 and relevant monographs (Prigogine and Defay, 1954; Guggenheim, 1967; Rowlinson, 1969; Haase, 1956; Haase and Schönert, 1969).

II. Chemical Species and Components

Thus far, we have referred to a "substance" without elaborating the meaning of this expression. We shall now investigate this problem more closely.

A *chemical species* is any kind of particle in the sense of chemistry, for example, an atomic species, an ionic species, or a molecular species.

* The symbol Z_k° often used formerly in the literature is discarded for its ambiguity.

On the other hand, a substance the amount of which can be varied independently is called a *component*. It is the same concept that occurs in the phase rule.

Thus, in a fluid (gaseous or liquid) mixture of chloroform and "nitrogen peroxide," there are three chemical species ($CHCl_3$, NO_2, N_2O_4), but only two components provided the chemical equilibrium

$$2NO_2 \rightleftharpoons N_2O_4$$

has been established. We may choose chloroform and nitrogen dioxide to be the components of the mixture. The amount of nitrogen tetroxide then follows automatically from the condition of chemical equilibrium and from the total amounts of the components.

Again, in an aqueous solution* of potassium chloride, there are four kinds† of chemical species (H_2O, KCl, K^+, Cl^-), but only two components (water and potassium chloride). Here, the conditions of electric neutrality and of the chemical equilibrium

$$KCl \rightleftharpoons K^+ + Cl^-$$

provide the two equations we need to compute the amounts of the four species present.

If a possible chemical reaction is "frozen," that is to say, if the chemical species do not react though they could do so in principle, then all the species are to be considered as components since their amounts now can change arbitrarily.

A system or a phase is called a one-component system or a pure phase if it contains a single component though the number of chemical species may exceed one (e.g., pure water or pure nitrogen dioxide). Otherwise, we refer to the system or phase as a multicomponent system or phase. In particular, a single phase consisting of two or three components is called a binary or ternary mixture, respectively.

Henceforth, we shall carefully distinguish between chemical species and components and not use the word "substance" any longer. But it should be stressed that the general considerations in Section I hold for both chemical species and components and that there are relations

* A liquid or solid mixture is often called a "solution," in particular when there is lack of symmetry with respect to the concentrations of the components or with respect to the states of aggregation of the pure components.

† We here ignore the possible dissociation or association of water and the possible hydration.

between, say, the chemical potentials of the components and those of the chemical species.

As an example of the relations just mentioned, let us discuss a particular binary liquid mixture: the solution of an electrolyte of the general formula $X_{\nu_+}Y_{\nu_-}$ in a neutral solvent. We have the dissociation equilibrium

$$X_{\nu_+}Y_{\nu_-} \rightleftharpoons \nu_+ X^{|z_+|+} + \nu_- Y^{|z_-|-} \tag{2.1}$$

where ν_+ and ν_- are the dissociation numbers and z_+ and z_- the charge numbers of the cations and anions, respectively (e.g., for $CaCl_2$ in H_2O, $\nu_+ = 1$, $z_+ = 2$, $\nu_- = 2$, and $z_- = -1$). If we denote the chemical potential of the solvent (component 1) and that of the electrolyte (component 2) by μ_1 and μ_2, respectively, we derive from Eq. (1.3) applied to the components

$$G = n_1\mu_1 + n_2\mu_2. \tag{2.2}$$

Here, G is the Gibbs function of the liquid mixture considered, and n_1 and n_2 are the (stoichiometric) amounts of substance of components 1 and 2 (solvent and electrolyte), respectively. Now, Eq. (1.3) continues to hold if we introduce the chemical species present in the solution. These are: solvent molecules (subscript L), undissociated electrolyte molecules (subscript u), cations (subscript +), and anions (subscript —). We thus have

$$G = n_L\mu_L + n_u\mu_u + n_+\mu_+ + n_-\mu_-, \tag{2.3}$$

where G is the same quantity as in Eq. (2.2), while n_i or μ_i now denotes the amount of substance or the chemical potential* of the chemical species i. Introducing the degree of dissociation α of the electrolyte,

* If we introduce the electrochemical potential η_i of the ionic species i ($i = +, -$),

$$\eta_i = \mu_i + z_i\mathfrak{F}\psi$$

where $\mathfrak{F}$ is the Faraday constant and ψ the electric potential, we find

$$n_+\eta_+ + n_-\eta_- = n_+\mu_+ + n_-\mu_- + (z_+n_+ + z_-n_-)\mathfrak{F}\psi.$$

Since the condition of electric neutrality in the interior of the solution (no space charges) requires that $z_+n_+ + z_-n_- = 0$, we conclude that

$$n_+\eta_+ + n_-\eta_- = n_+\mu_+ + n_-\mu_-.$$

Thus, the results are not affected by whether we use the electrochemical or the chemical potentials for the ionic species.

we obtain

$$n_L = n_1, \qquad n_u = (1-\alpha)n_2, \qquad n_+ = \alpha\nu_+ n_2, \qquad n_- = \alpha\nu_- n_2. \qquad (2.4)$$

The general condition of homogeneous chemical equilibrium (see p. 85), when applied to the reaction (2.1), leads to the following relation:

$$\mu_u = \nu_+\mu_+ + \nu_-\mu_-. \qquad (2.5)$$

We now insert Eqs. (2.4) and (2.5) into Eq. (2.3) and compare the result to Eq. (2.2). We then find

$$\mu_1 = \mu_L, \qquad (2.6)$$

$$\mu_2 = \mu_u, \qquad (2.7)$$

and thus, by virtue of Eq. (2.5),

$$\mu_2 = \nu_+\mu_+ + \nu_-\mu_-. \qquad (2.8)$$

If there is complete dissociation ($\alpha = 1$), the concept of the chemical potential μ_u of the undissociated part of the electrolyte loses its sense. But, in view of the relations

$$n_L = n_1, \qquad n_u = 0, \qquad n_+ = \nu_+ n_2, \qquad n_- = \nu_- n_2 \qquad (\alpha = 1),$$

following from Eq. (2.4), we derive from Eqs. (2.2) and (2.3)

$$\mu_1 = \mu_L, \qquad \mu_2 = \nu_+\mu_+ + \nu_-\mu_-.$$

Thus, the two equations (2.6) and (2.8) remain valid even in the case of complete dissociation.

It is by reasoning of this kind that we can always derive connections between thermodynamic quantities relating to components and those referring to the chemical species actually present in the gaseous, liquid, or solid phase considered.

III. Pure Gases

A pure gas is a gaseous phase consisting of a single component (Section II). All its thermodynamic properties can be derived from the molar Gibbs function $\bar{G}$ (equal to the chemical potential μ) when this is given in terms of the thermodynamic temperature T and the pressure P (Section I).

Usually, one starts from the *equation of state* (see p. 12), which in this simple case is a functional relationship of the type

$$\bar{V} = \bar{V}(T, P), \tag{3.1}$$

$\bar{V}$ denoting the molar volume of the gas. It is possible and convenient to write for pure gases

$$P\bar{V}/RT = 1 + (B/\bar{V}) + (C/\bar{V}^2) + \cdots. \tag{3.2}$$

Here, R is the gas constant, B the second virial coefficient, C the third virial coefficient, and so on. The coefficients $B, C, \ldots$ only depend on the temperature. The quantity $P\bar{V}/RT$ is usually called the "compression factor."

We now invert the series in Eq. (3.2) and thus find an expansion in powers of the pressure:

$$P\bar{V} = RT + BP + [(C - B^2)/RT]P^2 + \cdots. \tag{3.3}$$

This expression is more convenient than Eq. (3.2) if one uses the pressure as one of the independent variables, as we do here.

Equations (3.2) and (3.3) are combined results of experience, statistical theory, and the definition of temperature (see p. 9).

A corollary of Eq. (3.3) is the general limiting law

$$\lim_{P \to 0} P\bar{V} = RT. \tag{3.4}$$

A gas for which the equation of state has the simple form

$$P\bar{V} = RT \tag{3.5}$$

for nonzero pressures is said to be a *perfect gas* (or an "ideal gas"). Otherwise, it is called an *imperfect gas* (or a "real gas"). Strictly speaking, a perfect gas does not occur in nature, but often, at low pressures, the deviations from Eq. (3.5) are small as compared to the uncertainties of the measurements and it is then quite legitimate to refer to a perfect gas existing in practice.

Integrating the first equation in (1.10) between the limits $P^{\dagger}$ (standard pressure, 1 bar) and P (arbitrary pressure) and taking account of Eq. (3.3), we obtain for the molar Gibbs function $\bar{G}$ or the chemical potential μ of any gas

$$\bar{G} = \mu = \mu^* + RT \ln(P/P^{\dagger}) + BP + C'P^2 + \cdots, \tag{3.6}$$

where

$$C' \equiv (C - B^2)/2RT. \tag{3.7}$$

The quantity μ^*, depending on the temperature only, is said to be the *standard chemical potential.*

From Eqs. (1.10) and (3.6), we derive the corresponding expressions for the molar entropy $\bar{S}$ and the molar enthalpy $\bar{H}$:

$$-\bar{S} = (d\mu^*/dT) + R\ln(P/P^\dagger) + (dB/dT)P + (dC'/dT)P^2 + \cdots, \tag{3.8}$$

$$\bar{H} = \mu^* - T(d\mu^*/dT) + [B - T(dB/dT)]P + [C' - T(dC'/dT)]P^2 + \cdots. \tag{3.9}$$

The quantities $-d\mu^*/dT$ and $\mu^* - T(d\mu^*/dT)$ are called the *standard molar entropy* and the *standard molar enthalpy*, respectively. They both are functions of temperature.

The temperature dependence of the virial coefficients is found either from empirical equations of state or from statistical mechanics (see Section IV). The complicated dependence of μ^* on the temperature can only be established by means of the statistical theory (Guggenheim, 1967). The "conventional" values of μ^* used in tables for thermodynamic functions are based on the conventions described earlier and concerning the enthalpy (p. 28) and the entropy (p. 89).

For $B = 0$, $C' = 0, \ldots$, Eqs. (3.6), (3.8), and (3.9) reduce to the relations valid for a perfect gas. It is interesting to note that then $\bar{H}$ only depends on the temperature, while $\bar{G}$ and $\bar{S}$ still contain terms involving the pressure. Some other properties of perfect gases have already been derived in Chapters 1 and 2. In terms of molecular theory, a perfect gas is a system of noninteracting particles.

More details on the thermodynamic functions of imperfect gases may be found elsewhere (Guggenheim, 1967; Beattie, 1949; Haase, 1956).

IV. Gaseous Mixtures

A. General

A gaseous mixture is a gas composed of two or more components. As independent variables to describe the state of the gas, we use the thermodynamic temperature T, the pressure P, and the mole fractions

$x_2, x_3, \ldots, x_N$ of components 2, 3, ..., N. The mole fraction of component 1 is given by

$$x_1 = - \sum_{j=2}^{N} x_j, \tag{4.1}$$

as follows from Eq. (1.6).

A quantity typical for gaseous mixtures is the *partial pressure* p_i of component i, defined by

$$p_i \equiv P x_i, \qquad i = 1, 2, 3, \ldots, N. \tag{4.2}$$

We infer from Eqs. (4.1) and (4.2) that

$$\sum_{i=1}^{N} p_i = P, \tag{4.3}$$

so that the total pressure equals the sum of the partial pressures in any gaseous mixture.

As in the case of pure gases, we begin by discussing the *equation of state*, which here takes the form

$$\bar{V} = \bar{V}(T, P, x_2, x_3, \ldots, x_N), \tag{4.4}$$

where $\bar{V}$ denotes the molar volume of the gaseous mixture. Again, experience and statistical theory tell us that it is expedient to write

$$P\bar{V}/RT = 1 + (B/\bar{V}) + (C/\bar{V}^2) + \cdots \tag{4.5}$$

or

$$P\bar{V} = RT + BP + [(C - B^2)/RT]P^2 + \cdots. \tag{4.6}$$

These expressions are analogous to the relations (3.2) and (3.3). Again, R is the gas constant, but the virial coefficients denoted by B, C, ... depend on both temperature and composition ($x_2, x_3, \ldots, x_N$).

As we can deduce from either experimental results or from statistical theory (Lennard-Jones and Cook, 1927; Mayer, 1939; Fuchs, 1941), the second, third, ..., nth virial coefficient is a polynomial of second, third, ..., nth order in the mole fractions $x_1, x_2, x_3, \ldots, x_N$. Thus, we have for the second virial coefficient

$$B = \sum_{i=1}^{N} \sum_{j=1}^{N} B_{ij} x_i x_j \tag{4.7}$$

and for the third virial coefficient

$$C = \sum_{i=1}^{N} \sum_{j=1}^{N} \sum_{k=1}^{N} C_{ijk} x_i x_j x_k, \tag{4.8}$$

where the quantities $B_{ij} = B_{ji}$ and $C_{ijk} = C_{ikj} = C_{jik} = C_{jki} = C_{kij} = C_{kji}$ are functions of the temperature only.

As a simple example, we write down the expression for the second virial coefficient B of a binary gaseous mixture ($N = 2$):

$$B = B_{11} x_1^2 + 2B_{12} x_1 x_2 + B_{22} x_2^2, \tag{4.9}$$

which follows directly from Eq. (4.7). Obviously, B_{11} is the second virial coefficient of the pure gaseous component 1 ($x_1 = 1$, $x_2 = 0$), B_{22} that of the pure gaseous component 2 ($x_1 = 0$, $x_2 = 1$), and B_{12} a quantity characteristic of the mixture.

As implied by Eqs. (4.6)–(4.8), the molar volume $\bar{V}$ of a gas mixture can be written

$$\bar{V} = (RT/P) + \sum_{i=1}^{N} \sum_{j=1}^{N} B_{ij} x_i x_j + \Phi. \tag{4.10}$$

Here, Φ represents a power series in the pressure P and in the mole fractions $x_1, x_2, \ldots, x_N$, the exponents being positive integers and the coefficients of the powers depending on the temperature.

We infer from Eq. (1.11) that in any mixture the partial molar volume V_i of component i can be computed from the molar volume $\bar{V}$ if this is given in terms of temperature, pressure, and composition:

$$V_i = \bar{V} - \sum_j x_j (\partial \bar{V} / \partial x_j)_{T,P,x_k}. \tag{4.11}$$

In particular, for a binary mixture, we have [see Eq. (1.12)]

$$\begin{aligned} V_1 &= \bar{V} - x_2 (\partial \bar{V} / \partial x_2)_{T,P}, \\ V_2 &= \bar{V} - x_1 (\partial \bar{V} / \partial x_1)_{T,P} = \bar{V} + (1 - x_2)(\partial \bar{V} / \partial x_2)_{T,P}, \end{aligned} \tag{4.12}$$

where the identity $x_1 + x_2 = 1$ has been used.

Inspection of Eqs. (4.10) and (4.11) shows that the partial molar volume V_i of component i in any gaseous mixture must be of the general form

$$V_i = (RT/P) + \Psi_i, \tag{4.13}$$

where Ψ_i denotes a power series in the pressure and in the mole fractions, the exponents being nonnegative integers $(0, 1, 2, \ldots)$.

We derive from Eq. (1.8)

$$V_i = (\partial \mu_i / \partial P)_{T,x}.$$

Here, μ_i is the chemical potential of component i and x stands for all independent mole fractions. Integration between $P^\dagger$ (standard pressure) and P (arbitrary pressure) gives

$$\mu_i = \mu_i(P^\dagger) + \int_{P^\dagger}^{P} V_i \, dP. \tag{4.14}$$

Here, $\mu_i(P^\dagger)$ denotes the value of μ_i for $P = P^\dagger$. Combining Eqs. (4.13) and (4.14), we obtain

$$\mu_i = \mu_i(P^\dagger) + RT \ln(P/P^\dagger) + \int_{P^\dagger}^{P} \Psi_i \, dP, \tag{4.15}$$

where the dependence of $\mu_i(P^\dagger)$ on temperature and composition still has to be settled.

One of the fundamental facts borne out by experiment and by statistical mechanics is the statement that the limiting law

$$\mu_i \to \mu_i^* + RT \ln(Px_i/P^\dagger) \qquad \text{for} \quad P \to 0 \tag{4.16}$$

holds for any composition of a gaseous mixture. Here, μ_i^* is a function of temperature characteristic of the pure gaseous component i. It is, in fact, the quantity called μ^* in Section III. This can be verified by applying Eq. (4.16) to the pure component i ($x_i = 1$) and comparing the result to Eq. (3.6).

A more rigorous way of stating the limiting law (4.16) is the equation

$$\mu_i = \mu_i^* + RT \ln(Px_i/P^\dagger) + \varphi_i, \tag{4.17}$$

where φ_i denotes a function of temperature, pressure, and composition, vanishing at zero pressure.

We now split the integral in Eq. (4.15) into two parts,

$$\int_{P^\dagger}^{P} \Psi_i \, dP = \int_{P^\dagger}^{0} \Psi_i \, dP + \int_{0}^{P} \Psi_i \, dP, \tag{4.18}$$

and consider Eq. (4.13). We then conclude that the relation (4.15) is only consistent with Eq. (4.17) if the conditions

$$\varphi_i = \int_0^P \Psi_i \, dP = \int_0^P [V_i - (RT/P)] \, dP \tag{4.19}$$

and

$$\mu_i{}^* + RT \ln x_i = \mu_i(P^\dagger) - \int_0^{P\dagger} \Psi_i \, dP \tag{4.20}$$

are fulfilled.

Combining Eqs. (4.2), (4.17), and (4.19), we obtain the final formula

$$\mu_i = \mu_i{}^* + RT \ln(p_i/P^\dagger) + \int_0^P [V_i - (RT/P)] \, dP. \tag{4.21}$$

This is the basic relation for the derivation of thermodynamic properties from the equation of state.

With the help of Eqs. (1.8) and (1.9)], we find for the partial molar entropy S_i and the partial molar enthalpy H_i of component i

$$S_i = -(\partial \mu_i/\partial T)_{P,x}, \qquad H_i = \mu_i + TS_i. \tag{4.22}$$

Thus, there follows from Eqs. (4.19) and (4.21)

$$-S_i = (d\mu_i{}^*/dT) + R \ln(p_i/P^\dagger) + (\partial \varphi_i/\partial T)_{P,x}, \tag{4.23}$$

$$H_i = \mu_i{}^* - T(d\mu_i{}^*/dT) + \varphi_i - T(\partial \varphi_i/\partial T)_{P,x}. \tag{4.24}$$

The quantities $\mu_i{}^*$, $-d\mu_i{}^*/dT$, and $\mu_i{}^* - T(d\mu_i{}^*/dT)$ are called the *standard chemical potential*, the *standard partial molar entropy*, and the *standard partial molar enthalpy* of component i, respectively (compare Section III). They depend on the temperature only.

B. Perfect Gaseous Mixtures

We now suppose that the pressure is so low that experimentally the thermodynamic properties of the gaseous mixture are indistinguishable from those obtained for zero pressure; in other words, we assume that the function φ_i vanishes within the experimental errors. Then we have a *perfect gaseous mixture.*† We derive from Eqs. (1.4), (4.2), (4.6), (4.13), (4.17), (4.23), and (4.24), introducing the condition $\varphi_i = 0$,

$$P\bar{V} = RT, \qquad V_i = RT/P, \qquad p_i = (n_i/V)RT, \tag{4.25}$$

$$\mu_i = \mu_i{}^* + RT \ln(p_i/P^\dagger), \tag{4.26}$$

$$S_i = -(d\mu_i{}^*/dT) - R \ln(p_i/P^\dagger), \tag{4.27}$$

$$H_i = \mu_i{}^* - T \, d\mu_i{}^*/dT. \tag{4.28}$$

† We prefer this expression to the alternative name "ideal gaseous mixture" in view of the fact that an "ideal mixture" is a different type of mixture (p. 311).

Here, n_i is the amount of substance of component i, and V the volume of the mixture. It is remarkable that the partial molar enthalpy only depends on the temperature. Other properties of perfect gaseous mixtures have already been derived in Chapters 1 and 2. In terms of statistical theory, a perfect gaseous mixture is a system of noninteracting particles of different types.

It may occur that a gas at low pressures is not perfect, in the sense that the equations (4.25)–(4.28) apply to the components but that these formulas hold for the chemical species actually present. This may be even true for a pure gas. As an example, we mention pure gaseous nitrogen dioxide (NO_2), where there exists an association equilibrium

$$NO_2 \rightleftharpoons 2N_2O_4$$

leading to nitrogen tetroxide (N_2O_4). If this system is perfect in the sense that the relations (4.25)–(4.28) hold for the two chemical species (NO_2 and N_2O_4), it is obvious that the system seems to be an imperfect gas when it is treated like an ordinary one-component system. In practice, it might be difficult to separate these "chemical effects" from the "physical effect" of nonvanishing virial coefficients due to interactions between the molecules.

C. Slightly Imperfect Gaseous Mixtures

Any gaseous mixture that is not perfect is called an *imperfect gaseous mixture.*

If chemical reactions are known to occur within the gaseous mixture, it is expedient to apply the formulas to the chemical species actually present, as explained above. It is only for the sake of simplicity that we continue to use the components.

When the pressure in an imperfect gaseous mixture is low enough to render unimportant the contributions from the third and higher virial coefficients, then the system is said to be a *slightly imperfect gaseous mixture.* Thus, in this subsection, we only take account of terms involving the second virial coefficient.

Let us consider a *binary* (nonreacting), slightly imperfect gaseous mixture. Then, we derive from Eqs. (4.2), (4.6), (4.9), (4.12), (4.19), (4.21), (4.23), and (4.24)

$$P\bar{V} = RT + (B_{11}x_1^2 + 2B_{12}x_1x_2 + B_{22}x_2^2)P, \tag{4.29}$$

$$V_1 = (RT/P) + B_{11} + bx_2^2, \tag{4.30}$$

$$V_2 = (RT/P) + B_{22} + bx_1^2, \tag{4.31}$$

$$\mu_1 = \mu_1^* + RT\ln(Px_1/P^\dagger) + (B_{11} + bx_2^2)P, \tag{4.32}$$

$$\mu_2 = \mu_2^* + RT\ln(Px_2/P^\dagger) + (B_{22} + bx_1^2)P, \tag{4.33}$$

$$S_1 = -(d\mu_1^*/dT) - R\ln(Px_1/P^\dagger) - [(dB_{11}/dT) + (db/dT)x_2^2]P, \tag{4.34}$$

$$S_2 = -(d\mu_2^*/dT) - R\ln(Px_2/P^\dagger) - [(dB_{22}/dT) + (db/dT)x_1^2]P, \tag{4.35}$$

$$H_1 = \mu_1^* - T(d\mu_1^*/dT) + \{B_{11} - T(dB_{11}/dT) + [b - T(db/dT)]x_2^2\}P, \tag{4.36}$$

$$H_2 = \mu_2^* - T(d\mu_2^*/dT) + \{B_{22} - T(dB_{22}/dT) + [b - T(db/dT)]x_1^2\}P, \tag{4.37}$$

where

$$b \equiv 2B_{12} - B_{11} - B_{22}. \tag{4.38}$$

For $B_{11} = 0$, $B_{22} = 0$, $B_{12} = 0$, we recover the formulas for perfect gaseous mixtures given earlier.

We have now to investigate whether B_{12} and thus b can be computed from the virial coefficients B_{11} and B_{22} of the two pure gaseous components. If such a computation is possible, we are able to predict the thermodynamic properties of the binary gaseous mixture from those of the pure gases.

The simplest assumption is

$$b = 0, \tag{4.39}$$

which seldom, if ever, holds in practice. But it is interesting to note some implications of this conjecture. Thus, we conclude from Eqs. (4.32) and (4.33) that for $b = 0$ the chemical potentials μ_1 and μ_2 of the two components of the gaseous mixture take the form

$$\mu_i = \mu_i^* + RT\ln(P/P^\dagger) + B_{ii}P + RT\ln x_i, \qquad i = 1, 2. \tag{4.40}$$

If we proceed to the limit $x_i \to 1$ in Eq. (4.40), we find

$$\mu_i^\bullet = \mu_i^* + RT\ln(P/P^\dagger) + B_{ii}P, \qquad i = 1, 2, \tag{4.41}$$

where $\mu_i^\bullet$ is the chemical potential (or molar Gibbs function) of the pure gaseous component i at the given values of T and P (see Section I).

From Eqs. (4.40) and (4.41), there follows

$$\mu_i = \mu_i^\bullet + RT\ln x_i. \tag{4.42}$$

As we shall see later, this formula describes a type of gaseous mixture called an "ideal mixture," though it is still a slightly imperfect gaseous mixture. Furthermore, we infer from Eqs. (4.30), (4.31), (4.36), (4.37), and (4.39) that in a mixture of this kind the partial molar volumes and the partial molar enthalpies do not depend on the composition.

We conclude from Eq. (4.38) that the assumption $b = 0$ is equivalent to the simple combination rule

$$B_{12} = \tfrac{1}{2}(B_{11} + B_{22}), \tag{4.43}$$

corresponding to an arithmetic mean. Now, both experience and statistical theory lead to more accurate rules which we shall shortly outline.

In most cases, the temperature dependence of the quantities B_{11}, B_{22}, and B_{12} may be described by the relations

$$\begin{aligned} B_{11} &= \alpha_{11} - (\beta_{11}/T) - (\gamma_{11}/T^2), \\ B_{22} &= \alpha_{22} - (\beta_{22}/T) - (\gamma_{22}/T^2), \\ B_{12} &= \alpha_{12} - (\beta_{12}/T) - (\gamma_{12}/T^2), \end{aligned} \tag{4.44}$$

with the nine constants $\alpha_{11}, \ldots, \gamma_{12}$.

If we put $\gamma_{11} = \gamma_{22} = \gamma_{12} = 0$, the formulas (4.44) lead to the results following from the equation of state of van der Waals. The case $\beta_{11} = \beta_{22} = \beta_{12} = 0$ corresponds to the equation of state of Berthelot. In both cases, the constants are empirical. If we leave the relations (4.44) unchanged and still consider the constants to be empirical, then the results coincide with those implied by the equation of state of Beattie and Bridgeman. A number of combination rules have been proposed for the set of constants in (4.44); for example,

$$\alpha_{12} = \tfrac{1}{8}(\alpha_{11}^{1/3} + \alpha_{22}^{1/3})^3, \tag{4.45}$$

$$\beta_{12} = (\beta_{11}\beta_{22})^{1/2}, \tag{4.46}$$

$$\gamma_{12} = (\gamma_{11}\gamma_{22})^{1/2}. \tag{4.47}$$

The last two equations are combinations involving the geometric mean. Equation (4.45) is said to represent a Lorentz combination. Obviously, we can, by means of such relations, compute the thermodynamic properties of the mixture from those of the pure components.

From the point of view of certain models of the statistical theory, the constants α_{11}, β_{11}, γ_{11} may be replaced by expressions containing two molecular parameters (interaction energy and intermolecular distance) describing the interaction of a pair of molecules of component 1, while the constants $\alpha_{22}, \beta_{22}, \gamma_{22}$ and $\alpha_{12}, \beta_{12}, \gamma_{12}$ are then to be replaced by similar expressions referring to the interaction between a pair of molecules of component 2 and between a pair of molecules of different kind (1 and 2) in the mixture, respectively. Molecular theory also tells us how to compute the constants valid for the mixture from those relating to the pure components: a geometric-mean combination for the interaction energies and an arithmetic-mean combination for the intermolecular distances.

When the gases belong to a group of similar substances, we may apply the "principle of corresponding states" and use formulas containing "reduced quantities," that is to say, quantities divided by the critical temperature T_{c1} or T_{c2} of the pure component 1 or 2, by the critical molar volume $V^{\bullet}_{c1}$ or $V^{\bullet}_{c2}$ of the pure component 1 or 2, and so on. We define

$$T_{12} \equiv (T_{c1}T_{c2})^{1/2}, \qquad V_{12} \equiv \tfrac{1}{8}[(V^{\bullet}_{c1})^{1/3} + (V^{\bullet}_{c2})^{1/3}]^3,$$

where T_{12} and V_{12} have the dimensions of thermodynamic temperature and molar volume, respectively. If we plot $B_{11}/V^{\bullet}_{c1}$ versus T/T_{c1}, $B_{22}/V^{\bullet}_{c2}$ versus T/T_{c2}, and B_{12}/V_{12} versus T/T_{12}, we find that, for a pair of similar nonpolar substances, the three functions are represented by the same curve. Thus, B_{12} can be computed from B_{11} and B_{22} when the critical data of the pure components are available. This elegant procedure, due to Guggenheim and McGlashan (1951), is independent of analytic expressions such as (4.44).

In Table I, the second, third, fourth, and fifth columns give experimental values* of B_{11}, B_{22}, B_{12}, and b for the gaseous system methane (component 1) + n-butane (component 2), in the temperature range between 150°C and 300°C, due to Beattie and Stockmayer (1942). The sixth column, with the heading B_{12} ($b = 0$), gives B_{12} calculated from Eq. (4.43), equivalent to the assumption $b = 0$, which obviously is a bad approximation here. The last column, with the heading B_{12} (G.), gives B_{12} computed by Guggenheim and McGlashan according to the method just sketched, which obviously is a good approximation here.

More details on the thermodynamic properties of gaseous mixtures may be found elsewhere (Beattie, 1949; Haase, 1956).†

* The change of sign in B_{11} (second virial coefficient of pure gaseous methane) between 225°C and 250°C should be noted. Since, for a slightly imperfect pure gas the statement $B = 0$ is equivalent to the validity of the equation $P\overline{V} = RT$ [see Eq. (3.3)] at the temperature in question, this temperature is called the "Boyle temperature."

† For recent publications on gas mixtures, see Scott and Fenby (1969).

TABLE I

THE QUANTITIES B_{11}, B_{22}, B_{12}, AND b RELATED TO THE SECOND VIRIAL COEFFICIENT IN THE GASEOUS MIXTURE METHANE (1) + n-BUTANE (2) FOR DIFFERENT TEMPERATURES

Temperature (°C)	B_{11} ($cm^3 mol^{-1}$)	B_{22} ($cm^3 mol^{-1}$)	B_{12} ($cm^3 mol^{-1}$)	b ($cm^3 mol^{-1}$)	B_{12} ($b = 0$)[a] ($cm^3 mol^{-1}$)	B_{12} (G.)[b] ($cm^3 mol^{-1}$)
150	−11.4	−328.7	−81.6	176.9	−170	−83
175	−7.5	−287.3	−69.4	156.0	−147	−71
200	−4.0	−254.2	−60.4	137.4	−129	−62
225	−0.9	−224.5	−51.2	123.0	−113	−53
250	+1.9	−198.1	−42.0	112.2	−98	−45
275	+4.5	−176.0	−35.2	101.1	−86	−37
300	+6.8	−157.4	−29.2	92.2	−75	−31

[a] Computed from Eq. (4.43), equivalent to assuming $b = 0$.

[b] Computed by Guggenheim and McGlashan (1951) according to the method described in the text.

D. Activity Coefficients

We now return to the general formula (4.21) valid for the chemical potential μ_i of component i in any gaseous mixture. If we proceed to the limit $x_i \to 1$ at constant T and P, we obtain

$$\mu_i^\bullet = \mu_i^* + RT\ln(P/P^\dagger) + \int_0^P [V_i^\bullet - (RT/P)]\, dP. \tag{4.48}$$

Here, $\mu_i^\bullet$ and $V_i^\bullet$ denote the chemical potential (molar Gibbs function) and partial molar volume (molar volume) of the pure gaseous component i at the given values of temperature and pressure. Combining (4.21) with (4.48), we find

$$\mu_i - \mu_i^\bullet = RT\ln x_i + \int_0^P (V_i - V_i^\bullet)\, dP. \tag{4.49}$$

We define a quantity f_i called the *activity coefficient* of component i:

$$RT\ln f_i \equiv \int_0^P (V_i - V_i^\bullet)\, dP. \tag{4.50}$$

Thus, we have

$$\mu_i = \mu_i^\bullet + RT\ln x_i + RT\ln f_i. \tag{4.51}$$

The activity coefficients, in general, depend on temperature, pressure, and composition. Since for $x_i = 1$ the quantity μ_i equals $\mu_i^\bullet$ we derive from Eq. (4.51)

$$\lim_{x_i\to 1} f_i = 1, \tag{4.52}$$

a "normalization" implied by physical reasons.

For a *perfect gaseous mixture*, we obtain, according to Eqs. (4.2) and (4.26),

$$\mu_i = \mu_i^\bullet + RT\ln x_i, \qquad \mu_i^\bullet = \mu_i^* + RT\ln(P/P^\dagger), \tag{4.53}$$

and thus we have, in view of Eq. (4.51),

$$f_i = 1.$$

Here, the activity coefficients equal unity for all temperatures, pressures, and compositions.

Let us, in general, define an *ideal mixture* by the conditions

$$\mu_i = \mu_i^\bullet + RT\ln x_i \qquad \text{or} \qquad f_i = 1. \tag{4.54}$$

We immediately see from Eq. (4.53) that a perfect gaseous mixture is a special case of an ideal mixture where the dependence of $\mu_i^{\bullet}$ on the pressure is given by the second relation in Eq. (4.53). In Eq. (4.54), however, nothing is implied regarding the pressure dependence of $\mu_i^{\bullet}$. An example of an ideal mixture which is not a perfect gaseous mixture has already been discussed in the text following Eq. (4.42).

Thus, the activity coefficients measure the deviations between the real behavior of gaseous mixtures and that of ideal mixtures.*

We next consider a *binary, slightly imperfect gaseous mixture*. Comparing Eqs. (4.32) and (4.33) to Eq. (4.51), we derive

$$\mu_i^{\bullet} = \mu_i^* + RT\ln(P/P^{\dagger}) + B_{ii}P, \qquad i = 1, 2, \tag{4.55}$$

$$RT\ln f_i = bP(1 - x_i)^2, \qquad i = 1, 2. \tag{4.56}$$

These formulas may also be deduced by substituting Eqs. (4.30) and (4.31) into Eqs. (4.48) and (4.50). Again, we infer from Eq. (4.56) that a binary, slightly imperfect gaseous mixture with $b = 0$ represents an ideal mixture.

We conclude this section by stating some *general laws* concerning the activity coefficients in gaseous mixtures.

It follows from the definition of $V_i^{\bullet}$ that

$$\lim_{x_i \to 1} V_i = V_i^{\bullet}. \tag{4.57}$$

Let us choose the set of independent mole fractions x_j in such a way that x_i is the dependent mole fraction. Then, it is obvious from Eqs. (4.13) and (4.57) that the quantity in Eq. (4.50),

$$\int_0^P (V_i - V_i^{\bullet})\, dP = RT\ln f_i,$$

is a *power series* in both P and the x_j, the exponents being *positive integers*. Also, the lowest power in the x_j must be a quadratic one, such as x_2^2 for $\ln f_1$ in binary mixtures, x_2^2 or x_2x_3 or x_3^2 for $\ln f_1$ in ternary mixtures, and so on. This is implied by the structure of the expressions (4.10) and (4.11). A simple example is given by Eq. (4.56).

* There are other functions used in the literature, such as activities, fugacities, and fugacity coefficients. In particular, the fugacity coefficients describe the deviations from perfect gaseous mixtures. To avoid confusion and to be able to treat gaseous, liquid, and solid mixtures on the same footing, we only use the activity coefficients.

Taking account of the fact that in the case of reacting gaseous mixtures the chemical species and not the components are the fundamental parts of the system, we thus formulate the following general theorem:

At given temperature and pressure, the logarithm of the activity coefficient of any chemical species present in a gaseous mixture may be expressed in terms of the mole fractions of all the other species as a power series with positive integers as exponents, the sum of the exponents of each term being greater than unity.

We have restricted the wording of the theorem to constant pressure, though there is a similar theorem for the dependence of the activity coefficients on the pressure, the lowest power being P (see above). The reason for this restriction is that, as we shall see later, in liquid and solid mixtures there are similar general laws for the dependence of the activity coefficients on the composition at given temperature and pressure.

It should be stressed that these theorems are not implied by thermodynamics alone. They contain additional information from either experience or statistical mechanics. This can be verified for gaseous mixtures if one traces back the origin of the power series in $\ln f_i$ to Eqs. (4.13) and (4.10).

V. Pure Liquids

A. General

A pure liquid is a liquid phase consisting of a single component. All its thermodynamic properties can be derived from the molar Gibbs function $\bar{G}$ (equal to the chemical potential μ) when this is known as a function of thermodynamic temperature T and pressure P.

With regard to the last statement, there are two exceptions: glasses and liquid crystals.

A *glass* is an undercooled liquid which is internally metastable (see p. 88) and thus requires, besides T and P, one, two, or more independent variables to complete the thermodynamic description. These additional variables, called "internal parameters," are similar, in principle, to the extents of reaction of unknown chemical reactions.

A *liquid crystal* is an internally stable liquid (which may coexist with gaseous, liquid, or solid phases) that exhibits anisotropy. Thus, we have to introduce the stress components (see p. 13) in place of the pressure.

Both glasses and liquid crystals fall outside the scope of this chapter.

B. Behavior at High Temperatures

For an ordinary liquid far from absolute zero and far from critical conditions, we may write*

$$\bar{G} = \mu = a_0 + a_1 P - a_2 T + a_3 PT - a_4 P^2 - a_5 T^2 - a_6 P^2 T + a_7 PT^2 - a_8 T \ln(T/T^{\dagger}), \tag{5.1}$$

where

$$T^{\dagger} \equiv 1\text{K} \tag{5.2}$$

and $a_0, \ldots, a_8$ are positive empirical constants. Equation (5.1) represents —except the last term— a series expansion in powers of P and T. It is quite different from Eq. (3.6) valid for pure gases and containing a term involving the logarithm of the pressure.

Using Eqs. (1.10), we derive the expressions for the molar volume $\bar{V}$, the molar entropy $\bar{S}$, and the molar enthalpy $\bar{H}$, following from the empirical relation (5.1):

$$\bar{V} = a_1 - 2a_4 P + a_3 T - 2a_6 PT + a_7 T^2, \tag{5.3}$$

$$\bar{S} = a_2 + a_8 - a_3 P + 2a_5 T + a_6 P^2 - 2a_7 PT + a_8 \ln(T/T^{\dagger}), \tag{5.4}$$

$$\bar{H} = a_0 + a_1 P + a_8 T - a_4 P^2 + a_5 T^2 - a_7 PT^2. \tag{5.5}$$

Equation (5.3) is the equation of state of the liquid. Since $\bar{V}$ is always positive, Eq. (5.3) requires that

$$(a_1 + a_3 T + a_7 T^2) > (2a_4 P + 2a_6 PT). \tag{5.6}$$

As a matter of fact, at low pressures, the term a_1 in (5.3) exceeds considerably all the other terms in magnitude:

$$a_1 \gg | -2a_4 P + a_3 T - 2a_6 PT + a_7 T^2 |, \tag{5.7}$$

as borne out by experiment.

* Eq. (5.1) is an extension of the usual empirical approximations given in the literature.

We now introduce the isothermal compressibility $\varkappa$, the thermal expansivity β, and the molar heat capacity at constant pressure $\bar{C}_P$, defined by the relations (see p. 72)

$$\varkappa \equiv -(1/\bar{V})(\partial\bar{V}/\partial P)_T, \tag{5.8}$$

$$\beta \equiv (1/\bar{V})(\partial\bar{V}/\partial T)_P, \tag{5.9}$$

$$\bar{C}_P \equiv (\partial\bar{H}/\partial T)_P = T(\partial\bar{S}/\partial T)_P. \tag{5.10}$$

We then obtain from Eqs. (5.3)–(5.5)

$$\varkappa = (2a_4 + 2a_6 T)/(a_1 - 2a_4 P + a_3 T - 2a_6 PT + a_7 T^2), \tag{5.11}$$

$$\beta = (a_3 - 2a_6 P + 2a_7 T)/(a_1 - 2a_4 P + a_3 T - 2a_6 PT + a_7 T^2), \tag{5.12}$$

$$\bar{C}_P = a_6 + 2a_5 T - 2a_7 PT. \tag{5.13}$$

Since $\bar{C}_P$ is always positive, Eq. (5.13) requires that

$$(a_6 + 2a_5 T) > 2a_7 PT. \tag{5.14}$$

Actually, at low pressures, we find empirically

$$a_5 > a_7 P, \tag{5.15}$$

thus leading to an increase of $\bar{C}_P$ with increasing temperature.

Table II gives some examples (Rowlinson, 1969) of experimental values for $\bar{V}$, $\varkappa$, β, and $\bar{C}_P$. Although these quantities refer to varying pressure, since they have been measured along the saturation curve, they here nearly coincide with the quantities valid for atmospheric pressure (1 atm $\approx$ 1 bar).

The values of $\bar{V}$, $\varkappa$, β, and $\bar{C}_P$ are always positive—with the notable exception of β for water between 0°C and 4°C— and usually increase with increasing temperature. This is consistent with Eqs. (5.3) and (5.11)–(5.13) in view of the inequalities (5.6), (5.7), and (5.15).

At high pressures, the equation of state (5.3) must be replaced by a more accurate relation, for example, by the empirical formula due to Tait (1889),

$$(\bar{V}_0 - \bar{V})/\bar{V}_0 = AP/(B + P),$$

where $\bar{V}_0$ is the molar volume at zero pressure and A and B are positive parameters. The Tait equation gives an almost perfect fit of experimental data up to pressures of about 1000 bar.

TABLE II

MOLAR VOLUME $\bar{V}$, COMPRESSIBILITY $\varkappa$, THERMAL EXPANSIVITY β, AND MOLAR HEAT CAPACITY AT CONSTANT PRESSURE $\bar{C}_P$ FOR TWO PURE LIQUIDS AT DIFFERENT TEMPERATURES

Temperature (°C)	$\bar{V}$ / cm^3 mol^{-1}	$\varkappa$ / bar^{-1} $\times 10^4$	β / K^{-1} $\times 10^3$	$\bar{C}_P$ / J K^{-1} mol^{-1}
Water[a]:				
0.01[b]	18.019	0.508	−0.0685	75.99
10	18.021	0.478	0.0880	75.55
20	18.048	0.4586	0.207	75.35
40	18.158	0.4423	0.386	75.28
60	18.323	0.4448	0.523	75.38
80	18.538	0.4614	0.642	75.61
100	18.799	0.490	0.752	75.87
150	19.648	0.621	1.035	77.6
Carbon tetrachloride[c]:				
−22.96[b]	91.7	0.75	1.14	130
−10	93.1	0.821	1.16	130
0	94.23	0.890	1.18	130
10	95.35	0.960	1.196	131
20	96.50	1.035	1.219	132
30	97.69	1.120	1.242	132
50	100.17	1.33	1.292	133
70	102.87	1.59	1.346	134

[a] Critical temperature: 374.2°C. [b] Triple point. [c] Critical temperature: 283.2°C.

It should be mentioned that, except for the Tait equation with its limited scope to fit compressibility data, the expressions discussed thus far do not agree quantitatively with the experimental facts. Molecular theory, too, is far from being completely developed for liquids. A very good account of our present knowledge on liquids is given by Rowlinson (1969) and Egelstaff (1967).*

Near the critical point (where liquid and gas coincide), the equations given thus far cease to hold, in principle. What is needed here is a common description of liquid and gas (vapor), but we shall not go into details. We only mention that, at the critical point, the quantities $\varkappa$, β, and $\bar{C}_P$ tend toward infinity.

* See also Volume VIII, "Liquid State," of this series.

C. Behavior at Low Temperatures

The only liquid that remains internally stable down to absolute zero is helium. All other substances, if prevented from crystallization and thus undercooled, become glasses at low temperatures. Nevertheless, we shall investigate the behavior of an internally stable liquid at low temperatures.

We have to take account of two facts: the Nernst heat theorem (p. 86) and the rather strong temperature dependence of thermodynamic functions at low temperatures.

As explained in Chapter 1, Section XXIII, both the molar heat capacity $\bar{C}_P$ and the conventional value of the molar entropy $\bar{S}$ (for internally stable phases) vanish at $T = 0$:

$$\lim_{T\to 0} \bar{C}_P = 0, \qquad \lim_{T\to 0} \bar{S} = 0. \tag{5.16}$$

From Eq. (5.9) and from Chapter 1, Eq. (16.17), there follows

$$\begin{aligned}\bar{S}(T, P_{\text{II}}) - \bar{S}(T, P_{\text{I}}) &= (\partial \bar{S}/\partial P)_T(P_{\text{II}} - P_{\text{I}}) + \cdots \\ &= -\beta \bar{V}(P_{\text{II}} - P_{\text{I}}) + \cdots,\end{aligned}$$

where P_{I} and P_{II} are two arbitrary values of the pressure. Proceeding to the limit $T \to 0$ in the last expression and using (5.16), we find

$$\lim_{T\to 0} \beta = 0. \tag{5.17}$$

Thus, the thermal expansivity β vanishes at absolute zero, too.

Stipulating the validity of an expression such as Eq. (5.1), we impose the conditions (5.16) and (5.17) on Eqs. (5.4), (5.12), and (5.13). We then obtain

$$a_2 = a_3 = a_6 = a_8 = 0.$$

We insert this into Eq. (5.1) and expand the series to higher powers of T to take account of the strong temperature dependence mentioned. We then have, for constant pressure,

$$\bar{G} = \mu = b_1 + b_2 T^2 + b_3 T^3 + b_4 T^4, \tag{5.18}$$

where $b_1 \equiv a_0 + a_1 P - a_4 P^2$, $b_2 \equiv a_7 P - a_5$, and b_3 and b_4 are new empirical constants. Equation (5.18) will be discussed further in Section VII since it also holds for pure solids at very low temperatures.

D. Ionic Melts

A pure liquid may contain more than one chemical species. Examples are associated liquids, such as water, and molten electrolytes, such as pure liquid sodium chloride or pure liquid potassium hydroxide. Since a molten electrolyte mainly consists of ions, it will be called an *ionic melt*. The most important class of ionic melts are molten (fused) salts.

An ionic melt can only exist at elevated temperatures. Thus, its thermodynamic properties will again be described by Eq. (5.1), approximately. Here, the molar volume $\overline{V}$ has the order of magnitude of 10 to 10^2 $cm^3\ mol^{-1}$, the compressibility $\varkappa$ that of 10^{-6} to 10^{-5} bar^{-1}, the expansivity β that of 10^{-4} to 10^{-3} K^{-1}, the molar heat capacity $\overline{C}_P$ that of 10^2 J K^{-1} mol^{-1}. All these quantities again increase with increasing temperature [for more details, see the work of Janz (1967)].

For pure ionic melts (as well as for ionic melt mixtures), there is an additional method of determining thermodynamic functions: the measurement of the electromotive force of a galvanic cell containing a fused electrolyte in place of an electrolyte solution.

As our first example, we consider a galvanic cell consisting of molten lead (electrode), molten lead chloride (ionic melt), and graphite saturated with chlorine gas (electrode):

$$\text{Pb(liquid)} \mid \text{PbCl}_2\text{(liquid)} \mid \text{Cl}_2\ \text{(C)},$$

where temperature and pressure are supposed to be uniform. When we stipulate local heterogeneous equilibrium at the phase boundaries, then the cell is reversible and it is indeed a chemical cell. According to Chapter 1, Eq. (21.5), the electromotive force Φ of a chemical cell is related to the affinity A of the chemical reaction associated with the flow of the positive charge 1 Far (one faraday) passing though the cell from left to right:

$$\mathfrak{F}\Phi = A,$$

$\mathfrak{F}$ being the Faraday constant. In our example we have [see Chapter 1, Eq. (20.8)]

$$A = \tfrac{1}{2}\mu_{\text{Pb}} + \tfrac{1}{2}\mu_{\text{Cl}_2} - \tfrac{1}{2}\mu_{\text{PbCl}_2}.$$

Here, μ_{Pb}, μ_{Cl_2}, and μ_{PbCl_2} denote the chemical potentials (molar Gibbs functions) of liquid lead, gaseous chlorine, and liquid lead chloride, respectively. From the last two equations, we conclude that we are able to derive (conventional) values of μ_{PbCl_2} from the electromotive-force measurements provided μ_{Pb} and μ_{Cl_2} are known.*

A less familiar example is the following one. Let a pure ionic melt be part of a galvanic cell where two similar electrodes are at different heights in the earth's gravitational field or at different distances from the rotating axis in a centrifuge, thus giving rise to a pressure gradient. Such a galvanic cell is called a gravitational or centrifugal cell, respectively. If we assume that the electrodes are reversible for the cationic species (X) of the fused electrolyte ($X_{\nu_+}Y_{\nu_-}$), then the phase diagram of the cell may be written

$$|_{P_I}\ \mathrm{X(liquid\ or\ solid)}\ |_{P_I}\ \mathrm{X}_{\nu_+}\mathrm{Y}_{\nu_-}\mathrm{(liquid)}\ |_{P_{II}}\ \mathrm{X(liquid\ or\ solid)}\ |_{P_I}, \tag{5.19}$$

where the temperature is supposed to be uniform. The symbols P_I and P_{II} denote two values of the pressure P. A simple example is

$$|_{P_I}\ \mathrm{Ag(solid)}\ |_{P_I}\ \mathrm{AgNO_3(liquid)}\ |_{P_{II}}\ \mathrm{Ag(solid)}\ |_{P_I}, \tag{5.20}$$

with $\nu_+ = \nu_- = 1$. If we stipulate that there is local heterogeneous equilibrium at the phase boundaries, then the galvanic cell (5.19) or (5.20) is again a reversible cell, since nothing happens at zero electric current, but it is no longer a chemical cell. The cell (5.19) is one of the simplest types of galvanic cell conceivable: There are only two kinds of phases and two components in the whole system, which is symmetric except for the pressure differences. The electromotive force Φ of the cell (5.19) can be shown to be†

$$\begin{aligned} \Phi &= (M'/z_+\mathfrak{F}) \int_{P_I}^{P_{II}} [(1/\varrho) - (1/\varrho')]\, dP \\ &\approx (1/z_+\mathfrak{F})[(M'/M)\bar{V} - \bar{V}'](P_{II} - P_I). \end{aligned} \tag{5.21}$$

Here, M' is the molar mass of X, and M that of $X_{\nu_+}Y_{\nu_-}$; z_+ is the charge number of the cationic species; and $\bar{V}$ and $\bar{V}'$ are the molar volumes and ϱ and ϱ' the densities of the ionic melt and the electrode, respectively. Thus, the measurable electromotive force is determined by the difference

* The argument may be reversed: The thermodynamic properties of liquid metals (such as molten lead) can be derived from electromotive-force measurements provided the thermodynamic functions of the other phases are known. Another example refers to mercury: the molar Gibbs function of this liquid can be obtained from measurements on the chemical cell (21.1) given in Chapter 1.

† The rigorous derivation (Haase, 1969a) of Eq. (5.21) implies the combination of the condition for electrochemical equilibrium at the phase boundaries with the condition for sedimentation equilibrium in the phases supporting pressure gradients.

in reciprocal densities of the two phases present. For the cell (5.20), where $z_+ = 1$, $M' = M_{Ag}$, $M = M_{AgNO_3}$, Eq. (5.21) has been checked experimentally (Duby and Townsend, 1968). This is, of course, not a suitable method of measuring densities of ionic melts; but it is an interesting application of thermodynamics.

VI. Liquid Mixtures

A. General

A liquid mixture is a liquid phase containing two or more components. As independent variables to describe the state of the mixture, we again use the thermodynamic temperature T, the pressure P, and the mole fractions $x_2, x_3, \ldots, x_N$ of the components 2, 3, . . ., N. In some special cases, however, it is convenient to introduce other composition variables (see Sections VI,C and VI,D).

There exists a great variety of liquid mixtures. The binary liquid systems water + urea, benzene + polystyrene, water + sodium chloride, and potassium nitrate + silver nitrate are examples of low-molecular nonelectrolyte solutions (Section VI,B), of high-molecular nonelectrolyte solutions (Section VI,C), of electrolyte solutions (Section VI,D), and of ionic melt mixtures (Section VI,E), respectively.

The number of chemical species in a liquid mixture often exceeds the number of components. Thus, it is frequently convenient first to introduce the "true mole fractions" of the species actually present and then to proceed to the "stoichiometric mole fractions" of the components.

Usually, one compares the thermodynamic properties of the liquid mixture with those of the pure liquid components at the same temperature and pressure. We are thus interested in quantities such as the difference between the chemical potential μ_i of component i in the mixture and the chemical potential $\mu_i^{\bullet}$ of the pure liquid component i at the given values of T and P. It is convenient to use the dimensionless quantity

$$\psi_i \equiv (\mu_i - \mu_i^{\bullet})/RT, \tag{6.1}$$

where R denotes the gas constant.*

* $\mu_i^{\bullet}$ may refer to an unstable state. In the system water + urea, at 25°C, for example, pure liquid urea would be a hypothetical supercooled liquid. In the literature, we frequently find a quantity a_i called "activity" and usually, but not always, defined by $\ln a_i \equiv \psi_i$.

The simplest type of liquid mixture is an *ideal mixture* (sometimes called "perfect solution") defined by the relation

$$\psi_i = \ln x_i, \tag{6.2}$$

valid for any composition. Comparison of Eqs. (6.1) and (6.2) with Eq. (4.54) shows that Eq. (6.2) coincides with the definition of an ideal mixture already given for gaseous mixtures.

If it is definitely known that there are, in the liquid system considered, chemical species not identical with the components chosen, then it is expedient to apply the definition (6.2) to the chemical species actually present. Now, x_i denotes the true mole fraction of the chemical species i, while $\mu_i^{\bullet}$ in Eq. (6.1) may have a merely hypothetical meaning. We then call the liquid mixture a *true ideal mixture*. When, however, the definition (6.2) is given in terms of the components (x_i denoting the stoichiometric mole fraction), then the mixture is designated as a *stoichiometric ideal mixture*.

It is an important fact, borne out by experiment and statistical theory, that in any liquid (and solid) mixture, the asymptotic law

$$\mu_i \to \mu_i^{\infty} + RT \ln x_i$$

holds for any chemical species i if the mole fractions of all components (the "solutes") except one (the "solvent") tend toward zero. Here, x_i is the true mole fraction and μ_i^{∞} denotes a function of T and P which is different from $\mu_i^{\bullet}$ in Eq. (6.1), except for the solvent ($x_i = 1$ then denotes the pure liquid solvent so that $\mu_i = \mu_i^{\bullet} = \mu_i^{\infty}$).

A more rigorous way of stating this limiting law is the equation

$$\mu_i = \mu_i^{\infty} + RT \ln x_i + \lambda_i,$$

where λ_i is a function of temperature, pressure, and composition, vanishing at $x_1 = 1$ (x_1 is the stoichiometric mole fraction of solvent). This again implies that λ_i can be represented as a power series in the independent mole fractions, the exponents being nonnegative numbers which, however, are not necessarily integers.* If λ_i refers to a solvent species and is expressed in terms of the mole fractions of the solutes, then the sum of the exponents of each term must be greater than unity. (Of course, any component may be chosen to be the solvent.) This is a necessity since otherwise the dominating effect of the term $RT \ln x_i$

* Fractional exponents occur in electrolyte solutions.

near $x_1 = 1$ would be destroyed. Comparison to our formulations on p. 313 shows that such a general statement also applies to gaseous mixtures.*

We repeat that this general law (Haase, 1956) can only be justified by experience or statistical mechanics. It is not an implication of thermodynamics since, then, logarithmic terms and terms with negative exponents in the series expansion of λ_i with respect to the mole fractions would be possible (Haase, 1956).

All the *limiting laws for infinite dilution* can be readily derived from the above general statement (Haase, 1956).

An *ideal dilute solution* is any liquid (or solid) mixture that, within experimental error, exhibits the behavior characteristic for infinite dilution. Thus, any mixture may, on sufficient dilution, become an ideal dilute solution. On the other hand, an ideal mixture only occurs with certain types of components: the different kinds of particles (molecules, atome, or ions) present in the liquid system must be very similar (theoretically nearly identical), but then Eq. (6.2) holds for the whole range of compositions.

The ideal mixture (and sometimes the ideal dilute solution) is a useful standard system for defining quantities such as "activity coefficients" and "excess functions." We shall give the exact definitions in the following subsections since the different types of liquid mixtures require different treatments.

There is an enormous amount of publications on liquid mixtures. Scott and Fenby (1969) state that for the period 1966–1968 there are about 1500 papers on nonelectrolyte solutions alone. We are thus compelled to give a survey of fundamental principles and to select the examples in a rather personal way.

B. Low-Molecular Nonelectrolyte Solutions

A low-molecular nonelectrolyte solution is a liquid mixture containing nonionic components having small molar masses ("low-molecular-weight" nonelectrolytes). Examples are the binary liquid systems water + urea, methanol + benzene, and zinc + silver.

There is, of course, a continuous transition to other types of liquid mixtures. Thus, in the liquid system benzene + sulfur, above a certain

* The function λ_i only differs by a constant from $RT \ln f_i$ as defined in Eq. (4.50).

temperature, sulfur becomes a polymer. Again, the liquid system water + acetic acid is an electrolyte solution at low acid concentrations but practically a nonelectrolyte solution at high acid concentrations. Finally, liquid systems such as sodium + sodium chloride are examples for the transition from metallic nonelectrolyte solutions to ionic melt mixtures.

Usually, the *stoichiometric ideal mixture* (nearly realized in mixtures of isotopes and of isomers as well as in systems such as benzene + toluene) is taken as the standard system. All the thermodynamic properties of low-molecular nonelectrolyte solutions are then described as deviations from this standard mixture.

The simplest way of handling any mixture which is not a stoichiometric ideal mixture would be to assume a true ideal mixture. In the case of nonelectrolyte solutions, this conjecture is equivalent to the supposition of a number of association equilibria in a reacting mixture where the chemical potential μ_i of any chemical species i is given by Eqs. (6.1) and (6.2),

$$\mu_i = \mu_i^\bullet + RT \ln x_i . \tag{6.3}$$

This special type of true ideal mixture is called an *ideal associated mixture*. One of the drawbacks of this simple model is the fact that it cannot explain the phase separation (liquid–liquid equilibrium) frequently observed in nonelectrolyte solutions.*

In low-molecular nonelectrolyte solutions, we describe the deviations in the behavior of a real mixture from that of a stoichiometric ideal mixture in either of two ways.

The first way consists in introducing the *activity coefficient* f_i of component i defined by

$$\ln f_i \equiv \psi_i - \ln x_i, \qquad i = 1, 2, \ldots, N. \tag{6.4}$$

Thus, we have, in view of Eq. (6.1),

$$\mu_i = \mu_i^\bullet + RT \ln x_i + RT \ln f_i, \qquad i = 1, 2, \ldots, N. \tag{6.5}$$

* The proof is simple. Let the two liquid phases in mutual equilibrium be denoted by a prime (′) and a double prime (″). Then we have for any pair of coexisting phases $\mu_i' = \mu_i''$, and thus, on account of Eq. (6.3), $x_i' = x_i''$, valid for any species i that can pass the phase boundary. Since the pair of coexisting phases may be chosen close to the critical solution point (where the two liquid phases coincide), the equality $x_i' = x_i''$ holds indeed for all species present and thus means equal compositions of the two phases. But this contradicts the assumption of a phase separation.

There follows, for a stoichiometric ideal mixture,

$$f_i = 1 \qquad \text{(stoichiometric ideal mixture).} \tag{6.6}$$

This definition of activity coefficient coincides with that given for ideal gaseous mixtures [see Eq. (4.51)].

The second scheme of description (due to Scatchard) defines an *excess function* (superscript E):

$$y^{\mathrm{E}} \equiv y - y^{\mathrm{id}}, \tag{6.7}$$

where y is any function of state of the mixture and y^{id} the value of this function for a stoichiometric ideal mixture of the same temperature, pressure, and composition. In particular, there follows from Eqs. (6.5) and (6.6)

$$\mu_i^{\mathrm{E}} = RT \ln f_i, \qquad i = 1, 2, \ldots, N, \tag{6.7a}$$

μ_i^{E} being the excess chemical potential of component i.

If y is an extensive property, we call it Z. Accordingly, we denote the molar function of the mixture by $\bar{Z}$ (molar volume $\bar{V}$, molar enthalpy $\bar{H}$, molar entropy $\bar{S}$, molar Gibbs function $\bar{G}$, etc.). We thus have for any molar excess function

$$\bar{Z}^{\mathrm{E}} = \bar{Z} - \bar{Z}^{\mathrm{id}}. \tag{6.8}$$

In particular, the molar excess Gibbs function

$$\bar{G}^{\mathrm{E}} = \bar{G} - \bar{G}^{\mathrm{id}} \tag{6.9}$$

may be expressed in terms of the activity coefficients f_i of all the components 1, 2, . . ., N.

We derive from Eqs. (1.7), (6.5), and (6.6)

$$\bar{G} = \sum_{i=1}^{N} x_i \mu_i^{\bullet} + RT \sum_{i=1}^{N} x_i \ln(x_i f_i), \tag{6.10}$$

$$\bar{G}^{\mathrm{id}} = \sum_{i=1}^{N} x_i \mu_i^{\bullet} + RT \sum_{i=1}^{N} x_i \ln x_i . \tag{6.11}$$

Thus we find, on account of Eq. (6.9),

$$\bar{G}^{\mathrm{E}} = RT \sum_{i=1}^{N} x_i \ln f_i , \tag{6.12}$$

the relation desired. Both $\bar{G}^{\mathrm{E}}$ and f_i depend on temperature, pressure, and composition.

Considering the general thermodynamic formulas following from Eqs. (1.1) and (1.2),

$$\bar{V} = (\partial \bar{G}/\partial P)_{T,x}, \qquad \bar{S} = -(\partial \bar{G}/\partial T)_{P,x}, \qquad \bar{H} = \bar{G} + T\bar{S} \tag{6.13}$$

where the subscript x denotes constant composition, and taking account of Eqs. (6.8) and (6.9), we obtain

$$\bar{V}^{\mathrm{E}} = (\partial \bar{G}^{\mathrm{E}}/\partial P)_{T,x}, \quad \bar{S}^{\mathrm{E}} = -(\partial \bar{G}^{\mathrm{E}}/\partial T)_{P,x}, \quad \bar{H}^{\mathrm{E}} = \bar{G}^{\mathrm{E}} + T\bar{S}^{\mathrm{E}}, \tag{6.14}$$

$\bar{V}^{\mathrm{E}}$ denoting the molar excess volume, $\bar{S}^{\mathrm{E}}$ the molar excess entropy, $\bar{H}^{\mathrm{E}}$ the molar excess enthalpy.

It follows from Eqs. (6.11) and (6.13) that

$$\bar{V}^{\mathrm{id}} = \sum_{i=1}^{N} x_i V_i^{\bullet}, \tag{6.15}$$

$$\bar{S}^{\mathrm{id}} = \sum_{i=1}^{N} x_i S_i^{\bullet} - R \sum_{i=1}^{N} x_i \ln x_i, \tag{6.16}$$

$$\bar{H}^{\mathrm{id}} = \sum_{i=1}^{N} x_i H_i^{\bullet}. \tag{6.17}$$

Here, $V_i^{\bullet}$, $S_i^{\bullet}$, and $H_i^{\bullet}$ are the (partial) molar volume, entropy, and enthalpy of the pure liquid component i at the given temperature and pressure. Thus, for stoichiometric ideal mixtures, there is no change of volume or enthalpy on mixing, while the "molar entropy of mixing," equal to the last term on the right-hand side of Eq. (6.16), is a positive universal function of composition.

Let us, in general, consider the process of mixing the pure liquid components, at constant temperature and pressure, so as to form a liquid mixture. The increase $\Delta \bar{Z}$ of any molar function in this process is called the "molar function of mixing." We infer from Eqs. (6.8), (6.11), and (6.15)–(6.17)

$$\Delta \bar{V} = \bar{V} - \sum_{i=1}^{N} x_i V_i^{\bullet} = \bar{V}^{\mathrm{E}}, \tag{6.18}$$

$$\Delta \bar{H} = \bar{H} - \sum_{i=1}^{N} x_i H_i^{\bullet} = \bar{H}^{\mathrm{E}}, \tag{6.19}$$

$$\Delta \bar{S} = \bar{S} - \sum_{i=1}^{N} x_i S_i^{\bullet} = \bar{S}^{\mathrm{E}} - R \sum_{i=1}^{N} x_i \ln x_i, \tag{6.20}$$

$$\Delta \bar{G} = \bar{G} - \sum_{i=1}^{N} x_i \mu_i^{\bullet} = \bar{G}^{\mathrm{E}} + RT \sum_{i=1}^{N} x_i \ln x_i. \tag{6.21}$$

Thus, in particular, the molar enthalpy of mixing ("molar heat of mixing") equals the molar excess enthalpy. When $\bar{H}^E > 0$, we have "endothermic mixing"; when $\bar{H}^E < 0$, there is "exothermic mixing."

Experimental values of $\bar{V}^E$ and $\bar{H}^E$ may be directly obtained from density measurements and calorimetric determinations of the heat of mixing.

Experimental values of $\bar{G}^E$ can be derived from vapor pressures and vapor compositions or, less frequently, from freezing points combined with calorimetric data (see, for example, Goates and Sullivan (1958)]. For metallic solutions (liquid alloys), measurements on reversible cells may be used for the determination of $\bar{G}^E$, too (Wagner, 1952). For example, the electromotive force of the reversible cell (compare p. 318)

$$\text{Pb(liquid)} \mid \text{PbCl}_2\text{(liquid)} \mid \text{Pb} + \text{Cd(liquid)} \qquad (T, P \quad \text{constant})$$

is connected to the activity coefficient of lead in the liquid mixture (alloy) lead + cadmium.

The molar excess entropy $\bar{S}^E$ can be computed from $\bar{H}^E$ and $\bar{G}^E$ according to the last equation in (6.14). The first two relations in (6.14) provide a check for the consistency of the different kinds of measurements.*

We shall now confine the discussion to *binary*, low-molecular nonelectrolyte solutions.

Both experience and statistical theory tell us that it is possible and convenient to express the molar excess Gibbs function $\bar{G}^E$ as a polynomial in x_1 or x_2 for given temperature and pressure. Using the notation

$$x \equiv x_2, \qquad 1 - x \equiv x_1, \tag{6.22}$$

we may write the power series in the form (Guggenheim, 1937; Redlich and Kister, 1948; Scatchard, 1949)

$$\begin{aligned} \bar{G}^E &= x_1 x_2 [B_0 + B_1(x_1 - x_2) + B_2(x_1 - x_2)^2 + \cdots] \\ &= x(1 - x)[B_0 + B_1(1 - 2x) + B_2(1 - 2x)^2 + \cdots]. \end{aligned} \tag{6.23}$$

Here, $B_0, B_1, B_2, \ldots$ are empirical parameters, depending on T and P, which may be positive or negative. We see that $\bar{G}^E = 0$ for both $x = 0$ and $x = 1$ as implied by the definition of $\bar{G}^E$.

* Consistency and mathematical evaluation of experimental data are discussed in the following papers: Beatty and Calingaert (1934), Coulson and Herington (1948), Redlich and Kister (1948), and Jost and Röck (1954).

To derive the corresponding expressions for the activity coefficients f_1 and f_2 of the components 1 and 2, we apply Eq. (1.12) to the chemical potentials μ_1 and μ_2 and take account of the definitions (6.7), (6.7a), (6.9), and (6.22). We then discover that

$$\begin{aligned} RT \ln f_1 &= \mu_1{}^{\mathrm{E}} = \bar{G}^{\mathrm{E}} - x(\partial \bar{G}^{\mathrm{E}}/\partial x)_{T,P}, \\ RT \ln f_2 &= \mu_2{}^{\mathrm{E}} = \bar{G}^{\mathrm{E}} + (1-x)(\partial \bar{G}^{\mathrm{E}}/\partial x)_{T,P}. \end{aligned} \tag{6.24}$$

Inserting Eq. (6.23) into (6.24), we conclude that

$$\begin{aligned} RT \ln f_1 &= x^2[B_0 + B_1(4x-3) + B_2(2x-1)(6x-5) + \cdots], \\ RT \ln f_2 &= (1-x)^2[B_0 + B_1(4x-1) + B_2(2x-1)(6x-1) + \cdots]. \end{aligned} \tag{6.25}$$

Thus, we infer that the polynomial in (6.23) is equivalent to a series expansion given by Margules (1895).

We also note that, for constant temperature and pressure, $\ln f_1$ or $\ln f_2$ is a power series in x_2 or x_1 with positive integers as exponents, the lowest power being $x_2{}^2$ or $x_1{}^2$, respectively. This is a special case of a more general theorem already formulated on p. 321.

Combining Eqs. (6.14) and (6.23), we obtain

$$\begin{aligned} \bar{V}^{\mathrm{E}} = x(1-x)[&(\partial B_0/\partial P) + (\partial B_1/\partial P)(1-2x) \\ &+ (\partial B_2/\partial P)(1-2x)^2 + \cdots], \end{aligned} \tag{6.26}$$

$$\begin{aligned} \bar{S}^{\mathrm{E}} = -x(1-x)[&(\partial B_0/\partial T) + (\partial B_1/\partial T)(1-2x) \\ &+ (\partial B_2/\partial T)(1-2x)^2 + \cdots], \end{aligned} \tag{6.27}$$

$$\begin{aligned} \bar{H}^{\mathrm{E}} = x(1-x)\{&B_0 - T(\partial B_0/\partial T) + [B_1 - T(\partial B_1/\partial T)](1-2x) \\ &+ [B_2 - T(\partial B_2/\partial T)](1-2x)^2 + \cdots\}. \end{aligned} \tag{6.28}$$

We recall that $B_0, B_1, B_2, \ldots$ only depend on T and P.

Neglecting all terms involving $B_1, B_2, \ldots$, we find from the preceding equations

$$RT \ln f_1 = B_0 x^2, \qquad RT \ln f_2 = B_0(1-x)^2, \tag{6.29}$$

$$\bar{G}^{\mathrm{E}} = B_0 x(1-x), \tag{6.30}$$

$$\bar{V}^{\mathrm{E}} = (\partial B_0/\partial P)x(1-x), \tag{6.31}$$

$$\bar{S}^{\mathrm{E}} = -(\partial B_0/\partial T)x(1-x), \tag{6.32}$$

$$\bar{H}^{\mathrm{E}} = [B_0 - T(\partial B_0/\partial T)]x(1-x). \tag{6.33}$$

Thus, the molar excess functions $\bar{G}^E$, $\bar{V}^E$, $\bar{S}^E$, and $\bar{H}^E$ are all symmetric (parabolic) functions in x with a maximum or minimum at $x = \frac{1}{2}$. This type of mixture has been called a *simple mixture* (Guggenheim, 1967). Equations equivalent to (6.29) had already been used by Porter (1920). Later (Haase, 1951), the thermodynamic implications and the range of validity of Porter's formulas were realized. Comparing Eqs. (4.56) and (6.29), we conclude that binary, slightly imperfect gaseous mixtures may be considered to be a special case of simple mixtures.

Binary, low-molecular nonelectrolyte solutions with nonpolar components are frequently either accurately or approximately simple mixtures, in particular when the deviations from ideal behavior are small.

As an example, we take the liquid system carbon tetrachloride + cyclohexane. Here, $\bar{G}^E$ has been derived from vapor–liquid-equilibrium measurements (Scatchard *et al.* 1939) between 30°C and 70°C. Furthermore, $\bar{H}^E$ has been determined by calorimetric measurements (Adcock and McGlashan, 1954) between 10°C and 55°C, $\bar{V}^E$ by density measurements (Wood and Gray, 1952) between 15°C and 75°C. All these quantities practically refer to atmospheric pressure. If we write [compare Eq. (5.1)]

$$B_0 = a_0 + a_1 P - a_2 T - bT \ln(T/T^\dagger), \tag{6.34}$$

where $T^\dagger \equiv 1$ K and a_0, a_1, a_2, and b are empirical constants, then Eqs. (6.30)–(6.33) imply that

$$\bar{G}^E = x(1-x)[(a_0 + a_1 P - a_2 T - bT \ln(T/T^\dagger)], \tag{6.35}$$

$$\bar{V}^E = x(1-x)a_1, \tag{6.36}$$

$$\bar{S}^E = x(1-x)[a_2 + b + b \ln(T/T^\dagger)], \tag{6.37}$$

$$\bar{H}^E = x(1-x)(a_0 + a_1 P + bT). \tag{6.38}$$

As a matter of fact, these equations fit the experimental data when we use the values ($P = 1$ bar)

$$a_0 = 1176 \quad \text{J mol}^{-1},$$

$$a_1 = 0.0165 \quad \text{J mol}^{-1}\,\text{bar}^{-1} = 0.165 \quad \text{cm}^3\,\text{mol}^{-1},$$

$$a_2 = 14.18 \quad \text{J K}^{-1}\,\text{mol}^{-1},$$

$$b = -1.96 \quad \text{J K}^{-1}\,\text{mol}^{-1}.$$

Thus, in the temperature range investigated, all the molar excess functions are positive for the system carbon tetrachloride + cyclohexane.*

Most binary, low-molecular nonelectrolyte solutions, in particular those containing one or two polar components, exhibit a more complicated behavior ($B_1 \neq 0$, $B_2 \neq 0$, ...). Examples have already been given in Chapter 2 and may also be found in the literature (Haase, 1956; Rowlinson, 1969). We shall restrict the discussion to a few typical examples, relating to atmospheric pressure again.

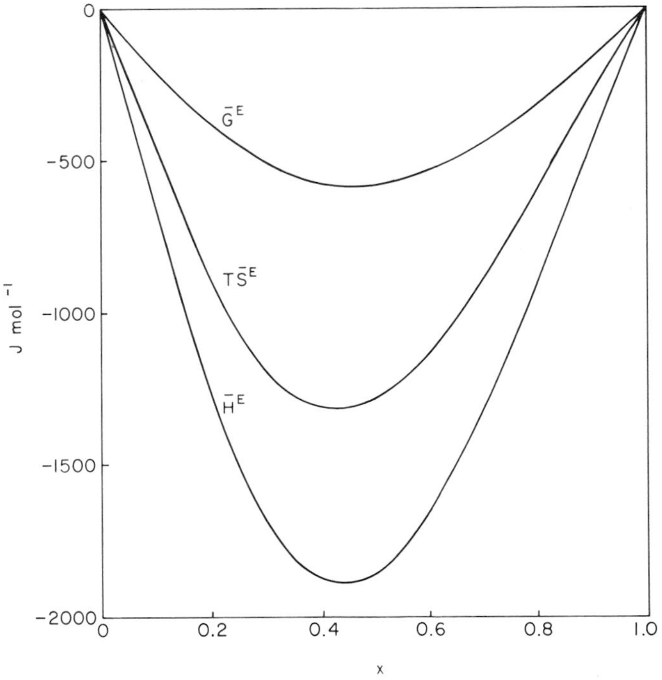

FIG. 1. Liquid system chloroform + acetone (Röck and Schröder, 1957) at 40°C: Molar excess Gibbs function $\bar{G}^E$, molar excess enthalpy $\bar{H}^E$, and product of thermodynamic temperature T and molar excess entropy $\bar{S}^E$ as functions of the mole fraction x of acetone.

Our first example is the liquid system chloroform + acetone at 40°C (Fig. 1). Here, $\bar{G}^E$, $\bar{H}^E$, and $\bar{S}^E$ are negative and no longer symmetric with respect to x.

Our second example is water + dioxan at 25°C (Fig. 2). Here, $\bar{H}^E$

* See also the recent data (Boublik *et al.*, 1969) on carbon tetrachloride + cyclopentane.

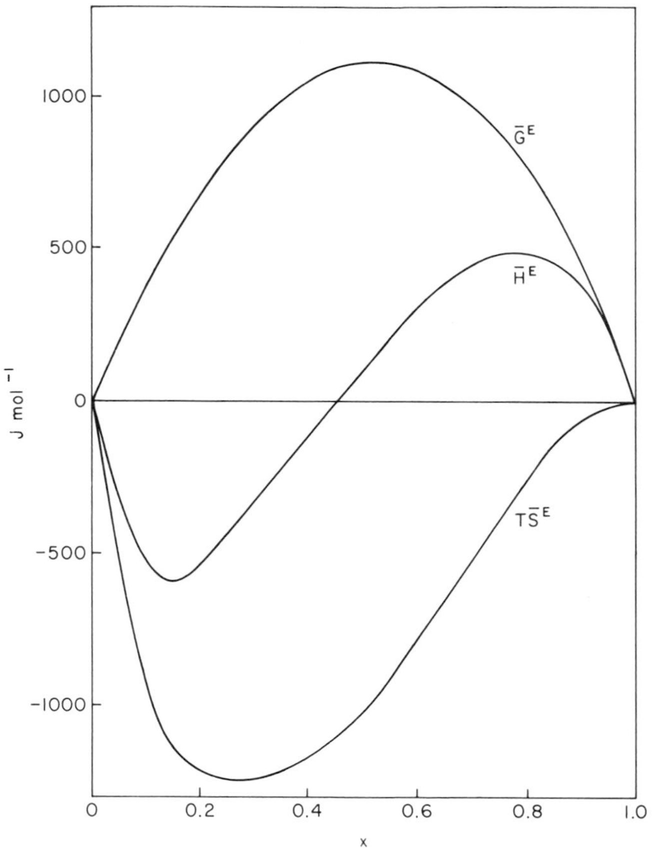

FIG. 2. Liquid system water + dioxan (Goates and Sullivan, 1958) at 25°C: $\bar{G}^E$, $\bar{H}^E$, and $T\bar{S}^E$ as functions of the mole fraction x of dioxan.

and $\bar{S}^E$ change signs, while $\bar{G}^E$ is positive over the whole range of compositions.

Our third example is triphenylmethane + sulfur at 190°C (Fig. 3). Here, similar statements apply as to the last system. It should be noted that in the binary liquid mixtures of sulfur with triphenylmethane, benzene, and toluene there are two regions of immiscibility (liquid–liquid equilibria) separated from each other by a region of complete miscibility (where the measurements leading to Fig. 3 have been carried out).

In these and other cases of complicated behavior, the experimental values of the excess functions can be fitted by Eq. (6.23) with the help of three or four parameters and a temperature dependence similar to that given for B_0 in Eq. (6.34). Thus, for triphenylmethane + sulfur, the experimental values approximately fit the equations following from Eqs. (6.23), (6.27), and (6.28) with three parameters B_0, B_1, and B_2 given by

$$B_0/(\mathrm{J\ mol^{-1}}) = 85{,}616 - 1263(T/T^{\dagger}) + 178(T/T^{\dagger})\ln(T/T^{\dagger}),$$

$$B_1/(\mathrm{J\ mol^{-1}}) = 119{,}043 - 1956(T/T^{\dagger}) + 277(T/T^{\dagger})\ln(T/T^{\dagger}),$$

$$B_2/(\mathrm{J\ mol^{-1}}) = 92{,}048 - 1507(T/T^{\dagger}) + 213(T/T^{\dagger})\ln(T/T^{\dagger}).$$

One of the most conspicuous features of the composition dependence of the molar excess functions in binary systems is the following rule: Though the functions $\bar{V}^{\mathrm{E}}(x)$, $\bar{H}^{\mathrm{E}}(x)$, and $\bar{S}^{\mathrm{E}}(x)$ may exhibit very complicated behavior (including S-shaped curves), the function $\bar{G}^{\mathrm{E}}(x)$ re-

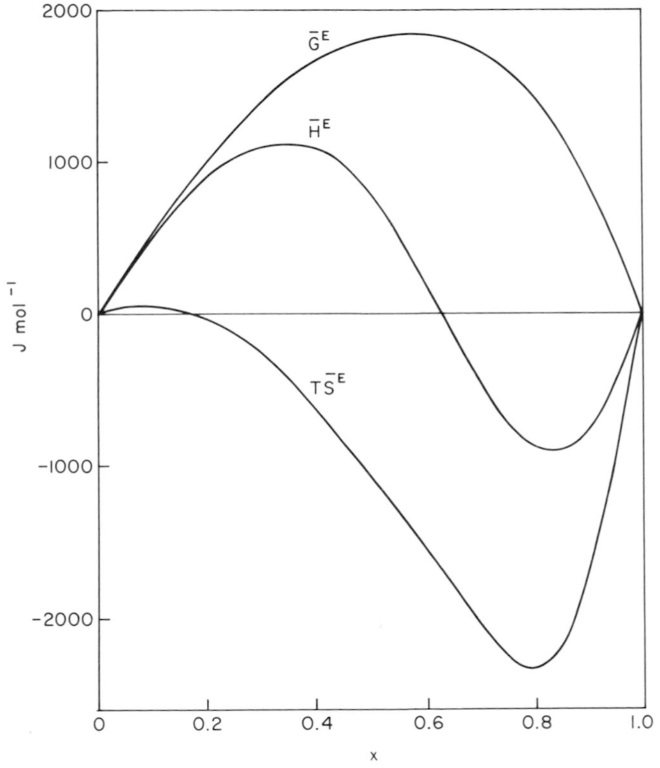

FIG. 3. Liquid system triphenylmethane + sulfur (Haase *et al.*, 1967) at 190°C: $\bar{G}^{\mathrm{E}}$, $\bar{H}^{\mathrm{E}}$, and $T\bar{S}^{\mathrm{E}}$ as functions of the mole fraction x of sulfur (S_8).

mains nearly symmetric and is still roughly represented by the formula $\bar{G}^{E} = B_0 x(1 - x)$. This is easily verified in the last three examples given. An inspection of numerous data shows indeed that the rule holds for nearly all binary, low-molecular nonelectrolyte solutions. We shall see later that this *symmetry rule* (Haase, 1950, 1951) is valid for other types of binary mixtures, too.

It may occur that in the expression for the molar entropy of mixing [see Eqs. (6.16), (6.20), and (6.22)],

$$\Delta\bar{S} = \Delta\bar{S}^{\text{id}} + \bar{S}^{E}, \tag{6.39}$$

with

$$\Delta\bar{S}^{\text{id}} = -R[x(\ln x) + (1 - x)\ln(1 - x)], \tag{6.40}$$

the absolute magnitude of the molar excess entropy $\bar{S}^{E}$ (which may be positive or negative) exceeds the molar entropy of mixing $\Delta\bar{S}^{\text{id}}$ for an ideal mixture (which is always positive), at least in the major part of the composition range. In the vicinity of $x = 0$ and $x = 1$ the logarithmic term in (6.39) and (6.40) becomes so important that here $\Delta\bar{S}$ nearly coincides with $\Delta\bar{S}^{\text{id}}$. The first example for $|\bar{S}^{E}| > \Delta\bar{S}^{\text{id}}$ is the system cyclohexane + methanol (Haase and Rehage, 1955; Haase, 1958), where $\bar{S}^{E}$ is positive. The second example for $|\bar{S}^{E}| > \Delta\bar{S}^{\text{id}}$ is the system water + triethylamine (Haase and Rehage, 1955; Haase, 1958), already discussed and represented graphically in Chapter 2. Here, $\bar{S}^{E}$ is negative and therefore $\Delta\bar{S}$ is negative, too,* except for the immediate neighborhood of $x = 0$ and $x = 1$, where $\Delta\bar{S}$ is again positive. The last effect cannot be seen in the figure since this would require an enlargement of the scale. It may be added that the signs of the molar entropy of mixing have a direct bearing on the type of liquid–liquid equilibria occurring in these systems (Haase and Rehage, 1955; Haase, 1958): There is a miscibility gap with an upper consolute point (upper critical solution temperature) in cyclohexane + methanol, while in water + triethylamine, we have a miscibility gap with a lower consolute point (lower critical solution temperature).

We finally point out that the statistical theory of nonelectrolyte solutions has made considerable progress in the last decades. There are excellent books on this topic (Guggenheim, 1952; Prigogine, 1967; Rowlinson, 1969).

* This also occurs in the liquid metallic systems magnesium + antimony and magnesium + bismuth at high temperatures (Vetter and Kubaschewski, 1953).

C. High-Molecular Nonelectrolyte Solutions

A high-molecular nonelectrolyte solution is a liquid mixture consisting of nonionic components which differ appreciably in molecular sizes. One component, called the "solvent," is a low-molecular-weight nonelectrolyte, while the other components, called the "solutes," are high-molecular-weight nonelectrolytes. The solutes may be, in particular, polymers or high polymers. For the sake of simplicity, we shall only consider a single solute, that is to say, we restrict the discussion to *binary* systems.

Examples of binary, high-molecular nonelectrolyte solutions are the liquid systems hexane + hexadecane (on the border between low-molecular and high-molecular solutions), benzene + polystyrene (a polymer or high polymer solution), and carbon tetrachloride + rubber (a high polymer solution).

Both experimental investigation and theoretical treatment of these solutions bear certain features alien to other classes of liquid mixtures. Thus, the thermodynamics of polymer solutions has grown into a field of its own. We therefore confine ourselves to general considerations and some simple examples. For more details, the reader is referred to the monographs (Miller, 1948; Mark and Tobolsky, 1950; Hildebrand and Scott, 1950; Flory, 1953; Tompa, 1956; Huggins, 1958).

The solvent (low-molecular-weight component) will be designated as component 1, the solute (high-molecular-weight component) as component 2. To describe the state of a binary, high-molecular nonelectrolyte solution, we use the thermodynamic temperature T, the pressure P, and a quantity x^* called "volume fraction" of the solute and defined by[†]

$$x^* \equiv rn_2/(n_1 + rn_2) = rx/[1 + (r - 1)x]. \tag{6.41}$$

Here, n_i, $i = 1, 2$, is the amount of substance of component i; r is the degree of polymerization of the solute (average number of monomers or of structural units in one solute molecule); and $x(= x_2)$ is the mole fraction of the solute. The mole fraction and volume fraction of the solvent are then $1 - x$ and $1 - x^*$, respectively.

[†] There are other quantities also called "volume fractions." These differ from x^* by the replacement of r in Eq. (6.41) by some other parameter such as the number of segments (or sites on a quasicrystalline lattice) in a polymer molecule or the ratio of the molar volume of the pure liquid solute to that of the pure liquid solvent.

For low-molecular nonelectrolyte solutions, the stoichiometric ideal mixture was taken to be the standard system. Such a mixture can be described by the expression for the Gibbs function G following from Eqs. (1.4) and (6.11). Confining ourselves to binary systems, we thus have

$$G^{\mathrm{id}} = n_1\mu_1^{\bullet} + n_2\mu_2^{\bullet} + RT\{n_1 \ln[n_1/(n_1 + n_2)] + n_2 \ln[n_2/(n_1 + n_2)]\}, \qquad (6.42)$$

the superscript "id" denoting an ideal mixture.

For binary, high-molecular nonelectrolyte solutions, we use another reference system defined by the following relation:

$$\begin{aligned} G^{\mathrm{st}} &= n_1\mu_1^{\bullet} + n_2\mu_2^{\bullet} + RT\{n_1 \ln[n_1/(n_1 + rn_2)] + n_2 \ln[rn_2/(n_1 + rn_2)]\} \\ &= n_1\mu_1^{\bullet} + n_2\mu_2^{\bullet} + RT[n_1 \ln(1 - x^*) + n_2 \ln x^*], \end{aligned} \qquad (6.43)$$

the superscript "st" denoting the new standard mixture. By means of Eqs. (1.1), (1.2) and (6.43), we find for the volume V^{st}, the entropy S^{st}, and the enthalpy H^{st}

$$\begin{aligned} V^{\mathrm{st}} &= n_1 V_1^{\bullet} + n_2 V_2^{\bullet}, \\ S^{\mathrm{st}} &= n_1 S_1^{\bullet} + n_2 S_2^{\bullet} - R[n_1 \ln(1 - x^*) + n_2 \ln x^*], \\ H^{\mathrm{st}} &= n_1 H_1^{\bullet} + n_2 H_2^{\bullet}, \end{aligned}$$

and for the quantities defined by Eq. (6.1)[†]

$$\psi_1^{\mathrm{st}} = \ln(1 - x^*) + [1 - (1/r)]x^*, \qquad (6.44a)$$

$$\psi_2^{\mathrm{st}} = \ln x^* - (r - 1)(1 - x^*). \qquad (6.44b)$$

We infer from these equations that, in the standard mixture, there is no change of volume and of enthalpy on mixing the pure liquid components[‡] at given temperature and pressure. It is only with respect to the entropy of mixing that there are differences between this type of mixture and an ideal mixture. For $r = 1$, the two classes of liquid systems coincide.

[†] We recall that it is usual to write $\psi_i = \ln a_i$, $i = 1, 2$, and to call a_i the "activity" of component i.

[‡] Quantities like $\mu_2^{\bullet}$, $V_2^{\bullet}$, $S_2^{\bullet}$, and $H_2^{\bullet}$ refer to the pure liquid solute supposed to be internally stable. If the solute is a polymer (or high polymer), it actually is a glass in the pure liquid state.

To have a brief term for the standard system described by Eqs. (6.43) and (6.44), we call it an *ideal athermal mixture* since it is certainly an athermal mixture (zero enthalpy of mixing) and it resembles an ideal mixture.

We now restrict the discussion to polymer solutions (including high polymer solutions).

Experience shows that polymer solutions containing solvent molecules similar in size and shape to the structural units of the polymer are approximately ideal athermal solutions.* Furthermore, the simplest statistical approach to binary, athermal liquid mixtures containing small and large molecules leads to the formulas given above, which were the earliest derived for polymer solutions from molecular theory (Flory, 1941, 1942, 1945; Huggins, 1941, 1942a,b,c).

We now introduce quantities to describe the deviations in the behavior of a real polymer solution from that of an ideal athermal mixture (Rehage, 1955; Haase, 1956).

We define, for any polymer solution,

$$\bar{\bar{Z}} = Z/(n_1 + rn_2), \tag{6.45}$$

a sort of "molar function" conjugate to any extensive property Z. We also define a kind of "molar excess function" [compare Eq. (6.8)]

$$\bar{\bar{Z}}^{\mathrm{E}} = \bar{\bar{Z}} - \bar{\bar{Z}}^{\mathrm{st}}, \tag{6.46}$$

where $\bar{\bar{Z}}^{\mathrm{st}}$ is the value of $\bar{\bar{Z}}$ for an ideal athermal mixture of the same temperature, pressure, and composition. On account of Eqs. (1.1) and (1.2), we have

$$\bar{\bar{V}}^{\mathrm{E}} = (\partial\bar{\bar{G}}^{\mathrm{E}}/\partial P)_{T,x^*}, \qquad \bar{\bar{S}}^{\mathrm{E}} = -(\partial\bar{\bar{G}}^{\mathrm{E}}/\partial T)_{P,x^*}, \qquad \bar{\bar{H}}^{\mathrm{E}} = \bar{\bar{G}}^{\mathrm{E}} + T\bar{\bar{S}}^{\mathrm{E}}, \tag{6.47}$$

similar to the relations (6.14).

To have a common basis for comparison with other liquid mixtures, we now introduce the activity coefficient f_1 or f_2 of the solvent or solute, respectively, according to Eq. (6.4):

$$\ln f_i \equiv \psi_i - \ln x_i, \qquad \ln f_i^{\mathrm{st}} \equiv \psi_i^{\mathrm{st}} - \ln x_i, \qquad i = 1, 2, \tag{6.48}$$

* Besides density, heat of mixing, and vapor pressure, the osmotic pressure is one of the most important measurable quantities in polymer solutions.

x_i being the mole fraction of component i. We immediately obtain

$$\ln f_i = \psi_i - \psi_i^{st} + \ln f_i^{st}, \qquad i = 1, 2. \tag{6.49}$$

Remembering that $x_1 = 1 - x$ and $x_2 = x$, we derive from Eqs. (6.41), (6.44), and (6.48)

$$\begin{aligned}\ln f_1^{st} &= \ln\{1 - [1 - (1/r)]x^*\} + [1 - (1/r)]x^* \\ &= 1 - \{1/[1 + (r - 1)x]\} - \ln[1 + (r - 1)x],\end{aligned} \tag{6.50a}$$

$$\begin{aligned}\ln f_2^{st} &= \ln[r - (r - 1)x^*] - (r - 1)(1 - x^*) \\ &= \ln r - \{(r - 1)(1 - x)/[1 + (r - 1)x]\} - \ln[1 + (r - 1)x].\end{aligned} \tag{6.50b}$$

Thus, we have expressed the activity coefficients of the two components of an ideal athermal mixture as functions of both the volume fraction x^* and the mole fraction x of the solute.

Combining Eqs. (1.1), (6.1), (6.41), (6.45), and (6.46), we find

$$RT(\psi_1 - \psi_1^{st}) = \bar{\bar{G}}^E - x^*(\partial \bar{\bar{G}}^E/\partial x^*)_{T,P}, \tag{6.51a}$$

$$RT(\psi_2 - \psi_2^{st}) = r\bar{\bar{G}}^E + r(1 - x^*)(\partial \bar{\bar{G}}^E/\partial x^*)_{T,P}. \tag{6.51b}$$

Comparison of the last relations with Eqs. (6.49) and (6.50) leads to the following formulas

$$\begin{aligned}\ln f_1 = 1 &- \{1/[1 + (r - 1)x]\} - \ln[1 + (r - 1)x] \\ &+ (1/RT)[\bar{\bar{G}}^E - x^*(\partial \bar{\bar{G}}^E/\partial x^*)_{T,P}],\end{aligned} \tag{6.52a}$$

$$\begin{aligned}\ln f_2 = \ln r &- \{(r - 1)(1 - x)/[1 + (r - 1)x]\} - \ln[1 + (r - 1)x] \\ &+ (r/RT)[\bar{\bar{G}}^E + (1 - x^*)(\partial \bar{\bar{G}}^E/\partial x^*)_{T,P}].\end{aligned} \tag{6.52b}$$

We are thus able to derive the functions $\ln f_1(x)$ and $\ln f_2(x)$ from the function $\bar{\bar{G}}^E(x^*)$ at given temperature and pressure.

Both experience and statistical theory tell us that it is possible and convenient to express $\bar{\bar{G}}^E$ as a polynomial in x^* in a way analogous to the series expansion (6.23) for binary, low-molecular nonelectrolyte solutions (Haase, 1956):

$$\bar{\bar{G}}^E = x^*(1 - x^*)[B_0 + B_1(1 - 2x^*) + B_2(1 - 2x^*)^2 + \cdots]. \tag{6.53}$$

Here, $B_0, B_1, B_2, \ldots$ are positive or negative empirical parameters depending on T and P. For $r = 1$ ($x^* = x$), the two power series (6.23) and (6.53) coincide. The corresponding expressions for $\bar{\bar{V}}^E$, $\bar{\bar{S}}^E$, and $\bar{\bar{H}}^E$ follow immediately from Eqs. (6.47) and (6.53).

When we insert Eq. (6.53) into Eqs. (6.52) and then express the right-hand sides of (6.52a) and (6.52b) in terms of x, taking account of Eq. (6.41), we obtain a polynomial in x, the exponents being positive integers. In particular, the lowest power in $\ln f_1$ is x^2. This is consistent with our previous statements concerning the composition dependence of the activity coefficients in gaseous mixtures (p. 313) and in low-molecular nonelectrolyte solutions (p. 327). This again implies that the limiting laws for infinite dilution hold for polymer solutions, too.

The quantity usually needed in practice is ψ_1 ($= \ln a_1$). Combining Eqs. (6.44a), (6.51a), and (6.53), we find

$$\psi_1 = \ln(1 - x^*) + [1 - (1/r)]x^* + (x^{*2}/RT)[B_0 + B_1(4x^* - 3) + B_2(2x^* - 1)(6x^* - 5) + \cdots]. \tag{6.54}$$

This formula is identical with Eq. (6.25) for $r = 1$ ($x^* = x$), as is easily seen from Eq. (6.48).

Neglecting all terms involving $B_1, B_2, \ldots$, we have a "generalized simple mixture" (see p. 328). We then derive from Eqs. (6.47), (6.53), and (6.54)

$$\psi_1 = \ln(1 - x^*) + [1 - (1/r)]x^* + (B_0/RT)x^{*2}, \tag{6.55a}$$

$$\bar{\bar{G}}^E = B_0 x^*(1 - x^*), \tag{6.55b}$$

$$\bar{\bar{V}}^E = (\partial B_0/\partial P)x^*(1 - x^*), \tag{6.55c}$$

$$\bar{\bar{S}}^E = -(\partial B_0/\partial T)x^*(1 - x^*), \tag{6.55d}$$

$$\bar{\bar{H}}^E = [B_0 - T(\partial B_0/\partial T)]x^*(1 - x^*). \tag{6.55e}$$

These equations are equivalent to a widely used formula proposed by Huggins (1955), where the quantity B_0/RT is (unfortunately) designated as μ and interpreted in terms of molecular theory.

Describing the dependence of B_0 on T and P in the same way as we did in Eq. (6.34) for low-molecular solutions (simple mixtures), we arrive at expressions for $\bar{\bar{G}}^E$, $\bar{\bar{V}}^E$, $\bar{\bar{S}}^E$, and $\bar{\bar{H}}^E$ analogous to those for $\bar{G}^E$, etc. in low-molecular solutions [see Eqs. (6.35)–(6.38)]. In particular, for a given pressure, we obtain

$$B_0 = b_0 - \beta_0 T - \gamma_0 T \ln(T/T^\dagger), \tag{6.56}$$

where b_0, β_0, and γ_0 are empirical constants. It follows from Eqs. (6.55) and (6.56) that

$$\bar{\bar{G}}^{\mathrm{E}} = x^*(1 - x^*)[b_0 - \beta_0 T - \gamma_0 T \ln(T/T^\dagger)], \quad (6.57a)$$

$$\bar{\bar{S}}^{\mathrm{E}} = x^*(1 - x^*)[\beta_0 + \gamma_0 + \gamma_0 \ln(T/T^\dagger)], \quad (6.57b)$$

$$\bar{\bar{H}}^{\mathrm{E}} = x^*(1 - x^*)(b_0 + \gamma_0 T). \quad (6.57c)$$

We conclude that b_0 only occurs in $\bar{\bar{H}}^{\mathrm{E}}$ and β_0 only in $\bar{\bar{S}}^{\mathrm{E}}$, while γ_0 is met with in both excess functions. It is only for the case $\gamma_0 = 0$ that we have a clear-cut division of B_0 into an "enthalpy term" and an "entropy term."

The polydispersity of the polymeric material, the high viscosities of solutions rich in polymer, and other factors are responsible for our lack of full information about the thermodynamic properties of polymer solutions, except for the region of high dilution. Nevertheless, the experimental data accumulated by now seem to indicate the following facts:

1. The equations (6.57) rarely, if ever, hold exactly.
2. The accuracy of the measurements is usually not high enough to justify the use of more than two parameters (B_0 and B_1) in Eqs. (6.53) and (6.54).
3. The temperature range covered by experiments is usually not wide enough to vindicate the use of more than the first two constants in relations of the type (6.56).
4. The symmetry rule formulated on p. 332 is also valid for polymer solutions provided we consider the functions $\bar{\bar{Z}}^{\mathrm{E}}(x^*)$ in place of the functions $\bar{Z}^{\mathrm{E}}(x)$.

In view of the first three facts, we have the following expressions containing four constants ($b_0, \beta_0, b_1, \beta_1$) at a given pressure [see Eqs. (6.47), (6.53), (6.54), and (6.56)]:

$$B_0 = b_0 - \beta_0 T, \qquad B_1 = b_1 - \beta_1 T, \quad (6.58a)$$

$$\psi_1 = \ln(1 - x^*) + [1 - (1/r)]x^* + \chi x^{*2}, \quad (6.58b)$$

$$\bar{\bar{G}}^{\mathrm{E}} = x^*(1 - x^*)(b_0 + b_1 - \beta_0 T - \beta_1 T - 2b_1 x^* + 2\beta_1 T x^*), \quad (6.58c)$$

$$\bar{\bar{S}}^{\mathrm{E}} = x^*(1 - x^*)(\beta_0 + \beta_1 - 2\beta_1 x^*), \quad (6.58d)$$

$$\bar{\bar{H}}^{\mathrm{E}} = x^*(1 - x^*)(b_0 + b_1 - 2b_1 x^*), \quad (6.58e)$$

with

$$\chi \equiv (b_0 - 3b_1 + 4b_1 x^* - \beta_0 T + 3\beta_1 T - 4\beta_1 T x^*)/RT, \quad (6.58f)$$

where the dimensionless quantity χ is also denoted by μ in the literature (see p. 337) though it obviously depends on both T and x^*.

As we infer from Eqs. (6.58), $\bar{\bar{H}}^E$ and $\bar{\bar{S}}^E$ do not depend on the temperature. The constants b_0 and b_1 only occur in $\bar{\bar{H}}^E$, the constants β_0 and β_1 only in $\bar{\bar{S}}^E$. Therefore, we may split the quantity χ into an "enthalpy term" χ_H and an "entropy term" χ_S:

$$\chi = \chi_H + \chi_S, \tag{6.59a}$$

where

$$\chi_H \equiv (1/RT)(b_0 - 3b_1 + 4b_1x^*), \tag{6.59b}$$

$$\chi_S \equiv -(1/R)(\beta_0 - 3\beta_1 + 4\beta_1x^*). \tag{6.59c}$$

We are now going to explain two empirical regularities observed in the literature (Huggins, 1958) concerning χ, χ_H, and χ_S.

The first rule refers to the linear dependence obtained when χ_H is plotted against χ_S. As a matter of fact, we derive from Eqs. (6.59b) and (6.59c), by elimination of x^*,

$$\chi_H RT\beta_1 = -\chi_S Rb_1 + b_0\beta_1 - b_1\beta_0. \tag{6.59d}$$

Thus, the alleged regularity follows directly from a thermodynamic analysis of the relations and definitions used in the description of the systems.

The second regularity concerns the composition dependence of χ, χ_H, and χ_S. We now apply the symmetry rule. According to Eqs. (6.58), this rule here simply states that

$$| b_0 - \beta_0 T | \gg | b_1 - \beta_1 T |,$$

while $| b_0 |$ is comparable to $| b_1 |$ and $| \beta_0 |$ is comparable to $| \beta_1 |$. We thus conclude from Eqs. (6.58f), (6.59b), and (6.59c) that the dependence of χ on composition is less pronounced than that of χ_H and χ_S. This is precisely what is found experimentally.†

Two systems investigated more thoroughly are toluene + polystyrene and cyclohexane + polystyrene. Here, measurements of vapor pressures (Schmoll and Jenckel, 1956), osmotic pressures (Rehage and Meys,

† The actual concentration dependence is more complicated than that implied by our formulas. But the argument given above can be easily generalized to expressions containing more than two parameters.

1958), and heats of mixing (Jenckel and Gorke, 1956) have been performed over a considerable range of compositions and temperatures.

The evaluation (Rehage and Meys, 1958) of the results shows that both systems exhibit positive values of $\bar{\bar{G}}^E$ and negative values of $\bar{\bar{S}}^E$. In toluene + polysterene, we have $\bar{\bar{H}}^E = 0$ (athermal mixture), in cyclohexane + polystyrene, $\bar{\bar{H}}^E > 0$ (endothermic mixing). The solutions in toluene can be approximately described by Eqs. (6.58) with $b_0 = b_1 = \beta_1 = 0$. Thus, there remains a single constant (β_0) occurring in $\bar{\bar{S}}^E$. The solutions in cyclohexane (with miscibility gap and upper consolute point) require all four constants in Eqs. (6.58). The functions $\bar{\bar{G}}^E$, $\bar{\bar{S}}^E$, and $\bar{\bar{H}}^E$ are slightly asymmetric with respect to x^*.

Recent data on the two systems mentioned are based on investigations of sedimentation equilibrium (Scholte, 1970a), and of light scattering (Scholte, 1970b). It should also be mentioned that modern research work on polymer solutions involves studies on liquid–liquid equilibrium (Koningsveld, 1968) and on swelling (Rehage, 1964).

D. Electrolyte Solutions

An electrolyte solution is a liquid mixture consisting of both nonelectrolytes ("solvents") and electrolytes. If one or more among the electrolytes are polymers, then we call the mixture a polyelectrolyte solution. For the sake of simplicity, we shall restrict the discussion to a *binary* system containing one solvent (component 1), such as water or methanol, and one electrolyte (component 2) yielding two kinds of ions upon dissociation, such as nitric acid or calcium chloride. For more complicated cases, the reader should consult the monographs (Harned and Owen, 1958; Robinson and Stokes, 1959).

As independent variables to describe the state of the electrolyte solution, we choose the thermodynamic temperature T, the pressure P, and either the (stoichiometric) mole fraction x ($= x_2$) of the electrolyte or the molality m of the electrolyte defined by

$$m \equiv n_2/M_1 n_1 = x/M_1(1 - x). \tag{6.60}$$

It follows that

$$x = M_1 m/(1 + M_1 m). \tag{6.61}$$

Here, n_1 or n_2 denotes the (stoichiometric) amount of substance of the solvent or electrolyte, respectively, and M_1 is the molar mass of the solvent. The unit universally employed for m is the SI unit mol kg^{-1}. By means of the abbreviation

$$m^\dagger \equiv 1 \quad \text{mol kg}^{-1},$$

we can use the quantity $m/m^\dagger$ as a dimensionless composition variable. According to Eq. (6.60) or (6.61), the unit to be chosen for M_1 is then kg mol^{-1}.

The chemical species actually present in the solution, besides solvent molecules, are the cations (species +), the anions (species −), and the undissociated electrolyte molecules (species u). Introducing the molality m_i of the species i,

$$m_i \equiv n_i/M_1 n_1, \qquad i = +, -, \text{u},$$

n_i being the (true) amount of substance of the species i, we derive [see Eqs. (2.4)]

$$m_+ = \nu_+ \alpha m, \qquad m_- = \nu_- \alpha m, \qquad m_\text{u} = (1 - \alpha)m. \tag{6.62}$$

Here, α is the degree of dissociation of the electrolyte, while ν_+ and ν_- are the dissociation numbers of the cations and anions, respectively. The condition of electric neutrality requires that

$$z_+\nu_+ + z_-\nu_- = 0, \tag{6.63}$$

where z_+ and z_- are respectively the charge number of the cations and that of the anions.

Let us denote the chemical potential of a component ($k = 1, 2$) or of an electrolytic species ($k = +, -, \text{u}$) by μ_k. We then have, in view of Eqs. (2.5) and (2.8),

$$\mu_2 = \mu_\text{u} = \nu_+\mu_+ + \nu_-\mu_-. \tag{6.64}$$

According to Chapter 1, Eq. (19.2), we obtain from Eq. (6.60)

$$(\partial\mu_1/\partial m)_{T,P} = -M_1 m(\partial\mu_2/\partial m)_{T,P}. \tag{6.65}$$

This relation establishes a connection between the composition dependence of the chemical potential of the solvent and that of the chemical potential of the electrolyte.

We introduce the following quantities:

$$\nu \equiv \nu_+ + \nu_-, \qquad \nu_\pm{}^\nu \equiv \nu_+^{\nu_+}\nu_-^{\nu_-}, \tag{6.66a}$$

$$\mu_i{}^\ominus \equiv \lim_{m\to 0} [\mu_i - RT\ln(m_i/m^\dagger)], \qquad i = +, -, \mathrm{u}, \tag{6.66b}$$

$$\ln \gamma_i \equiv [(\mu_i - \mu_i{}^\ominus)/RT] - \ln(m_i/m^\dagger), \qquad i = +, -, \mathrm{u}, \tag{6.66c}$$

$$\mu_2{}^\ominus \equiv \nu_+\mu_+{}^\ominus + \nu_-\mu_-{}^\ominus, \qquad \ln K_\mathrm{m} \equiv (\mu_\mathrm{u}{}^\ominus - \mu_2{}^\ominus)/RT, \tag{6.66d}$$

$$\gamma_\pm{}^\nu \equiv \gamma_+^{\nu_+}\gamma_-^{\nu_-}, \qquad \gamma \equiv \alpha\gamma_\pm, \tag{6.66e}$$

R denoting the gas constant again. The quantities $\mu_i{}^\ominus$ and $\mu_2{}^\ominus$ are standard values of the chemical potentials (in the molality scale), depending on T and P only; γ_i is the "practical activity coefficient" of species i, while $\gamma_\pm$ is known as the "practical mean ionic activity coefficient" and γ should be designated as the "stoichiometric practical mean ionic activity coefficient," but will more briefly be called the *conventional activity coefficient* of the electrolyte. It is γ which is always measurable and which is found in all modern tables on electrolyte solutions. The activity coefficients are dimensionless and depend on temperature, pressure, and composition. The meaning of K_m will appear later.

We should like to emphasize that $\mu_i{}^\ominus$ and thus $\mu_2{}^\ominus$ are finite and nonzero. This follows from the validity of the limiting laws for infinite dilution, which imply that (see p. 321)

$$\mu_i \to \mu_i{}^\ominus + RT\ln(m_i/m^\dagger) \qquad \text{for} \quad m \to 0, \qquad i = +, -, \mathrm{u},$$

or, more rigorously [see Eq. (6.66c)],

$$\mu_i = \mu_i{}^\ominus + RT\ln(m_i/m^\dagger) + RT\ln\gamma_i, \qquad \lim_{m\to 0}\gamma_i = 1, \; i = +, -, \mathrm{u}. \tag{6.67a}$$

Thus, we conclude from Eq. (6.66e) that

$$\lim_{m\to 0}\gamma_\pm = 1, \qquad \lim_{m\to 0}\gamma = 1, \tag{6.67b}$$

since $\alpha \to 1$ for $m \to 0$. We infer that γ measures the deviations in the behavior of a real electrolyte solution from that of a (completely dissociated) ideal dilute solution.

Combining Eqs. (6.62), (6.64), and (6.66), we find

$$\mu_2 = \mu_2{}^\ominus + \nu\, RT\ln(\nu_\pm m\gamma/m^\dagger), \tag{6.68}$$

$$K_\mathrm{m} = [\nu_\pm{}^\nu/(1-\alpha)](m/m^\dagger)^{\nu-1}(\gamma^\nu/\gamma_\mathrm{u}). \tag{6.69}$$

Equation (6.68) is the general expression for the chemical potential of the electrolyte in the solution, while Eq. (6.69) represents the generalized law of mass action for the dissociation equilibrium of a single electrolyte that produces two kinds of ions. The dimensionless quantity K_m, depending on T and P only, is called the *dissociation constant* (in the molality scale).

We now consider the solvent. The dimensionless quantity [see Eq. (6.1)]

$$\varphi \equiv -\psi_1/\nu M_1 m = (\mu_1^{\bullet} - \mu_1)/\nu RTM_1 m \tag{6.70}$$

is said to be the *osmotic coefficient*. It is a measurable function of T, P, and m. In modern work on electrolyte solutions, φ is usually given besides γ. Since for $m = 0$ the chemical potential μ_1 of the solvent in the solution coincides with the chemical potential $\mu_1^{\bullet}$ of the pure liquid solvent, Eq. (6.70) requires that

$$\lim_{m \to 0} (m\varphi) = 0.$$

The limiting value of φ for $m \to 0$ will be derived below.

For an electrolyte that dissociates in several stages (such as sulfuric acid in water), γ and φ are still given by Eqs. (6.68) and (6.70); the former relation is then to be interpreted as the definition of γ; the numbers to be used for ν and $\nu_{\pm}$ are those obtained for dissociation at infinite dilution. Thus, for aqueous sulfuric acid with the dissociation equilibria

$$H_2SO_4 \rightleftharpoons H^+ + HSO_4^-,$$
$$HSO_4^- \rightleftharpoons H^+ + SO_4^{--},$$

we have to consider the overall equilibrium

$$H_2SO_4 \rightleftharpoons 2H^+ + SO_4^{--}$$

in order to find the values for ν and $\nu_{\pm}$; these are, on account of Eq. (6.66a),

$$\nu = 3, \qquad \nu_{\pm} = 4^{1/3}.$$

Of course, Eq. (6.69) ceases to hold. This relation is to be replaced by two equations referring to the two actual dissociation equilibria.

These or similar considerations apply to all electrolytes having two ion constituents, such as H_2SO_4 and H_3PO_4 ($\nu = 4$, $\nu_{\pm} = 9^{1/4}$). Electrolytes with three or more ion constituents, such as $NaHSO_4$ and K_2HPO_4, cannot be treated so simply since there are three or more ionic species at infinite dilution [for more details, see Prue (1966)].

There follows from Eqs. (6.65), (6.68), and (6.70)

$$d(m\varphi) = dm + m\,d(\ln \gamma), \qquad T, P \quad \text{constant},$$

or

$$d(\ln \gamma) = d\varphi + [(\varphi - 1)/m]\,dm, \qquad T, P \quad \text{constant}.$$

Integration between the limits $m = 0$ ($\gamma = 1$) and m (arbitrary value of the molality) gives

$$\varphi = 1 + (1/m) \int_0^m m\,d(\ln \gamma), \tag{6.71a}$$

or

$$\ln \gamma = \varphi - 1 + \int_0^m [(\varphi - 1)/m]\,dm. \tag{6.71b}$$

Here, the assumption $\varphi = 1$ for $m = 0$ has been used in Eq. (6.71b), while Eq. (6.71a) does not anticipate this result still to be proved.

We infer from Eq. (6.67b) that $\ln \gamma$ can be represented by a power series in m with nonnegative exponents (compare p. 321). If we only retain the lowest power, we find for high dilution:

$$\ln \gamma = bm^r, \qquad r > 0, \tag{6.72}$$

b and r being constants for given temperature and pressure. (We shall see later that actually $r = \frac{1}{2}$.) Equations (6.71a) and (6.72) combine to give

$$\varphi - 1 = [br/(r + 1)]m^r = [r/(r + 1)] \ln \gamma. \tag{6.73}$$

Thus, we have

$$\lim_{m \to 0} \varphi = 1, \tag{6.74}$$

the result desired and anticipated in Eq. (6.71b).

We define, according to Eq. (6.1),

$$\psi_2 \equiv (\mu_2 - \mu_2^{\bullet})/RT. \tag{6.75}$$

Here, $\mu_2^{\bullet}$ denotes the chemical potential of the pure liquid electrolyte. This is a real quantity for acids such as hydrogen chloride or nitric acid and a hypothetical quantity for salts like solium chloride or silver nitrate. We obtain from Eqs. (6.68) and (6.75)

$$\nu \ln(\nu_{\pm} m\gamma/m^{\dagger}) = \psi_2 + C_2, \tag{6.76}$$

with

$$C_2 \equiv (\mu_2^{\bullet} - \mu_2^{\theta})/RT. \tag{6.77}$$

The dimensionless quantity C_2 depends on T and P.

Another function that will be used later is denoted by Γ and defined by

$$\Gamma \equiv (1 - x)\psi_1 + x\psi_2. \tag{6.78}$$

We conclude from Eqs. (1.7), (1.12), (6.1), and (6.78) that

$$\Gamma = [\bar{G} - (1 - x)\mu_1^{\bullet} - x\mu_2^{\bullet}]/RT, \tag{6.79}$$

$$\psi_1 = \Gamma - x(\partial\Gamma/\partial x)_{T,P}, \qquad \psi_2 = \Gamma + (1 - x)(\partial\Gamma/\partial x)_{T,P}, \tag{6.80}$$

where $\bar{G}$ denotes the molar Gibbs function of the solution. Comparing Eqs. (6.21) and (6.79), we see that Γ is equivalent to $\Delta\bar{G}/RT$ for binary nonelectrolyte solutions.

From Eq. (6.80), we conclude that, for a treatment symmetric in both components (suitable for systems like water + nitric acid), it is expedient to use the mole fraction x of the electrolyte as the composition variable and to start from the function $\Gamma(T, P, x)$ and then to derive the expressions for ψ_1, ψ_2, φ, and γ by applying Eqs. (6.80), (6.70), and (6.76) successively.

The relations (6.70), (6.71), (6.76), and (6.78) show how the various quantities may be determined experimentally. If the electrolyte is volatile, both ψ_1 and ψ_2 and thus Γ can be obtained from partial vapor pressures. From ψ_1, we find φ and hence γ. Thus, by combining ψ_2 and γ, we can derive the constant C_2. When the electrolyte, however, is nonvolatile, we only obtain ψ_1 and hence φ and γ from vapor pressures, but not ψ_2 and C_2. (These quantities are then computed indirectly from the analytical expression for Γ.) In both cases, ψ_1 and φ may also be derived from freezing-point determinations combined with calorimetric data, while γ can also be found from electromotive-force measurements on chemical cells containing the electrolyte in question provided suitable electrodes exist.

According to Eqs. (6.66e) and (6.69), known values of γ and of α (derived from spectroscopic data, for example) lead to values of $\gamma_{\pm}$, γ_{u}, and K_{m}.

The considerations thus far apply to an electrolyte solution of any composition. We now proceed to investigate the behavior at high dilution and then revert to the subject of concentrated solutions.

For extremely dilute electrolyte solutions, the "Debye–Hückel limiting law" holds. This famous law, originally derived by Debye and Hückel (1923) by means of a simplified statistical treatment, has been confirmed by both more rigorous molecular theory (Mayer, 1950; Poirier, 1953; Kirkwood and Poirier, 1954) and by experience. Let us denote the Avogadro constant by L, the elementary charge (proton charge) by e, the Boltzmann constant by k, the density of the solvent by ϱ, the permittivity of vacuum by ε_0, and the dielectric constant (relative permittivity) of the solvent by ε. Then the Debye–Hückel limiting law (for a single electrolyte with two ionic species) may be written as

$$\ln \gamma = a\zeta(m/m^{\dagger})^{1/2}, \tag{6.81}$$

where

$$\zeta \equiv z_+ z_- [\tfrac{1}{2}(z_+{}^2\nu_+ + z_-{}^2\nu_-)]^{1/2}, \tag{6.81a}$$

$$a \equiv (2\pi L\varrho m^{\dagger})^{1/2}(e^2/4\pi\varepsilon_0\varepsilon kT)^{3/2}. \tag{6.81b}$$

The dimensionless quantity a is called the "Debye–Hückel constant." It only contains properties of the pure liquid solvent, besides universal constants. Thus, we have for the solvent water, that is to say, for aqueous solutions, at 25°C,

$$a = 1.176. \tag{6.81c}$$

Equation (6.81) has the mathematical form of Eq. (6.72) with $r = \frac{1}{2}$. Thus, we immediately derive from Eqs. (6.73) and (6.81)

$$\varphi - 1 = \tfrac{1}{3} \ln \gamma = \tfrac{1}{3}a\zeta(m/m^{\dagger})^{1/2}. \tag{6.82}$$

This formula holds in the same range as Eq. (6.81).

We now consider the functions ψ_1, ψ_2, and Γ in terms of x. With the help of Eq. (6.60) or (6.61), we obtain for an extremely dilute solution ($x \ll 1$)

$$M_1 m = x.$$

Combining this with Eqs. (6.70), (6.76), (6.78), (6.81), and (6.82), we find

$$\psi_1 = -\nu x + \tfrac{1}{3}B_0 x^{3/2}, \tag{6.83a}$$

$$\psi_2 = -C + \nu \ln x - B_0 x^{1/2}, \tag{6.83b}$$

$$\Gamma = \nu x \ln x - (C + \nu)x - \tfrac{2}{3}B_0 x^{3/2}, \tag{6.83c}$$

with

$$B_0 \equiv -\nu a\zeta(M^\dagger/M_1)^{1/2}, \qquad C \equiv C_2 - \nu \ln \nu_\pm + \nu \ln(M_1/M^\dagger), \qquad (6.83\text{d})$$
$$M^\dagger \equiv 1/m^\dagger = 1 \quad \text{kg mol}^{-1}.$$

Here, in view of the condition $x \ll 1$, all powers in x beyond $x^{3/2}$ have been dropped. The value of B_0 for aqueous solutions at 25°C can be derived from Eqs. (6.81c) and (6.83d):

$$B_0 = -17.53\nu\zeta/2. \qquad (6.83\text{e})$$

Equations (6.83) apply to the same range of compositions as do the classical expressions (6.81) and (6.82) of the Debye–Hückel limiting law.

There are numerous formulas, partly theoretical and partly empirical, that have been proposed to describe the thermodynamic properties of less dilute solutions (Harned and Owen, 1958; Robinson and Stokes, 1959; Guggenheim, 1966). These relations are still restricted to a composition range which is small as compared to the range of existence of electrolytic solutions. In particular, solvent and electrolyte are not treated on the same footing. As a matter of fact, in most molecular theories, the ionic species are considered to be embedded in a structureless continuum with the properties of the pure solvent. Furthermore, the physical and mathematical background of the molecular models is still being debated (Résibois, 1968). Thus, strictly speaking, only the Debye–Hückel limiting law is accepted by all authors.

We shall shortly develop an empirical scheme of description (Haase, 1963) covering the whole range between pure liquid solvent ($x = 0$) and pure liquid electrolyte ($x = 1$).

We start with the following expression for the function $\Gamma(x)$:

$$\Gamma = (1 - x)\ln(1 - x) + \nu x \ln x - (C + \nu - 1)x$$
$$- \tfrac{2}{3}B_0 x^{3/2} + \tfrac{1}{2}B_1 x^2 + \tfrac{2}{5}B_2 x^{5/2} + \tfrac{1}{3}B_3 x^3 + \tfrac{2}{7}B_4 x^{7/2} + \cdots, \qquad (6.84)$$

where C and B_0 have the same meaning as in Eqs. (6.83), and B_1, B_2, B_3, B_4, ... are empirical dimensionless parameters depending on T and P only.

Since Γ must vanish for both $x = 0$ and $x = 1$, we have the condition

$$\tfrac{1}{2}B_1 + \tfrac{2}{5}B_2 + \tfrac{1}{3}B_3 + \tfrac{2}{7}B_4 + \cdots = \nu - 1 + C + \tfrac{2}{3}B_0, \qquad (6.85)$$

implying a relation between the parameters. Thus, if the series expansion is carried on to the term involving B_5, for example, we have four independent parameters when C is known. In view of Eq. (6.83d), C is related to C_2 and this is obtained from partial vapor pressures provided the electrolyte is volatile.

The polynomial (6.23) used for binary nonelectrolyte solutions is formally a special case of Eq. (6.84) if we take account of the fact that there is no dissociation ($\nu = 1$) and that fractional powers do not exist ($B_0 = B_2 = B_4 = \cdots = 0$). As a matter of fact, we derive from Eqs. (6.84) and (6.85), if we only retain terms involving C and B_1,

$$\Gamma = (1 - x)\ln(1 - x) + x \ln x - Cx(1 - x),$$

and this is identical with Eq. (6.30) describing a "simple mixture," as can be easily verified by inspection of Eqs. (6.21) and (6.79).

Inserting Eq. (6.84) into Eqs. (6.80) and (6.76) and using Eqs. (6.60) and (6.83d), we find

$$\begin{aligned}\psi_1 = \ln(1 - x) - (\nu - 1)x + \tfrac{1}{3}B_0 x^{3/2} - \tfrac{1}{2}B_1 x^2 - \tfrac{3}{5}B_2 x^{5/2} \\ - \tfrac{2}{3}B_3 x^3 - \tfrac{5}{7}B_4 x^{7/2} \cdots,\end{aligned} \tag{6.86}$$

$$\begin{aligned}\psi_2 + C - \nu \ln x + \nu \ln(1 - x) = \nu \ln \gamma \\ = \nu \ln(1 - x) - B_0 x^{1/2} + (B_1 - \nu + 1)x + (\tfrac{1}{3}B_0 + B_2)x^{3/2} \\ + (B_3 - \tfrac{1}{2}B_1)x^2 + (B_4 - \tfrac{3}{5}B_2)x^{5/2} - \tfrac{2}{3}B_3 x^3 - \tfrac{5}{7}B_4 x^{7/2} \cdots.\end{aligned} \tag{6.87}$$

Since $\ln(1 - x)$ can be developed into powers of x, Eq. (6.87) represents a series expansion of $\ln \gamma$ in x, the exponents of the powers being nonnegative numbers.

At first sight, Eq. (6.86) seems to contradict our general statement on p. 321 regarding the lowest power in the development of the thermodynamic functions of the solvent with respect to the mole fraction of the solute (electrolyte). But we should remember that the expression to be investigated is $\psi_1 - \ln x_L$, where x_L denotes the true mole fraction of the solvent, given by

$$x_L = (1 - x)/(1 + \beta x), \qquad \beta \equiv (\nu - 1)\alpha.$$

Thus, we have

$$\ln(1 - x) = \ln x_L + \ln(1 + \beta x) = \ln x_L + \beta x - \tfrac{1}{2}(\beta x)^2 \cdots.$$

In view of Eqs. (6.60) and (6.69), there follows

$$1 - \alpha = bx^{\nu-1} + \text{higher powers of } x,$$

b being a positive constant for given temperature and pressure. We obtain

$$\ln(1 - x) = \ln x_L + (\nu - 1)(1 - bx^{\nu-1} \cdots)x \cdots.$$

Combining this with Eq. (6.86), we find

$$\psi_1 - \ln x_L = (1 - \nu)bx^{\nu} + \tfrac{1}{3}B_0 x^{3/2} \cdots.$$

Since $\nu > 1$, the requirement is fulfilled.

Comparing Eqs. (6.84), (6.86), and (6.87) to Eqs. (6.81) and (6.83), we conclude that, for extreme dilution, the Debye–Hückel limiting law also follows from our formulas. In particular, from Eqs. (6.83d) and (6.87), we obtain for constant temperature and pressure

$$[\partial(\ln \gamma)/\partial x^{1/2}]_{x=0} = -B_0/\nu \qquad \text{or} \qquad [\partial(\ln \gamma)/\partial(m/m^{\dagger})^{1/2}]_{m=0} = a\zeta,$$

which is the rigorous form of the limiting law.

For all systems for which experimental data are available in a wide range of compositions, the preceding equations produce a good fit when three, four, or five independent parameters are used (Haase *et al.*, 1963a; Haase, 1965). The parameters for aqueous solutions of 17 electrolytes at 25°C (and 1 bar) have been tabulated (Haase, 1965). It also should be noted that the miscibility gaps (Haase *et al.*, 1963b) in the case of the systems H_2O + HCl, H_2O + HBr, and H_2O + HI can almost correctly be reproduced by our equations (Haase *et al.*, 1963a). We shall only give two examples here.

Our first example is the system H_2O + HCl at 25°C (Fig. 4). Though the function $\ln \gamma(m)$ exhibits a minimum and a maximum, Eq. (6.87) fits the curve quite well if four independent parameters are employed.

Our second example is the system $H_2O + HClO_4$ at 25°C (Fig. 5). Here, only a minimum appears in the curve for $\ln \gamma(m)$ since the composition range investigated is smaller than in the first example. The curve is fitted almost perfectly, three free parameters being required this time.

For aqueous perchloric acid, there also exist values for the degree of dissociation α derived from spectroscopic data (Fig. 6). From this, we obtain $\gamma_{\pm}$, γ_u, and K_m, using Eqs. (6.66e) and (6.69). A plot of the three quantities γ, $\gamma_{\pm}$, and γ_u against composition at 25°C (Fig. 7) shows the

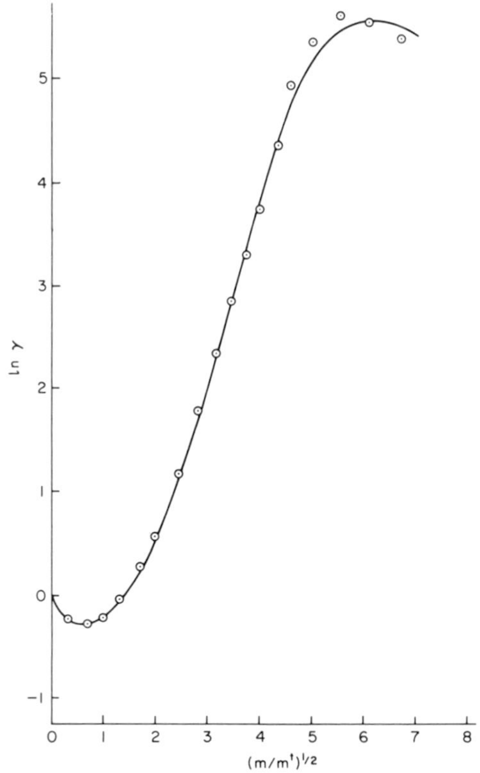

FIG. 4. Liquid system water + hydrogen chloride at 25°C: logarithm of conventional activity coefficient γ versus the square root of the molality m of the acid ($m^\dagger = 1$ mol kg^{-1}). Circles indicate measured values (Haase *et al.*, 1963b); solid curve calculated (Haase *et al.*, 1963a) from Eq. (6.87) with $B_1 = 191.5$,

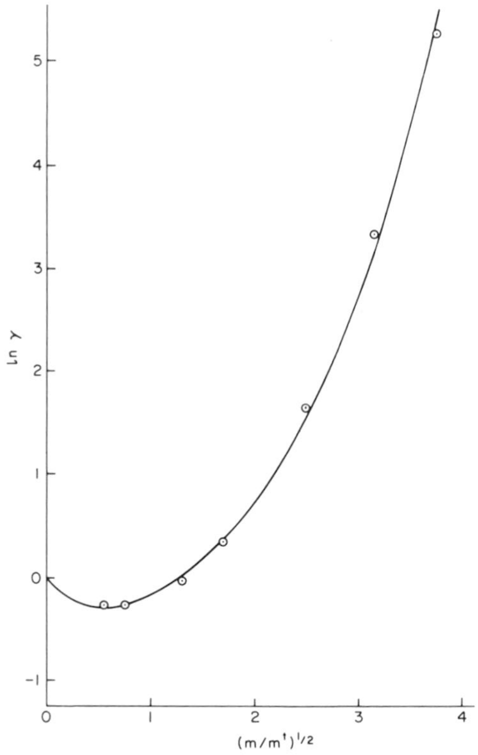

FIG. 5. Liquid system water + perchloric acid at 25°C: logarithm of conventional activity coefficient γ versus the square root of the molality m ($m^\dagger = 1$ mol kg^{-1}). Circles indicate measured values (Haase *et al.*, 1965); solid curve calculated (Haase *et al.*, 1965) from Eq. (6.87) with $B_1 = 161.7$, $B_2 = -449.4$,

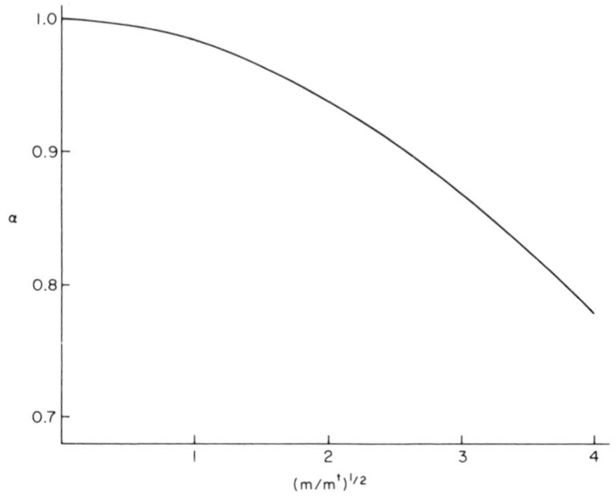

FIG. 6. (*above*). Liquid system water + perchloric acid at 25°C: degree of dissociation α of the acid versus the square root of the molality m from nuclear magnetic resonance data (Hood and Reilly, 1960) ($m^\dagger = 1$ mol kg^{-1}).

FIG. 7. (*right*). Liquid system water + perchloric acid at 25°C: logarithms of the conventional activity coefficient γ (——), of the mean activity coefficient $\gamma_\pm$ (– – –), and of the activity coefficient γ_u of the undissociated part of the electrolyte (–·–) versus the square root of the molality m derived from various experimental data (Haase *et al.*, 1965) ($m^\dagger = 1$ mol kg^{-1}).

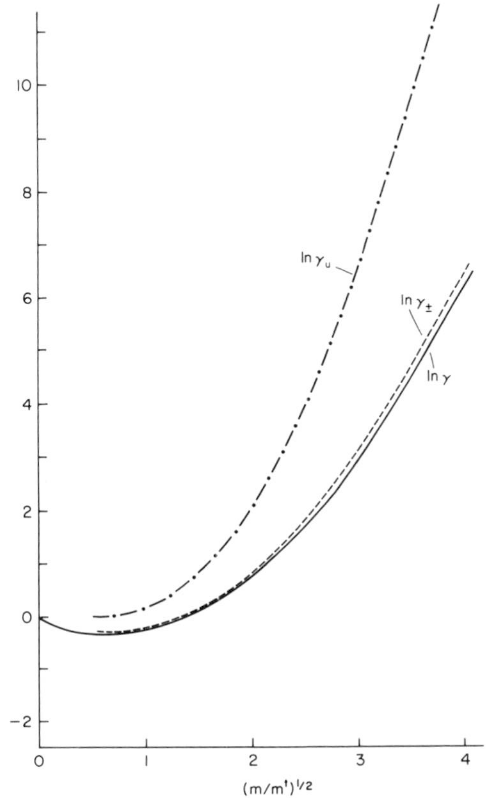

behavior of the three activity coefficients. The value of K_m obtained (Haase *et al.*, 1965) for this system at 25°C is $K_m = 34.5$.

These examples refer to solutions of "strong" electrolytes. For solutions of "weak" electrolytes, such as aqueous acetic acid, the investigation of a wide range of compositions would only reveal that these systems behave like nonelectrolyte solutions. The determination of the dissociation constant here requires different methods (Prue, 1966).

E. Ionic Melt Mixtures

An ionic melt mixture is a liquid mixture entirely composed of electrolytes. The most familiar examples are molten salt mixtures. For the sake of brevity and simplicity, we only discuss *binary* ionic melts containing *three ion constituents*, such as $KNO_3 + AgNO_3$ or $PbCl_2 + PbBr_2$. For more details and for other types of systems, the reader is referred to the literature (Sundheim, 1964; Blander, 1964; Lumsden, 1966; Bloom, 1967; Janz, 1967; Mamantov, 1969).

The two electrolytes forming the binary ionic melt are called component 1 and component 2. The ionic species only occurring in component 1 or 2 is designated as species *a* or *b*, respectively. The ionic species common to both components (such as NO_3^- in $KNO_3 + AgNO_3$ or Pb^{++} in $PbCl_2 + PbBr_2$) is designated as species *c*. We denote the number of ions of kind *a* or *c* produced by one molecule of component 1 by ν_a or ν_c, respectively, and the number of ions of kind *b* or *c* produced by one molecule of component 2 by ν_b or ν_c', respectively. We have for the system NaCl (1) + $MgCl_2$ (2), for example, $\nu_a = \nu_{Na^+} = 1$, $\nu_b = \nu_{Mg^{++}} = 1$, $\nu_c = \nu_{Cl^-} = 1$, $\nu_c' = \nu'_{Cl^-} = 2$.

The species actually present in an ionic melt are mainly ions, including complex ions, but there are possibly neutral molecules, too. Thus, the relations between the true amounts of the chemical species and the stoichiometric (measurable) amounts of the components may be very complicated.

The macroscopic state of the binary system is described by the thermodynamic temperature T, the pressure P, and the stoichiometric mole fraction x ($= x_2$) of component 2. The true mole fractions x_a, x_b, x_c of the ionic species *a*, *b*, *c* may, in principle, be expressed in terms of the independent mole fraction x. But these expressions are simple only for complete dissociation and absence of ionic complexes. We then obtain

$$x_a = (\nu_a/\nu)(1 - x), \quad x_b = (\nu_b/\nu)x, \quad x_c = (\nu_c/\nu) + [(\nu_c' - \nu_c)/\nu]x, \tag{6.88a}$$

with

$$\nu \equiv \nu_1 + (\nu_2 - \nu_1)x, \qquad \nu_1 \equiv \nu_a + \nu_c, \qquad \nu_2 \equiv \nu_b + \nu_c'. \tag{6.88b}$$

Introducing the limiting values of the true mole fractions in the pure liquid components,

$$\begin{aligned} x_a^\circ &\equiv \lim_{x\to 0} x_a, & x_c^\circ &\equiv \lim_{x\to 0} x_c, \\ x_b^\circ &\equiv \lim_{x\to 1} x_b, & x_c^{\circ\prime} &\equiv \lim_{x\to 1} x_c, \end{aligned} \tag{6.89}$$

we derive from Eqs. (6.88)

$$x_a^\circ = \nu_a/\nu_1, \qquad x_c^\circ = \nu_c/\nu_1, \qquad x_b^\circ = \nu_b/\nu_2, \qquad x_c^{\circ\prime} = \nu_c'/\nu_2, \tag{6.90}$$

$$\begin{aligned} x_a/x_a^\circ &= \nu_1(1-x)/[\nu_1 + (\nu_2 - \nu_1)x], \\ x_b/x_b^\circ &= \nu_2 x/[\nu_1 + (\nu_2 - \nu_1)x], \end{aligned} \tag{6.91a}$$

$$\begin{aligned} x_c/x_c^\circ &= \nu_1[1 + (r-1)x]/[\nu_1 + (\nu_2 - \nu_1)x], \\ x_c/x_c^{\circ\prime} &= \nu_2[1 + (r'-1)(1-x)]/[\nu_1 + (\nu_2 - \nu_1)x], \end{aligned} \tag{6.91b}$$

with

$$r \equiv \nu_c'/\nu_c, \qquad r' \equiv 1/r, \tag{6.91c}$$

where ν_1 and ν_2 are the total numbers of ions in one molecule of component 1 and 2, respectively.

For the limiting values of the chemical potentials μ_a, μ_b, and μ_c of the ionic species in the pure liquid components, we use a notation analogous to that employed for the true mole fractions in Eq. (6.89):

$$\begin{aligned} \mu_a^\circ &\equiv \lim_{x\to 0} \mu_a, & \mu_c^\circ &\equiv \lim_{x\to 0} \mu_c, \\ \mu_b^\circ &\equiv \lim_{x\to 1} \mu_b, & \mu_c^{\circ\prime} &\equiv \lim_{x\to 1} \mu_c, \end{aligned} \tag{6.92}$$

valid for any binary ionic melt.

The chemical potentials μ_1 and μ_2 of the components 1 and 2 are related to the chemical potentials μ_a, μ_b, and μ_c of the ionic species a, b, and c by the general equations (compare p. 299)

$$\mu_1 = \nu_a\mu_a + \nu_c\mu_c, \qquad \mu_2 = \nu_b\mu_b + \nu_c'\mu_c. \tag{6.93}$$

If we denote the chemical potential of the pure liquid component 1 or 2 by $\mu_1^\bullet$ or $\mu_2^\bullet$, respectively, we obtain, in view of Eq. (6.92),

$$\mu_1^\bullet = \nu_a\mu_a^\circ + \nu_c\mu_c^\circ, \qquad \mu_2^\bullet = \nu_b\mu_b^\circ + \nu_c'\mu_c^{\circ\prime}. \tag{6.94}$$

Combining Eqs. (6.93) and (6.94), we find

$$\begin{aligned}\psi_1 &\equiv (\mu_1 - \mu_1^\bullet)/RT \\ &= [\nu_a(\mu_a - \mu_a^\circ)/RT] + [\nu_c(\mu_c - \mu_c^\circ)/RT],\end{aligned} \tag{6.95a}$$

$$\begin{aligned}\psi_2 &\equiv (\mu_2 - \mu_2^\bullet)/RT \\ &= [\nu_b(\mu_b - \mu_b^\circ)/RT] + [\nu_c'(\mu_c - \mu_c^{\circ\prime})/RT],\end{aligned} \tag{6.95b}$$

where ψ_1 and ψ_2 are identical to the logarithms of the quantities called "activities" in the literature and R is the gas constant.

The components of an ionic melt may be chosen to be as similar as desired. Therefore, it is convenient to consider the *true ideal mixture* (p. 321) as the reference system:

$$\mu_i = \mu_i^\bullet + RT \ln x_i, \qquad i = a, b, c, \tag{6.96}$$

$\mu_i^\bullet$ denoting the chemical potential of the (hypothetical) pure liquid species i. On account of Eqs. (6.89), (6.92), and (6.95), we have for our binary system

$$\psi_1 = \nu_a \ln(x_a/x_a^\circ) + \nu_c \ln(x_c/x_c^\circ), \tag{6.97a}$$

$$\psi_2 = \nu_b \ln(x_b/x_b^\circ) + \nu_c' \ln(x_c/x_c^{\circ\prime}), \tag{6.97b}$$

the description of the true ideal mixture in terms of the measurable quantities ψ_1 and ψ_2.

A standard type of ionic melt, however, is useful only if ψ_1 and ψ_2 can be given explicitly in terms of the independent variables T, P, and x. We thus introduce a *standard ideal mixture* by the requirement that the true ideal mixture described by Eqs. (6.97) shall not contain any neutral molecules or ionic complexes. Then the relations (6.91) apply and we derive from Eqs. (6.97)

$$\begin{aligned}\psi_1^* &= \nu_1 \ln \nu_1 + \nu_a \ln(1 - x) - \nu_1 \ln[\nu_1 + (\nu_2 - \nu_1)x] \\ &\quad + \nu_c \ln[1 + (r - 1)x],\end{aligned} \tag{6.98a}$$

$$\begin{aligned}\psi_2^* &= \nu_2 \ln \nu_2 + \nu_b \ln x - \nu_2 \ln[\nu_2 + (\nu_1 - \nu_2)(1 - x)] \\ &\quad + \nu_c' \ln[1 + (r' - 1)(1 - x)],\end{aligned} \tag{6.98b}$$

the superscript asterisk denoting the standard ideal mixture. Taking account of Eqs. (1.7), (6.13), and (6.19)–(6.21), we have

$$\begin{aligned} \Delta\bar{G}^* &= RT[(1-x)\psi_1^* + x\psi_2^*], \\ \Delta\bar{S}^* &= -R[(1-x)\psi_1^* + x\psi_2^*], \\ \Delta\bar{H}^* &= 0, \end{aligned} \tag{6.99}$$

where $\Delta\bar{G}$, $\Delta\bar{S}$, and $\Delta\bar{H}$ are the molar Gibbs function of mixing, the molar entropy of mixing, and the molar enthalpy of mixing ("molar heat of mixing"), respectively.

It is interesting to note that, for binary ionic melts of the type NaCl + KCl or NaCl + NaBr, the molecular model developed by Temkin (1945) for ideal ionic melts coincides with the results of Eqs. (6.98) and (6.99). We then have

$$\nu_1 = \nu_2 = 2, \qquad \nu_a = \nu_b = \nu_c = \nu_c' = r = r' = 1,$$
$$\psi_1^* = \ln(1-x), \qquad \psi_2^* = \ln x,$$

and it is indeed only for this simple type of binary ionic melt that the model is applicable. Another requirement for the validity of the model is that the sizes of the noncommon ions should be nearly equal. Actually, however, even systems such as KCl + AgCl show appreciable deviations from this ideal behavior (see Fig. 8). Nevertheless, the standard ideal mixture defined by Eqs. (6.98) and (6.99) represents a useful reference system.

To describe the deviations in the behavior of a real binary ionic melt from that of a standard ideal mixture, we use the molar *excess function* $\bar{Z}^E$ defined by

$$\bar{Z}^E = \bar{Z} - \bar{Z}^* = \Delta\bar{Z} - \Delta\bar{Z}^*.$$

Here, $\bar{Z}$ or $\Delta\bar{Z}$ is the molar function of state or the molar function of mixing for any mixture and $\bar{Z}^*$ or $\Delta\bar{Z}^*$ the value of $\bar{Z}$ or $\Delta\bar{Z}$ for a standard ideal mixture of the same temperature, pressure, and composition (compare p. 324). Thus, in view of Eqs. (1.7), (6.13), (6.95), and (6.99), we obtain for the molar excess Gibbs function $\bar{G}^E$

$$\bar{G}^E/RT = (1-x)(\psi_1 - \psi_1^*) + x(\psi_2 - \psi_2^*), \tag{6.100}$$

and for the molar excess entropy $\bar{S}^E$ and the molar excess enthalpy $\bar{H}^E$

$$\bar{S}^E = -(\partial\bar{G}^E/\partial T)_{P,x}, \qquad \bar{H}^E = \Delta\bar{H} = \bar{G}^E + T\bar{S}^E, \tag{6.101}$$

$\Delta\bar{H}$ being the molar heat of mixing of the real melt.

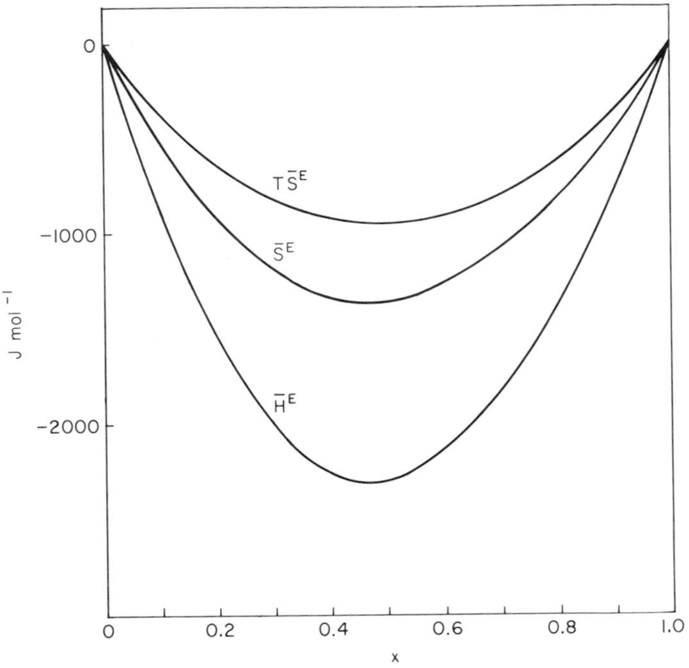

FIG. 8. Liquid system silver chloride (AgCl) + potassium chloride (KCl) at 650°C: molar excess Gibbs function $\bar{G}^E$, molar excess enthalpy $\bar{H}^E$, and product of thermodynamic temperature T and molar excess entropy $\bar{S}^E$ as functions of the mole fraction x of KCl (after Richter, 1969a).

Since ψ_1 and ψ_2 may be found from determinations of partial vapor pressures, of electromotive forces of chemical cells, etc., and $\bar{H}^E$ may be measured directly, we can derive $\bar{G}^E$, $\bar{S}^E$, and $\bar{H}^E$ from experimental data.

The experimental results for about 50 binary systems indicate (Dijkhuis *et al.*, 1968; Richter, 1969a) that $\bar{G}^E$, $\bar{S}^E$, and $\bar{H}^E$ can be represented by Eqs. (6.23), (6.27), and (6.28) and that one, two, or three parameters are required. Thus, for the excess functions of binary ionic melts, the same polynomials may be used as in the case of binary, low-molecular nonelectrolyte solutions.

These series expansions, in conjunction with the relations (6.98) and (6.100), again guarantee the validity of the general laws for infinite dilution (p. 321). These limiting laws, when applied to experimental situations, have a complicated form since in ionic melts there occur chemical species that are common to "solvent" and "solute." This may be seen, for example, in the generalizations of Henry's and Raoult's

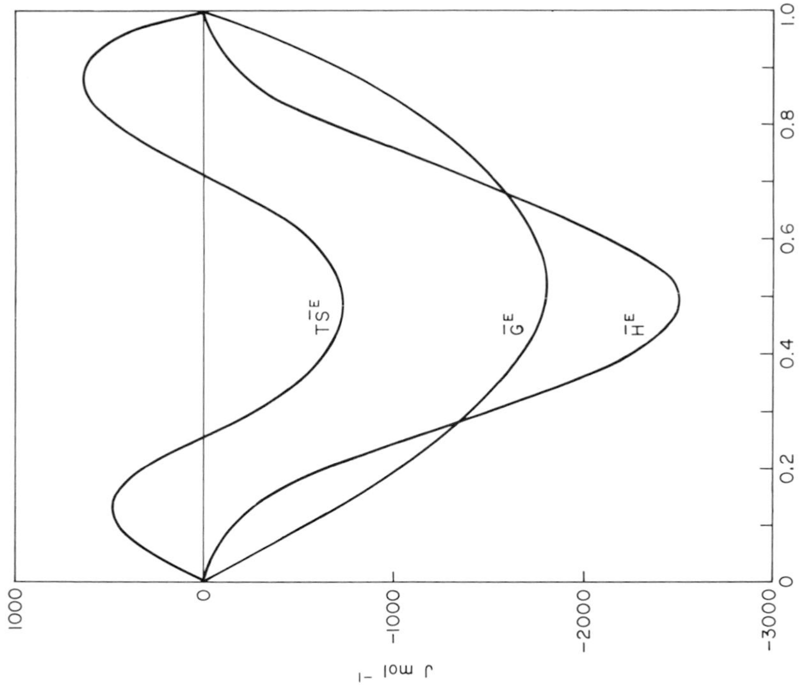

Fig. 10. Liquid system lead chloride ($PbCl_2$) + sodium chloride (NaCl) at 600°C: $\bar{G}^E$, $\bar{H}^E$, and $T\bar{S}^E$ as functions of the mole fraction x of NaCl (after Richter, 1969a).

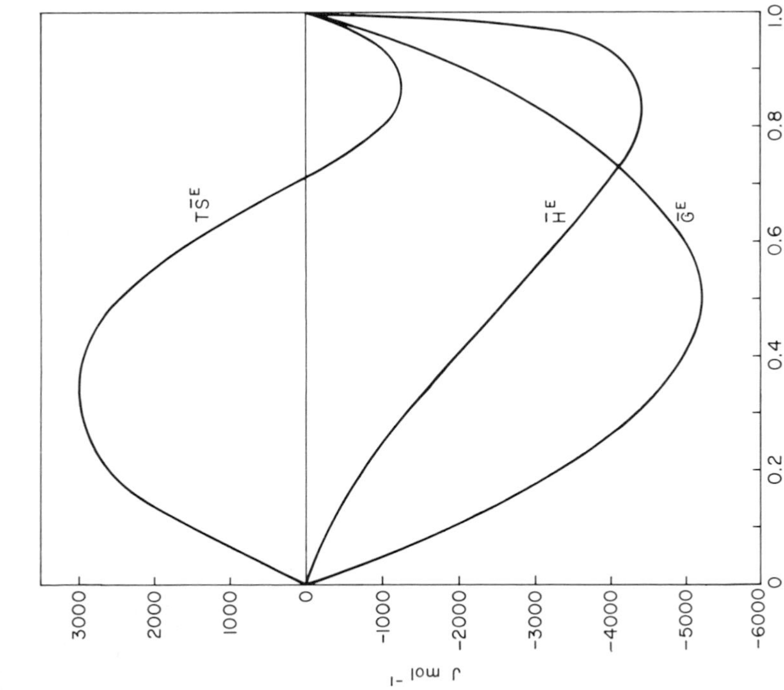

Fig. 9. Liquid system lead chloride ($PbCl_2$) + potassium chloride (KCl) at 600°C: $\bar{G}^E$, $\bar{H}^E$, and $T\bar{S}^E$ as functions of the mole fraction x of KCl (after Richter, 1969a).

limiting laws for the partial vapor pressures (Richter, 1968; Haase, 1969b) and in the generalization of van't Hoff's limiting law for the melting-point depression (Haase, 1952; Sinistri, 1961; Haase and Schönert, 1969).

The fact that, by combination of Eqs. (6.98) and (6.100) with the polynomial (6.23), there results the proper form of the limiting laws for infinite dilution is one more reason for introducing the standard ideal mixture as a reference system and not to use the stoichiometric ideal mixture any longer (Richter, 1969a,b; Haase, 1969b,c,d).

An inspection of the experimental data again reveals that the *symmetry rule* first mentioned on p. 332 also holds for binary ionic melts (Richter, 1969a). Three examples are given in Figs. 8–10. In the system $PbCl_2$ + NaCl at 600°C (Fig. 10), there are even three stationary points (two maxima and one minimum) in the function $\bar{S}^E(x)$; nevertheless, the function $\bar{G}^E(x)$ is still nearly symmetric.

VII. Pure Solids

A pure solid is a solid phase consisting of a single component. We exclude both anisotropic bodies and internally metastable phases, such as crystals of CO, NO, N_2O, and H_2O at very low temperatures. Then, all the thermodynamic properties of the solid can be derived from the molar Gibbs function $\bar{G}$ when this is known as a function of thermodynamic temperature T and pressure P.

Solids exhibit a crystalline structure. The type of lattice varies between, say, that of an elementary metal or of a molecular compound and that of an ionic crystal. Though our knowledge on the structure of crystals is more detailed than that on the structure of liquids, the quantitative success of statistical theory in computing thermodynamic functions of solids is still rather limited.*

Experience shows that, in principle, Eq. (5.18) holds for internally stable solids at very low temperatures and constant pressure. We rewrite Eq. (5.18) as

$$\bar{G} = c - bT^2 - b'T^3 - aT^4, \tag{7.1}$$

where a, b, and b' are empirical positive constants, c being fixed by the convention for the enthalpy mentioned on p. 28. Now, statistical theory, when applied to certain simple models, tells us that $b = b' = 0$. This is

* See also Volume X, "Solid State," of this series.

usually confirmed by measurements. But for metals, both experience and molecular theory taking account of the contribution of the electrons, seem to lead to the statement $b \neq 0$, $b' = 0$. Though this is not quite certain (Keesom and Kok, 1932, 1933; Silvidi and Daunt, 1950; Münster, 1956) we shall, for the sake of generality, only drop the term involving b' in Eq. (7.1). We thus write

$$\bar{G} = c - bT^2 - aT^4 \tag{7.2}$$

as our final formula.

According to Eqs. (1.10), (5.10), and (7.2), the molar entropy $\bar{S}$, the molar enthalpy $\bar{H}$, and the molar heat capacity at constant pressure $\bar{C}_P$ are given by*

$$\bar{S} = 2bT + 4aT^3, \tag{7.3}$$

$$\bar{H} = c + bT^2 + 3aT^4, \tag{7.4}$$

$$\bar{C}_P = 2bT + 12aT^3. \tag{7.5}$$

For the classical case $b = 0$, Eq. (7.3) or (7.5) is called the "Debye T^3-law".

At intermediate temperatures, the Debye–Einstein theory gives a rough fit of the experimental data. A critical discussion of this approximation may be found elsewhere (Blackmann, 1942; Münster, 1956; Guggenheim, 1967).

At high temperatures, Eq. (5.1) holds approximately. Here, the molar heat capacity $\bar{C}_P$ has the order of magnitude of 10 J K^{-1} mol^{-1}, the molar volume that of 10 to 10^2 cm^3 mol^{-1}, the isothermal compressibility that of 10^{-6} bar^{-1}, the thermal expansivity that of 10^{-6} to 10^{-5} K^{-1}. All these quantities again increase with increasing temperature. The limiting value of $\bar{C}_P$ for very high temperatures, at least for monatomic crystals, is roughly $3R$ (R is the gas constant). This is said to be the "Dulong–Petit rule."

VIII. Solid Mixtures

A solid mixture (or solid solution) is a solid phase consisting of two or more components. The requirements for the existence of such mixtures are more stringent than in the case of liquid mixtures. The phenomena

* At absolute zero, both $\bar{C}_P$ and the thermal expansivity vanish for any phase of fixed content. The statement $\bar{S} \to 0$ for $T \to 0$ holds for any pure phase which is internally stable.

of partial miscibility or of immiscibility of the components are common in solid systems.

Examples relating to the comparatively few binary systems forming a complete series of solid solutions are given in Table III, partly taken from the compilations by Hildebrand and Scott (1950). Besides these "substitutional solid solutions," there are also "interstitial solid solutions," such as iron + carbon or platinum + hydrogen.

TABLE III

COMPONENTS OF BINARY SYSTEMS FORMING A COMPLETE SERIES OF SOLID SOLUTIONS

First component	Second component	Remarks
Ar	Kr	Face-centered cubic lattice
Ar	CH_4	Face-centered cubic lattice
Kr	CH_4	Face-centered cubic lattice
$N_2(\alpha)$	$CO(\alpha)$	Cubic lattice
$N_2(\beta)$	$CO(\beta)$	Hexagonal lattice
K	Rb	Body-centered cubic lattice
$Ti(\alpha)$	$Zr(\alpha)$	Hexagonal close-packed lattice
Cu	Ni	Face-centered cubic lattice, nearly ideal mixture
$Co(\beta)$	Ni	Face-centered cubic lattice
$Co(\beta)$	$Fe(\gamma)$	Face-centered cubic lattice
Au	Cu	Face-centered cubic lattice
Au	Ag	Face-centered cubic lattice, nearly ideal mixture
$TiCl_4$	$TiBr_4$	Ideal mixture
$(C_6H_5)_4Sn$	$(C_6H_5)_4Pb$	Ideal mixture
$(C_6H_5)_3N$	$(C_6H_5)_3P$	Ideal mixture
$(C_6H_5)_2S$	$(C_6H_5)_2Se$	Ideal mixture
C_6H_5Cl	C_6H_5Br	Ideal mixture
NaCl	AgCl	Ionic solid solution
KCl	KBr	Ionic solid solution

The wide range covered by the concept of a solid mixture would require a treatment discriminating between the different types of mixtures: there are, for example, solid metallic solutions (solid alloys) or solid solutions of organic compounds as well as ionic crystals with several components. But in view of the difficulties of the experimental methods (vapor-pressure determinations, electromotive-force measurements, calorimetric investigation, etc.), only few reliable data exist and these mainly refer to metallic systems. Thus, it is universal praxis to treat solid mixtures on an equal footing with nonelectrolyte solutions. In particular, one uses the stoichiometric ideal mixture as the reference system and defines activity coefficients and excess functions in the same way as for low-molecular nonelectrolyte solutions.

There are examples of binary solid solutions that are, within experimental error, ideal mixtures: chlorobenzene + bromobenzene, tetraphenyl-tin + tetraphenyl-lead, and some other systems. We recall that ideal mixtures never exhibit miscibility gaps.

One of the few nonideal solid mixtures thoroughly studied is the binary system zinc + silver in the temperature range between 850 K and 1200 K and within the limits of existence of the solid solutions. Here, spectrophotometric measurements have been performed to compute the vapor pressure of zinc. The results (Scatchard and Westlund, 1953) can be represented by Eqs. (6.23), (6.27), and (6.28), where two parameters and altogether three constants are required. The excess entropy is positive and considerable in magnitude: its maximum value is more than twice as great as the maximum of the entropy of mixing for an ideal mixture.

Two further examples of binary nonideal solid solutions may be mentioned: the system argon + krypton (Walling and Halsey, 1958) between 70 and 96 K, where partial vapor pressures have been measured, and the system tetramethylmethane + carbon tetrachloride (Chang and Westrum, 1965), forming a continuous series of "plastically crystalline" solid solutions in the vicinity of 230 K, where the heats of mixing have been derived from heat-capacity determinations and data on the liquid mixtures.

For lack of a sufficient number of exact and direct measurements, most considerations on solid solutions involve phase behavior (liquid–solid and solid–solid equilibria). Analytic expressions for thermodynamic functions, such as those discussed in the earlier sections on liquid mixtures, can hardly be derived from such considerations alone.

More information on solid mixtures may be found in the literature concerning experimental methods and results (Hildebrand and Scott, 1950; Wagner, 1952; Kubaschewski and Evans, 1955) and statistical theory (Münster, 1956). It should also be stressed that in solid alloys such interesting phenomena occur as order–disorder transformations (Münster, 1956) and intermetallic compounds (Wallace, 1964).

Nomenclature

$\bar{C}_P$	molar heat capacity at constant pressure
$\mathfrak{F}$	Faraday constant
f_i	activity coefficient of component *i*
G	Gibbs function
$\bar{G}$	molar Gibbs function
$\bar{G}^{\mathrm{E}}$	molar excess Gibbs function
H	enthalpy
$\bar{H}$	molar enthalpy
$\bar{H}^{\mathrm{E}}$	molar excess enthalpy
H_i	partial molar enthalpy of component *i*
$H_i^{\bullet}$	(partial) molar enthalpy of pure component *i*
K_m	dissociation constant
M_i	molar mass of component *i*
m	molality
n	amount of substance
n_i	amount of substance of component *i* or chemical species *i*
P	pressure
p_i	partial pressure
R	gas constant
S	entropy
$\bar{S}$	molar entropy
$\bar{S}^{\mathrm{E}}$	molar excess entropy
S_i	partial molar entropy of component *i*
$S_i^{\bullet}$	(partial) molar entropy of pure component *i*
T	thermodynamic temperature
V	volume
$\bar{V}$	molar volume
$\bar{V}^{\mathrm{E}}$	molar excess volume
V_i	partial molar volume of component *i*
$V_i^{\bullet}$	(partial) molar volume of pure component *i*
x_i	mole fraction of component *i*
x^*	volume fraction of polymer
z_+, z_-	charge numbers
α	degree of dissociation
β	thermal expansivity
γ	activity coefficient for electrolyte solutions
$\varkappa$	compressibility
μ	chemical potential
μ_i	chemical potential of component *i*
$\mu_i^{\bullet}$	chemical potential of pure component *i*
ν_+, ν_-	dissociation numbers for electrolyte solutions
$\nu_a, \nu_b, \nu_c, \nu_c'$	dissociation numbers for ionic melts
φ	osmotic coefficient
ψ_i	logarithm of activity of component *i*

References

Adcock, D. S., and McGlashan, M. L. (1954). *Proc. Roy. Soc. Ser. A* **226**, 266.
Beattie, J. A. (1949). *Chem. Rev.* **44**, 141.
Beattie, J. A., and Stockmayer, W. H. (1942). *J. Chem. Phys.* **10**, 473.
Beatty, H. A., and Calingaert, G. (1934). *Ind. Eng. Chem.* **26**, 904.
Blackman, M. (1942). *Rep. Progr. Phys.* **8**, 11.
Blander, M., ed. (1964). "Molten Salt Chemistry." Wiley (Interscience). New York.
Bloom, H. (1967). "The Chemistry of Molten Salts." Benjamin, New York.
Boublik, T., Lam, V. T., Murakani, S., and Benson, G. C. (1969). *J. Phys. Chem.* **73**, 2356.
Chang, E. T., and Westrum, E. F., Jr. (1965). *J. Phys. Chem.* **69**, 2176.
Coulson, E. A., and Herington, E. F. G. (1948). *Trans. Faraday Soc.* **44**, 629.
Debye, P., and Hückel, E. (1923). *Phys. Z.* **24**, 185.
Dijkhuis, C., Dijkhuis, R., and Janz, G. J. (1968). *Chem. Rev.* **68**, 253.
Duby, P., and Townsend, Jr., H. E. (1968). *J. Electrochem. Soc.* **115**, 605.
Egelstaff, P. A. (1967). "An Introduction to the Liquid State." Academic Press, New York.
Flory, P. J. (1941). *J. Chem. Phys.* **9**, 660.
Flory, P. J. (1942), *J. Chem. Phys.* **10**, 51.
Flory, P. J. (1945), *J. Chem. Phys.* **13**, 453.
Flory, P. J. (1953). "Principles of Polymer Chemistry." Cornell Univ. Press, Ithaca, New York.
Fuchs, K. (1941). *Proc. Roy. Soc. Ser. A* **179**, 408.
Goates, J. R., and Sullivan, R. J. (1958). *J. Phys. Chem.* **62**, 188.
Guggenheim, E. A. (1937). *Trans. Faraday Soc.* **32**, 151.
Guggenheim, E. A. (1952). "Mixtures." Oxford Univ. Press (Clarendon), London and New York.
Guggenheim, E. A. (1966). "Applications of Statistical Mechanics." Oxford Univ. Press (Clarendon), London and New York.
Guggenheim, E. A. (1967). "Thermodynamics." North-Holland Publ., Amsterdam.
Guggenheim, E. A., and McGlashan, M. L. (1951). *Proc. Roy. Soc. Ser. A* **206**, 448.
Haase, R. (1950). *Z. Phys. Chem. (Leipzig)* **194**, 217.
Haase, R. (1951). *Z. Elektrochem.* **55**, 29.
Haase, R. (1952). *C. R. Reunion Chim. Phys., 2nd, Paris, 1952*, p. 131.
Haase, R. (1956). "Thermodynamik der Mischphasen." Springer, Berlin.
Haase, R. (1958). *Ber. Bunsenges. Phys. Chem.* **62**, 1043.
Haase, R. (1963). *Z. Phys. Chem. (Frankfurt am Main)* **39**, 360.
Haase, R. (1965). *Angew. Chem.* **77**, 517.
Haase, R. (1969a). "Thermodynamics of Irreversible Processes." Addison-Wesley, Reading, Massachusetts.
Haase, R. (1969b). *Z. Phys. Chem. (Frankfurt am Main)* **63**, 95.
Haase, R. (1969c). *J. Phys. Chem.* **73**, 1160.
Haase, R. (1969d). *J. Phys. Chem.* **73**, 4023.
Haase, R., and Rehage, G. (1955). *Ber. Bunsenges. Phys. Chem.* **59**, 994.
Haase, R., and Schönert, H. (1969). Solid-liquid equilibrium. "The International Encyclopedia of Physical Chemistry and Chemical Physics," Vol. 1, Topic 13. Pergamon, New York.

Haase, R., Naas, H., and Dücker, K.-H. (1963a). *Z. Phys. Chem.* (*Frankfurt am Main*) **39**, 383.

Haase, R., Naas, H., and Thumm, H. (1963b). Z. Phys. Chem. (Frankfurt am Main) **37**, 210.

Haase, R., Dücker, K.-H., and Küppers, H. A. (1965). *Ber. Bunsenges. Phys. Chem.* **69**, 97.

Haase, R., Rehse, M., and Daelman, C. (1967). *Monatsh. Chem.* **98**, 922.

Harned, H. S., and Owen, B. B. (1958). "The Physical Chemistry of Electrolytic Solutions." Reinhold, New York.

Hildebrand, J. H., and Scott, R. L. (1950). "The Solubility of Nonelectrolytes." Reinhold, New York.

Hood, G. C., and Reilly, C. A. (1960). *J. Chem. Phys.* **32**, 127.

Huggins, M. L. (1941). *J. Chem. Phys.* **9**, 440.

Huggins, M. L. (1942a). *J. Phys. Chem.* **46**, 151.

Huggins, M. L. (1942b). *Ann. N. Y. Acad. Sci.* **41**, 11.

Huggins, M. L. (1942c). *J. Amer. Chem. Soc.* **64**, 1712.

Huggins, M. L. (1955). *J. Polym. Sci.* **16**, 209.

Huggins, M. L. (1958). "Physical Chemistry of High Polymers." Wiley, New York.

Janz, G. J. (1967). "Molten Salts Handbook." Academic Press, New York.

Jenckel, E., and Gorke, K. (1956). *Ber. Bunsenges. Phys. Chem.* **60**, 579.

Jost, W., and Röck, H. (1954). *Chem. Eng. Sci.* **3**, 17.

Keesom, W. H., and Kok, J. A. (1932). *Commun. Kamerlingh Onnes Lab. Univ. Leiden* **219d**.

Keesom, W. H., and Kok, J. A. (1933). *Commun. Kamerlingh Onnes Lab. Leiden* **232d**.

Kirkwood, J. G., and Poirier, J. C. (1954). *J. Phys. Chem.* **58**, 591.

Koningsveld, R. (1968). "Advances in Colloid and Interface Science," Vol. II. Elsevier, Amsterdam.

Kubaschewski, O., and Evans, E. L. (1955). "Metallurgical Thermochemistry." Pergamon, New York.

Lennard-Jones, J. E., and Cook, W. R. (1927). *Proc. Roy. Soc. Ser. A* **115**, 334.

Lumsden, J. (1966). "Thermodynamics of Molten Salt Mixtures." Academic Press, New York.

Mamantov, G., ed. (1969). "Molten Salts." Marcel Dekker, New York.

Margules, M. (1895). *Sitzungsber. Akad. Wiss. Wien Math. Naturwiss. Kl.* **104**, 1243.

Mark, H., and Tobolsky, A. V. (1950). "Physical Chemistry of High Polymeric Systems." Wiley (Interscience), New York.

Mayer, J. E. (1939). *J. Phys. Chem.* **43**, 71.

Mayer, J. E. (1950). *J. Chem. Phys.* **18**, 1423.

Miller, A. R. (1948). "The Theory of Solutions of High Polymers." Oxford Univ. Press (Clarendon), London and New York.

Münster, A. (1956). "Statistische Thermodynamik." Springer, Berlin.

Poirier, J. C. (1953). *J. Chem. Phys.* **21**, 974.

Porter, A. W. (1920). *Trans. Faraday Soc.* **16**, 336.

Prigogine, I. (1967). "The Molecular Theory of Solutions." North-Holland Publ., Amsterdam.

Prigogine, I., and Defay, R. (1954). "Chemical Thermodynamics." Longmans, Green, New York.

PRUE, J. E. (1966). Ionic equilibria. "The International Encyclopedia of Physical Chemistry and Chemical Physics," Vol. 3, Topic 15. Pergamon, New York.
REDLICH, O., and KISTER, A. T. (1948). *Ind. Eng. Chem.* **40**, 345.
REHAGE, G. (1955). *Z. Elektrochem.* **59**, 78.
REHAGE, G. (1964). *Kolloid-Z* **196**, 97.
REHAGE, G., and MEYS, H. (1958). *J. Polym. Sci.* **30**, 271.
RÉSIBOIS, P. M. V. (1968). "Electrolyte Theory." Harper, New York.
RICHTER, J. (1968). *Ber. Bunsenges. Phys. Chem.* **72**, 681.
RICHTER, J. (1969a). *Z. Naturforsch. A* **24**, 835.
RICHTER, J. (1969b). *Z. Naturforsch. A* **24**, 447.
ROBINSON, R. A., and STOKES, R. H. (1959). "Electrolyte Solutions." Butterworths, London.
RÖCK, H., and SCHRÖDER, W. (1957). *Z. Phys. Chem. (Frankfurt am Main)* **11**, 41.
ROWLINSON, J. S. (1969). "Liquids and Liquid Mixtures." Butterworths, London.
SCATCHARD, G. (1949). *Chem. Rev.* **44**, 7.
SCATCHARD, G., and WESTLUND, Jr., R. A. (1953). *J. Amer. Chem. Soc.* **75**, 4189.
SCATCHARD, G., WOOD, S. E., and MOCHEL, J. M. (1939). *J. Amer. Chem. Soc.* **61**, 3206.
SCHMOLL, K., and JENCKEL, E. (1956). *Ber. Bunsenges. Phys. Chem.* **60**, 756.
SCHOLTE, TH. G. (1970a). *J. Polym. Sci. (Part A-2)* **8**, 141.
SCHOLTE, TH. G. (1970b). *European Polym. J.* **6**, 1063.
SCOTT, R. L., and FENBY, D. V. (1969). *Annu. Rev. Phys. Chem.* **20**, 111.
SILVIDI, A. A., and DAUNT, J. G. (1950). *Phys. Rev.* **77**, 125.
SINISTRI, C. (1961). *Z. Phys. Chem. (Frankfurt am Main)* **30**, 349.
SUNDHEIM, B. R., ed. (1964). "Fused Salts." McGraw-Hill, New York.
TEMKIN, M. (1945). *Acta Physicochim. U.R.S.S.* **20**, 411.
TOMPA, H. (1956). "Polymer Solutions." Academic Press, New York.
VETTER, F. A., and KUBASCHEWSKI, O. (1953). *Ber. Bunsenges. Phys. Chem.* **57**, 243.
WAGNER, C. (1952). "Thermodynamics of Alloys." Addison-Wesley, Reading, Massachusetts.
WALLACE, W. E. (1964). *Annu. Rev. Phys. Chem.* **15**, 109.
WALLING, J. F., and HALSEY, G. D., Jr. (1958). *J. Phys. Chem.* **62**, 752.
WOOD, S. E., and GRAY, J. A. (1952). *J. Amer. Chem. Soc.* **74**, 3729.

Chapter 4

Gas–Liquid and Gas–Solid Equilibria at High Pressures, Critical Curves, and Miscibility Gaps

E. U. FRANCK

I. Introduction

The equilibrium vapor density of a solid or liquid substance can be enhanced considerably if this substance is in contact with another gas which is highly compressed. This enhancement of vapor density may amount to several orders of magnitude, and the phenomena can be discussed as solubilities of the solid or liquid material in dense gases as solvents. The range of investigation of the gaseous solvents has to be extended to supercritical temperatures where the density can be widely varied without phase separation. Pressures up to several thousand bars may be required to obtain "liquidlike" densities of the solvent gas at such temperatures.

The study of gas–solid and gas–liquid solubility equilibria is of practical importance in various fields. Technical extraction processes utilize gaseous solvents (Prausnitz, 1969). High pressure gas-chromatography can be a useful analytical method (Schneider and Bartmann, 1969). The solvent power of the compressed gas often can be easily adjusted by changing the pressure. Solubility of inorganic solids in high pressure steam is important for power plants, and more concentrated supercritical aqueous solutions are used for synthetic crystal growing. The unusual miscibility of dense supercritical water with various nonpolar compounds may lead to new chemical applications. Numerous inorganic compounds are transported as high pressure gaseous solutions during the hydrothermal formation of mineral deposits within the Earth's crust (Helgeson, 1969; Barnes and Czamanske, 1967). In addition, the investigation of high pressure gas solubility reveals valuable information about molecular interactions.

The recent development of experimental methods for high pressure–high temperature research and the availability of new construction materials has caused a considerable increase in available information on high pressure gaseous solutions. Not only can the solubility be determined by quantitative chemical analysis and thermodynamic measurements, but visual observations and spectrophotometric and electrochemical determinations at maintained high pressures and temperatures provide detailed data on intermolecular associations and ionization within the fluid solutions.

II. General Description

This discussion will be mainly restricted to two-component systems consisting of a compressed gas 1 and a solid or liquid, condensed substance 2. The mole fraction x_2 in the gas phase is subject to four different influences: the external hydrostatic pressure exerted by gas 1 which increases the free energy of phase 2, the nonideality of the pure saturated vapor of 2, the vapor-phase interactions of 1 and 2 molecules, and, finally, the possible solubility of gas 1 in the condensed phase 2. These influences will be discussed in this sequence. The third effect, i.e., the gas-phase interaction between unlike particles, is particularly important. The last mentioned situation, i.e., solubility in both condensed and gas phases, will be described in terms of the critical phenomena in two-component systems in Section III.

A. Effect of Pressure on the Solid or Liquid Phase

In theory, there are two ways to measure the vapor pressure of a solid or liquid under an external hydrostatic pressure: with an ideal membrane, semipermeable to the vapor of (2), or with an atmosphere of an "inert" gas (1) which in practice, however, may dissolve in the solid or liquid. The second case is of interest here, and if the solubility in the condensed phase is neglected, the description is simple.

The fugacity f_2 of component 2 under the total hydrostatic pressure P follows from

$$\mu_2^{s0} + \int_{P_2^0}^{P} V_2^{s}\, dP = \mu_2^{g0} + RT \ln f_2. \tag{2.1}$$

Here μ_2^{s0} is the chemical potential of 2 in the solid or liquid state at normal saturation pressure P_2^0; μ_2^{g0} is the chemical potential of 2 in the gaseous state at the usual standard pressure; V_2^s stands for the molar volume of 2 in the solid or liquid state. If V_2^s can be considered as independent of pressure up to P and if 1 and 2 behave as ideal gases, then the fugacities can be replaced by the gas pressures and

$$\ln(P_2/P_2^0) = (V_2^s/RT)P. \tag{2.2}$$

Here P_2 is the partial pressure of 2 in equilibrium with the condensed phase 2 under the total pressure P, and it is assumed that $P_2^0 \ll P$. As long as the assumptions mentioned are valid, Eq. (2.2) can be replaced by

$$\ln(x_2/x_2^0) = (V_2^s/RT)P, \tag{2.3}$$

where x_2 is the actual mole fraction of 2 in the gas phase and x_2^0 is the mole fraction which component 2 could have if its concentration in the gaseous mixture were that of the pure saturated vapor.

The increases of vapor pressure or mole fraction, i.e., the ratios P_2/P_2^0 and x_2/x_2^0 are plotted in Fig. 1 for 25°C and 500°C and for two typical values of the volume per mole V_2^s. The ratio P_2/P_2^0 is often called the Poynting factor. It is certainly quite small, i.e., below 2, at conditions where the assumptions of gas ideality and of pressure independence of V_2^s apply. Thus it is almost impossible to determine the Poynting factor accurately experimentally since the nonideality contributions are of comparable magnitude.

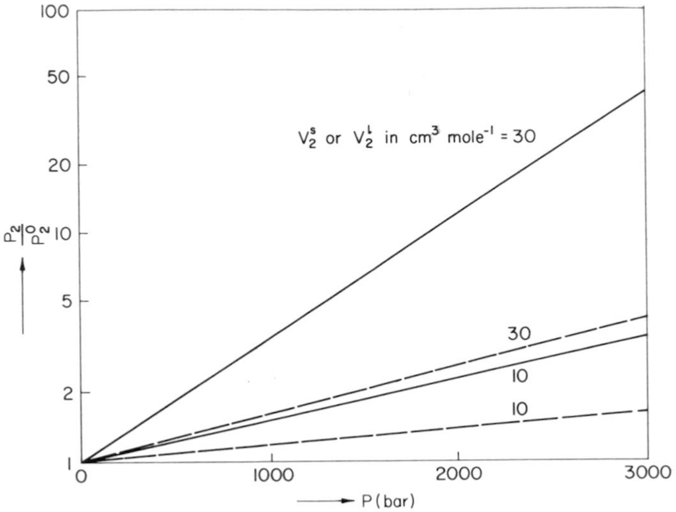

FIG. 1. Increase of vapor pressure P_2/P_2° of a solid or liquid substance 2 caused by an external pressure P at 25°C (——) and at 500°C (- - -). V_2^s and V_2^l are the molar volumes of 2.

B. Solvent–Solute Interaction in the Gas Phase

Particles 1 and 2 in the gas phase at elevated pressures will in most cases deviate from ideal behavior. This deviation can be described in a formal way, using the fugacity coefficient φ_2 of component 2:

$$\mu_2^{s^0} + \int_{P_2^0}^{P} V_2^{\,s}\, dP = \mu_2^{g^0} + RT \ln(Px_2\varphi_2/P^0). \qquad (2.4)$$

In analogy to Eq. (2.3) one obtains

$$\ln(x_2/x_2^0) = (V_2^s/RT)P - \ln \varphi_2. \qquad (2.5)$$

Since x_2/x_2^0 gives the enhancement of the mole fraction of 2 in the gas phase, an "enhancement factor" E is sometimes used (Prausnitz, 1969):

$$E = (1/\varphi_2) \exp(V_2^s P/RT). \qquad (2.6)$$

The fugacity coefficient can be expressed using the partial molar volume V_2

$$\ln \varphi_2 = (1/RT) \int_{P^0}^{P} (V_2 - (RT/P)\, dP; \qquad V_2 = (\partial V/\partial n_2)_{T,P,n_1} \qquad (2.7)$$

Which gives, combined with Eq. (2.5),

$$\ln(x_2/x_2^0) = (V_2^s P/RT) - (1/RT)\int_{P^0}^{P} (V_2 - RT/P)\, dP; \qquad (2.8)$$

assuming that $V_2 = f(P)$ obeys the ideal gas law up to the standard pressure P°. If the actual value of V_2 is V_2^{real} and the "ideal" value V_2^{ideal}, Eq. (2.8) would read:

$$\ln(x_2/x_2^0) = (V_2^s P/RT) - (1/RT)\int_{P^0}^{P} (V_2^{\text{real}} - V_2^{\text{ideal}})\, dP. \qquad (2.9)$$

Equation (2.9) shows clearly, that an increase of x_2 will occur, whenever V_2^{real} is smaller than V_2^{ideal} in the gaseous mixture, i.e., whenever the attraction between particles 1 and 2 predominates. For the more common case, when the volumetric data are expressed in pressure-explicit form, Eq. (2.10) for φ_2 has to be used

$$\ln \varphi_2 = (1/RT)\int_{V}^{\infty} [(\partial P/\partial n_2)_{TVn_1} - (RT/V)]\, dV - \ln z, \qquad (2.10)$$

where

$$z = PV/RT \qquad (2.11)$$

is the "compressibility factor" of the gas mixture. It will be the purpose of Sections III and IV to use specific expressions for φ_2 in combination with Eq. (2.5).

III. Detailed Description—Weak Interaction in the Gas Phase

If the solubility of gas 1 in the solid or liquid condensed phase of component 2 can be neglected, the quantitative calculation of the gas phase solubility depends mainly on sufficient information of the fugacity φ_2 in Eq. (2.5), i.e., on the partial molar volume V_2. Provided that x_2-values and the total density in the gas phase are not too high and if the intermolecular 1–2 interaction is mainly of the van der Waals-type, an equation of state of the virial form for V_2 is appropriate. Treatments of this kind have been developed and discussed repeatedly (Eucken and Bressler, 1928; Braune and Strassmann, 1929; Robin and Vodar, 1953; Ewald *et al.*, 1953; Franck, 1956). A recent and very thorough discussion of this group of gaseous solutions has been given by Prausnitz (1969, pp. 163–176).

Using the first three terms, the virial expansion in powers of reciprocal volume for the compressibility factor of a mixture z_{mix} is

$$z_{\text{mix}} = PV/RT = 1 + B_{\text{mix}}/V + C_{\text{mix}}/V^2. \tag{3.1}$$

For m components, the second and third virial coefficients are

$$B_{\text{mix}} = \sum_{i=1}^{m} \sum_{j=1}^{m} x_i x_j B_{ij}, \tag{3.2}$$

$$C_{\text{mix}} = \sum_{i=1}^{m} \sum_{j=1}^{m} \sum_{k=1}^{m} x_i x_j x_k C_{ijk}, \tag{3.3}$$

and for binary mixtures

$$B_{\text{mix}} = x_1^2 B_{11} + 2x_1 x_2 B_{12} + x_2^2 B_{22}, \tag{3.4}$$

$$C_{\text{mix}} = x_1^3 C_{111} + 3x_1^2 x_2 C_{112} + 3x_1 x_2^2 C_{122} + x_2^3 C_{222}. \tag{3.5}$$

Equation (2.10) combined with Eqs. (2.12), (2.15), and (2.16) gives the fugacity coefficient for component 2

$$\ln \varphi_2 = (2/V)(x_2 B_{22} + x_1 B_{12}) + (3/2V^2)(x_2^2 C_{222} + 2x_1 x_2 C_{122} + x_1^2 C_{112}) - \ln z_{\text{mix}}. \tag{3.6}$$

Inserting Eq. (2.17) in Eq. (2.5) and converting the total pressure P into reciprocal molar volume, one obtains:

$$\ln(x_2/x_2^0) = (V_2^s - 2B_{12})(1/V) + (V_2^s B_{11} - \tfrac{3}{2} C_{112})(1/V^2). \tag{3.7}$$

Only the first term on the right side does not disappear if the gas 1 is perfect. Of the virial coefficients, B_{11} and B_{12} are negative for all the systems discussed here. The third virial coefficients are probably positive. Thus the first term in Eq. (3.7) is positive, the second negative. Additional terms in $1/V^3$, etc. have been neglected because the corresponding coefficients can not be calculated.

The binary virial coefficient B_{12} can be determined from experimental x_2/x_2^0-ratios, using the slope of Eq. (2.18) at zero density. It is obvious, however, that x_2-values can be predicted to some extent if the virial coefficients are known. The "pure" virial coefficients B_{11}, B_{22}, C_{111}, C_{222}, etc. can be derived from experimental PVT-data. Several procedures have been developed to derive the "mixed" coefficients B_{12}, C_{112}, etc. from properties of the pure components. They can not be surveyed here in detail. One possibility is, to use the expression obtained from statistical mechanics for the B and C coefficients as functions of the intermolecular potential $\Gamma_{12}(r)$, where r is the intermolecular distance. For B_{12} the expression is

$$B_{12} = 2\pi N_0 \int_0^\infty [1 - \exp(-\Gamma_{12}(r)/kT)]\, dr \tag{3.8}$$

and analogous but more complicated relations are valid for C_{112} and C_{122} (see, for example, Prausnitz, 1969). To calculate B_{12} by Eq. (3.8) a function $\Gamma_{12}(r)$ has to be inserted. One relation suitable among others, is the Lennard–Jones function

$$\Gamma_{12}(r) = 4\varepsilon_{12}[(\sigma_{12}/r)^{12} - (\sigma_{12}/r)^6] \tag{3.9}$$

with

$$\sigma_{12} = \tfrac{1}{2}(\sigma_1 + \sigma_2) \tag{3.10}$$

and

$$\varepsilon_{12} = (\varepsilon_1 \times \varepsilon_2)^{1/2} \tag{3.11}$$

where ε is the minimum potential energy and σ the collision diameter, i.e., the distance where $\Gamma(r) = 0$. The Lennard–Jones constants ε and σ for the pure components can often be taken from existing tables (see Curtiss, 1967, Hirschfelder *et al.*, 1964). As an example Fig. 2 shows the results which Robin and Vodar (1953) have obtained by this method with the experimentally determined solubility of methanol in several inert gases (Kritchevsky and Koroleva, 1941). Solubility of the inert gases in the liquid methanol phase was not taken into account. Application of Eq. (3.7) was restricted to the first right-hand term, i.e., to second virial coefficients. The density scale of the abscissa is in amagat units, i.e., in ratios of the true molar volume in the gas phase to the standard molar volume.

The agreement between the calculated straight curves and the experimental points in Fig. 2 is good. But this can not be expected for higher mole fractions x_2 and stronger interactions. Ewald *et al.* (1953) have used Eq. (3.7) including the second right-hand term. The necessary Lennard–Jones constants were derived from the critical temperatures and critical molar volumes. One example, the solubility of solid CO_2 in air for three temperatures is shown in Fig. 3. The improvement caused by the inclusion of the V^{-2}-term of Eq. (3.7) is obvious. For other systems, however, e.g., for naphthalene–hexachloroethane and *p*-chloro-iodo-benzene in ethylene the virial treatment of Eq. (3.7) is not sufficient. An improved procedure will be shown below.

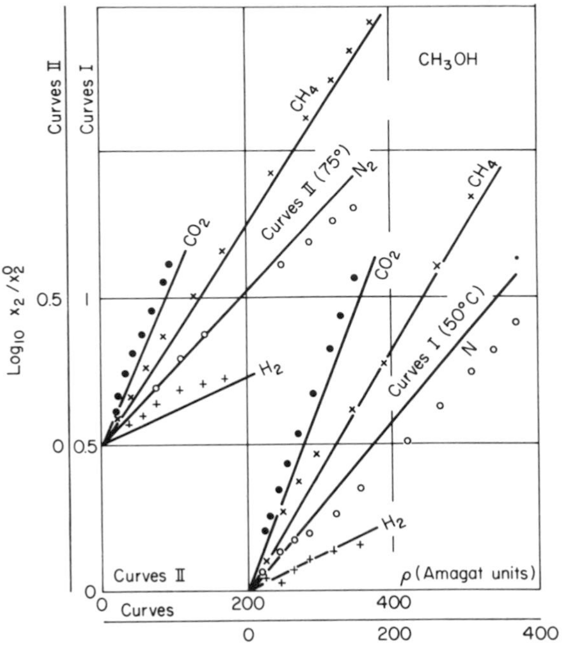

FIG. 2. Solubility of liquid CH_3OH in H_2, N_2, CH_4 and CO_2 at 50°C and 75°C as a function of gas density after Robin and Vodar (1953). Measurements from Krichevski and Koroleva (1941).

Another way of representing solubilities in compressed gases is shown in Fig. 4 (from Prausnitz, 1969) with more recent data on the system solid carbon dioxide and air by Webster (1952). The mole fraction $x(CO_2)$ in the gas phase as a function of pressure exhibits a minimum. The data can be well represented with the inclusion of third virial coefficients. It is actually a ternary system, and the virial coefficients were calculated using the Kihara potential. The positions of the mole fraction minimums are of practical interest, e.g., for finding the optimum conditions for freezing-out processes. Figure 4 shows that the minimum is already found if only second virial coefficients are considered. Reuss and Beenakker (1956) and Hinckley and Reid (1964) have derived expressions for the minimum coordinates in the x_2–P-diagram, neglecting the Poynting factor:

$$x_2(\text{minimum}) = -5.44 B_{12} P_2^0 / RT \tag{3.12}$$

and

$$P(\text{minimum}) = -RT/(B_{11} + B_{12}), \tag{3.13}$$

where the various quantities have the same meaning as in Eqs. (2.4) and (3.4).

At very high densities in the gas phase the virial method can not be applied. A good equation of state, however, can be applied to calculate the fugacity coefficient φ_2 in Eq. (2.5). The Redlich–Kwong equation Eq. (3.14) has been used successfully for this purpose.

$$z = PV/RT = V/(v - b) - a/RT^{3/2}(V + b); \qquad (3.14)$$

a and b are constants which can also be evaluated for mixtures using suitable combination rules. Chueh and Prausnitz (1967) have given tables of such constants. Figure 5 demonstrates the success of this procedure with the system naphtalene–ethylene.

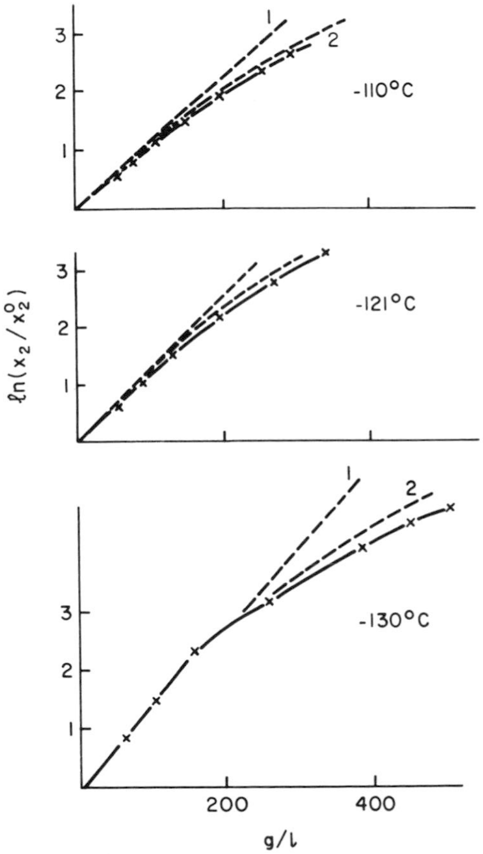

FIG. 3. Solubility of solid CO_2 in air as function of gas density after Ewald *et al.* (1953). Measurements from Webster (1952).

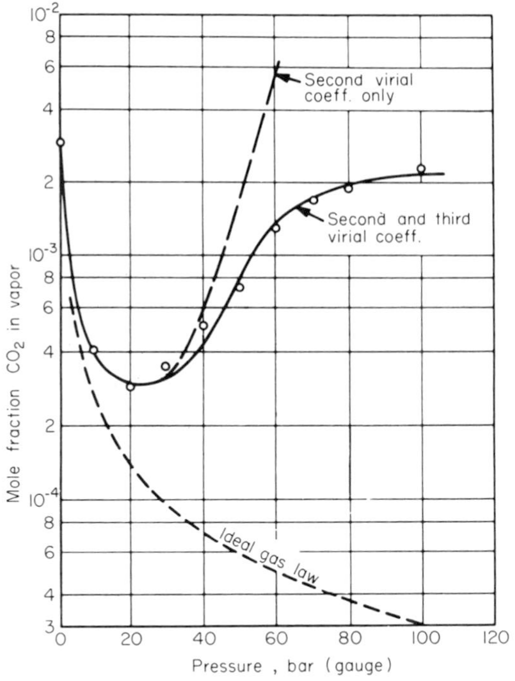

FIG. 4. Solubility of solid CO_2 in air as a function of gas pressure after Prausnitz (1969). Measurements from Webster (1952).

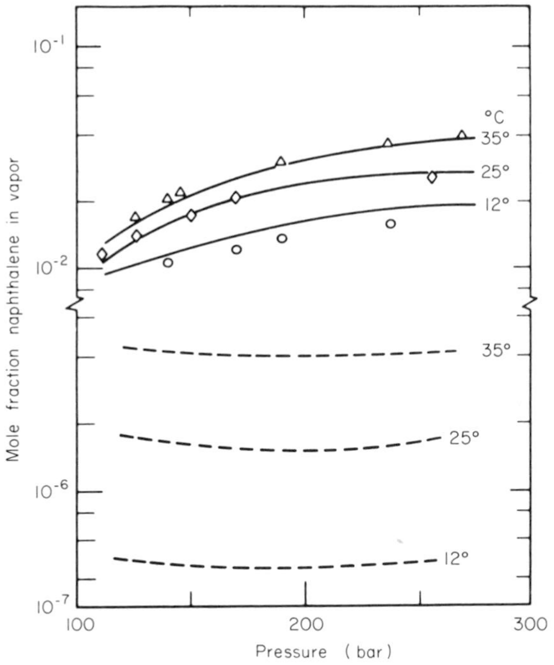

FIG. 5. Solubility of solid naphtalene in ethylene as a function of pressure after Prausnitz (1969). Measurements from Diepen and Scheffer (1948). Vapor-phase fugacities from Redlich–Kwong equation (——) and ideal-gas law (– – – –).

The influence of solubility of component 1 in the condensed phase of component 2 has been neglected above. If the mole fraction y_1 of 1 in the solid or liquid 2 is small, it can be accounted for by using Henry's constant H_{12} which must be determined experimentally or estimated from a dependable correlation. For a sparingly soluble gas the accuracy of H_{12} is not critical. The general situation of high pressure liquid gas equilibria at elevated temperature and the critical phenomena in such systems are discussed below.

IV. Detailed Description—Strong Interaction in the Gas Phase

A. Solubility Equations

The treatment of solubility in compressed gases based on the virial concept is of little value if the interactions 1–2 in the gas phase are too strong to be considered as of the van der Waals-type. This is to be expected for large and very polarizable molecules 2, for highly polar particles and ions, and for systems where weak types of genuine bonds, like hydrogen bonds, can occur in the gas phase. Examples are large aromatic compounds in gaseous olefines or systems with a dense gas phase of hydrogen chloride or gaseous hydrogen chloride–water vapor mixtures. They can be used as solvents for gas phase crystal synthesis of inorganic compounds or even of pure gold (Rabenau, 1969). The most important group are the aqueous supercritical solutions. The enhancement factor E (see Eq. (2.6)) for the solution of some high melting inorganic materials in steam can attain many orders of magnitude, although the mole fraction x_2 may still be much smaller than unity.

One way to discuss such mixtures with strong interaction is to assume a well defined solvation of particles 2 by those of component 1 according to a "chemical" equation:

$$\text{"2"} + n \times \text{"1"} = \text{"}21_n\text{"} \tag{4.1}$$

where it is assumed, that n can increase, with rising pressure of "1" from 0 to a maximum value m, which is the maximum "association number." The analytically determinable mole fraction X_2 of the solute would then be

$$X_2 = \sum_{n=0}^{m} x_{21_n} \tag{4.2}$$

if x_1 is still close to unity. It shall be assumed further that the association of each of the particles 1 with one particle 2 is independent and accompanied by the same amount of enthalpy ΔH_{21} and entropy ΔS_{21} at a given temperature and pressure. This means that only one "solvation layer" around each particle 2 is considered. If apart from this association all particles in the gas phase would behave ideally, the following relation would hold

$$\ln(x_{21_n}/x_2) = n \ln(K_{21}/V) \tag{4.3}$$

where

$$\ln K_{21} = (\Delta H_{21}/RT) - (\Delta S_{21}/R) \tag{4.4}$$

with K_{21} the equilibrium constant or the ratio of the partition functions for the reaction of one mole of component 1 reacting with one mole of 21_n to form 21_{n+1}; K_{21} has the dimension of volume per mole. The mole fraction of nonsolvated particles x_2 is related to x_2^0 the mole fraction found if the condensed phase were unaffected by the total pressure P through the relation

$$x_2/x_2^0 = \exp(V_2^s P/RT) \tag{4.5}$$

which was given as Eq. (2.3) above. Using the definition of X_2 Eq. (4.2), one obtains

$$\ln(X_2/x_2^0) = (V_2^s P/RT) + \ln\left[\sum_{n=0}^{m} (K_{21}/V)^n\right]. \tag{4.6}$$

Since actually nonideal behavior is expected in the gas phase, Eq. (4.6) has to be extended. The virial expansion is used again with second and third coefficients. Since only x_1 is assumed to be high and all other mole fractions x_{21_n} from $n = 0$ to $n = m$ are expected to be comparatively small, only interactions between 1–1 and 1–21_n interactions need to be allowed for by the respective virial coefficients. In accordance with the procedure used to derive Eq. (3.7) one obtains instead of Eq. (4.3), including Eq. (4.4):

$$\ln(x_{21_n}/x_2^0) = (V_2^s P/RT) + \ln(K_{21}/V)^n + (2/V)(B_{2,1} - B_{21_n,1}) + (3/2V^2)(C_{2,11} - C_{21_n,11}) + n[(2B_{11}/V) + (3C_{111}/2V^2)] \tag{4.7}$$

It is practical to introduce abbreviated symbols for the 1–2, 1–21_n and 1–1 interactions, namely:

$$2B_{11}/V + 3C_{111}/2V^2 = A_{11} \tag{4.8}$$

and

$$(2/V)(B_{2,1} - B_{21_n,1}) + (3/2V^2)(C_{2,11} - C_{21_n,11}) = A_{21_n}; \qquad (4.9)$$

A_{11} and A_{21_n} are dimensionless functions of the total molar volume V of the gas mixture.

The terms A_{21_n} allow for that part of intermolecular interaction between unlike molecules which does not lead to association. Introducing A_{11} and A_{21_n} changes Eq. (4.7) to

$$\ln(x_{21_n}/x_2^0) = (V_2^s P/RT) + n \ln(K_{21}/V) + nA_{11} + A_{21_n} \qquad (4.10)$$

and instead of Eq. (4.6) we obtain:

$$\ln(X_2/x_2^0) = (V_2^s P/RT) + \ln \sum_{n=0}^{m} [((K_{21}/V) \exp A_{11})^n \exp A_{21_n}]. \qquad (4.11)$$

The evaluation of the sum in Eq. (4.11) is impeded by the terms A_{21_n}, which are actually unknown. Two simple limiting situations can be considered, however:

I. $B_{2,1} \approx B_{21_n,1}$ and $C_{2,11} \approx C_{21_n,11}$ (4.12)

for all values of n. This would make $A_{21_n} = 0$ and eliminate the last factor in the sum of Eq. (4.11).

II. $B_{2,1} - B_{21_n,1} \approx -nB_{11}$ and $C_{2,11} - C_{21_n,11} \approx -nC_{111}$ (4.13)

for all values of n. This would make $A_{21_n} = -nA_{11}$ and eliminate both factors, $\exp(A_{11})$ and $\exp(A_{21})$ in the sum of Eq. (4.11).

Using the simplifications I and II one obtains from Eq. (4.11):

$$\ln(X_2/x_2^0) = (V_2^s P/RT) + \ln \sum_{n=0}^{m} [(K_{21}/V) \exp A_{11})]^n \qquad (4.14)$$

or

$$\ln(X_2/x_2^0) = (V_2^s P/RT) + \ln \sum_{n=0}^{m} (K_{21}/V)^n. \qquad (4.15)$$

The sums can now be evaluated to give:

$$\ln(X_2/x_2^0) = (V_2^s P/RT) + \ln\{1 - [(K_{21}/V) \exp A_{11})]^{m+1}\} - \ln[1 - (K_{21}/V) \exp A_{11}] \qquad (4.16)$$

and

$$\ln(X_2/x_2^0) = (V_2^s P/RT) + \ln[1 - (K_{21}/V)^{m+1}] - \ln[1 - (K_{21}/V)]. \tag{4.17}$$

Again two kinds of conditions in the gas phase can be considered: The density and the intermolecular interaction may be weak, which means that $V \gg K_{21}$ and that $\exp A_{11}$ is close to unity. Then

$$\ln(X_2/x_2^0) \approx (V_2^s + K_{21})1/V. \tag{4.18}$$

This is the same form of solubility equation which was derived already above in Section II as Eq. (3.7), taking only the first right-hand term of that equation. The association constant for 2–1 pairs in Eq. (4.18) replaces the binary virial coefficient $2B_{12}$ in Eq. (3.7) which is in accordance with the common interpretation of the second virial coefficient. Figure 2 has shown, that measured solubilities indeed confirm Eq. (4.18) in certain cases. If, however, gas densities and 2–1 interactions are strong, one will expect, that

$$(K_{21}/V) \exp A_{11} \gg 1 \tag{4.19}$$

and from Eq. (4.16)

$$\ln(X_2/x_2^0) \approx (V_2^s P/RT) + mA_{11} + m \ln(K_{21}/V) \tag{4.20}$$

or from Eq. (4.17):

$$\ln(X_2/x_2^0) \approx (V_2^s P/RT) + m \ln(K_{21}/V). \tag{4.21}$$

The last equations, (4.20) and (4.21), predict a proportionality between the logarithm of the solubility $\ln(X_2/x_2^0)$ and the logarithm of the total gas density $\ln(1/V)$ if the 2–1 association in the gas phase proceeds to a state of saturation, characterized by the association number m. As will be shown below, there are examples of experimental solubilities which over wide ranges of gas density confirm Eq. (4.21). In order to predict solubilities from Eq. (4.21), knowledge of the maximum association number m is required. In some cases m can be predicted from available chemical information. Sterical considerations can also be of assistance. Otherwise m has to be derived from experimental results within a region, where Eq. (4.21) is obeyed.

Equation (4.11) and the following relations are inconsistent in so far as the total pressure P in the Poynting term on the right side has not been expressed by a suitable function of the reciprocal volume. This would make the expressions less simple, however, and the Poynting term is of relatively little influence in cases where Eqs. (4.20) and (4.21) are valid. Thus even the use of an estimated value for the pressure will often be sufficient.

It is plausible that this "association" concept can in principle be extended to include ionic ionization of the particles "2" within a dense gas phase of polar molecules "1." Extensive measurements of electrolytic conductance show that many hydroxides, acids and salts do form ions in supercritical aqueous solutions (see below). The procedure described above has been extended to include the influence of ion solvation on the solubility (Franck, 1956). Numerical evaluation of the resulting equations, however, evidently requires detailed information about the ionization equilibria and the activity coefficients of the ionic species in the dense gas phase.

B. Solubility Examples

In Section III it has already been shown that the solubility of naphtalene in dense ethylene cannot be sufficiently described by the virial treatment and Fig. 5 demonstrated the success of the application of the Redlich–Kwong equation to calculate the required fugacity coefficient of naphtalene (Prausnitz, 1969). Figure 6 shows, that Eq. (4.20) is

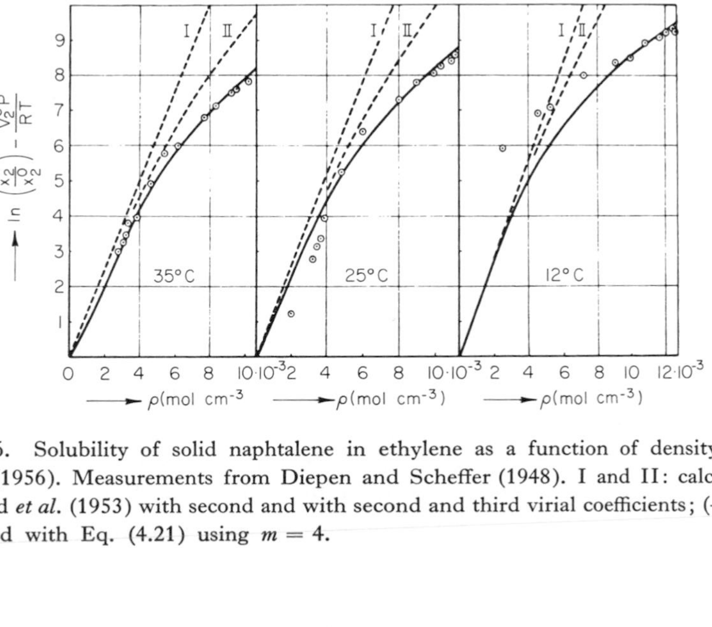

Fig. 6. Solubility of solid naphtalene in ethylene as a function of density after Franck (1956). Measurements from Diepen and Scheffer (1948). I and II: calculated by Ewald *et al.* (1953) with second and with second and third virial coefficients; (——): calculated with Eq. (4.21) using $m = 4$.

also capable of describing these measurements. Certainly, m and the association equilibrium constant K_{21} had to be adjusted. It appears, however, that a single value $m = 4$ can be used for the whole region of temperature and density; K_{21} is equal to 850 cm³ mol⁻¹, 760 cm³ mol⁻¹ and 720 cm³ mol⁻¹ for 12, 25, and 35°C, respectively. From this temperature dependence of K_{21}, an energy of association at constant density for one ethylene molecule with one naphtalene molecule of $\Delta E = -1100$ cal mol⁻¹ can be derived. Both, the m and the ΔE values appear plausible and should facilitate the estimation of such quantities for analogous systems.

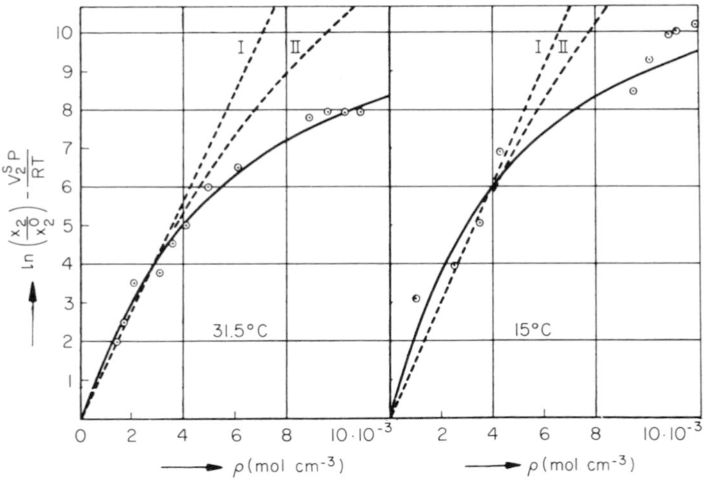

FIG. 7. Solubility of solid p-chloro-iodo-benzene in ethylene as a function of density after Franck (1956). Measurements from Ewald (1955). I and II: calculated by Ewald *et al.* (1953) with second and with second and third virial coefficients; (——): calculated with Eq. (4.21), using $m = 3$.

Another system of organic compounds which has been investigated is p-chloro-iodo-benzene and ethylene (Fig. 7). Here a single value of $m = 3$ has been used. K_{21} is 2000 and 1400 cm³/mole for 15 and 31.5°C. One obtains $\Delta E = -3800$ cal mol⁻¹ in this case, although the derivation from data at only two temperatures may not be justified.

Good examples for the application of Eqs. (4.20) and (4.21) are the solutions of solid KCl and SiO_2 in supercritical stam, as long as the solute mole fractions are still small. In these cases the virial treatment according to Eq. (4.19) is completely inadequate. Figure 8 gives the comparison with KCl-solubility measurements of Jasmund (1952); $m = 4$

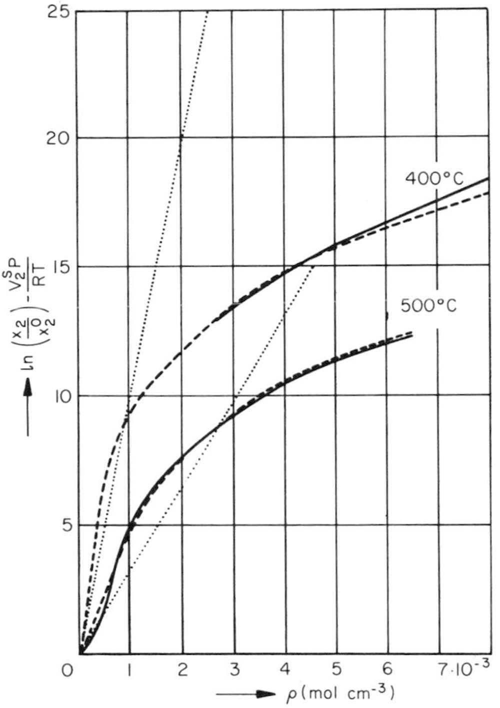

FIG. 8. Solubility of solid KCl in supercritical water vapor as a function of density after Franck (1956); (——): experimental curve from Jasmund (1952), (......): initial slope according to Eq. (4.19), (– – – –): calculated with Eq. (4.21), using $m = 4$.

has been used and ΔE was found equal to 11.6 kcal mol^{-1}. This last value should be the molar energy of hydration for one water molecule to one of the highly polar KCl molecules. Figure 9 gives data on the solubility of quartz in supercritical steam according to measurements of Kennedy (1950). At the high densities the agreement between experimental and calculated data is not quite as good as in Fig. 8. One has to consider, however, that a constant value of $m = 2$ has been used in the whole region. It may be that at high densities a lower mean number of association is to be used because of some agglomeration of SiO_2 units. In Fig. 9, $\ln X_2$ instead of $\ln(X_2/x_2^0)$ was plotted because the normal saturation pressure of quartz is not sufficiently well known. The solubility of a great number of inorganic oxides and hydroxides in supercritical gaseous water up to pressures of several kilobar has been investigated by Glemser and Wendlandt (1963, 1964, 1966). Often the observed solubility confirms the logarithmic law of Eq. (4.21). A par-

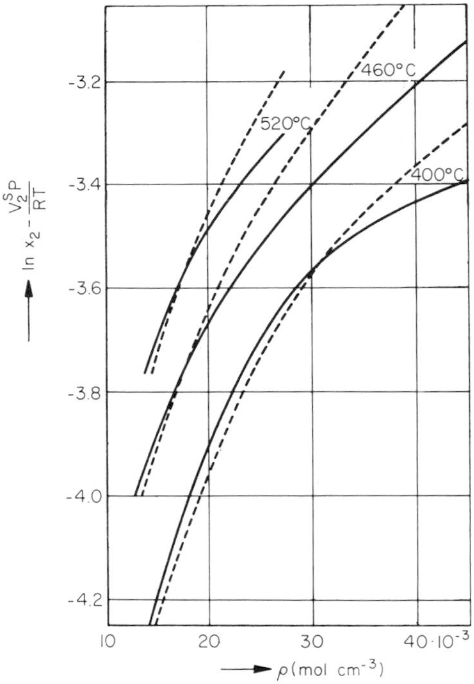

FIG. 9. Solubility of quartz in supercritical water vapor as a function of density after Franck (1956); (——): experimental curve from Kennedy (1950), (– – – –): calculated with Eq. (4.21), using $m = 4$.

ticularly good example is the solubility of GeO_2 (Glemser and Wendlandt, 1966), shown in Fig. 10. Many solubility measurements in supercritical steam, especially with sparingly soluble inorganic compounds of geological interest, were made by Morey and co-workers (see Morey, 1957). Barnes and Czamanske (1967) have accumulated and reviewed numerous data on solubilities and transport of ore minerals in supercritical water. Additional data on aqueous systems of geological importance are given by Ellis (1967) and by Roedder (1967). A very general survey on the thermodynamics of hydrothermal systems at elevated temperatures and pressures has been made by Helgeson (1969). The thermodynamic characteristics of hydrothermal solutions up to about 300°C (the critical temperature of water is 374°C) have been thoroughly discussed by Cobble (1966) and his results can be of use for dense gaseous solutions also. Extensive studies of salt solubilities, especially at higher concentrations up to about 350°C have been made by Marshall and co-workers (1968, 1969).

Franck (1961) has given a general discussion of supercritical water as an electrolytic solvent. Very often *PVT*-data of dense supercritical water are necessary for the solubility calculations. Recent data were presented by Maier and Franck (1966) to 850°C and 6 kbar, by Köster and Franck (1969) to 600°C and 10 kbar and by Burnham *et al.* (1969) to 900°C and 9 kbar.

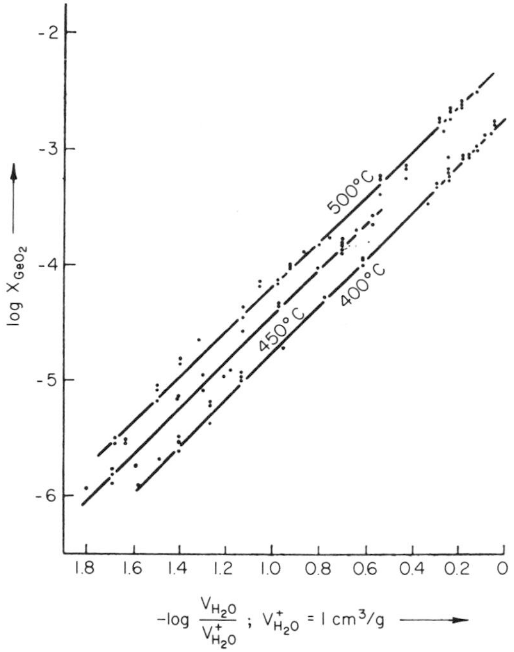

FIG. 10. Solubility of solid GeO_2 in supercritical water vapor as a function of the logarithm of the water density after Glemser *et al.* (1966); V_{H_2O} = specific volume of the water.

If the partial molar volume V_2 of the solute in dense gaseous phases is known, Eq. (2.8) provides a straightforward way to calculate the solubility. Experimental data on pressure–volume–temperature relations as a function of the mole fraction x_2 of solutions of solids in dense gases are limited, however. Copeland *et al.* (1953) have investigated the H_2O–NaCl system to 400°C and 300 bar. Very high negative partial molal volumes of sodium chloride have been found at the highest temperatures, which can be expected because of the hydration discussed above. The same system was studied by Khaibullin and Borisow (1965) and by Lemmlein and Klevtsov (1961). The latter group obtained *PVTx*-data to 500°C, 1750 bar and 30 weight percent NaCl in the one phase fluid region. Within the experimental accuracy the isochores $P = P(T)_{V,x}$ were linear in the region covered. This behavior, which is known for many pure substances can facilitate the extrapolation of $PVTx$-data. Rodnyanski and Galinker (1955) have made measurements with the H_2O–KCl system. The ionic dissociation of electrolytes becomes important in supercritical gaseous water at densities higher than about 0.5 gm cm^{-3}. At such densities ion hydration can influence solubilities already to a considerable degree (Franck, 1956). Extensive studies of such gaseous aqueous electrolyte solutions have been made by several groups, mainly using conductivity measurements. They were extended to about 800°C and to about 5 kbar (see as examples Franck, 1956a,b,c; Quist *et al.*, 1963; Quist and Marshall, 1968; Dunn and Marshall, 1969; Ritzert and Franck, 1968; and an extensive review by Marshall, 1969). The dissociation constant of KCl at 400°C and at water densities of 0.6 gm cm^{-3} and 1.0 gm cm^{-3} are 2.5×10^{-3} and 18×10^{-3} mol $\times$ l^{-1}. At 750°C and a water density of 0.6 gm cm^{-3} a value of 1.4×10^{-3} mol l^{-1} has been found (Ritzert and Franck, 1968). Such values are typical for a number of alkali halides and hydroxides and several simple compounds of second group bivalent cations. Some conductance determinations have been carried out to 1000°C and 12 kbar with alkali chlorides (Mangold and Franck, 1969) which give the degree of association at such conditions. Apparently 80% of the electrolyte of 0.01 molal solutions of KCl and LiCl water at 1000°C and a density of 0.7 gm cm^{-3} are dissociated to ions.

Measurements of the optical absorption in the visible and ultraviolet spectral ranges to 500°C and 6 kbar show that several ions of heavy metals, especially cobalt and nickel can exist as well defined aquo-complexes in the dense supercritical gas phase (Lüdemann and Franck, 1967, 1968). Supercritical solutions of bivalent cobalt chloride for example form tetrahedral neutral dichloro-diaquo complexes and six-coordinated octahedral mixed chloro-aquo-complexes. At 500°C and a water density of 0.9 gm cm^{-3} about 20% of the cobalt exists as tetrahedral complexes. This proportion increases with increasing temperature and with decreasing water density. Addition of high concentrations of alkali halides to the gaseous solutions appear to favor the formation of tetrahedral complexes. Analogous behavior has been found for supercritical nickel solutions. The considerable consequences which complex formation can have for the solubility and transport of ore forming minerals in dense gaseous phases have been extensively discussed (Helgeson, 1964).

V. Liquid–Gas Critical Phenomena in Binary Systems

A comprehensive discussion of the phenomena of solubility of solids and liquids in dense gases has to consider the complete phase diagram which includes also the solubility of the "light" compound in the condensed phases. These diagrams are three-dimensional for two component systems. The discussion of this section will be restricted to such systems and variables T, P, x will be used. Recent surveys of the thermodynamics and characteristics of solid–fluid and fluid–fluid phase equilibria at high pressures have been given by Rowlinson (1969), Prausnitz (1969), Tsiklis (1968), Schneider (1966, 1970), Franck (1961), Tödheide (1966, 1970).

Two-component systems exhibit a critical curve which extends in the PTx-diagram from the critical point of the first pure component to the critical point of the second pure component. If the critical temperatures of the two partners are not too different and if there are not too great differences in molecular seize and polarity between them, the critical curve is uninterrupted. Figure 11a demonstrates this behavior schematically; T_{t_1} and T_{t_2} are the tripel point temperatures and T_{c_1} and T_{c_2} the critical temperatures of the components 1 and 2. The solid curves leading to T_1, T_2, C_1 and C_2 are the sublimation pressure and vaporization pressure curves of 1 and 2. The dotted line between C_1 and C_2 is the projection of the critical curve on the PT-lane; Q is a quadruple point for the four phases–solid S_1, solid S_2, liquid L, and gas G. Mutual solubility in the solid phases has been excluded. The finely dotted, almost vertical, lines indicate the melting pressure curves. The crossed curves are projections of three phase surfaces S_1, S_2L, S_1LG, S_2LG on the PT-plane. Numerous and well investigated systems have phase diagrams as in Fig. 11a, for example, combinations of simple hydrocarbons like ethane-heptane (see Rowlinson, 1969). The projections of such critical curves may have maximum or minimum pressures and also maximum and minimum temperatures. Since the present discussion is mainly concerned with components of very different properties, the emphasis of the following sections will be on systems with more complex phase diagrams.

A. The Solid–Liquid–Gas Three-Phase Surface and the Liquid–Gas Critical Curve

For systems where the lower boiling component has a critical temperature lower than the tripel point temperature of the higher boiling

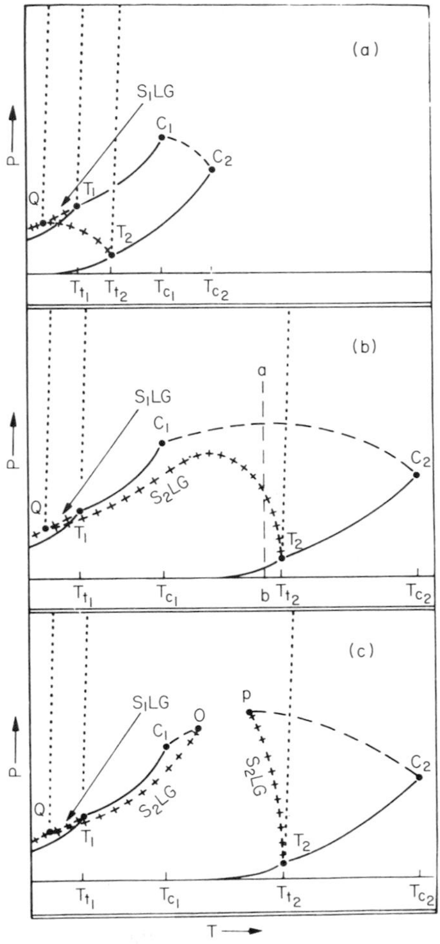

FIG. 11. Projection of the PTx-phase diagram of a binary system on the PT-plane.

component $T_{c_1} < T_{t_2}$, the S_2LG three-phase equilibrium surface may extend to temperatures and pressures approaching the critical point of "1." This is demonstrated by Fig. 11b. An example for such behavior is the system H_2O–NaCl. To discuss its properties an isothermic cross section between T_{c_1} and T_{t_2} at a–b in Fig. 11b is examined. The resulting Px-diagram is shown in Fig. 12. One point on the critical curve is C. The points A and B are on the three phase surface S_2LG. The left part (fully drawn) of the curve A–E is a solubility curve of solid "2" in the gas "1." Figure 13 is a diagram of four isotherms similar to that of Fig. 12, drawn from experimental data for H_2O–NaCl from Sourirajan and

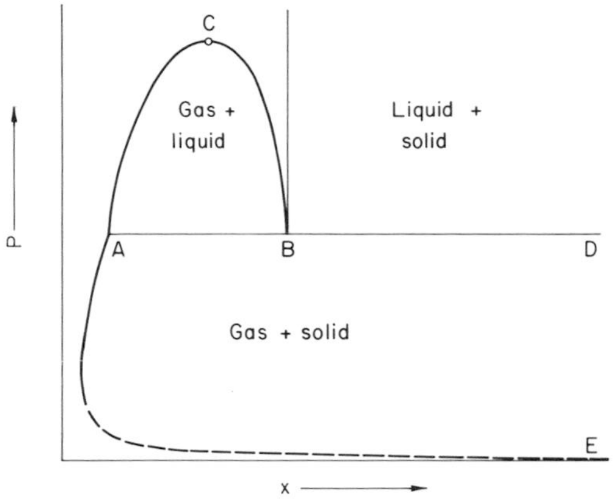

FIG. 12. *Px*-cross section of the *PTx*-phase diagram at a–b between T_{c_1} and T_{t_2} of Fig. 11b.

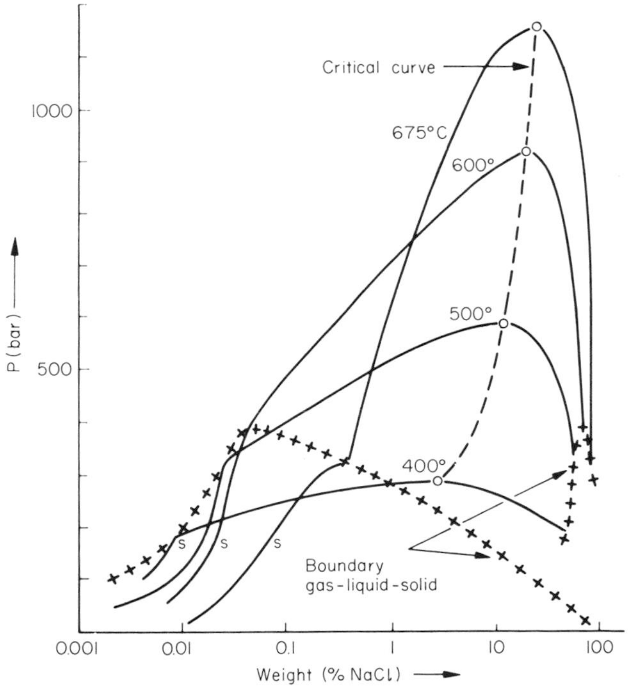

FIG. 13. Isotherms in the pressure-composition diagram of the H_2O–NaCl-system from measurements of Sourirajan and Kennedy (1962).

Kennedy (1962). Figure 13 is distorted because of the logarithmic scale for the mole fraction. The curve branches denoted by letters s are the solubility curves of solid NaCl in compressed steam corresponding to the left side of the A–E branch in Fig. 12. The two crossed curves are the loci of points of the character A and B (Fig. 12) for the NaCl–H_2O-system projected on a T–x–plane. It follows from Fig. 13, that the vapor pressure of a saturated aqueous NaCl solution has a maximum of about 400 bar at 600°C. It is believed that in the NaCl–H_2O system the critical curve, part of which is shown in Fig. 13, will continue uninterrupted to the pure fluid NaCl.

If the solubility of solid "2" is very small, it is possible, that the three-phase surface S_2LG interferes with the liquid–gas critical curve, which will be interrupted at two "critical end points," O and P in Fig. 11c. Between O and P the critical curve and the three phase surface disappear. An example is the system SiO_2–H_2O. Figure 14 is a diagram from measurements of Kennedy and co-workers (1962) which shows the upper critical end point UCEP of this system, where the projection of the three phase surface quartz–liquid solution–gas ends at 1080°C and 9.8 kbar. This point corresponds to the point P in Fig. 11c. The solubility of SiO_2 in supercritical water at 500°C and 1 kbar is only 0.26 weight percent, that is by about a factor of 200 less than the solubility of NaCl. An extensive discussion of the quartz solubility at high temperatures and pressures is given by Holland (1967).

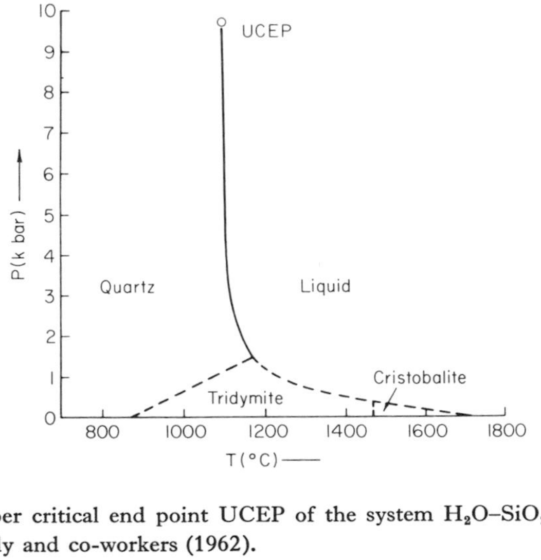

FIG. 14. Upper critical end point UCEP of the system H_2O–SiO_2 from measurements of Kennedy and co-workers (1962).

TABLE I

INORGANIC COMPOUNDS IN WATER[a]

Uninterrupted critical curve		Interrupted critical curve	
NaCl	K_2CO_3	LiF	Na_2SO_4
NaBr	$K_4P_2O_7$	Li_2SO_4	K_2SO_4
NaJ	RbCl	NaF	Tl_2SO_4
$Na_20.2B_2O_3$	CsCl	Na_2CO_3	SiO_2
KF	Cs_2SO_4	$Na_4P_2O_7$	
KCl	$CaNO_3$		
KJ	$PbCl_2$		

[a] According to measurements of Keevil (1942) and Morey and Chen (1956).

Table I gives a list of inorganic compounds with small volatility which in combination with water have either a uninterrupted or an interrupted critical curve. It is based on measurements of Keevil (1942) and Morey and Chen (1956).

B. Gas–Gas Immiscibility

Critical curves of binary systems can also be interrupted because of the interference with an immiscibility range in the liquid region. This is schematically shown in Fig. 15. One branch of the critical curve begins at the critical point C_1 of the lower boiling component and ends at a lower critical end point LCEP. The other branch of the critical curve begins at C_2 and can either exhibit a minimum temperature at higher pressures (possibility a in Fig. 15) or preceed directly to higher temperatures and higher pressures (possibility b). Such critical curves indicate that certain systems can separate into two fluid phases at temperatures higher than the critical temperature of the less volatile component. This phenomenon has been predicted and discussed already by van der Waals (1894) and Kamerlingh-Onnes *et al.* (1907), who suggested this behavior is "immiscibility in the gas phase." Although the dense fluids concerned are not gases in the ordinary sense, this term will be used here too because it has been widely accepted. Figure 15 shows, that the isotherms which define the two-phase fluid range have a narrowed portion which gradually leads to separation into two isotherms with increasing temperature. These

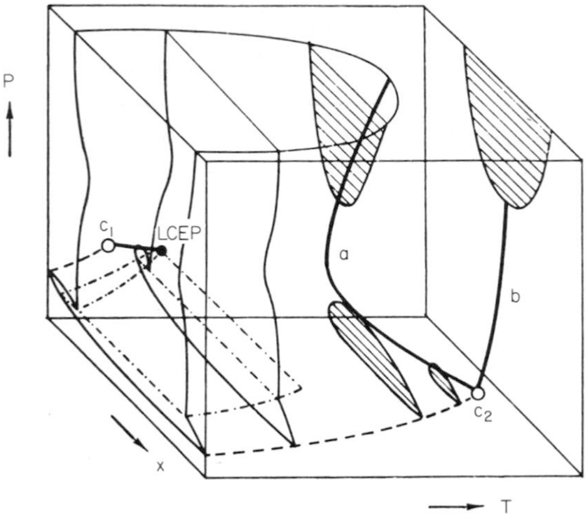

FIG. 15. Two types of critical curves, a and b, which result from an interference of an immiscibility in the liquid region with the gas–liquid critical curve; C_1, C_2: Critical points of the pure components. LCEP: lower critical end point.

divided isothermal cross sections are shaded in Fig. 15. Krichevski and Bolshakov (1941) were the first to find immiscibility in the gas phase of type a (Fig. 15) experimentally for the system NH_3–N_2. The results were confirmed by Lindroos and Dodge (1952). Type *b* behavior was found for the first time by Tsiklis (1952) for He–NH_3. Schneider (1970) gives a compilation of gas–gas equilibria which have been investigated experimentally. About 25 of these systems have critical curves of type a with temperature minima. Of this group around 20 have either ammonia or water as the less volatile component. Another 16 systems have been investigated which have critical curves of type b. All except one are combinations with helium. The exception is Ar–H_2O (Lentz and Franck, 1969).

Several typical *Px*-diagrams with experimentally determined isotherms are shown in Fig. 16 and 17. The behavior of the NH_3–N_2 system to 16 kbar is demonstrated by Fig. 16. The separation of the isotherms occurs at the temperature minimum of 87°C. The data are from several publications of Krichevski, Tsiklis and co-workers (1941) (1952) (1967). The system CO_2–H_2O belongs also to type a, it is of considerable interest for geochemistry because both compounds are major components of hydrothermal fluids. Figure 17 contains the experimental isotherms (Tödheide and Franck, 1963). The critical curve has a minimum temperature at 266°C and 2.5 kbar. This means, that for example at 300°C

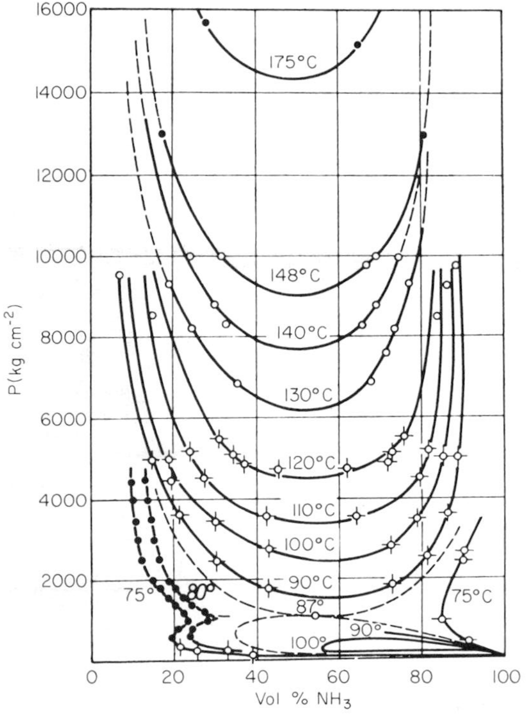

FIG. 16. Isotherms which are the boundaries of the two-phase region in a pressure-composition diagram for the system NH_3–N_2 from measurements of Krichevski and Bolshakow (1941), Krichevski and Korolewa (1941), Tsiklis (1952), and Tsiklis and Rott (1967).

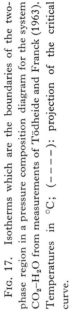

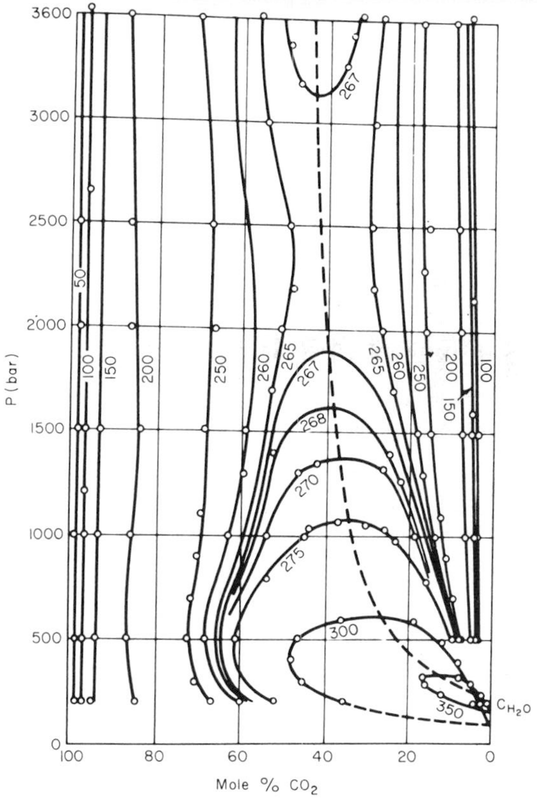

FIG. 17. Isotherms which are the boundaries of the two-phase region in a pressure composition diagram for the system CO_2–H_2O from measurements of Tödheide and Franck (1963). Temperatures in °C; (– – – –): projection of the critical curve.

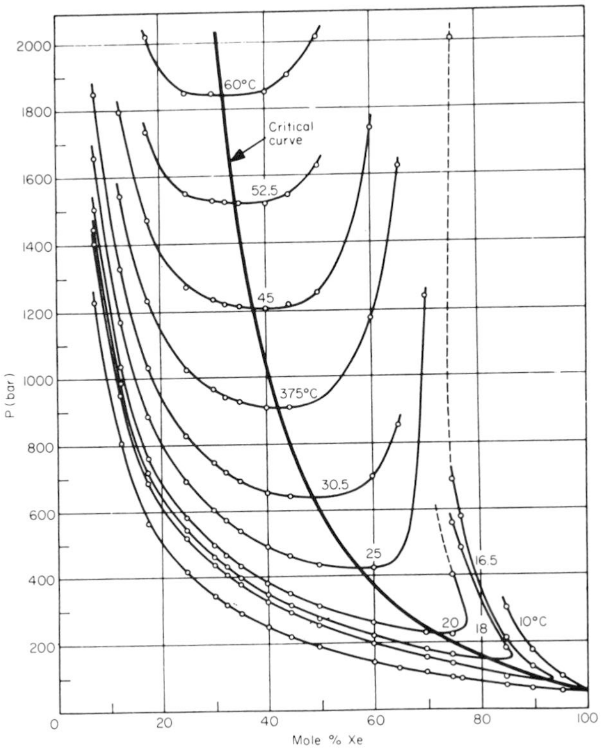

FIG. 18. Isotherms which are the boundaries of the two-phase region in a pressure composition diagram for the system He–Xe from measurements of De Swaan Aarons and Diepen (1966).

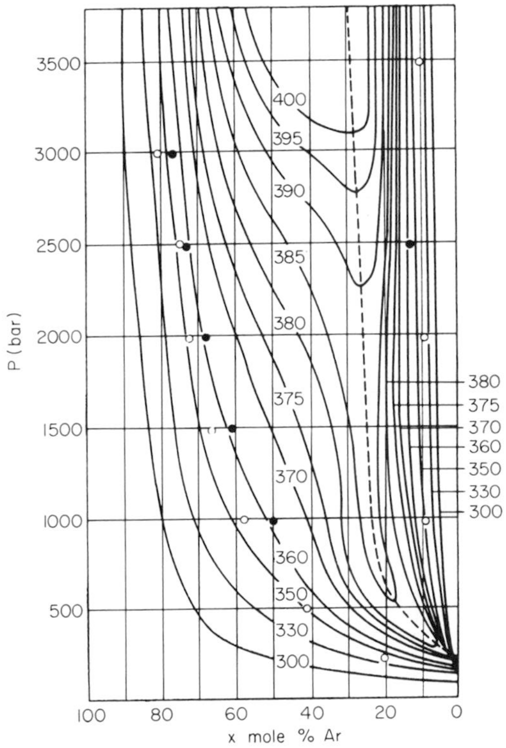

FIG. 19. Isotherms which are the boundaries of the two-phase region in a pressure-composition diagram for the system Ar–H_2O from measurements of Lentz and Franck (1969). For 350 and 360°C experimental points from Tsiklis and Prochorov (1966) are indicated; (-- – –): critical curve.

and pressures of 2 or 3 kbar complete miscibility of CO_2 and H_2O exists in fluid phases of liquidlike density.

Figures 18 and 19 give two examples of type b behavior (see Fig. 15). Isothermal cross sections and the upper branch of the critical curve for the He–Xe system (De Swaan Arons and Diepen, 1966) are shown in Fig. 18. There is no minimum in the critical curve and all isotherms have a pressure minimum. Another example is the Ar–H_2O system (Lentz and Franck, 1969), Fig. 19. A smaller number of experimental points obtained earlier by Tsiklis and Prochorov (1966) seemed to indicate divided isotherms and a minimum temperature in the critical curve. The more extensive recent investigation proved that this is not the case, however, and that the Ar–H_2O system belongs to type b. It can be presumed, that H_2O–N_2 and similar systems have phase diagrams analogous to Ar–H_2O.

A compilation of *PT*-projections of critical curves is shown in Fig. 20. The curve for the NH_3–H_2O system is of the uninterrupted type. To the right of each of these curves is the region of complete miscibility. The curves H_2O–benzene (Danneil *et al.*, 1967) and H_2O–ethane (Alwani and Schneider, 1969) show, that even hydrocarbons and water can form dense fluid mixtures at all compositions beyond 300 or 400°C. The

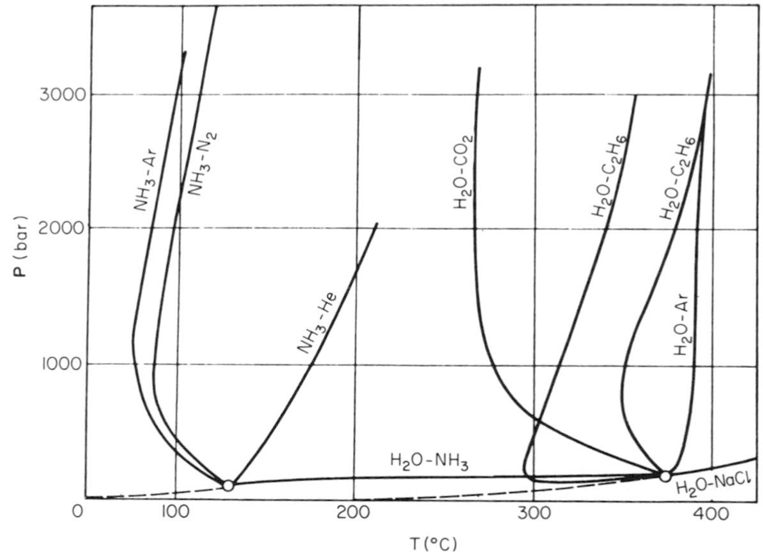

Fig. 20. Several experimentally determinal critical curves of the types a and b (see Fig. 15). Only the H_2O–NH_3-curve is uninterrupted.

H_2O–benzene curve is typical also for a number of other combinations of water with aromatic hydrocarbons (Schneider, 1970). The He–Ar system has recently been found of the type a group by Street (1969) who gives an interesting survey of six critical curves found for binary He-containing systems (Fig. 21). When proceeding from He–H_2 to He–Xe, i.e., to greater size and polarizability of the second partner the curves shift from type a to type b.

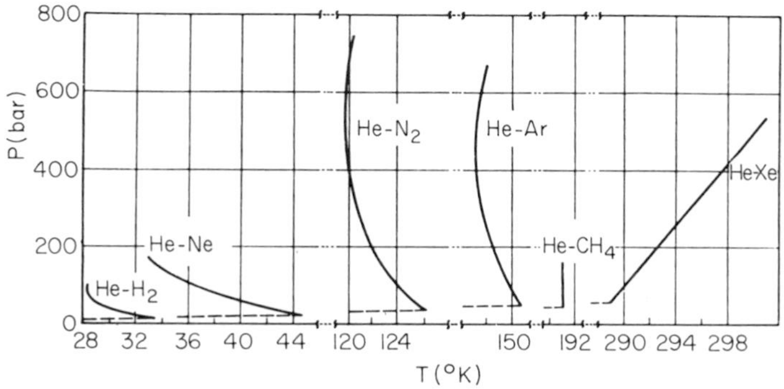

FIG. 21. *PT*-projections of critical curves of binary systems containing He. After Street (1969).

It has been pointed out above, that the gas-phase immiscibility can be explained as a combination of liquid–liquid immiscibility with the liquid–gas phase separation. This has been discussed in detail (Rowlinson, 1969) and Fig. 22 gives a schematic diagram from the extensive investigation of Schneider (1970). In each of the six cases Fig. 22a–f, it is assumed, that the liquid–liquid upper critical solution temperature increases with pressure. It merges in Fig. 22b,d, and f with liquid–gas critical curves of different shapes. For five of these six types of diagrams have examples been found experimentally which have hydrocarbons combined with CO_2 or H_2O. The LLG-curves are projections of the liquid–liquid–gas three phase surface on the *PT*-phase. Figure 23 gives a whole set of critical curves determined by Schneider *et al.* (1967) for combinations of carbon dioxide with *n*-alkanes from methane to *n*-hexadecane. This last compound shows the critical curve of type b, Fig. 22. If the mutual solubility in the liquid phase becomes even smaller, the pressure maximum in the critical curves of Fig. 22b and Fig. 23 will disappear and the critical curve will approach the dotted line in Fig. 22b, which is the type of several curves in Fig. 20.

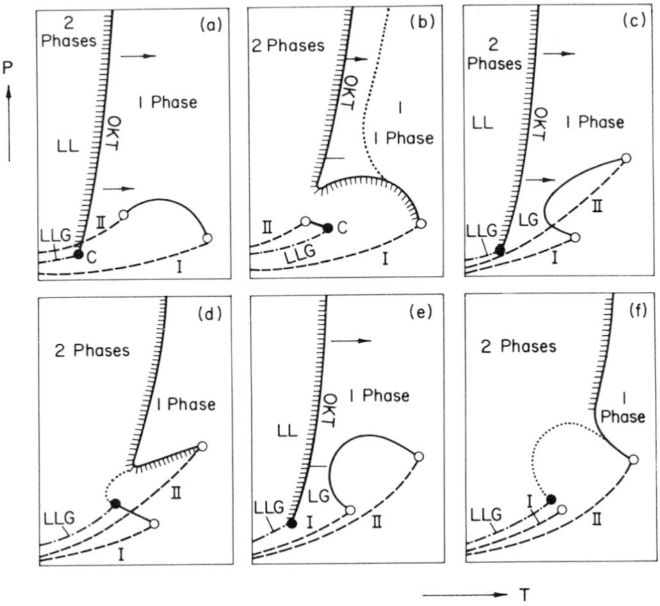

FIG. 22. *PT*-projections of several types of phase diagrams of binary mixtures of hydrocarbons combined with either CO_2 or H_2O after Schneider (1970). I: Hydrocarbon. II: CO_2 or H_2O. Examples: type a, *n*-octane–CO_2; type b, *n*-hexadecane–CO_2; type c, naphtalene–H_2O; type d, benzene–H_2O; type e, no hydrocarbon–H_2O system known; type f, ethane–H_2O.

The criteria for critical points in systems with gas–gas equilibria are analogous to those for other fluid–fluid equilibria. If the Gibbs free energy G is plotted as a function of the mole fraction x_1, the conditions for the critical $G(x)$ isotherm at the critical point are

$$(\partial^2 G/\partial x^2)_c = 0, \tag{5.1}$$

$$(\partial^3 G/\partial x^3)_c = 0, \tag{5.2}$$

$$(\partial^4 G/\partial x^4)_c = 0. \tag{5.3}$$

From Eqs. (5.1)–(5.3) and the condition

$$(\partial^2 G/\partial x^2)_{T,P} \geq 0 \tag{5.4}$$

for the stability or metastability of the homogeneous one phase state, additional relations for the curvature of the curves of the isothermal and isobaric enthalpy $H(x)$ and volume $V(x)$ at the critical point can be

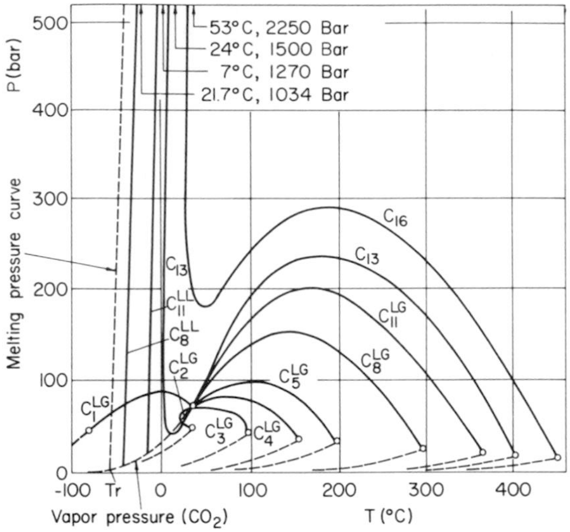

FIG. 23. Critical curves for several binary systems, which are combinations of carbon dioxide with *n*-alkanes from methane to *n*-hexadecane; LL: liquid–liquid critical immiscibility curve; LG: liquid–gas critical immiscibility curve.

derived. For the branches of the critical curves which belong to gas–gas phase equilibria the conditions

$$(\partial^2 H/\partial x^2)_c < 0 \tag{5.5}$$

and

$$(\partial^2 V/\partial x^2)_c < 0 \tag{5.6}$$

should hold. The last Eq. (5.6) could be verified with experimentally determined *PVT*-data for Ar–H_2O (Lentz and Franck, 1969) and benzene–H_2O (Alwani and Schneider, 1969). Eq. (5.5) could not yet been confirmed experimentally.

Several approaches have been made to predict the tendency of a system for gas–gas phase separation. Kaplan (1968) calculated the Hildebrand solubility parameter D from the critical data of each substance

$$D^2 = 3RT_c/2V_c \tag{5.7}$$

and used the difference ΔD^2 of the two components as a criterion for the phase separation. By comparison of ΔD^2 within series of analogous systems Kaplan (1968) could predict gas–gas immiscibility for a number of not yet investigated binary combinations.

Several authors have used the van der Waals equation and have applied suitable relations for concentration dependence and combination rules of the two constants a and b of this equation. Thus Temkin (1959) found as conditions for gas–gas phase separation

$$b_{22} \geq 0.42b_{11} \tag{5.8}$$

$$a_{22} < 0.053a_{11}. \tag{5.9}$$

These are the van der Waals constants for the pure substances. The less volatile component has the indices 22. The criteria proved to be relatively successful, for example, for He containing systems, (see also Alwani and Schneider, 1969). A very thorough and extensive theoretical investigation has been made by Van Konynenburg and Scott (1968), to find by model calculations what types of critical curves are produced by systematic variations of the van der Waals constants for the binary mixtures.

Based on the principle of corresponding states and the Lennard–Jones intermolecular potential, Zandbergen *et al.* (1967) have calculated the free energy G of mixtures as a function of P, T, and x. The procedure has been critically discussed by Prausnitz (1969). It gives good qualitative results for the He–Xe-system.

The discussion of gas–gas phase separation is at the present time mainly descriptive. The basic thermodynamic relevance and the technical applicability of this phenomenon, however, will stimulate continuing effort in this field. More experimental data on excess values of thermodynamic functions and spectrophotometric and other molecular information from the dense gase phase at supercritical conditions would be very desirable.

References

Alwani, Z., and Schneider, G. M. (1969). *Ber. Bunsenges. Phys. Chem.* **73**, 294.

Barnes, H. L., and Czamanske, G. K. (1967). Solubility and transport of ore materials, *in* "Geochemistry and Hydrothermal Deposits" (H. L. Barnes, ed.). Holt, New York.

Braune, H., and Strassmann, F. (1929). *Z. Phys. Chem.* **A143**, 225.

Burnham, C. W., Holloway, J. R., Davis, N. F. (1969). *Amer. J. Sci.* **297-A**, 70.

Chueh, P. L., and Prausnitz, J. M. (1967). *A. J. Ch. E. J.* **13**, 896.

Cobble, J. W. (1966). *Science* **152**, 1479.

Copeland, C. S., Silverman, J., and Benson, S. W. (1953). *J. Amer. Phys.* **21**, 12.

Curtiss, C. F. (1967). "A Treatise in Physical Chemistry," Academic Press, New York.

Danneil, A., Tödheide, K. and Franck, E. U. (1967). *Chem. Ing. Tech.* **39**, 816.

De Swaan Arons, J., and Diepen, G. A. M. (1966). *J. Chem. Phys.* **44**, 2322.
Diepen, G. A. M., and Scheffer, F. E. C. (1948). *J. Amer. Chem. Soc.* **70**, 4085.
Dunn, L. A., and Marshall, W. L. (1969). *J. Phys. Chem.* **73**, 723.
Ellis, A. J. (1967). The Chemistry of some explored geothermal systems, *in* "Geochemistry and Hydrothermal Deposits" (H. L. Barnes, ed.). Holt, New York.
Eucken, A., and Bressler, F. (1928). *Z. Phys. Chem.* **A134**, 230.
Ewald, A. H. (1955). *Trans. Faraday Soc.* **51**, 347.
Ewald, A. H., Jepson, W. B. and Rowlinson, J. S. (1953). *Discuss. Faraday Soc.* **15**, 238.
Franck, E. U. (1956). *Z. Phys. Chem.* (*Frankfurt am Main*) **6**, 345.
Franck, E. U. (1956a). *Z. Phys. Chem.* (*Frankfurt am Main*) **8**, 92.
Franck, E. U. (1956b). *Z. Phys. Chem.* (*Frankfurt am Main*) **8**, 107.
Franck, E. U. (1956c). *Z. Phys. Chem.* (*Frankfurt am Main*) **8**, 192.
Franck, E. U. (1961). *Angew. Chem.* **73**, 309.
Glemser, O., and Wendlandt, H. G. (1963). "Advances in Inorganic Chemistry and Radiochemistry" (H. J. Eméleus and A. G. Sharpe, eds.), Vol. 5. Academic Press, New York.
Glemser, O., and Wendlandt, H. G. (1964). *Ang. Chem.* (*Int. Ed.*) **3**, 47.
Glemser, O., Stöcker, U., and Wendlandt, H. G. (1966). *Ber. Bunsenges. Phys. Chem.* **70**, 1129.
Helgeson, H. C. (1964). "Complexing and Hydrothermal Ore Deposition," Pergamon Press, Oxford.
Helgeson, H. C. (1969). *Amer. J. Sci.* **267**, 729.
Hinckley, R. B., and Reid, R. C. (1964). *A. J. Ch. E. J.* **10**, 416.
Hirschfelder, J. O., Curtiss, C. F., and Bird, R. B. (1964). "The Molecular Theory of Gases and Liquids," 2nd ed. Wiley, New York.
Holland, H. D. (1967). Gangue minerals in hydrothermal deposits *in* "Geochemistry of Hydrothermal Ore Deposits" (H. L. Barnes, ed.). Holt, New York.
Jasmund, K. (1952). *Heidelberg. Beitr. Mineral. Petrogr.* **3**, 380.
Kamerlingh-Onnes, H., and Keesom, W. H. (1907). *Proc. Roy. Acad. Sci. Amst.* **9**, 786; **10**, 231.
Kaplan, R. (1968). *Amer. Inst. Chem. Eng. J.* **14**, 821.
Keevil, N. B. (1942). *J. Amer. Chem. Soc.* **64**, 841.
Kennedy, G. C. (1950). *Econ. Geol.* **45**, 629.
Kennedy, G. C., Wasserburg, G. J., Heard, H. C., and Newton, R. C. (1962). *Amer. J. Sci.* **260**, 501.
Khaibullin, J. K., and Borisow, N. M. (1965). *Russ. J. Phys. Chem.* **39**, 361.
Köster, H., and Franck, E. U. (1969). *Ber. Bunsenges. Phys. Chem.* **73**, 716.
Krichevski, J. R., and Bolshakow, P. (1941). *Acta Phys. Chem. USSR*, **14**, 353.
Krichevski, J. R., and Korolewa (1941). *Acta Physicochim. USSR* **15**, 327.
Lemmlein, G. G., and Klevtsov, P. V. (1961). *Geochimia* No. 2, 133; *Geochemistry USSR* No. 2, 148.
Lentz, H., and Franck, E. U. (1969). *Ber. Bunsenges. Phys. Chem.* **73**, 28.
Lindroos, A. E., and Dodge, B. F. (1952). *Chem. Eng. Progr. Symp. Ser.* **48**, 10.
Lüdemann, H.-D. (1968). *Ber. Bunsenges. Phys. Chem.* **72**, 514.
Lüdemann, H.-D., and Franck, E. U. (1967). *Ber. Bunsenges. Phys. Chem.* **71**, 455.
Maier, S., and Franck, E. U. (1966). *Ber. Bunsenges. Phys. Chem.* **70**, 639.
Mangold, K., and Franck, E. U. (1969). *Ber. Bunsenges. Phys. Chem.* **73**, 21.

MARSHALL, W. L. (1968). *Rev. Pure Appl. Chem.* **18**, 167.
MARSHALL, W. L. (1969). Correlations in aqueous electrolyte behavior to high temperatures and pressures *in* The Record of Chemical Progress, June 1969 issue.
MOREY, G. W. (1957). *Econ. Geol.* **52**, 225.
MOREY, G. W., and CHEN, W. T. (1956). *J. Amer. Chem. Soc.* **78**, 4249.
PRAUSNITZ, J. M. (1969). "Molecular Thermodynamics of Fluid Phase Equilibria." Prentice Hall, Englewood Cliffs, New Jersey.
QUIST, A. S., and MARSHALL, W. L. (1968). *J. Phys. Chem.* **72**, 684.
QUIST, A. S., FRANCK, E. U., JOLLEY, H. R., and MARSHALL, W. L. (1963). *J. Phys. Chem.* **67**, 2453.
RABENAU, A., RAU, H., ROSENSTEIN, G. (1969), *Angew. Chemie* **81**, 148.
REUSS, J., and BEENAKKER, J. J. M. (1956). *Physica* **22**, 869.
RITZERT, G., and FRANCK, E. U. (1968). *Ber. Bunsenges. Phys. Chem.* **72**, 798.
ROBIN, S., and VODAR B. (1953). *Discuss. Faraday Soc.* **15**, 233.
RODNYANSKI, J. M., and GALINKER, J. S. (1955). *C. R. Acad. Sci. Russ.* **105**, 115.
ROEDDER, E. (1967). Fluid inclusions as samples of ore fluids, *in* "Geochemistry and Hydrothermal Deposits" (H. L. Barnes, ed.). Holt, New York.
ROWLINSON, J. S. (1969). "Liquids and Liquid Mixtures," 2nd. ed., Butterworth, London and Washington, D.C.
SCHNEIDER, G. M. (1966). *Ber. Bunsenges. Phys. Chem.* **70**, 497.
SCHNEIDER, G. M. (1970). Gas-Gas-Gleichgewichte, *in* "Topics in Current Chemistry," Vol. 13, p. 559. Springer, New York.
SCHNEIDER, G. M., and BARTMANN, D. (1969). *Ber. Bunsenges. Phys. Chem.* **73**, 917.
SCHNEIDER, G. M., ALWANI, Z., HEIM, W., HORVATH, E., and FRANCK, E. U. (1967). *Chem. Ing. Tech.* **39**, 649.
SOURIRAJAN, S., and KENNEDY, G. C. (1962). *Amer. J. Sci.* **260**, 115.
STREET, W. B. (1969). *Trans. Faraday Soc.* **65**, 696.
TEMKIN, M. J. (1959). *Russ. J. Phys. Chem.* **33**, 275.
TÖDHEIDE, K. (1966). *Ber. Bunsenges. Phys. Chem.* **70**, 1022.
TÖDHEIDE, K. (1970). *Naturwissenschaften* **57**, 72.
TÖDHEIDE, K., and FRANCK, E. U. (1963). *Z. Phys. Chem. Frankfurt* **37**, 387.
TSIKLIS, D. S. (1952). *Dokl. Acad. Nauk SSSR* **86**, 993, 1159.
TSIKLIS, D. S. (1968). "Handbook of Techniques in High Pressure Research and Engineering." Plenum Press, New York.
TSIKLIS, D. S., and PROCHOROW, W. M. (1966). *J. Phys. Chem.* **40**, 2335.
TSIKLIS, D. S., and ROTT, L. A. (1967). *Russ. Chem. Rev.* **36**, 351.
VAN DER WAALS, J. D. (1894). Zittinsversl. Kon. Acad. v. Wetensch. Amsterdam 133.
VAN KONYNENBURG, P., and SCOTT, R. L. (1968). Ph. D. Thesis. Chemistry Dept., Univ. of California, Los Angeles.
WEBSTER, T. J. (1952). *Proc. Roy. Soc.* **A 214**, 61.
ZANDBERGEN, P., KNAAP, H. F. P., and BEENAKKER, J. J. M. (1967). *Physica* **33**, 379.

Chapter 5

Thermodynamics of Matter in Gravitational, Electric, and Magnetic Fields

HERBERT STENSCHKE

I. Thermodynamics of Matter in Gravitational Fields and in Rotational Frames of Reference

A. MATTER IN EXTERNAL FIELDS

We first consider matter in states of equilibrium placed in an external field $\Phi(\mathbf{r})$. By "external field," we mean that the field is not modified by the presence of the matter on which the fields are acting. The gravitational field is the foremost example of this type of field, whereas the electric and magnetic fields are, in most cases, determined not only by their external sources, but also by the distribution of matter in space.

1. *Phases in Gravitational fields*

A physical system is called a phase if all its volume elements are described by the same thermodynamic potential; that is, such that all physical variables depend in the same way on each other. In a gravitational field, however, different parts of the phase are generaly subject to different conditions; thus, the field has the effect of making the phase spatially inhomogeneous. If the phase contains only one type of particle representing an independent variable (particle number), one may describe it by the chemical potential $\mu(T, p, \mathbf{r})$, where the vector $\mathbf{r}$ denotes the location of the volume elements.

2. *Equilibrium in Gravitational fields*

The energy density in the presence of a gravitational field $\Phi(\mathbf{r})$ is increased with respect to the field-free case by the amount

$$W_{\text{grav}} = m(\mathbf{r})\Phi(\mathbf{r}), \tag{1.1}$$

where $m(\mathbf{r})$ denotes the mass density. In the case of gravitation, the variable $m(\mathbf{r})$ occurring if the field is applied does not represent an independent degree of freedom, because $m(\mathbf{r})$ is, in all practical cases, proportional to the particle density $n(\mathbf{r})$:

$$m(\mathbf{r}) = m_0 n(\mathbf{r}), \tag{1.2}$$

where m_0 is the mass of one particle. Hence, a gravitational field has no other effect than to add the term $m_0\Phi(\mathbf{r})$ to the chemical potential:

$$\mu(T, p, \mathbf{r}) = \mu(T, p) + m_0\Phi(\mathbf{r}). \tag{1.3}$$

Equilibrium with respect to entropy and particle-number exchange of any two subsystems (situated at $\mathbf{r}$ and $\mathbf{r}'$, respectively) implies that temperature and chemical potential are constant over the whole system:

$$T(\mathbf{r}) = T(\mathbf{r}') \tag{1.4}$$

$$\mu(T, p, \mathbf{r}) = \mu(T, p, \mathbf{r}'). \tag{1.5}$$

Since μ depends explicitly on $\mathbf{r}$, the equilibrium condition (1.5) leads to a dependence of the pressure on the space coordinates and the temperature according to

$$\mu(T, p(T, \mathbf{r})) + m_0 \Phi(\mathbf{r}) = \mu(T, p(T, \mathbf{r}')) + m_0 \Phi(\mathbf{r}') = \text{const.} \quad (1.6)$$

The entropy density, which is a function of temperature and pressure, then becomes, in general, a function of $\mathbf{r}$ also.

3. *Pressure and Entropy Distribution of a Perfect Gas and an Incompressible Fluid*

Taking the derivative of the equilibrium condition with respect to $\mathbf{r}$, one obtains

$$v(\mathbf{r})\, dp = m_0\, d\Phi(\mathbf{r}). \quad (1.7)$$

[Here, we have used $\partial\mu(T, p, \mathbf{r})/\partial p = v(\mathbf{r})$.]

If we apply Eq. (1.7) to a perfect gas ($v = kT/p$), we get for the pressure distribution in the field

$$p(\mathbf{r}) = p(\mathbf{r}') \exp\{-(m_0/kT)[\Phi(\mathbf{r}') - \Phi(\mathbf{r})]\}. \quad (1.8)$$

In the case of an incompressible fluid ($v = \text{const}$), one obtains, correspondingly,

$$p(\mathbf{r}) - p(\mathbf{r}') = (m_0/v)[\Phi(\mathbf{r}') - \Phi(\mathbf{r})]. \quad (1.9)$$

The derivative of Eq. (1.6) with respect to the temperature yields the following expression for the entropy per particle $-\partial\mu(T, p, \mathbf{r})/\partial T$:

$$s(T, \mathbf{r}) - s(T, \mathbf{r}') = v(\mathbf{r}')\, \partial p(T, \mathbf{r}')/\partial T - v(\mathbf{r})\, \partial p(T, \mathbf{r})/\partial T, \quad (1.10)$$

which, for a perfect gas, reads

$$s(T, \mathbf{r}) - s(T, \mathbf{r}') = (m_0/T)[\Phi(\mathbf{r}') - \Phi(\mathbf{r})]. \quad (1.11)$$

Due to Eq. (1.11), the equilibrium specific heat also becomes a function of $\mathbf{r}$:

$$C(T, \mathbf{r}) - C(T, \mathbf{r}') = -(m_0/T)[\Phi(\mathbf{r}') - \Phi(\mathbf{r})]. \quad (1.12)$$

For an incompressible fluid, the pressure does not depend explicitly on the temperature, so that, in the equilibrium state, the entropy and the specific heat are constant in space.

4. *Mixture of Perfect Gases in a Gravitational Field*

The pressure of a mixture of perfect gases is the sum of the partial pressures of the components,

$$p = p_1 + p_2. \tag{1.13}$$

In a gravitational field, p_1 and p_2, and, consequently, the total pressure p, become functions of the space coordinates according to the condition of equilibrium for each component:

$$\begin{aligned} \mu_1(T, p_1(T, \mathbf{r})) + m_{01}\Phi(\mathbf{r}) &= \mu_1(T, p_1(T, \mathbf{r}')) + m_{01}\Phi(\mathbf{r}') \\ \mu_2(T, p_2(T, \mathbf{r})) + m_{02}\Phi(\mathbf{r}) &= \mu_2(T, p_2(T, \mathbf{r}')) + m_{02}\Phi(\mathbf{r}'). \end{aligned} \tag{1.14}$$

Since both substances are assumed to be perfect gases, we can use Eq. (1.8) to obtain the ratio of the partial pressures:

$$\frac{p_1(\mathbf{r})}{p_2(\mathbf{r})} = \frac{p_1(\mathbf{r}')}{p_2(\mathbf{r}')} \exp\left\{- \frac{(m_{01} - m_{02})[\Phi(\mathbf{r}) - \Phi(\mathbf{r}')]}{kT}\right\}. \tag{1.15}$$

In consequence of Eq. (1.15), the ratio of the concentrations depends in the same way on the absolute difference in the atomic masses.

5. *Chemical Reactions*

A chemical reaction is described by the equation

$$\sum_{i=1}^{n} \nu_i B_i = 0, \tag{1.16}$$

where the B_i represent the components of the reaction and the ν_i are the stoichiometric coefficients. The equilibrium composition $N_1(T, p)$, $N_2(T, p)$, ..., $N_n(T, p)$ is determined by the equation

$$\sum_{i}^{n} \nu_i \mu_i(T, p, N_1(T, p), N_2(T, p), \ldots, N_n(T, p)) = 0 \tag{1.17}$$

and the conservation of the total particle number

$$\sum_{i=1}^{n} \nu_i N_i(T, p) = 0. \tag{1.18}$$

If the reaction takes place in a gravitational field, the chemical potential

μ_i is related to the chemical potential μ_i^0 in the absence of the field by

$$\mu_i(T, p, N_1, N_2, \ldots, N_n, \mathbf{r}) = \mu_i^0(T, p, N_1, N_2, \ldots, N_n) + m_i \Phi(\mathbf{r}). \quad (1.19)$$

With Eq. (1.19), the equilibrium condition (1.17) reads

$$\sum_{i=1}^{n} \nu_i \mu_i^0(T, p, N_1(T, p), N_2(T, p), \ldots, N_n(T, p)) + \Phi(\mathbf{r}) \sum_{i=1}^{n} \nu_i m_i = 0. \quad (1.20)$$

The last term in Eq. (1.20) vanishes because of the conservation of the total mass. Hence, the equilibrium condition is the same as in the field-free case. It should be remarked, however, that the equilibrium composition $N_1, N_2, \ldots, N_n$ itself becomes spatially inhomogeneous, since the pressure p varies in space according to Eq. (1.5).

B. Thermodynamics of Rotating Systems

A uniform translation with nonrelativistic velocity $\mathbf{v}$ does not influence the internal thermodynamic state of a physical system. Its Gibbs free energy as function of the velocity is then given by the "Lagrange function"

$$G(T, p, \mathbf{v}) = G(T, p) - \tfrac{1}{2} M \mathbf{v}^2, \quad (1.21)$$

where M denotes the total mass of the system, which in the nonrelativistic limit does not depend on T and p. Therefore, a translation only adds a constant energy to the thermodynamic potential. Hence, it can be omitted.

The situation is different if we consider phases which rotate with constant angular velocity $\boldsymbol{\omega}$. For, only in the case in which the phase is a rigid body is the Gibbs potential affected by a constant rotational energy $(-\tfrac{1}{2}\theta\boldsymbol{\omega}^2)$; in general, the rotational energy must be described by means of the centrifugal field

$$W(\mathbf{r}) = -\tfrac{1}{2} m (\boldsymbol{\omega} \times \mathbf{r})^2.$$

In addition, phases displaying quantum effects, such as He II and superconductors, are essentially influenced by the Coriolis vector potential

$$\mathbf{A}(\boldsymbol{\omega}) = \boldsymbol{\omega} \times \mathbf{r},$$

which has no effect on the equilibrium states of dissipative phases.

In order to demonstrate this state of affairs, it is convenient to start with the Schrödinger equation of N particles; i.e., with the Hamiltonian

$$H = (1/2m) \sum_{i=1}^{N} p_i^2 + V(\mathbf{r}_1, \ldots, \mathbf{r}_N) + U(\mathbf{r}_1, \ldots, \mathbf{r}_N). \qquad (1.22)$$

The first and second terms of H, representing the kinetic energy and the interatomic potentials, respectively, are invariant under rotation. The term $U(\mathbf{r}_1, \ldots, \mathbf{r}_N)$ denotes an external potential which keeps the particles inside a cylinder. If this vessel rotates with angular velocity $\boldsymbol{\omega}$, the potential transforms as follows:

$$U(\mathbf{r}_1, \ldots, \mathbf{r}_N; \boldsymbol{\omega}) = \left\{\exp\left[-i\frac{\mathbf{L}\cdot\boldsymbol{\omega}}{\hbar}t\right]\right\} U(\mathbf{r}_1, \ldots, \mathbf{r}_N) \exp\left[-i\frac{\mathbf{L}\cdot\boldsymbol{\omega}}{\hbar}t\right] \qquad (1.23)$$

where $\mathbf{L}$ is the angular momentum. According to Eqs. (1.22)–(1.23), the Schrödinger equation of the rotating system reads

$$\left\{\frac{1}{2m}\sum_i p_i^2 + V(\mathbf{r}_1, \ldots, \mathbf{r}_N) + \left[\exp\left(-i\frac{\mathbf{L}\cdot\boldsymbol{\omega}}{\hbar}t\right)\right]\right.$$
$$\left.\times\; U(\mathbf{r}_1, \ldots, \mathbf{r}_N) \exp\left(+i\frac{\mathbf{L}\cdot\boldsymbol{\omega}}{\hbar}t\right)\right\}\psi = ih\frac{\partial\psi}{\partial t}. \qquad (1.24)$$

Stationary solutions of Eq. (1.24) can only be obtained in the rotating frame of reference. There, the wave function

$$\psi' = \psi \exp[i(\mathbf{L}\cdot\boldsymbol{\omega}/\hbar)t] \qquad (1.25)$$

obeys the equation

$$H'\psi' = i\hbar\, \partial\psi'/\partial t, \qquad (1.26)$$

where

$$H' = H - \boldsymbol{\omega}\cdot\mathbf{L}, \qquad (1.27)$$

obviously does not depend on time explicitly. In terms of the Coriolis potentials

$$A_i(\boldsymbol{\omega}) = \boldsymbol{\omega} \times \mathbf{r}_i \qquad (1.28)$$

and the centrifugal potential

$$W(\mathbf{r}_1, \ldots, \mathbf{r}_N) = -\sum_{i=1}^{N} \tfrac{1}{2}m(\boldsymbol{\omega}\times\mathbf{r}_i)^2, \qquad (1.29)$$

the transformed Hamiltonian H' takes the form

$$H' = \frac{1}{2m} \sum_{i=1}^{N} (\mathbf{P}_i^{\omega})^2 + V(\mathbf{r}_1, \ldots, \mathbf{r}_N) + U(\mathbf{r}_1, \ldots, \mathbf{r}_N) + W(\mathbf{r}_1, \ldots, \mathbf{r}_N) \tag{1.30}$$

where we have introduced the canonical momenta

$$\mathbf{P}_i^{\omega} = \mathbf{p}_i - mA_i(\boldsymbol{\omega}). \tag{1.31}$$

Contrary to ordinary momenta, the momenta $\mathbf{P}_i^{\omega}$ do not commute. This has different consequences for quantum systems as compared with the classical limit. Turning first to the latter, one realizes that, in order to calculate the thermodynamic potential of a rotating system, one has to insert into the partition function the energy spectrum of $H + W$ instead of the spectrum of H. One then obtains a Gibbs density

$$\mu_{\omega}(T, p) = \mu(T, p) + \langle W(\mathbf{r}_1, \ldots, \mathbf{r}_N) \rangle, \tag{1.32}$$

where the thermodynamic average $\langle W \rangle$ is a function of the space coordinates

$$W(\mathbf{r}) = \langle W(\mathbf{r}_1, \ldots, \mathbf{r}_N) \rangle = -\tfrac{1}{2} m(\boldsymbol{\omega} \times \mathbf{r})^2. \tag{1.33}$$

Hence, in the classical limit, a system rotating with angular velocity $\boldsymbol{\omega}$ feels only the centrifugal effect of the rotation, and, thus, is equivalent to a system in a gravitational field $\Phi(\mathbf{r}) = W(\mathbf{r})$.

For quantum systems, on the other hand—we are thinking of systems which are able to carry nondissipative currents—the fact that the momenta $\mathbf{P}_i^{\omega}$ do not commute becomes important. This is plausible, considering the fact that quantum mechanics here is essentially equivalent to subjecting the canonical momenta to Sommerfeld's quantization rule:

$$\oint \mathbf{p}_i \, d\mathbf{r}_i = n_i h, \qquad n_i = 0, 1, 2, \ldots.$$

In case of the existence of nondissipative currents, these conditions again imply a quantization condition for the circulation of the macroscopic current density, which, in turn, classifies the superfluid flow according to quantum numbers of the circulation. Hence, it may happen, and, in fact, it does, that the states of lowest energy of a rotating superfluid such as He II contain a lattice of vertices, each quantized in units of h/m, as predicted by Feynman (1955). The centrifugal effect in He II is negligible compared with the effect of forming vertices. In rotating superconductors, the formation of a lattice of vertices occurs only at angular velocities larger than a certain value ($\omega_{c1} = 10^7 H_{c1}$ G^{-1} sec^{-1}).

II. Static Electric and Magnetic Fields

1. *Electric and Magnetic Variables*

The electromagnetic properties of substances are described by four vector fields; the electric field $\mathbf{E}(\mathbf{r}, t)$, the electric induction field $\mathbf{D}(\mathbf{r}, t)$, the magnetic field $\mathbf{H}(\mathbf{r}, t)$, and the magnetic induction field $\mathbf{B}(\mathbf{r}, t)$. Maxwell's equations relate these fields to the charge density $\varrho(\mathbf{r}, t)$ and the current density $\mathbf{j}(\mathbf{r}, t)$. We confine ourself to the *static* case; i.e., all charges and all currents are constant in time. In this case, Maxwell's equations separate into two independent sets, so that electrostatics and magnetostatics can be considered distinct phenomena.

Electrostatics:

$$\operatorname{curl} \mathbf{E}(\mathbf{r}) = 0 \tag{2.1}$$

$$\operatorname{div} \mathbf{D}(\mathbf{r}) = 4\pi\varrho(\mathbf{r}). \tag{2.2}$$

Magnetostatics:

$$\operatorname{curl} \mathbf{H}(\mathbf{r}) = 4\pi/c\mathbf{j}(\mathbf{r}) \tag{2.3}$$

$$\operatorname{div} \mathbf{B}(\mathbf{r}) = 0. \tag{2.4}$$

In addition, there is, in each case, another equation, called a constitutive relation, connecting **E** and **D**, and **H** and **B**, respectively, characterizing the properties of the matter under consideration.

The fact that we confine ourselves to the static case is not only due to the simplification of the mathematics, but also has an important thermodynamic basis. In the case of fields varying in time, the electric energy is always converted into magnetic energy and vice versa. This process, however, gives rise to dissipation of energy, which again affects the minimum and maximum principles of the thermodynamic potentials. Strictly speaking, the methods of equilibrium thermodynamics are only sufficient to describe matter in static fields, and, if currents are present, they have to be nondissipative.

The description of the behavior of matter in electric and magnetic fields is more complicated than a description of systems in gravitational fields because gravitational fields as considered in Section I constitute "external" fields in the sense that they are determined by external sources only and are not modified by the matter present. In electric and magnetic fields, matter, due to its being polarized and magnetized, is an additional source of the field which modifies the externally applied field. Furthermore, this modification depends on the shape of the specimen situated in the field.

For the description of systems in fields, it is most convenient to choose as variables not the local fields $\mathbf{E}(\mathbf{r})$, $\mathbf{H}(\mathbf{r})$, $\mathbf{D}(\mathbf{r})$ and $\mathbf{B}(\mathbf{r})$, but the fields $\mathbf{E}_e$ and $\mathbf{H}_e$ (assumed to be homogeneous in space) before the specimen was introduced and the total electric moment $|\mathbf{P}$ as well as the total magnetic moment $|\mathbf{M}$ of the body.

2. *The Energy of Dielectric and Magnetic Materials in Given External Fields*

In terms of the fields **E** and **D**, the electric energy of a dielectric specimen is given by

$$W_{el} = (1/4\pi) \int d^3\mathbf{r} \int_0^{\mathbf{D}(\mathbf{r})} \mathbf{E}(\mathbf{r}, \mathbf{D}')\, d\mathbf{D}', \tag{2.5}$$

where the first integration extends over the whole space, including the part which is not occupied by the dielectric body. The induction field **D** is given by

$$\mathbf{D}(\mathbf{r}) = \mathbf{E}(\mathbf{r}) + 4\pi\mathbf{P}(\mathbf{r}), \tag{2.6}$$

where $\mathbf{P}(\mathbf{r})$ is the dielectric polarization (the density of the dipole moment). If we introduce the "external" field $\mathbf{E}_e$ which is the field generated by the external sources if no dielectric specimen is present, Eq. (2.5) takes the form

$$W_{el} = (1/8\pi) \int \mathbf{E}_e^2(\mathbf{r})\, d^3\mathbf{r} + \int d^3\mathbf{r} \int \mathbf{E}_e(\mathbf{r})\, d\mathbf{P}', \tag{2.7}$$

which has been derived by Heine (1956). In case of a homogeneous field $\mathbf{E}_e(\mathbf{r}) = \mathbf{E}_e$, formula (2.7) reduces to

$$W_{el} = (1/8\pi)\mathbf{E}_e^2 V + \int \mathbf{E}_e\, d|\mathbf{P}, \tag{2.8}$$

where

$$|\mathbf{P} = \int d^3r\mathbf{P}(\mathbf{r}) \tag{2.9}$$

is the total dielectric dipole moment of the specimen. The first term in (2.8) represents the field energy of the empty space, so that the matter is responsible for the second one. From (2.8), one infers that, in a homogeneous external field $\mathbf{E}_e$ the differential electric energy of a dielectric system by change of the dipole moment $|\mathbf{P}$ can be written

$$dW_{el} = \mathbf{E}_e\, d|\mathbf{P}. \tag{2.10}$$

The analogous expressions in case of a magnetic specimen in an external magnetic field $\mathbf{H}_e(\mathbf{r})$ are given by

$$W_{\text{mag}} = (1/4\pi) \int d^3\mathbf{r} \int_0^{\mathbf{B}(\mathbf{r})} \mathbf{H}(\mathbf{r})\, d\mathbf{B}'$$
$$= (1/8\pi) \int d^3\mathbf{r}\, \mathbf{H}_e^2(\mathbf{r}) + \int d^3\mathbf{r}\, \mathbf{H}_e(\mathbf{r})\, d|\mathbf{M}, \quad (2.11)$$

and, in case $\mathbf{H}_e$ is homogeneous,

$$dW_{\text{el}} = \mathbf{H}_e\, d|\mathbf{M}, \qquad |\mathbf{M} = \int d^3r \mathbf{M}(\mathbf{r}), \quad (2.12)$$

where $|\mathbf{M}$ is the total magnetic dipole moment.

3. *The Dipole Moment of a Dielectric Ellipsoid in a Homogeneous Electric Field*

The problem of calculating by electrostatics, i.e., by Eqs. (2.1)–(2.2), the polarization of a specimen in an applied electric field is, in general, very complicated. Only in the special case that the material has the shape of an ellipsoid and is described by a constant permeability ε so that

$$D(\mathbf{r}) = \varepsilon E(\mathbf{r}) \quad (2.13)$$

can the solution be given in analytic form. If the charge density is zero everywhere, the electric field can be written as gradient of a scalar function $\phi(\mathbf{r})$ which fulfills the differential equation

$$\nabla \cdot [\varepsilon \nabla \phi(\mathbf{r})] = 0 \quad (2.14)$$

and the following boundary conditions at the surface:

$$\phi_{\text{body}} = \phi_{\text{vacuum}}, \qquad \varepsilon\left(\frac{\partial \phi}{\partial \mathbf{n}}\right)_{\text{body}} = \left(\frac{\partial \phi}{\partial \mathbf{n}}\right)_{\text{vacuum}}, \quad (2.15)$$

where $(\partial\phi/\partial\mathbf{n})$ denotes the derivative of ϕ, normal to the interface.

Besides the boundary conditions (2.15), the solution of Eq. (2.14) must only obey the asymptotic condition

$$\phi(\mathbf{r}) \xrightarrow[|\mathbf{r}|\to\infty]{} \mathbf{E}_e \cdot \mathbf{r}, \quad (2.16)$$

where $\mathbf{E}_e$ is the applied electric field, which is assumed to be homogeneous. Inside of an ellipsoid with principal axes a, b, and c, the field $\mathbf{E}_i$ is then also homogeneous and given by (see, for example, Stratton (1941))

$$E_{ix} = \frac{E_{ex}}{1 + \frac{1}{2}abc(\varepsilon - 1)A_x}, \quad A_x = \int_0^\infty \frac{ds}{(s+a^2)[(s+a^2)(s+b^2)(s+c^2)]^{1/2}} \tag{2.17}$$

$$E_{iy} = \frac{E_{ey}}{1 + \frac{1}{2}abc(\varepsilon - 1)A_y}, \quad A_y = \int_0^\infty \frac{ds}{(s+b^2)[(s+a^2)(s+b^2)(s+c^2)]^{1/2}} \tag{2.18}$$

$$E_{iz} = \frac{E_{ez}}{1 + \frac{1}{2}abc(\varepsilon - 1)A_z}, \quad A_z = \int_0^\infty \frac{ds}{(s+c^2)[(s+a^2)(s+b^2)(s+c^2)]^{1/2}}. \tag{2.19}$$

The field $\mathbf{E}_i$ is parallel to the applied field if $\mathbf{E}_e$ has the direction of one of the principal axes of the ellipsoid.

Using the definitions of the dielectric polarization (2.6) and the electric dipole moment (2.9), one obtains a linear relation between $|\mathbf{P}$ and the applied field $\mathbf{E}_e$

$$|\mathbf{P} = \boldsymbol{\alpha} \cdot \mathbf{E}_e. \tag{2.20}$$

The susceptibility $\boldsymbol{\alpha}$ then is a tensor:

$$\boldsymbol{\alpha} = \begin{pmatrix} \frac{(\varepsilon - 1)V}{1+[3(\varepsilon-1)V/8]A_x} & 0 & 0 \\ 0 & \frac{(\varepsilon - 1)V}{1+[3(\varepsilon-1)V/8]A_y} & 0 \\ 0 & 0 & \frac{(\varepsilon - 1)V}{1+[3(\varepsilon-1)V/8]A_z} \end{pmatrix}. \tag{2.21}$$

where V denotes the volume of the ellipsoid.

A relation analogous to Eq. (2.20) also holds between magnetic dipole moment $|\mathbf{M}$ and applied magnetic field $\mathbf{H}_e$ for a ellipsoidal magnetic specimen.

III. The Thermodynamic Potentials of Dielectric and Magnetic Substances

The thermodynamic treatment of dielectric systems is based on the assumption that each system is characterized by a single-valued function

$$Y = Y(x_1, \ldots, x_n; |\mathbf{P}), \tag{3.1}$$

the thermodynamic potential of the system. The function defines the states of the system for given values of the nonelectric variables $x_1, \ldots, x_n$ and the electric dipole moment $|\mathbf{P}$. The variables $x_1, \ldots, x_n$, and $|\mathbf{P}$ characterize all forms of energy the system is able to exchange with other systems according to the relation

$$dY = \sum_i \xi_i \, dx_i + \boldsymbol{\xi}_{|\mathbf{P}} \cdot d\,|\mathbf{P}. \tag{3.2}$$

The term $\boldsymbol{\xi}_{|\mathbf{P}} \cdot d|\mathbf{P}$ is the electric work. Comparison with formula (2.10) shows that $\boldsymbol{\xi}_{|\mathbf{P}}$ is the applied electric field $\mathbf{E}_e$.

The variables $\boldsymbol{\xi}_i$, which are called conjugates of x_i with respect to Y are, in general, functions of $x_1, \ldots, x_n, |\mathbf{P}$:

$$\xi_i = \partial Y(x_1, \ldots, x_n; |\mathbf{P})/\partial x_i = \xi_i(x_1, \ldots, x_n; |\mathbf{P}). \tag{3.3}$$

The electric field $\mathbf{E}_e$ also depends generally on all coordinates

$$\mathbf{E}_e = \partial Y(x_1, \ldots, x_n; |\mathbf{P})/\partial |\mathbf{P} = \mathbf{E}_e(x_1, \ldots, x_n; |\,\mathbf{P}). \tag{3.4}$$

From Eqs. (3.3)–(3.4), there follows trivially

$$\partial \xi_i(x_1, \ldots, x_n; |\mathbf{P})/\partial x_j = \partial \xi_j(x_1, \ldots, x_n; |\mathbf{P})/\partial x_i \tag{3.5}$$

and

$$\partial \xi_i(x_1, \ldots, x_n; |\mathbf{P})/\partial |\mathbf{P} = \partial \mathbf{E}_e(x_1, \ldots, x_n; |\mathbf{P})/\partial x_i. \tag{3.6}$$

Because of Eqs. (3.5)–(3.6), all $\xi_i(x_1, \ldots, x_n; |\mathbf{P})$ can be calculated from the state equation

$$\mathbf{E}_e = \mathbf{E}_e(x_1, \ldots, x_n; |\mathbf{P})$$

if the ξ_i are known for vanishing dipole moment:

$$\xi_i(x_1, \ldots, x_n; |\mathbf{P}) = \xi_i(x_1, \ldots, x_n; 0) + \frac{\partial}{\partial x_i} \int_0^{|\mathbf{P}} \mathbf{E}_e(x_1, \ldots, x_n; |\mathbf{P}')\, d|\mathbf{P}'. \tag{3.7}$$

In case the matter can be described by a susceptibility $\chi(x_1, \ldots, x_n)$,

$$\mathbf{E}_e = [1/\chi(x_1, \ldots, x_n)]\,|\mathbf{P}, \tag{3.8}$$

one gets

$$\xi_i(x_1, \ldots, x_n; |\mathbf{P}) - \xi_i(x_1, \ldots, x_n; 0) = -\frac{\partial \chi(x_1, \ldots, x_n)/\partial x_i}{\chi^2(x_1, \ldots, x_n)}\,|\mathbf{P}^2. \tag{3.9}$$

If one chooses instead of the variables $x_1, \ldots, x_{k-1}, x_k, x_{k+1}, \ldots, x_n, |\mathbf{P}$ the set $x_1, \ldots, x_{k-1}, \xi_k, x_{k+1}, \ldots, x_n, |\mathbf{P}$ as independent variables (ξ_k is the conjugate to x_k), one obtains the thermodynamic potential $Y^{[x_k]}$ pertaining to the new variables by Legendre transforming Y, i.e., by the formula

$$\begin{aligned} &Y^{[x_k]}(x_1, \ldots, x_{k-1}, \xi_k, x_{k+1}, \ldots, x_n; |\mathbf{P}) \\ &= Y(x_1, \ldots, x_{k-1}, x_k, x_{k+1}, \ldots, x_n; |\mathbf{P}) \\ &- x_k \frac{\partial}{\partial x_k} Y(x_1, \ldots, x_{k-1}, x_k, x_{k+1}, \ldots, x_n; |\mathbf{P}), \end{aligned} \tag{3.10}$$

where, on the right-hand side, x_k is to be replaced by

$$x_k = x_k(x_1, \ldots, x_{k-1}, \xi_k, x_{k+1}, \ldots, x_n; |\mathbf{P}).$$

Accordingly, if one replaces $|\mathbf{P}$ by its conjugate, i.e., by the applied field $\mathbf{E}_e$, Y has to be replaced by

$$Y(x_1, \ldots, x_n; \mathbf{E}_e) = Y - |\mathbf{P} \frac{\partial Y}{\partial |\mathbf{P}} = Y(x_1, \ldots, x_n; |\mathbf{P}) - |\mathbf{P} \cdot \mathbf{E}_e, \tag{3.11}$$

where $|\mathbf{P} = |\mathbf{P}(x_1, \ldots, x_n; \mathbf{E}_e)$.

The conjugate of $\mathbf{E}_e$ with respect to $Y(x_1, \ldots, x_n; \mathbf{E}_e)$ is the negative dipole moment

$$\boldsymbol{\xi}_{\mathbf{E}_e} = \partial Y(x_1, \ldots, x_n; \mathbf{E})/\partial \mathbf{E}_e = -|\mathbf{P}(x_1, \ldots, x_n; \mathbf{E}_e). \tag{3.12}$$

The relations analogous to formulae (3.5)–(3.6) then read

$$\partial \xi_i(x_1, \ldots, x_n; \mathbf{E}_e)/\partial x_j = \partial \xi_j(x_1, \ldots, x_n; \mathbf{E}_e)/\partial x_i \tag{3.13}$$

$$\partial \xi_i(x_1, \ldots, x_n; \mathbf{E}_e)/\partial \mathbf{E}_e = \partial\, |\mathbf{P}(x_1, \ldots, x_n; \mathbf{E}_e)/\partial x_i. \tag{3.14}$$

The above considerations are immediately applicable to the magnetic case if one replaces the electric field $\mathbf{E}_e$ by the external magnetic field $\mathbf{H}_e$ and the electric dipole moment $|\mathbf{P}$ by the magnetic dipole moment $|\mathbf{M}$.

IV. Applications

This section deals with the dependence of thermal properties of simple systems on external electric and magnetic fields. We are particularly interested in equilibrium states which can be found as extrema of thermodynamic functions. This excludes energy-dissipating processes, such as thermal conductivity and others. Moreover, the applied fields have to be constant in time.

A. Electrostriction and Magnetostriction

Choosing as independent variables the temperature T, the pressure p, the particle number N, and the external electric field $\mathbf{E}_e$, one of the relations (3.14) reads

$$\partial V(T, p, N, \mathbf{E}_e)/\partial \mathbf{E}_e = -\partial |\mathbf{P}(T, p, N, \mathbf{E}_e)/\partial p. \tag{4.1}$$

This equation states that a change in the external field is accompanied by a change in volume V if the total dipole moment depends on pressure. This effect is called electrostriction. In case the dielectric is described by a susceptibility $\chi_e(T, p, N)$, i.e.,

$$|\mathbf{P} = \chi_e(T, p, N)\mathbf{E}_e, \tag{4.2}$$

the change in volume is proportional to E_e^2:

$$\Delta V = V(\mathbf{E}_e) - V(0) = -\frac{1}{2}\,\frac{\partial \chi_e(T, p, N)}{\partial p}\,\mathbf{E}_e^2. \tag{4.3}$$

Magnetic substances exhibit in magnetic fields an analogous effect which is called magnetostriction. In terms of a magnetic susceptibility $\chi_m(T, p, N)$, the change in volume is, accordingly,

$$\Delta V = V(\mathbf{H}_e) - V(0) = -\frac{1}{2}\,\frac{\partial \chi_m(T, p, N)}{\partial p}\,\mathbf{H}_e^2. \tag{4.4}$$

B. Compressibility

If the susceptibility of a dielectric substance depends on pressure, the compressibility $\varkappa = (1/v)\,\partial V/\partial P$ is affected by an applied electric field. It is

$$\varkappa(E_e) - \varkappa(0) = -\frac{1}{2}\,\frac{\mathbf{E}_e^2}{V}\,\frac{\partial^2 \chi_e(T, p, N)}{\partial p^2}. \tag{4.5}$$

Specific Heat at Constant Magnetic Field and Constant Magnetic Moment

The entropy in the presence of a magnetic field is given by

$$S(T, p, N, \mathbf{H}_e) = S(T, p, N, 0) + \frac{\partial}{\partial T}\int_0^{\mathbf{H}_e} |\mathbf{M}(T, p, N, \mathbf{H}')\,d\mathbf{H}' \tag{4.6}$$

and the specific heat, accordingly, by

$$C_{\mathbf{H}_e} = C_{\mathbf{H}_e=0} + T\frac{\partial^2}{\partial T^2}\int_0^{\mathbf{H}_e} |\mathbf{M}(T, p, N, \mathbf{H}')\, d\mathbf{H}' \tag{4.7}$$

which, in case

$$|\mathbf{M}(T, p, N, \mathbf{H}_e) = \chi_m(T_e\, p, N)\mathbf{H}_e, \tag{4.8}$$

simplifies to

$$C_{\mathbf{H}_e} = C_{\mathbf{H}_e=0} + \frac{T\mathbf{H}_e^2}{2}\,\frac{\partial^2\chi_m(T, p, N)}{\partial T^2}. \tag{4.9}$$

Similar considerations lead to the specific heat at constant magnetic moment:

$$C_{|\mathbf{M}} = C_{|\mathbf{M}=0} - T\frac{\partial^2}{\partial T^2}\int_0^{|\mathbf{M}} \mathbf{H}_e(T, p, N, |\mathbf{M}')\, d|\mathbf{M}' \tag{4.10}$$

and, if Eq. (4.8) holds, then

$$C_{|\mathbf{M}} = C_{|\mathbf{M}=0} - \frac{T\,|\mathbf{M}^2}{2}\,\frac{\partial^2\chi_m^{-1}(T, p, N)}{\partial T^2}. \tag{4.11}$$

The difference $C_{|\mathbf{M}} - C_{\mathbf{H}_e}$ is calculated by means of the following identities

$$\frac{\partial S(T, \mathbf{H}_e)}{\partial T} = \frac{\partial S(T, |\mathbf{M})}{\partial T} + \frac{\partial S(T, |\mathbf{M})}{\partial T}\,\frac{\partial |\mathbf{M}(\mathbf{H}_e, T)}{\partial T}$$

$$\frac{\partial S(T, |\mathbf{M})}{\partial |\mathbf{M}} = \frac{\partial \mathbf{H}_e(T, |\mathbf{M})}{\partial T}$$

$$C_{|\mathbf{M}} - C_{\mathbf{H}_e} = T\frac{\partial \mathbf{H}_e(T, |\mathbf{M})}{\partial T}\,\frac{\partial |\mathbf{M}(\mathbf{H}_e, T)}{\partial T}. \tag{4.12}$$

By use of Jacobian identities, Eq. (4.12) can be brought into the form

$$C_{\mathbf{M}} - C_{\mathbf{H}_e} = -T\left[\frac{\partial |\mathbf{M}(T, \mathbf{H}_e)}{\partial T}\right]^2 \Big/ \frac{\partial \mathbf{M}(T, \mathbf{H}_e)}{\partial \mathbf{H}_e}, \tag{4.13}$$

and, if Eq. (4.8) holds, then

$$C_{|\mathbf{M}} - C_{\mathbf{H}_e} = -\frac{T\mathbf{H}_e^2}{\chi}\left[\frac{\partial\chi(T, P, N)}{\partial T}\right]^2, \tag{4.14}$$

a result which can be obtained directly from Eqs. (4.9) and (4.11).

C. The Magnetocaloric Effect

An adiabatic change of the external field and of the magnetic moment of a magnetic substance is accompanied by a change in its temperature. By use of Jacobian determinants,

$$\frac{\partial T(\mathbf{H}_e, S)}{\partial \mathbf{H}_e} = \frac{\partial(T, S)}{\partial(\mathbf{H}_e, S)} = \frac{\partial(T, S)}{\partial(\mathbf{H}_e, T)} \frac{\partial(\mathbf{H}_e, T)}{\partial(\mathbf{H}_e, S)}$$

$$= -\frac{\partial S(T, \mathbf{H}_e)}{\partial \mathbf{H}_e} \frac{T}{C_\mathbf{H}} = -\frac{\partial \,|\, \mathbf{M}(T, \mathbf{H}_e)}{\partial T} \frac{T}{C_{\mathbf{H}_e}}, \quad (4.15)$$

and, in the same way,

$$\frac{\partial T(|\, \mathbf{M}, S)}{\partial \,|\, \mathbf{M}} = -\frac{\partial \mathbf{H}_e(T, |\, \mathbf{M})}{\partial T} \frac{T}{C_{|\mathbf{M}}}.$$

Hence, magnetic materials with positive $\partial|\mathbf{M}/\partial T$ or $\partial\mathbf{H}_e/\partial T$ can be used to lower temperatures by adiabatically removing the external field or the magnetic moment. We shall consider this cooling method in Section V,C.

D. Chemical Reactions in Magnetic and Electric Fields

In Section I,A, we found that, due to the conservation of total mass, the equilibrium condition of a chemical reaction in a gravitational field is of the same form as in the field-free case. Since there exists no analogous conservation of the magnetic and electric moments, we expect that the equilibrium condition of a chemical reaction depends generally on applied magnetic and electric fields.

The magnetic moment of most substances is proportional to the particle number

$$|\mathbf{M}(T, p, N, \mathbf{H}_e) = N\mathbf{m}(T, p, \mathbf{H}_e). \quad (4.16)$$

Therefore, the chemical potential of a substance in the presence of an external magnetic field is given by

$$\mu(T, p, N, \mathbf{H}_e) = \mu(T, p, N, 0) - \int_0^{\mathbf{H}_e} \mathbf{m}(T, p, \mathbf{H}_e)\, d\mathbf{H}_e. \quad (4.17)$$

Let us consider a chemical reaction of the type

$$\nu\mathrm{A} \rightleftharpoons \mathrm{B} \quad (4.18)$$

and assume that substance A (for example, atomic hydrogen) is paramagnetic; substance B (molecular hydrogen) is, however, magnetically inert. If the reaction (4.18) then takes place in a magnetic field, the equilibrium condition reads

$$\nu\mu_A(T, p, N, \mathbf{H}_e) = \mu_B(T, p, N, \mathbf{H}_e).$$

Using Eq. (4.17), one obtains

$$\nu\mu_A(T, p, N, 0) - \nu \int_0^{\mathbf{H}_e} \mathbf{m}_A(T, p, \mathbf{H}')\, d\mathbf{H}' = \mu_B(T, p, N, 0). \quad (4.19)$$

Since, in our example, $\mathbf{m}(T, p, \mathbf{H}_e)$ is a positive quantity, the magnetic field shifts the equilibrium such that substance A is favored. Assuming that both substances can be considered to be perfect gases, the equilibrium of the reaction is described by an equilibrium constant $K(T, p)$, so that the molar concentrations

$$x_A = N_A/(N_A + N_B) \qquad \text{and} \qquad x_B = N_B/(N_A + N_B)$$

obey the mass action law

$$x_A{}^\nu x_B = K(T, p) = \exp\{-(1/kT)[\nu\mu_A(T, p) - \mu_B(T, p)]\}. \quad (4.20)$$

By means of Eq. (4.19), we obtain

$$x_A{}^\nu(\mathbf{H}_e)x_B(\mathbf{H}_e) = x_A{}^\nu(0)x_B(0) \exp\left[-(\nu/kT) \int_0^{\mathbf{H}_e} \mathbf{m}(T, p, \mathbf{H}')\, d\mathbf{H}'\right]. \quad (4.21)$$

This relation states that the equilibrium is influenced considerably by an applied field in favor of substance A if the magnetic energy per atom is of the order of the thermal energy, i.e., if

$$\int_0^{\mathbf{H}_e} \mathbf{m}(T, p, \mathbf{H}')\, d\mathbf{H}' \approx kT. \quad (4.22)$$

This remark is thus only of theoretical significance, since the magnetic fields necessary to produce a detectable effect would be prohibitive. Considering now a chemical reaction involving three dielectric substances A, B, and C,

$$\nu_A A + \nu_B B \rightleftharpoons \nu_C C, \quad (4.23)$$

the mass action law in the presence of an applied field reads

$$x_A^{\nu_A}(\mathbf{E}_e)x_B^{\nu_B}(\mathbf{E}_e)x_C^{\nu_C}(\mathbf{E}_e) = x_A^{\nu_A}(0)x_B^{\nu_B}(0)x_C^{\nu_C}(0) \exp(-\Delta W_{el}/kT), \quad (4.24)$$

where

$$x_A = \frac{N_A}{N_A+N_B+N_C}, \qquad x_B = \frac{N_B}{N_A+N_B+N_C}, \qquad x_C = \frac{N_C}{N_A+N_B+N_C}$$

and

$$\Delta W_{el} = \int_0^{\mathbf{E}_e} [\nu_A \mathbf{p}_A(T, p, \mathbf{E}') + \nu_B \mathbf{p}_B(T, p, \mathbf{E}') - \nu_C \mathbf{p}_C(T, p, \mathbf{E}')] \, d\mathbf{E}'.$$

The quantities $\mathbf{p}_A$, $\mathbf{p}_B$, and $\mathbf{p}_C$ denote the dipole moments per particle of substances A, B, and C, respectively. Thus, in equilibrium, electric fields favor substances with positive dipole moments.

V. The Magnetic Properties of Matter

A. Classification of Magnetic Materials

We distinguish between ferromagnetic and nonferromagnetic materials. In the latter case, the magnetic moment depends linearly on an applied weak field

$$|\mathbf{M} = \chi \mathbf{H}_e, \qquad |\mathbf{M} = 0 \quad \text{if} \quad \mathbf{H}_e = 0. \tag{5.1}$$

If χ is positive (negative) the material is called paramagnetic (diamagnetic).

Ferromagnetics, antiferromagnetics, and ferrimagnetics, forming the other class, exhibit the pecularity of spontaneous magnetization, i.e., their magnetic moment may be different from zero even if there is no applied field. Whereas the paramagnetic behavior can be described by superposing the effects of practically independent electrons, atoms, or groups of atoms, ferromagnetism is a typical example of a cooperative phenomenon.

B. Diamagnetic Materials

The properties of diamagnetic materials can frequently be obtained by the linear response of noninteracting electrons (freely moving in a metal or localized near an atom) to an applied field. The negative sign of the susceptibility is due to induced currents, the magnetic field of which tries to weaken the external field. At first glance, the negative sign of the diamagnetic susceptibility seems to contradict the general conditions of thermodynamic stability. According to these conditions, all second derivatives of a thermodynamic potential must have a definite sign. In particular, the magnetic susceptibility should be positive, so that one is tempted to conclude that diamagnetic systems are unstable. This conclusion, however, is premature, since one has to keep in mind that, according to Eq. (2.11), a magnetic system is always coupled with the magnetic field in the vacuum. Hence, the thermodynamic potential of the total magnetic system is given by

$$Y_{\text{total}}(x_1, \ldots, x_n; \mathbf{H}_e) = Y_{\text{matter}}(x_1, \ldots, x_n; \mathbf{H}_e) + (V/8\pi)\mathbf{H}_e^2. \quad (5.2)$$

The last term in (5.2) does not play any role in processes in which one of the variables $x_1, \ldots, x_n$ is changed (thermal process), but this term becomes important in case the applied field is changed. The stability condition upon Y_{total} then gives as the range of the magnetic susceptibility

$$\frac{\partial^2 Y_{\text{total}}(x_1, \ldots, x_n; \mathbf{H}_e)}{\partial \mathbf{H}_e^2} > 0: \qquad -\frac{V}{4\pi} < \chi < \infty. \quad (5.3)$$

Hence, the diamagnetic region

$$-V/4\pi < \chi < 0 \quad (5.4)$$

is stabilized by the vacuum.

An extreme form of diamagnetism is the Meissner–Ochsenfeld effect in superconductors, where $\chi = -V/4\pi$. We turn to this peculiarity in Section V,E, where the magnetic properties of superconductors are discussed.

C. Paramagnetic Materials

The sign of the paramagnetic susceptibility is positive because the magnetic properties of these materials result from permanent magnetic moments of the molecules or of the freely moving electrons. In case of perfect paramagnetism, these moments are independently lined up by an applied magnetic field. Thermal motion prevents a complete alignment, so that the paramagnetic susceptibility is strongly temperature-dependent. In order to calculate the free energy of a perfect paramagnetic, let us consider a system of N independent elementary magnetic moments occupying states with the orientational energy

$$\varepsilon_m = g\mu_B m H_e, \qquad m = -j, -j+1, \ldots, j-1, j, \quad (5.5a)$$

where μ_B is the Bohr magneton,

$$\mu_B = 0.9273 \times 10^{-20} \quad \text{erg G}^{-1}, \tag{5.5b}$$

and g is the gyromagnetic ratio. From the excitation spectrum, we obtain the sum of states $Z(T, H_e)$ and the free energy $F(T, N, H_e)$:

$$Z(T, H_e) = Z(H_e/T) = \sum_{m=-j}^{m=j} \exp(-\varepsilon_m/kT) = \sum_{m=-j}^{+j} \exp(-m\mu_B g H_e/kT) \tag{5.6}$$

and

$$F(T, N, H_e) = -NkT \log Z(H_e/T). \tag{5.7}$$

The magnetic moment $|M$ is then given by

$$|M(T, N, H_e) = -\frac{\partial F(T, N, H_e)}{\partial H_e} = \frac{Nk}{Z(H_e/T)} \frac{\partial Z(H_e/T)}{\partial(H_e/T)}. \tag{5.8}$$

We observe that the magnetic moment of perfect paramagnetics is a function of H_e/T only. Equivalent to this statement is that the energy $U(T, N, |M)$ does not depend on the magnetic moment:

$$\partial U(T, N, |M)/\partial |M = 0. \tag{5.9}$$

Hence, changing the magnetic moment does not affect the isothermal energy, but only influences the entropy of the system. In the special case $j = \frac{1}{2}$, the sum of states, the free energy, and the magnetic moment are given by

$$Z_{1/2}(H_e/T) = 2 \cosh(g\mu_B H_e/2kT), \tag{5.10}$$

$$F_{1/2}(H_e/T, N) = -NkT \log[2 \cosh(g\mu_B H_e/2kT)], \tag{5.11}$$

$$|M_{1/2}(H_e/T, N) = \tfrac{1}{2} N g\mu_B \tanh(g\mu_B H_e/2kT).$$

In the case of classical elementary dipole moments, we have to consider the limit

$$g\mu_B \to 0, \quad j \to \infty \quad \text{while} \quad \alpha = g\mu_B j = \text{const};$$

we then obtain Langevin's result

$$F_\infty(H_e/T, N) = -NkT \log[(kT/\alpha H_e) \sinh(\alpha H_e/kT)], \tag{5.12}$$

$$|M_\infty(H_e/T, N) = N\alpha L(\alpha H_e/kT), \tag{5.13}$$

where

$$L(x) = \coth x - \frac{1}{x}$$

is called the Langevin function. In the limit $\alpha H \ll kT$, Eq. (5.13) reduces to Curie's law

$$| M(H_e/T, N) = (N\alpha^2/3kT)H_e \tag{5.14}$$

which neglects the effect of saturation:

$$| M(H_e/T, N) \leq N\alpha. \tag{5.15}$$

Cooling by Adiabatic Demagnetization

Besides the adiabatic separation of ^3He–^4He mixtures, the magnetocaloric effect of paramagnetic substances is used for cooling purposes below 0.5°K. With this method, temperatures of 10^{-3}°K and below have been achieved. To discuss this method, we consider a perfect paramagnetic system in thermal equilibrium with the phonon system of a condensed body. The total entropy is then given by

$$S(T, H_e) = S_0(T) + S_{\text{para}}(T, H_e), \tag{5.16}$$

$$S_0(T) = AT^3, \tag{5.17}$$

where A is a constant. From the free energy (5.7), we obtain the entropy of the spin system:

$$\begin{aligned} S_{\text{para}}(T, H_e) &= -\frac{\partial F}{\partial T} \\ &= Nk\left\{\log Z\left(\frac{H_e}{T}\right) - \frac{H_e}{T}\left[\frac{\partial Z(H_e/T)}{\partial(H_e/T)} \Big/ Z\left(\frac{H_e}{T}\right)\right]\right\}. \end{aligned} \tag{5.18}$$

The entropy of perfect paramagnetic material depends only on H_e/T, just as its magnetic moment does. In order to keep the entropy (5.16) of the adiabatically isolated system constant, its temperature must decrease if the magnetic field is turned off. This method of cooling is most effective if the paramagnetic entropy S_{para} dominates S_0. In this approximation, the condition of constant entropy implies

$$H_e/T = \text{const}, \qquad T_f = (H_{e_f}/H_{e_i})T_i. \tag{5.19}$$

The indices i and f denote the initial and the final states of the process of adiabatic demagnetization, respectively. The method is limited by the fact that H_{ef} cannot be smaller than a certain value determined by the interaction between magnetic moments, which causes the paramagnetic material to deviate from its perfect behavior. This interaction can be taken into account by introducing an effective internal magnetic field h which constitutes a lower limit of H_{ef} and thus a limitation of the magnetic method of cooling. This internal field is of the order of 100 G for electronic magnetic moments and 1 G for nuclear magnetic moments. For this reason, the lowest temperatures (10^{-6}°K) have been reached by nuclear adiabatic demagnetization.

D. Ferromagnetics near the Curie Point

The phenomenon of a spontaneous magnetization can only be understood by taking the interaction between elementary magnetic moments into account. Pauli's principle leads generally to a difference in the energy between electrons of parallel or antiparallel spins. Consequently, an interaction between magnetic moments $\mathbf{m}_j$ and $\mathbf{m}_k$ results which is of the form

$$V = J\mathbf{m}_j \cdot \mathbf{m}_k. \tag{5.20}$$

The coupling is either ferromagnetic or antiferromagnetic according to the sign of the exchange integral J. In both cases, below a critical temperature

$$T_{\mathrm{c}} \approx J/k, \tag{5.21}$$

the magnetic system becomes ordered: below T_{c}, a spontaneous magnetization occurs. The transition to the ordered phase shows the characteristics of a phase transition of second order; that is, the transition is not accompanied by latent heat, and consequently does not allow processes of superheating and undercooling. The specific heat exhibits a jump and possibly a singularity at T_{c}.

An interaction of the form (5.20) between magnetic moments constitutes a severe complication of the quantum-statistical problem of summing the partition function. The thermodynamic potential can be found in closed form only for simplified models such as the Ising model. Instead of following this approach from first principles, we take advantage of the fact that the ferromagnetic transition is a phase transition of second order. Thus, following Landau (1958), we shall formulate a phenomenological theory of ferromagnetism near the Curie point which is based essentially on thermodynamic stability considerations, as shown by Falk (1968) and Stenschke and Falk (1968). Thermodynamic functions have to be single-valued, so that, in the following, we deal only with ferromagnetics consisting of a single Weiss domain.

1. *Thermodynamics of the Phase Transition*

The temperature dependence of the spontaneous magnetization in the neighborhood of T_c is of the following form:

$$M_s = \begin{cases} m \mid \tau \mid^{\beta}, & \tau \leq 0 \\ 0 & \tau < 0, \end{cases} \qquad \tau = (T - T_c)/T_c. \tag{5.22}$$

The exponent β is, as many experiments show, of the order of 1/3. Above T_c, where the spontaneous magnetization vanishes, the system exhibits a paramagnetic behavior. At T_c the magnetic susceptibility diverges according to

$$\chi = \left[\frac{\partial H(\tau, M)}{M}\right]^{-1}_{M=M_s} = \left[\frac{\partial^2 F(\tau, M)}{\partial M^2}\right]^{-1} = \chi_0 \mid \tau \mid^{-\gamma}. \tag{5.23}$$

Experiments show that $\gamma \approx 4/3$. The divergence of the susceptibility means that, at $\tau = 0$, the second derivative of the thermodynamic potential with respect to the magnetization vanishes. Hence, the phase limit at $\tau = 0$ is an instability of the paramagnetic and the ferromagnetic phases. The phase transition is of second order if the two phases are not separated by a region of instability. That is, both phases must be stable up to and on the phase limit. Since the second derivative of the thermodynamic potential vanishes there, the potential has to have a minimum of higher order at T_c, so that the following stability conditions must hold:

$$\begin{aligned} &\left[\frac{\partial^{\nu} F(\tau, M)}{\partial M}\right]_{M=M_s, \tau=0} = 0, \qquad \nu = 2, 3, \ldots, 2n+1, \\ &\left[\frac{\partial^{2n+2} F(\tau, M)}{\partial^{2n+2} M}\right]_{M=M_s, \tau=0} > 0, \qquad n = 1, 2, 3, \ldots. \end{aligned} \tag{5.24}$$

If the relations (5.24) were not fulfilled, the phase limit $\tau = 0$ would not be a point of stability of the paramagnetic as well as of the ferromagnetic phase. It can, therefore, never be reached; hence, the phase transition has to occur before reaching the phase limit. But this would contradict our basic assumption that the ferromagnetic transition is a phase transition of second order.

2. *Landau Theory*

In Landau's theory of phase transitions of second order, the free energy density $f(\tau, M)$ is expanded in a series of the following form:

$$f_L(\tau, M) = f_0(\tau) + \sum_{\mu=1}^{n+1} \alpha_{2\mu}^{(n)}(\tau) M^{2\mu}. \tag{5.25}$$

In order that the stability conditions (5.24) be fulfilled, the following conditions result

$$\begin{aligned} \alpha_{2\mu}^{(n)}(0) &= 0, \qquad \mu = 1, \ldots, n, \\ \alpha_{2n+2}^{(n)}(0) &> 0. \end{aligned} \tag{5.26}$$

The equilibrium value of M is obtained by minimizing the free energy density (5.25),

$$\frac{\partial f_L(\tau, M)}{\partial M} = M_s \sum_{\mu=1}^{n+1} 2\mu \alpha_{2\mu}^{(n)}(\tau) M_s^{2\mu-2} = 0. \tag{5.27}$$

Inserting the equilibrium value $M_s(\tau)$ back into the expansion (5.25) and requiring that all terms contribute equally to the jump in the specific heat which is derived from Eq. (5.25), one obtains, in lowest order, the temperature dependence of the coefficients $\alpha_{2\mu}^{(n)}(\tau)$:

$$\alpha_{2\mu}^{(n)}(\tau) = \alpha_{2\mu 0}^{(n)} \tau^{2(n+1-\mu)/(n+1)}. \tag{5.28}$$

With these results, we are in a position to express the critical indices β and γ and the index δ (defined below) of the ferromagnetic transition in terms of the classification number n. Comparing with experiment, one obtains agreement for $n = 2$: The exponents β and γ as defined in Eqs. (5.22)–(5.23) are then given by

$$\beta = 1/3, \qquad \gamma = 4/3. \tag{5.29}$$

The index δ of the critical isotherm, defined by

$$H(\tau = 0) = \text{const}\, M^{\delta}, \tag{5.30}$$

turns out to be

$$\delta = 5. \tag{5.31}$$

we add to the free energy density (5.25) the surface energy

$$f_s = \lambda^2(\nabla M)^2, \tag{5.32}$$

where λ is a constant, the Landau theory allows us to calculate the magnetization as a function of the space coordinates. The surface tension (5.32), which measures the energy associated with spatial changes of the magnetization, defines a correlation length ξ which varies with temperature according to

$$\xi(\tau) = \lambda/(|\alpha_{20}^{(n)}|)^{1/2} = \xi_0 |\tau|^{-2/3}. \tag{5.33}$$

This length gives the order of the thickness of the Bloch walls.

3. *Ferromagnetics in an Applied Magnetic Field*

In the presence of an external field, the Landau free energy density is given by

$$f(\tau, H_e, M) = f_L(\tau, M) - H_e \cdot M. \tag{5.34}$$

The equilibrium magnetization is then determined by the equation

$$\partial f_L(\tau, M)/\partial M = H_e. \tag{5.35}$$

The equilibrium value M_s obeying Eq. (5.35) is everywhere nonzero. If one inserts M_s into the second derivative of Eq. (5.34), one observes that

$$\left[\frac{\partial^2 f(\tau, H_e, M)}{\partial M^2}\right]_{M=M_s} > 0$$

everywhere. Hence, the magnetic instability is removed, so that, strictly speaking there is no ferromagnetic phase transition in an applied field.

E. The Magnetic Properties of Superconductors

1. *Perfect Diamagnetism*

The magnetic characteristic of the superconducting phase of a metal is its perfect diamagnetism. Below a critical field H_{c1}, the magnetic permeability vanishes, so that the magnetic moment of a long superconducting cylinder in a longitudinal field $\mathbf{H}_e$ is given by

$$|\mathbf{M} = -(V/4\pi)\mathbf{H}_e, \qquad H_e \leq H_{c1}(T), \quad T < T_c, \tag{5.36}$$

with V the volume of the metal and T_c the transition temperature in vanishing field. Equation (5.36) states that a magnetic field which is less than H_{c1} cannot penetrate into the superconductor (Meissner–Ochsenfeld effect).

The thermodynamic potential $G_s(T, \mathbf{H}_e)$ of the superconducting phase of a metal in an external field $\mathbf{H}_e$ is connected with the potential $G_s(T, 0)$ in zero field by the formula

$$G_s(T, \mathbf{H}_e) = G_s(T, 0) - \int_0^{\mathbf{H}_e} |\mathbf{M}\, d\mathbf{H}_e' = G_s(T, 0) + (V/8\pi)\mathbf{H}_e^2. \quad (5.37)$$

The critical field $H_c(T)$ (in general, not identical with H_{c1}) is defined by the condition that

$$G_s(T, \mathbf{H}_e) = G_N(T, \mathbf{H}_c), \quad (5.38)$$

where G_N is the thermodynamic potential of the metal in the normal state.

The superconducting phase is energetically favored for $H_e \leq H_c(T)$, the normal phase for $H_e \geq H_c(T)$. Since the normal phase of a metal is magnetically inert ($\mu \approx 1$), Eq. (5.38) reads

$$G_s(T, 0) = G_N(T, 0) - (V/8\pi)H_c^2(T). \quad (5.39)$$

The lowering of the thermodynamic potential of the superconducting phase as compared with the potential of the normal phase is, according to Bardeen *et al.* (1957), explained by the formation of electron pairs, so-called Cooper pairs, below T_c. The interaction favoring these pair states is an electron-lattice-electron interaction, which at distances greater than the Debye screening length, represents the dominant interaction between the conduction electrons. Due to this interaction, the electrons "condense" into the Cooper-pair state, which, in turn, has the effect of lowering the thermodynamic potential of the superconducting phase by an amount proportional to the density of Cooper paris $n_s(T)$, so that

$$G_s(T) = G_N(T) - V\alpha(T)n_s(T). \quad (5.40)$$

In the vicinity of T_c, the condensation energy per Cooper pair has the temperature dependence

$$\alpha(T) = \alpha_0(T_c - T)/T_c, \qquad \alpha_0 > 0. \quad (5.41)$$

The Cooper-pair density itself varies near T_e according to

$$n_s(T) = \begin{cases} n_0(T_c - T)/T_c, & T < T_c, \\ 0, & T > T_c. \end{cases} \tag{5.42}$$

Taking Eqs. (5.40)–(5.42) into account, we have that the transition into the superconducting phase, which, in zero magnetic field, is a phase transition of second order, is accompanied by a jump in the specific heat.

2. *London Equations*

The magnetic properties of superconductors can be described by an equation which relates the current carried by the Cooper pairs to the vector potential $\mathbf{A}(\mathbf{r})$ [$\mathbf{H}(\mathbf{r}) = \text{curl}\, \mathbf{A}(\mathbf{r})$]. This equation has the form

$$\mathbf{J}_s = en_s(T)\mathbf{V}_s(\mathbf{r}, T) = \frac{e\hbar}{m_e}\left[\nabla\phi(\mathbf{r}) - \frac{2e}{\hbar c}\mathbf{A}(\mathbf{r})\right]n_s(T), \tag{5.43}$$

where m_e and e denote the electron's mass and charge. The vector field $\mathbf{V}_s(\mathbf{r}, T)$ represents the spatial distribution of the velocity of the Cooper pairs, which, in case of vanishing magnetic field, is an irrotational vector field:

$$\text{curl}\, \mathbf{V}_s(\mathbf{r}, T) = 0 \qquad \text{if} \quad \mathbf{H}(\mathbf{r}) = 0. \tag{5.44}$$

The scalar function $\phi(\mathbf{r})$ in Eq. (5.43) represents the potential of $\mathbf{V}_s(\mathbf{r}, T)$. Equation (5.44) suggests a description of the moving Cooper pairs as a fluid governed by a Navier–Stokes equation which, however, does not contain dissipation even if the fluid should have nonvanishing viscosity. In his famous paper on the superfluidity of liquid helium, Landau (1941) has shown that a quantum fluid fulfills the condition (5.44) and, moreover, that such a system can be described in terms of a macroscopic wave function

$$\psi = [n_s(T)]^{1/2} \exp[i\phi(\mathbf{r})]. \tag{5.45}$$

From Eq. (5.43) one obtains London's equations

$$\text{curl}(\Lambda\mathbf{j}_s) = -(1/c)\mathbf{H}(\mathbf{r}), \tag{5.46}$$

$$(\partial/\partial t)(\Lambda\mathbf{j}_s) = \mathbf{E}(\mathbf{r}), \tag{5.47}$$

where

$$\Lambda = m_e/2e^2 n_s. \tag{5.48}$$

Together with Maxwell's equation

$$\text{curl}\,\mathbf{H} = (4\pi/c)\mathbf{j},$$

London's first equation (5.46) leads to

$$\lambda_L{}^2\, \nabla^2\mathbf{H}(\mathbf{r}) = \mathbf{H}(\mathbf{r}), \qquad \lambda_L{}^2 = \Lambda c^2/4\pi = m_e c^2/8\pi e^2 n_s. \tag{5.49}$$

The length λ_L is the London penetration length. Its physical meaning is most simply demonstrated by the one-dimensional solution of Eq. (5.49):

$$H(x) = H_0 \exp(-x/\lambda_L), \tag{5.50}$$

which shows that the magnetic field penetrates into the superconductor only within a surface sheet of thickness $\lambda_L \approx 10^{-5}$ cm. Thus, superconductors with thickness less than λ_L do not exhibit the Meissner–Ochsenfeld effect.

3. *Fluxoid Quantization*

The integral

$$\Phi_\Gamma = \oint_\Gamma [\mathbf{A}(\mathbf{r}) + c\Lambda\mathbf{j}]\, d\mathbf{s} \tag{5.51}$$

around a closed curve Γ is called the fluxoid pertaining to Γ. Applying Londons equations (5.46)–(5.47) and the equation

$$\partial\mathbf{H}/\partial t = -c\,\text{curl}\,\mathbf{E},$$

one easily shows that Φ_Γ is constant in time. Inserting $\mathbf{A} + c\Lambda\mathbf{j}_s$ from Eq. (5.43) into the definition of the fluxoid, one obtains

$$\Phi_\Gamma = (\hbar c/2e) \oint_\Gamma \nabla\phi(\mathbf{r})\, d\mathbf{s}. \tag{5.52}$$

In evaluating this integral, one has to distinguish between the two cases, that the curve Γ lies in a simply or a multiply connected piece of superconductor. In the first case, the function $\phi(\mathbf{r})$ is single-valued, so that the fluxoid vanishes. In the second, as for a superconducting ring, the function $\phi(\mathbf{r})$ may change by $2\pi n$ (n = integer) in order to leave $\psi(\mathbf{r})$ single-valued if one makes a single traversal of Γ. Thus, any fluxoid is a multiple of an elementary fluxoid quantum:

$$\Phi_\Gamma = n(hc/2e) \approx n(2 \times 10^{-7}) \quad \text{G cm}^2. \tag{5.53}$$

If one chooses the path of integration such that the current density vanishes on it, the fluxoid is identical with the magnetic field enclosed, namely,

$$\Phi_\Gamma = \iint \mathbf{H}(\mathbf{r})\, d\mathbf{f}. \tag{5.54}$$

For a superconductor which is thicker than the penetration length λ_L, the condition of choosing Γ such that Eq. (5.54) holds can always be fulfilled by taking Γ to be any closed loop inside the superconductor and not touching the surface layer in which the shielding supercurrents flow. By measuring the magnetic field frozen in a superconducting hollow cylinder, the fluxoid quantization has been observed experimentally by Doll and Näbauer (1961) and Deaver and Fairbank (1961).

The numerical value of the fluxoid quantum leads to a charge $2e$ for the carriers responsible for superconductivity, thus supporting the Cooper-pair concept. The mass of a Cooper pair has also been determined by an analogous measurement of the elementary "rotational fluxoid" quantum. As shown in Section I,B, the rotation of a quantum system corresponds to a Coriolis potential $2m_e(\boldsymbol{\omega} \times \mathbf{r})$ which replaces the vector potential $(2e/c)\mathbf{A}(\mathbf{r})$ in the above considerations. The rotational fluxoid is then, according to Eq. (5.43),

$$\Phi_\Gamma{}^{\omega} = \oint \left[\boldsymbol{\omega} \times \mathbf{r} + \frac{\mathbf{j}_s}{2en_s(T)}\right] d\mathbf{s}$$

$$= \frac{\hbar}{2m_e} \oint \nabla\phi(\mathbf{r})\, d\mathbf{s} = \frac{h}{2m_e} n = n(0.57) \quad \text{cm}^2\,\text{sec}^{-1}. \tag{5.55}$$

The experimental verification of the quantum $h/2m_e$ of the rotational fluxoid by Zimmerman and Mercereau (1965) again gives support to the pair concept.

4. *Type-II Superconductors*

Besides the penetration length λ_L, there is another length characteristic for a superconductor. The length ξ is a healing length which describes the distance at which the density n_s of the Cooper pairs reaches its equilibrium value if it is somewhere forceably put to zero. The length ξ is connected with an energy form similar to a surface tension; hence, it is proportional to $(\nabla n_s)^2/n_s$.

There are two groups of superconductors distinguished by the value

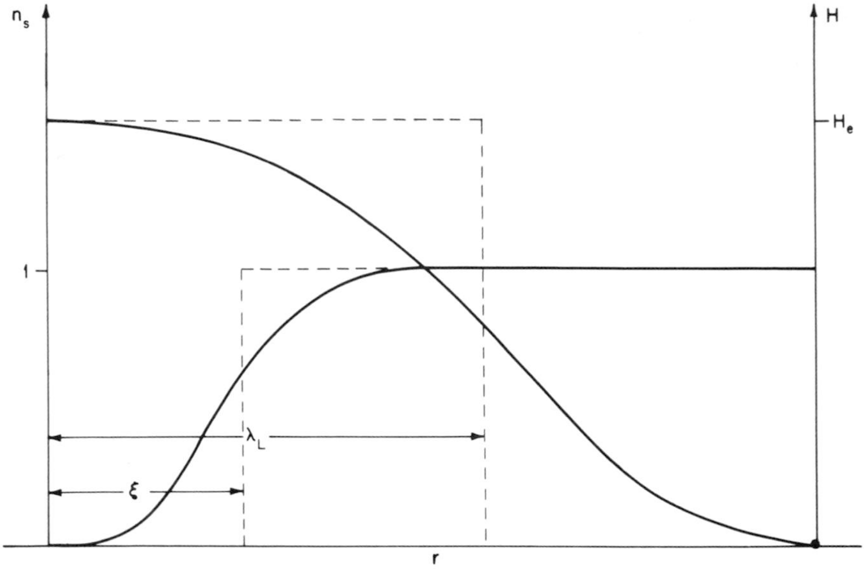

FIG. 1. Variation of the Cooper-pair density n_s and the magnetic field H if n_s is forceably put zero at $r = 0$.

of the ratio $\varkappa = \lambda_L/\xi$, called the Ginzburg–Landau parameter (1950) namely,

$$\varkappa < 1/\sqrt{2}: \quad \text{type-I superconductors,}$$

$$\varkappa > 1/\sqrt{2}: \quad \text{type-II superconductors.}$$

Type-I superconductors exhibit a Meissner–Ochsenfeld effect for all magnetic fields below the critical field H_c. If the applied field exceeds H_c, the superconductor experiences a phase transition into the normal phase.

Type-II superconductors also exhibit a Meissner–Ochsenfeld effect for external fields H_e smaller than a certain field H_{c1}. In contrast to type-I superconductors, there is, however, a second critical field H_{c2}, so that an external field $\mathbf{H}_e$ for which $H_{c1} < H_e < H_{c2}$ holds penetrates into the superconductor.

To estimate H_{c1}, we consider the situation sketched in Fig. 1. The Cooper-pair density is zero at $r = 0$ and reaches the equilibrium value within the healing length ξ. The magnetic field, on the other hand, takes the value of the external field at $r = 0$ and decreases to its equilibrium value, zero, within the characteristic length λ_L. In order to decide whether this situation is stable or unstable, we estimate the energies involved.

For this purpose, we represent the functions $n_s(r)$ and $H(r)$ approximately by square-well functions. The region $0 \leq r \leq \xi$ where the superconductor is normal contributes a deficiency of condensation energy of the amount

$$g_1 = (1/8\pi)[H_c^2(T) - H_e^2)]\pi\xi^2. \tag{5.56}$$

In the superconducting ring $\xi \leq r \leq \lambda_L$, the magnetization is approximately zero, so that an energy of the amount

$$g_2 = (1/8\pi)H_e^2(\lambda_L^2 - \xi^2)\pi \tag{5.57}$$

FIG. 2. Contour diagram of the Cooper-pair density n_s for the free-energy-minimum solution of the Ginzburg–Landau equations just below H_{c2}. (After W. H. Kleiner, L. M. Roth, and S. H. Autler (1964), *Phys. Rev.* **133A**, 1226.)

is gained. If λ_L is larger than ξ, the balance

$$g_1 + g_2 = (1/8)[\xi^2 H_c^2(T) - \lambda_L^2 H_e^2]$$

is negative when $H_e > (\lambda_L/\xi)H_c(T) = \varkappa H_c(T)$, so that the Meissner state is unstable for these values of the external magnetic field.

Abrikosov (1957) has solved the Ginzburg–Landau equations for superconductors with

$$\varkappa = \lambda_L/\xi > 1/\sqrt{2}.$$

He found that, for external fields H_e within the limits

$$H_{c1} = H_c\varkappa(\log \varkappa - 0.27) \tag{5.58}$$

$$H_{c2} = H_c\varkappa\sqrt{2}, \tag{5.59}$$

the magnetic field strength in the superconductor is distributed in the form of a two-dimensional lattice of flux lines. The magnetic field and the Cooper-pair density are then doubly periodic functions in the coordinates perpendicular to the applied field (Fig. 2). At $H = H_{c2}$, a transition into the normal state of the metal takes place, since, for this field strength, the distance between neighboring flux lines is of the order

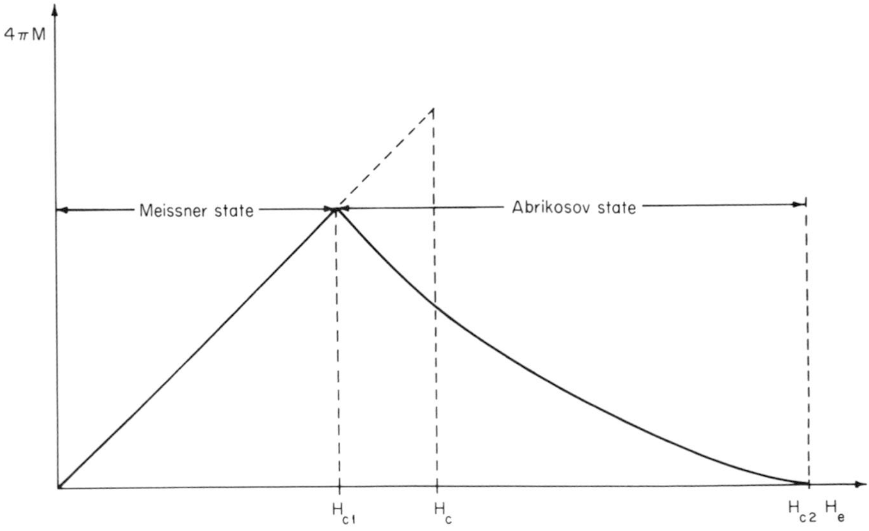

FIG. 3. The magnetization M as function of the applied magnetic field H_e for a type-II superconductor.

of the healing length ξ. This transition is a phase transition of second order. According to Abrikosov's solution, the magnetization of a type-II superconductor as function of the applied field is as given in Fig. 3.

Acknowledgment

The author would like to thank Professor G. Falk for helpful discussions and a critical reading of the manuscript.

General References

Callen, H. B. (1960). "Thermodynamics." Wiley, New York.

Falk, G. (1968). "Theoretische Physik," Vol. II. Springer, Berlin.

Guggenheim, E. A. (1967). "Thermodynamics." North-Holland Publ. Amsterdam.

Kittel, C. (1966). "Introduction to Solid State Physics." Wiley, New York.

Landau, L. D., and Lifshitz, E. M. (1958). "Statistical Physics." Pergamon, New York.

Landau, L. D., and Lifshitz, E. M. (1960). "Electrodynamics of Continuous Media." Pergamon, New York.

Lynton, A. B. (1962). "Superconductivity." Wiley, New York.

Pippard, A. B. (1957). "Elements of Classical Thermodynamics." Cambridge Univ. Press, London and New York.

Special References

Abrikosov, A. A. (1957). *Soviet Phys. JETP (English Transl.)* **5**, 1174.

Bardeen, J., Cooper, L. N. and Schrieffer, J. R., (1957). *Phys. Rev.* **108**, 1175.

Deaver, H. S., and Fairbank W. M. (1961). *Phys. Rev. Letters* **7**, 43.

Doll, R., and Näbauer, M. (1961). *Phys. Rev. Letters* **7**, 51.

Feynman, R. P. (1955). *Proc. Intern. Conf. Low Temp. Phys.*, 1955, **1**, Chapter 2. North-Holland Publ., Amsterdam.

Ginzburg, V. L., and Landau, L. D. (1950). *Soviet Phys. JETP (English Transl.)* **19**, 1064.

Heine, V. (1956). *Proc. Cambridge Phil. Soc.* **52**, 546.

Landau, L. D. (1941). *J. Phys., USSR* **5**, 71.

London, F. (1950). "Superfluids," Vol. I. Dover, New York.

Stenschke, H. (1968). *Z. Physik* **216**, 456.

Stenschke, H., and Falk, G. (1968).

Stratton, J. A. (1941). "Elektromagnetic Theory." McGraw-Hill, New York.

Zimmerman, J. E., and Mercereau, J. E. (1965). *Phys. Rev. Letters* **14**, 887.

Chapter 6

The Third Law of Thermodynamics

J. WILKS

I. Introduction

The third law of thermodynamics is an additional axiom of thermodynamics, which cannot be derived from the first and second laws. In its simplest form, it states that the entropy of a system in internal thermodynamic equilibrium tends to zero at absolute zero. The third law correlates a wide range of varied phenomena, and there are no known exceptions to its validity.

The third law has several fields of application. The vanishing of the entropy, taken together with other standard thermodynamic relationships, implies that many properties of matter have a common form of behavior at low temperatures. By comparing entropies derived from measurements and the third law, with values calculated from statistical mechanics, we obtain important information on the thermal motion both of gases and of solids at low temperatures. The law also enables us to predict the conditions of equilibrium in chemical reactions. In particular, it permits the determination of equilibrium conditions solely from calorimetric measurements.

The third law may be expressed in a more detailed and useful form by noting that the entropy of a system may arise in several different ways. For example, the total entropy of gaseous oxygen is made up of contributions from the translational and rotational motions, from the disorder of the magnetic moments, and from the mixing of the constituent isotopes. If the interactions between the various constituent parts of a system are weak, as is often the case, the total entropy will be very close to the sum of the various contributions calculated independently. We may then regard the entropy as the sum of component entropies each associated with a different type of disorder within the system.

The different contributions to the entropy are often said to arise from different "subsystems" of the main assembly. However, although a paramagnetic solid may be said to consist of vibrational and spin subsystems, it is difficult to identify a subsystem associated with an entropy of mixing. Therefore, we prefer to define each of the various modes in which entropy may arise as an *aspect* of the system. We now state the third law by saying that, "The contribution to the entropy of a system by each aspect which is in internal thermodynamic equilibrium tends to zero at absolute zero."

The third law is the direct descendant of Nernst's heat theorem (Nernst, 1906), which stated that the entropy change in a chemical reaction tends to vanish as the temperature approaches absolute zero. The validity and range of application of this theorem were at first the subject of much discussion, but the points at issue were eventually resolved, principally by W. F. Giaugue and G. N. Lewis in America, and F. E. Simon in Europe. Finally, Simon (1927) put forward a new formulation of the heat theorem, which, in a slightly modified form (Simon, 1937), is now the third law of thermodynamics. The early history of the law has been described by Simon (1956), who gives an interesting study of the development of the ideas behind the law. It should be noted, however, that the details of some of the early controversies are not the best introduction to the subject for the modern reader.

We now proceed to consider applications of the law. Quite commonly, systems remain in complete internal equilibrium down to the lowest temperatures. Therefore we shall first consider systems in equilibrium, and defer discussion of aspects not in equilibrium until Section VII. (As we discuss in Section VIII, aspects associated with nuclear properties do not usually affect either the physical properties of a system or its equilibrium condition. Therefore, to begin with, we shall ignore aspects associated with the orientation of nuclear spins and with the mixing of isotopic species.)

II. The Third Law and Low-Temperature Physics

We first restrict ourselves to systems in complete internal equilibrium, so that the entropy of the system tends to zero at absolute zero. The most important generalization of this result is that all specific heats tend to zero at absolute zero. This may be seen by writing the specific heat at, say, constant pressure in the form

$$C_p = T\left(\frac{\partial S}{\partial T}\right)_p = \left(\frac{\partial S}{\partial(\ln T)}\right)_p. \tag{2.1}$$

As T tends to zero, $\ln T$ tends to minus infinity and S tends to zero, hence C_p tends to zero. Similar proofs follow for C_v, and other specific heats.

It is well known experimentally that all specific heats tend to zero, as is shown for some typical solids in Fig. 1. These particular specific heats are associated with the thermal vibrations of the crystal lattice and have been treated theoretically by Einstein, Debye, Born, Blackman, and others (see, for example, Blackman, 1955). For a dielectric solid in which the specific heat arises solely from the vibrations of the lattice, it is possible to define a characteristic temperature, the Debye θ, such that, below about $\theta/50$, the specific heat vanishes as the cube of the temperature, its magnitude being given by the Debye relation $C = 464(T/\theta)^3$ cal deg^{-1} mol^{-1}.

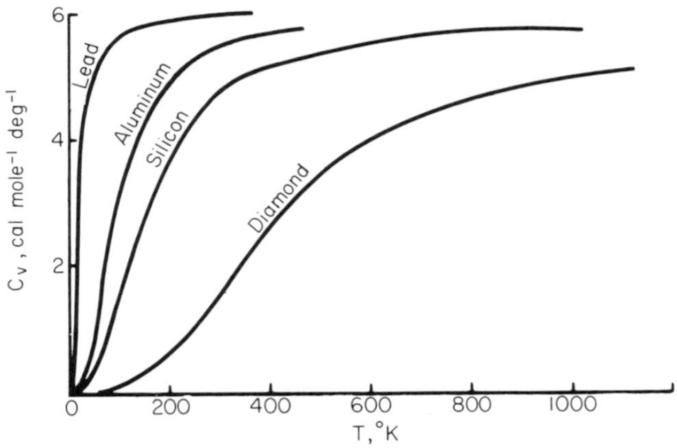

FIG. 1. The specific heats of some typical solids.

At room temperature, the specific heat of a metal is not significantly greater than that of a dielectric, but at helium temperatures, there is an appreciable difference. The electrons appear to make an additional contribution to the specific heat which is proportional to the absolute temperature, so that the total specific heat may be written

$$C_{\rm m} = aT^3 + bT \tag{2.2}$$

or

$$C_{\rm m}/T = aT^2 + b, \tag{2.3}$$

where aT^3 is the specific heat of the lattice and bT that of the electrons. Figure 2 shows plots of $C_{\rm m}/T$ against T^2 for copper and silver. The good straight line is firm evidence in favor of the above expression for the specific heat, while the slope and intercept of the line give the magnitude of each component. Not only does the specific heat of the whole system tend to zero, but the lattice and electronic contributions vanish independently. Hence, we see that the entropy of each aspect vanishes separately.

It is important to realize that the third law gives no direct indication of the temperature region in which the specific heat begins to fall below its classical value, for this depends very much on the substance in question. For many substances, the Debye characteristic temperature θ is of the order of 300°K, so that, by about 6°K, the specific heat is very small and may be readily extrapolated to absolute zero. However, this is not always the case. For example, there are some paramagnetic salts with lattice

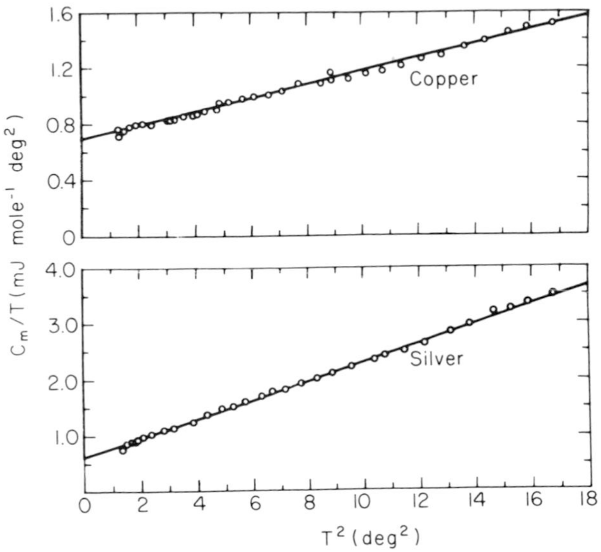

FIG. 2. The specific heat C_m of copper and silver, plotted as C_m/T against T^2. (After Corak *et al.* (1955).)

specific heats which are apparently quite similar, but for which no such extrapolation is possible because of their magnetic properties. Figure 3 shows the specific heat of such a salt, iron ammonium alum. Although the lattice specific heat is negligibly small below 1 °K, there is still a specific heat associated with the magnetic interactions. This shows a large co-operative peak at 26 m°K, and does not fall off until even lower temperatures.

Various miscellaneous consequences of the third law follow from the use of Maxwell's thermodynamic relations. For example, the fourth relation applied to magnetic materials states that

$$(\partial M/\partial T)_H = (\partial S/\partial H)_T, \tag{2.4}$$

where M and H are the magnetic moment and magnetic field. The magnetic susceptibility per unit volume is defined as

$$\varkappa = M/VH,$$

where V is the volume. It follows from the third law that

$$\lim_{T\to 0}(\partial\varkappa/\partial T)_H = 0; \tag{2.5}$$

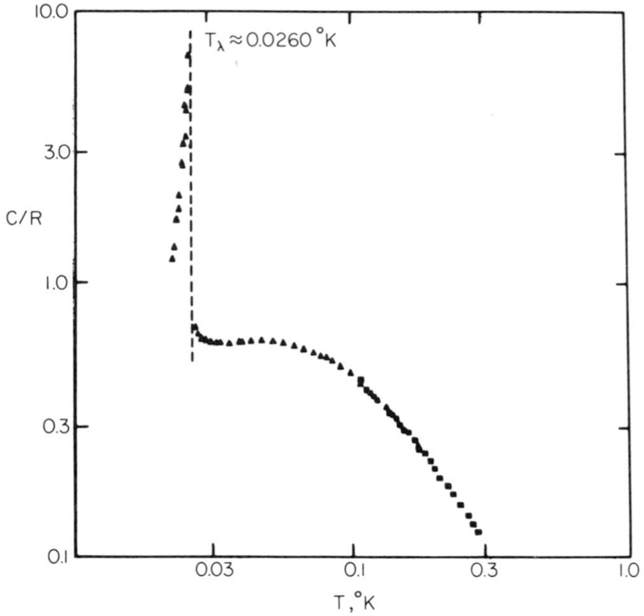

FIG. 3. The molar specific heat C of iron ammonium alum, expressed as the ratio C/R, where R is the gas constant. (After Vilches and Wheatley (1966).)

hence, the susceptibility of all magnetic substances becomes independent of temperature at the lowest temperatures. Although magnetism may arise in several different ways, this limiting behavior is always observed. For example, many paramagnetic materials undergo a transition at the so-called Curie point and take up a more ordered arrangement that is either ferromagnetic or antiferromagnetic. This transition to a more ordered state is accompanied by an excess specific heat, like that for iron ammonium alum shown in Fig. 3. As we have mentioned previously, the third law gives no indication as to the temperature of the transition to the ordered state, which depends entirely on the strength of the interactions between the magnetic moments. Thus, the Curie point of some paramagnetic salts is less than 0.01°K, while for iron it is about 1000°K.

The fourth Maxwell relation applied to the equation of state tells us that

$$(\partial V/\partial T)_p = -(\partial S/\partial p)_T, \tag{2.6}$$

so that, by applying the third law, we find

$$\lim_{T\to 0}(\partial V/\partial T)_p = 0, \tag{2.7}$$

that is, all expansion coefficients tend to zero at absolute zero. As in the case of magnetic ordering, the temperature below which the coefficient tends to zero is very variable. The expansion coefficient of diamond decreases steadily from a temperature of at least 500°K, whereas the expansion coefficient of liquid ^{3}He does not begin to fall off finally until below 0.2°K.

A metal in the superconducting state may be restored to the normal state by applying a certain critical magnetic field H_c whose magnitude is a function of temperature. A straightforward application of the second law (e.g., Kuper (1968)) gives the entropy difference between unit volume of normal and superconducting material as

$$S_n - S_s = -\frac{H_c}{4\pi}\left(\frac{\partial H_c}{\partial T}\right)_p . \tag{2.8}$$

Figure 4 shows values of H_c for mercury plotted against T. We see that $(\partial H_c/\partial T)_p$ tends to zero at absolute zero, implying that the entropy difference between the normal and superconducting states vanishes in accord with the third law.

The transition from the normal to the superconducting state is accompanied by a very small change in volume, which has been measured by Lasarew and Sudovstov (1949), whose results are shown in Fig. 5.

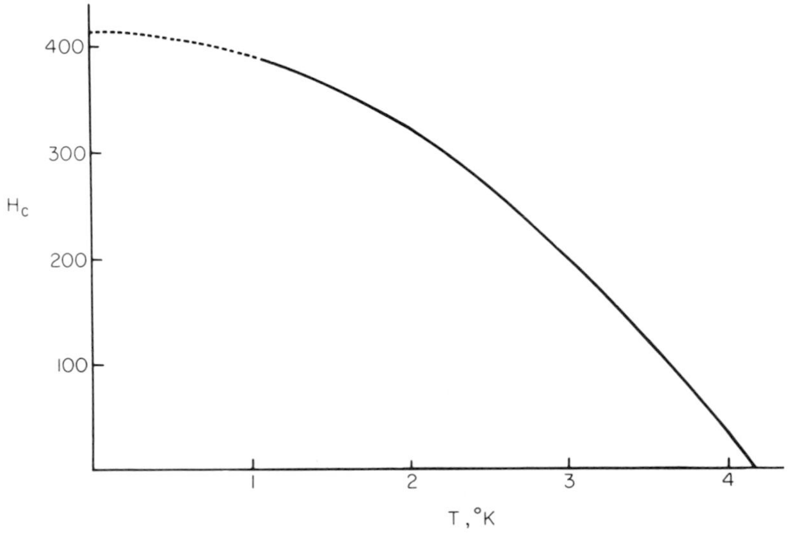

FIG. 4. The critical magnetic field for mercury. (After Misener (1940).)

At the lowest temperatures, ΔV appears to become independent of T; this follows from the thermodynamic relation

$$\Delta V = V \frac{H_c}{4\pi} \left(\frac{\partial H_c}{\partial p} \right)_T \tag{2.9}$$

together with the previous result that H_c becomes independent of T.

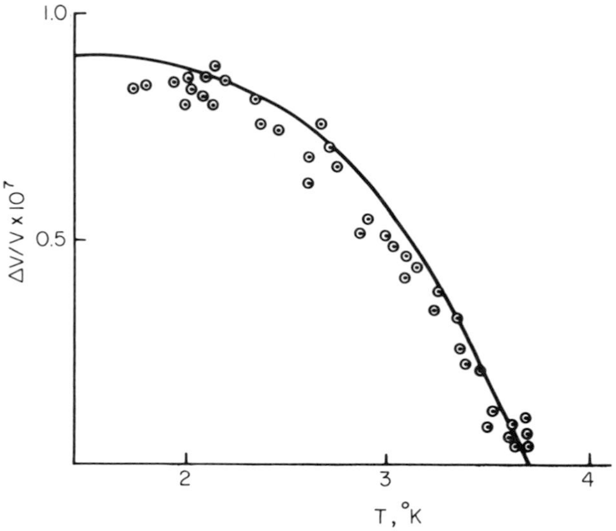

FIG. 5. The change in volume of tin at the superconducting transition; the points are measured values, and the full line is calculated from Eq. (2.9). (After Lasarew and Sudovstov (1949).)

Finally, we mention that thermoelectric effects are observed to vanish as the temperature tends to zero. That is, the emf dE in a thermocouple circuit with junctions at temperatures T and $T + dT$ tends to zero as T goes to zero. This result follows from the original Thomson treatment of thermoelectricity, which implies that the coefficient dE/dT is proportional to the differences in the entropies of transfer at the two junctions. According to the third law, these entropies of transfer vanish at absolute zero, and hence also the thermal emf's (see, for example, Chapter 28 of Lewis and Randall (1961)). [Note, however, that the Thomson relations are more properly obtained by the methods of irreversible thermodynamics (e.g., Bazarov (1964)). The entropy of transfer then appears only as the ratio of two coefficients, and it has yet to be shown rigorously that the third law can be applied to this quantity.]

III. Liquid Helium

Because the binding energy of liquid helium is very small, its structure is determined to an unusual extent by the zero-point energy of the atoms (see, for example, Wilks (1967)). Both liquid ^3He and ^4He have such large molar volumes that, under their saturated vapor pressure, they remain liquid down to the very lowest temperatures. Hence, they are the only fluids which demonstrate the consequences of the third law.

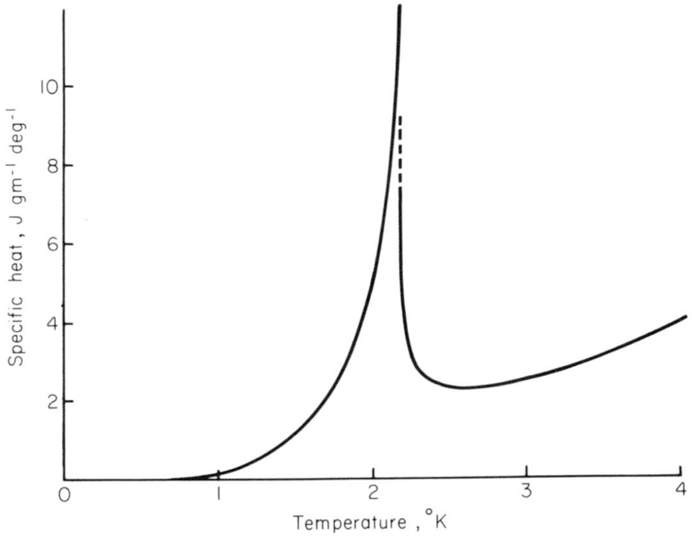

FIG. 6. The specific heat of liquid ^4He.

At about 3°K, both liquid ^3He and ^4He have a specific heat of the same order as that of a classical liquid. However, as the temperature drops, these specific heats decrease, either steadily, as in liquid ^3He, or abruptly at a lambda-type anomaly, as in liquid ^4He (Fig. 6). Below about 50 m°K the specific heat of liquid ^3He falls off approximately as the temperature T (Abel *et al.*, 1966), and the specific heat of liquid ^4He below 0.6°K decreases as T^3 (Wiebes *et al.*, 1957).

The third Maxwell thermodynamic relation applied to the surface tension σ of a liquid states that

$$(\partial\sigma/\partial T)_A = -(\partial S/\partial A)_T, \tag{3.1}$$

where A is the area of the surface. It follows from the third law that the

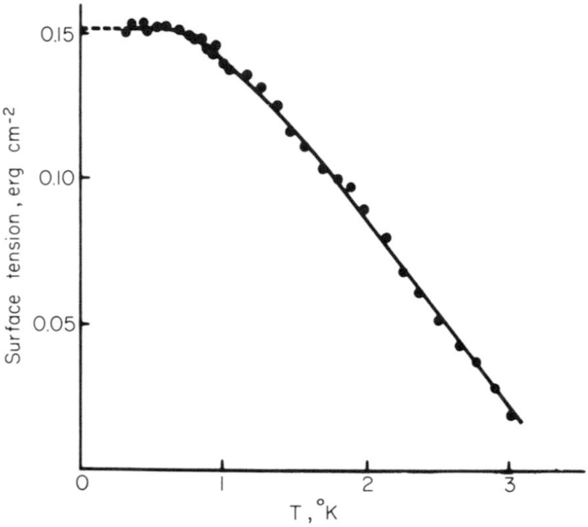

FIG. 7. The surface tension of liquid ^3He. (After Zinov'eva (1955).)

surface tension will tend to a constant value at low temperatures. Measurements on liquid ^3He (Fig. 7) and on liquid ^4He (Allen and Misener, 1938) are in good accord with this prediction.

Although neither liquid ^3He nor ^4He solidify under their saturated vapor pressure, they do so under an external pressure. Figure 8 shows the melting pressure for ^4He as a function of temperature. (The melting curve of ^3He is more complex, and is discussed in Section VIII.) Below about 1°K, the melting pressure has an almost constant value, as we would expect from the Clausius–Clapeyron relation

$$dp/dT = \Delta S/\Delta V, \tag{3.2}$$

where ΔS and ΔV are the entropy and volume changes on melting. According to the third law, the entropy of both liquid and solid tend to zero, so that ΔS tends to zero. Measurements show that the volume change tends to a constant and finite value, hence dp/dT tends to zero. In fact, as dp/dT varies approximately as T^8, the entropy difference ΔS vanishes much more rapidly than the entropy of either the liquid or solid. Thus, the measurements shown in Fig. 8 confirm the remarkable result that, below about 1°K, the liquid and solid states along the melting line have virtually the same entropy.

As entropy is a measure of disorder, the above results imply that liquid ^4He below about 1°K is in almost as ordered a state as the solid.

This conclusion, though unexpected, is well supported by experimental evidence. Below 0.6°K, the magnitude of specific heat is given by Deby's theory for a *solid* (assuming that only longitudinal modes of vibration are possible because of the low shear viscosity of the liquid). By inserting the measured velocity of longitudinal sound waves, we obtain a value for the specific heat in good agreement with that observed. Likewise, the thermal conductivity of the liquid below 0.6°K is very similar to what we would expect for a crystalline solid.

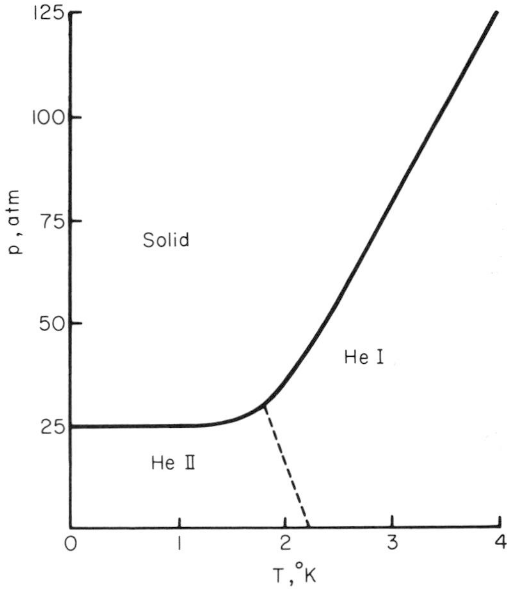

FIG. 8. The melting pressure of ^{4}He. (After Simon and Swenson (1950).)

Given that the heat motion in liquid ^{4}He below 1°K is very similar to that in the solid, it is not surprising that liquid ^{4}He turns out to have other unusual properties. The specific heat exhibits a large anomaly of the cooperative type at 2.17°K, the so-called lambda point (Fig. 6); above this temperature, helium is not greatly different from many other liquids, whereas, below the lambda point, its properties are unique. Experiments on the flow through extremely fine channels imply that the helium has virtually no viscosity and that its linear velocity of flow is greater in *narrower* channels. The liquid will also support second sound, that is, standing waves of temperature entirely analogous to standing waves of pressure.

Thermodynamics gives no indication of the microscopic mechanisms which are responsible for the entropy of liquid helium tending to zero at absolute zero. For an explanation, we must turn to quantum statistics. A perfect gas obeying either Bose or Fermi statistics eventually falls into a degenerate state at low temperatures, and the entropy tends to zero. Of course, the interactions between the atoms in liquid helium are so strong that we can hardly describe it as a perfect gas. Yet the main features of liquid ^{3}He are quite well described by a gas model—in particular, this treatment accounts for the linear specific heat. The position regarding liquid ^{4}He is more complicated. The theory of a Bose gas predicts a specific heat varying as $T^{3/2}$, and a much fuller treatment is necessary to account for the observed T^3 relationship in liquid ^{4}He. For further details, see (Wilks, 1967).

We sometimes associate an increase in the spatial disorder of a system with an increase in its entropy. Yet the fact that liquid and solid helium along the melting curve have almost the same entropy gives a warning against a too ready application of this concept. In general, the entropy of a system increases considerably on passing from the solid to the liquid phase, and this may usually be associated with the liquid being in a more disordered state than the solid. On the other hand, in liquid helium, the de Broglie wavelengths of the molecules are comparable with the molecular spacing, and no intuitive concept of spatial disorder is possible. The behavior of solutions of ^{3}He in liquid ^{4}He is a case in point.

Above 1°K, liquid ^{3}He is completely miscible with liquid ^{4}He in all proportions, and the entropy of mixing makes a large contribution to the entropy of the solution. Thus, for a solution consisting of equal parts of the two isotopes, the entropy of mixing per mole is of the order of R, the gas constant. The third law states that this entropy must vanish, and a phase separation is observed to occur below about 0.8°K (Fig. 9). As the temperature falls further, the upper and lower phases become richer in ^{3}He and ^{4}He, respectively, and the entropy of mixing decreases, in accord with the third law.

Yet it must be appreciated that the third law does not predict the existence of a phase separation; it only states that the entropy of mixing must vanish. A phase separation into two pure components is an obvious way of ensuring that the entropy vanishes, but in view of the unusual behavior of liquid ^{4}He, there is no *a priori* reason for supposing this to be the only way. In fact, measurements at the lowest temperatures show that the lighter phase tends to become pure ^{3}He, but that the heavier phase tends to a limiting condition of ^{4}He containing about 6% of ^{3}He as solute.

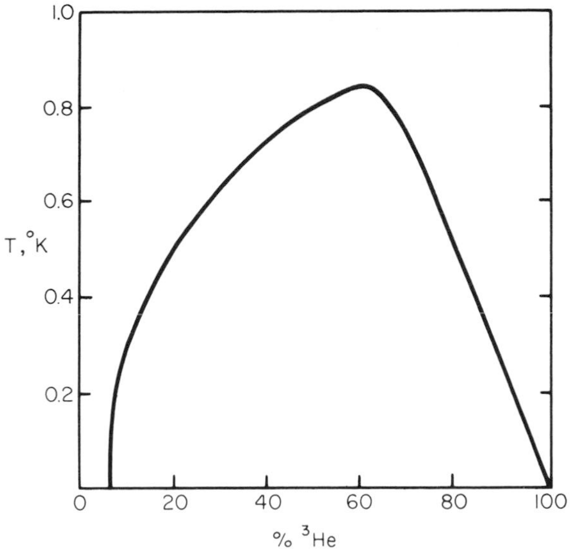

FIG. 9. The phase separation in ^3He–^4He mixtures. Liquid ^3He and ^4He are only miscible in all proportions in the upper part of the phase diagram; otherwise, they form two phases whose constitution may be read off the curve. (After Fairbank and Walters (1958) and Edwards *et al.* (1965).)

At first sight, it appears that these ^3He atoms will contribute a nonzero entropy of mixing, but this is not so, because we are dealing with fully degenerate quantum liquids. For a further discussion of this point, see De Bruyn Ouboter and Beenakker (1961).

IV. The Third Law and the Determination of Entropy

The vanishing of the specific heats (Section II) is of great importance, for it permits the use of absolute zero as a reference level for all thermodynamic calculations. The definition of entropy is based on the second law of thermodynamics and is only sufficient to define differences of entropy. Thus, if S_T and S_0 are the entropies of a substance at temperature T and zero,

$$S_T - S_0 = \int_0^T \frac{C\,dT}{T}. \tag{4.1}$$

Because, and only because, the specific heat tends to zero at absolute zero, the integral is finite. Hence, by stating that S_0 is zero, the third law allows us to allot a unique value to the entropy at any other temperature, namely:

$$S_T = \int_0^T \frac{C\,dT}{T}. \tag{4.2}$$

To evaluate the integral, we require values of the specific heat down to absolute zero, but it is sufficient to measure them to a temperature from which the values may be extrapolated to zero. However, if any aspect of a system still has any appreciable entropy which is not rapidly decreasing, this is a clear indication that measurements to lower temperatures are necessary. Note that, although a specific heat C may become small at low temperatures, errors in extrapolation may have large effects on the entropy, which depends on the ratio C/T.

A verification of this method of determining entropies concerns the behavior of allotropic solids. These solids exist in more than one crystal form, two examples being the graphite and diamond structure of carbon and the three allotropic forms of sulphur. At any given temperature, one allotropic state will have a lower free energy than the others, and will therefore be the stable form. The other states are metastable, the atoms are in a position of stable equilibrium with respect to small displacements, and there is a well-defined crystal structure. However, these metastable states will eventually transform into the more stable state of lowest energy, but, as high potential barriers have to be crossed, this may take an extremely long time. Thus, we have the possibility of studying different forms of the same solid over a wide range of temperature.

Consider two allotropic forms, α and β, which are in equilibrium at a temperature $T_{\alpha\beta}$ and which have entropies S_α and S_β. The difference in entropy between the two forms at this temperature ($\Delta S_{\alpha\beta} = S_\alpha - S_\beta$) may be directly determined from the measured heat of transformation. The third law gives an entirely independent method of calculating $\Delta S_{\alpha\beta}$, for, from Eq. (4.2),

$$\Delta S_{\alpha\beta} = S_\alpha - S_\beta = \int_0^{T_{\alpha\beta}} \frac{C_\alpha\,dT}{T} - \int_0^{T_{\alpha\beta}} \frac{C_\beta\,dT}{T}. \tag{4.3}$$

Thus, by measuring the specific heats down to low temperatures, we may obtain values for $\Delta S_{\alpha\beta}$ which should be equal to those determined directly. (These values are often given in entropy units E.U., which have the dimensions of cal deg^{-1} mol^{-1}.)

Sufficient information is available for making calculations for tin (Lange, 1924), sulphur (Eastman and McGavock, 1937), cyclohexanol

(Kelley, 1929), and phosphine (Stephenson and Giauque, 1937). In all these cases, the two values for the entropy difference agree to within the errors of measurement. As an example, we quote the the figures for phosphine, which are of particular interest, as this compound exists in three polymorphic forms α, β, and γ. The α and β forms are in equilibrium at 49.43°K and the measured entropy of transition is 3.757 e.u. The entropy of the β form is found by measuring the specific heat down to about 15°K, and then extrapolating to zero using the Debye theory of specific heats. On cooling the α form, it changes to the γ form at 30.29°K; its entropy is therefore determined by measuring (a) its specific heat down to this temperature, (b) the entropy of transformation at 30.29°K, and

TABLE I

THE ENTROPY OF α- AND β-PHOSPHINE AT 49.43°K[a]

S_β	E.U.	S_α	E.U.
0–15°K (Debye)	0.338	0–15°K Debye[b]	0.495
15–49.43°K	4.041	15–30.29°K[b]	2.185
		$\Delta S(\gamma \rightarrow \alpha)$	0.647
		30.29–49.43°K	4.800
Total	4.379		8.127

[a] From Stephenson and Giauque (1937).
[b] γ-Phosphine.

(c) the specific heat of the γ form down to 15°K. The results are shown in Table I. Thus, $S_\alpha - S_\beta$ is equal to 3.748 e.u. at 49.43°K, in good agreement with the directly determined value of 3.757 e.u.

The same method of calculating entropies may be extended to liquids and gases. We may deduce the entropy of a gas at the boiling point from the specific heats of the solid and liquid, together with the latent heats of melting and vaporization and any other transitions. As an example of such a calculation, Table II shows figures for the entropy of nitrogen gas at the boiling point. Note that, as solid nitrogen exhibits a phase transition at 35.61°K, it is essential to measure the heat of transition at 35.61°K and the specific heat to considerably lower temperatures, in order to obtain a reliable extrapolation to absolute zero.

TABLE II

THE ENTROPY OF NITROGEN GAS AT THE BOILING POINT[a]

Solid	0–10° (Debye)	0.458 E.U.
	10–35.61°	6.034
	Transition 35.61°	1.536
	35.61–63.14°	5.589
	Melting	2.729
Liquid	63.14–77.32°	2.728
	Vaporization	17.237
		36.31

[a] From Giauque and Clayton (1933).

V. Statistical Mechanics and the Third Law

We have seen in Section IV that the entropy of a gas may be derived from thermal measurements by using the third law. This entropy may also be calculated by the quite different methods of statistical mechanics. In their simplest form, the basic axioms of statistical mechanics state that each quantum-mechanical state has equal statistical weight, and that the entropy may be written:

$$S = k \ln W, \tag{5.1}$$

where k is a constant and W the number of complexions associated with a given state of the system. Hence, by methods given in any standard text, we may derive an expression for the entropy of a perfect gas in terms of the atomic and molecular constants.

The statistical method gives an exact expression for the entropy of a perfect gas, that is, a gas with negligible interactions between the molecules. Entropies calculated in this way are not directly comparable with those deduced from calorimetric measurements on *real* gases. The entropy of a real gas differs from that of its ideal counterpart because of departures from the ideal gas laws. This entropy difference may be readily calculated from the virial coefficients (see, for example, Wilson (1957)), and is usually quite small. Thus, for nitrogen gas at the boiling point, the difference is only 0.22 E.U. in a total entropy of about 36 E.U.

Having made the above correction, we may now compare the entropies

TABLE III

CALCULATED AND MEASURED ENTROPIES OF GASES

Gas	Temperature	Entropy (E.U.)		Reference
		Statistical	Thermal*	
A	B.P.	30.87	30.85	a
O_2	B.P.	40.68	40.70	b
N_2	B.P.	36.42	36.53	c
Cl_2	B.P.	51.55	51.56	d
HCl	B.P.	41.45	41.3	e
HD	298.1°K	34.35	34.45	f

* Corrected for nonideality.
[a] Clusius and Frank (1943).
[b] Giauque and Johnston (1929).
[c] Giauque and Clayton (1933).
[d] Giauque and Powell (1939).
[e] Giauque and Wiebe (1928).
[f] Clusius *et al.* (1937).

obtained calorimetrically and statistically. The second column of figures in Table III shows the entropies of several gases calculated from the statistical expression for a perfect gas, together with values obtained from calorimetric measurements plus a small correction for nonideality. The table shows that the statistical and thermal values are in close agreement, even though statistical mechanics does not make the explicit assumption that the entropy tends to zero at absolute zero. It therefore appears that the third law is a consequence of the axioms of statistical mechanics.

The fundamental axiom of statistical mechanics, Eq. (5.1), implies that the entropy of a perfect gas is zero at absolute zero. The quantum-mechanical treatment of a gas leads to an array of quantized energy levels. As the temperature falls, the system is more likely to be found in states corresponding to the lower levels. Hence, the number of complexions is reduced, and, likewise, the entropy. At absolute zero, only the lowest level will be occupied, the term $k \ln W$ vanishes, and the entropy is zero.

We should note that the value of entropy *zero* at absolute zero in the above statistical treatments comes about as the result of a particular choice of the form of the basic axiom, Eq. (5.1). This axiom derives from the fact that the equilibrium condition for an isolated system is a state of both maximum entropy and maximum probability. It therefore follows that the entropy and probability are in some way related, say, as $S = f(W)$. In fact, the additive property of entropies and the multiplicative property of probabilities are such that the only possible relation must be of the form

$$S = k \ln W + S_0, \tag{5.2}$$

where k and S_0 are two constants which are independent of the value of W (see, for example, Wilks (1961)). The value of k is identified by noting that Eq. (5.2) gives the equation of state of a perfect gas of N particles to be $pV = NkT$, so that k must be the gas constant per particle, and S_0 is taken as zero.

At first sight, there appears to be no compelling reason why we should put $S_0 = 0$. Equation (5.2) is sufficient to account for the experimental observations that varying the parameters of a system near absolute zero does not change the entropy, so that coefficients such as $(\partial S/\partial H)_p$ and $(\partial S/\partial V)_p$ tend to zero. To ensure these results, it is only necessary that, at absolute zero, all states of a given system should have the *same* entropy, say S_0. The form of Eq. (5.2) is also sufficient to ensure that the specific heat

$$C = T\left(\frac{\partial S}{\partial T}\right) = T\left(\frac{\partial S}{\partial(\ln T)}\right) \tag{5.3}$$

should vanish. As the temperature approaches zero, the term ln T tends to infinity, so the specific heat vanishes provided only that S remains finite. Thus, the physical content of the third law is obtained by saying that the entropy of a given system at absolute zero has a value S_0 which is characteristic of the system, but does not depend on any external parameter. For example, the agreement between the experimental and statistical values of the entropies of gases shown in Table III would still be valid if we added zero-point values S_0 to both the statistical and thermal entropies of each gas.

The most appropriate value for the constant S_0 has been discussed by Schrödinger (1952), who considers any thermodynamic system whose condition may be varied by changing the external parameters. The system might, for example, be a metal which is brought into the superconducting condition by varying a magnetic field, a liquid which is solidified by pressure, or a mixture of different chemical species which is cooled in different chemical combinations. It follows from the derivation of Eq. (5.2) that, if we cool a system down toward absolute zero under different conditions, specified by different external parameters, the entropy must always tend to the same value S_0. Moreover, this entropy is undetectable, since it is not affected by any of the usual parameters of the system. In particular, it will not contribute to the entropy determined calorimetrically, nor will it be of any consequence in determining the equilibrium constants of a chemical reaction. Hence, there is no occasion when we need to know its value, and Schrödinger concludes that the obvious procedure is to set it equal to zero.

The methods of statistical mechanics result in the vanishing of the entropy of a perfect gas. This result has also been obtained for a wide variety of systems, including crystal lattices, superconductors, liquid ^{3}He and ^{4}He, and magnetic material. As yet, however, there is no *general* proof that the entropy of all systems in equilibrium vanishes at absolute zero. Nor is it easy to see how a more general proof is to be obtained, in view of the wide variety of methods used to treat the different systems. (It is even less easy to form a general picture of how the entropy falls off toward absolute zero. In a simple magnetic system with a limited number of Zeeman levels with separation ΔE, the entropy falls off in the temperature region defined by $kT \sim \Delta E$. On the other hand, the entropy of a perfect gas of molecular weight 4 and of the same density as liquid helium decreases rapidly below the Bose condensation temperature of 3.14°K, even though the spacing of the energy levels is of the order of 10^{-15}°K. In fact, the details of the decline in the entropy of any system are determined by the form of the quantum distribution functions and by the density of states in the particular system in question. For a further discussion of this point, see ter Haar (1966).)

Viewed from a statistical point of view, the vanishing of the entropy at absolute zero is a consequence of the quantization of the energy levels of a system, together with the basic axioms of statistical mechanics. We conclude that there is no new principle enshrined in the third law which does not follow from quantum statistics. The value of the third law is that it encapsulates a particular aspect of quantum statistics in a graphic form, which can be very readily applied to a wide variety of physical situations.

VI. Chemical Equilibrium

It follows from the second law that any system maintained at a constant temperature T and pressure p is in thermal equilibrium when its Gibbs free energy G is a minimum. We now apply this result to a chemical reaction, which we represent schematically as

$$\alpha A + \beta B \rightleftharpoons \gamma C + \delta D, \tag{6.1}$$

where α, β, γ, and δ are the number of moles of the species A, B, C, and D. If the four species are placed in an isothermal isobaric container, their equilibrium concentrations will be such as to make the total free energy a minimum. The condition for equilibrium is simply

$$\Delta G = 0, \tag{6.2}$$

where ΔG is the increase in the total free energy of the system as the reaction proceeds from left to right. That is, we can calculate the equilibrium condition if we can find ΔG.

From standard thermodynamic relationships, we know that

$$\Delta G = \Delta H - T\,\Delta S, \tag{6.3}$$

where H and S are the enthalpy and entropy of the whole system. The quantity ΔH is the heat of reaction at constant pressure, and may be found calorimetrically. This heat of reaction may be found either by a direct measurement at the temperature T, or by a measurement at some more convenient temperature T', together with the relation

$$\Delta H = \Delta H' + \int_{T'}^{T} (-\alpha C_A - \beta C_B + \gamma C_C + \delta C_D)\, dT, \tag{6.4}$$

where C_A, C_B, etc. are the specific heats at constant pressure of the individual species. To find ΔS, we need values for the entropies associated with each side of the reaction equation. Measurements of the specific heat C of a substance, together with the relation $dS = C\,dT/T$, show how the entropy varies as a function of temperature. We can thus determine how ΔS and ΔG vary with temperature. Hence, given one value of ΔG, or one position of equilibrium, we can find all the rest. However, the second law is not sufficient to specify the absolute value of an entropy, and hence does not lead to an absolute value for ΔG. The third law, by stating that the entropy of each component is zero at temperature zero, leads to absolute values for the entropies, and hence allows us to calculate ΔG from purely calorimetric measurements.

As an example of the method, we now outline its application to a relatively simple reaction, the polymorphic transition

$$\text{graphite} \rightleftharpoons \text{diamond}$$

The heat of transformation of the reaction may be found by calorimetry, and a measurement of the specific heats down to low temperatures then gives the entropy and free-energy changes in the reaction. It is found that, under atmospheric pressure, the free energy of diamond is greater than that of graphite at all temperatures. Graphite is therefore the stable form. However, higher pressures favor the production of diamond. If ΔG_{T0} is the free energy change in the reaction at temperature T and at atmospheric (or zero) pressure, it follows from the relation

$$[\partial(\Delta G)/\partial p]_T = \Delta V, \tag{6.5}$$

that, at pressure p, the value of ΔG is

$$\Delta G_{Tp} = \Delta G_{T0} + \int_0^p \Delta V \, dp. \tag{6.6}$$

The molar volume of diamond is less than that of graphite, so the term under the integral is negative. Hence, by applying sufficient pressure, ΔG will eventually become negative, and diamond will be the stable allotrope.

The pressure p corresponding to the equilibrium line at temperature T is found by solving the equation

$$\Delta G_{T0} + \int_0^p \Delta V \, dp = 0. \tag{6.7}$$

Figure 10 shows the equilibrium line as calculated by Berman and Simon (1955) before the synthesis of diamond; it is, in fact, quite close to the line subsequently determined experimentally by Bundy *et al.* (1961). Under pressures of more than about 15,000 atm, diamond is the stable form at room temperature, but the rate of transition from graphite is virtually zero. It is therefore necessary to go to higher temperatures (and, consequently, higher pressures) to achieve a finite rate of reaction.

Some early work on this reaction illustrates the importance of measuring the specific heats to sufficiently low temperatures in order to obtain the correct values for the entropies. Bridgman (1947) heated some small diamonds to about 3000°K at pressures between 15,000 and 30,000 atm. At the lower pressures, the stones were completely graphitized, but at 30,000 atm, they were practically unchanged. We now know that this latter result was due to the pressure reducing the reaction rate, but Bridgman believed it to indicate that the conditions were close to the equilibrium line. However, the equilibrium curve could only pass near

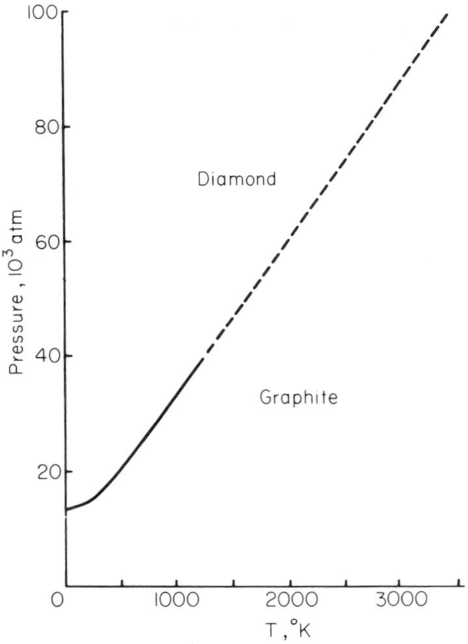

FIG. 10. The full line gives calculated values for the graphite–diamond equilibrium curve up to 1200°K, the dotted line represents a linear extrapolation. (After Berman and Simon (1955).)

Bridgman's points if the entropy of the diamond was actually about $R \ln 2$ greater than that determined calorimetrically. Terms of this order are known to arise from transitions in other substances at low temperatures, and therefore careful measurements on the specific heat and entropy of diamond were made down to 0.4°K. These were sufficient to show that no transition occurred at any temperature above a few hundredths of a degree, so that there was no reason for amending the equilibrium diagram (Berman and Simon, 1955). Thus, measurements at 0.4°K may, via the third law, have a direct bearing on reactions at 3000°K.

The above procedure to deduce the equilibrium conditions for a simple polymorphic transition may be generalized to discuss any form of chemical equilibrium. However, we shall refer only to one case of historical interest, the determination of the equilibrium constants of gas reactions. These constants are of considerable practical importance, but are difficult to measure directly over a wide range of pressure and temperature. It is therefore desirable to deduce them from thermodynamic information such as specific heats and heats of reaction which are more readily available. In fact, the equilibrium constant is directly related to the change in the Gibbs free energy during the reaction (see, for example, Chapter 15 of Lewis and Randall (1961). Hence, the third law, by indicating the value of ΔS, again leads to the equilibrium conditions.

There is not much information available to make direct comparisons of measured and calculated values of the equilibrium constants for gas reactions, because of the difficulty of obtaining accurate values by direct measurement. However, as an example, we quote the results of Newton and Dodge (1934), who evaluated the equilibrium constant for the reaction

$$CO + 2H_2 \rightleftharpoons CH_3OH$$

at temperatures of 225, 250, and 276°C, and 3-atm pressure from purely thermal data. They obtained results within a few parts per cent of values determined experimentally; bearing in mind the logarithmic relation between ΔG and the equilibrium constant, the agreement is satisfactory.

It is of interest to note that the third law was originally put forward by Nernst, as the heat theorem, in order to calculate the equilibrium constants of gas reactions. In spite of the scarcity of specific heat measurements at low temperatures, Nernst was able to predict quite good values of the equilibrium constants from calorimetric data. Today, however, atomic and molecular constants may be found with a high degree of precision, so accurate values of gas entropies are best obtained by statistical methods. Thus, the third law has rather outlived its usefulness so far as gas reactions are concerned, although it still provides a useful check on the calculated values of the entropies, as discussed in the following sections.

VII. Internal Equilibrium in Solids

In the previous sections, we have only discussed systems which have been in complete internal equilibrium. We now consider the nature of internal equilibrium, and the situation which arises when one or more aspects of a system are not in equilibrium.

Equilibrium may be defined in two ways. First, an equilibrium state is one whose thermodynamic functions do not change with time. Alternatively, according to the second law of thermodynamics, the equilibrium state of an isothermal isobaric system is that with the minimum value of the Gibbs free energy. Although these two definitions are consistent, many systems which are apparently stable in time have a free energy greater than the minimum possible. For example, diamond under normal conditions has a larger free energy than graphite, but diamond is metastable because large energy barriers have to be overcome before the atoms can rearrange themselves into the graphite lattice.

Other examples of systems with free energies greater than the minimum are those metallic alloys consisting of a disordered solution of two or more components. Theoretical studies show that the free energy of these alloys is reduced if the component atoms position themselves in a regular array to form a superlattice (see, for example, Wilson (1957)). Solid solutions do, in fact, form superlattices, provided that the reduction in potential energy is sufficient to overcome the disruptive and randomizing effect of the thermal motion. Hence, we often observe a transition from a high-temperature disordered state to a superlattice state at a sufficiently low temperature. In some alloys, the decrease in the free energy to be expected on forming a superlattice is small, so that we expect the transition to occur at a correspondingly low temperature. The rate of diffusion of the atoms may then be so low that the disordered phase remains *frozen-in* as an apparently stable phase down to the lowest temperatures.

Several features of the internal equilibrium of solids are illustrated by some classic studies of the behavior of glycerol. Below the melting point, which is in the region of room temperature, the most stable form of glycerol is a crystalline solid, but it also exists in another apparently stable form as a supercooled liquid. This liquid has a free energy appreciably greater than that of the solid, but if precautions are taken to avoid nucleation, it may be cooled to the lowest temperatures without any onset of crystallization. It is therefore possible to measure the specific heat of both the solid and the supercooled liquid over a wide range of temperatures. The results obtained (Fig. 11) show that there is a sharp drop in the specific heat of the supercooled liquid in a narrow temperature interval termed the *transition region.*

To understand the form of the specific heat curves, we note that the viscosities of glycerol and other supercooled liquids increase very rapidly as the temperature falls. Although the transition occurs at different temperatures in different materials, the viscosity in the transition region is usually of the order of 10^{13} poise. It follows that the stress required to maintain a constant strain will relax to $1/e$ of its value in the order of seconds or minutes. As the viscosity changes rapidly, we find that, at temperatures just above the transition region, the atoms have all the mobility characteristic of a liquid. At temperatures just below the transition region, they cannot change their configurations in any reasonable time, and the configurations are said to be *frozen in*. The only observable modes of motion are then the vibrations characteristic of a solid.

The specific heat of crystalline glycerol (curve *a* in Fig. 11) arises from the vibrations of the atoms about their mean position in the lattice. At relatively high temperatures, the specific heat of the supercooled liquid (curve *b*) is greater than that of the solid because more complex types of motion are possible in the liquid state. Then, at about 180°K, the specific heat of the liquid falls rapidly to a value close to that of the crystalline solid. This low-temperature state of supercooled glycerol is not a crystalline solid, because it contains all the spatial disorder characteristics of a liquid which has been abruptly frozen-in, as may be shown by X-ray studies. This quite different state is termed a *glass*, that is, a liquid which has become so viscous that configurational changes are no longer possible *in reasonable times*. Hence, these changes cannot contribute to the specific heat when measured in the normal way.

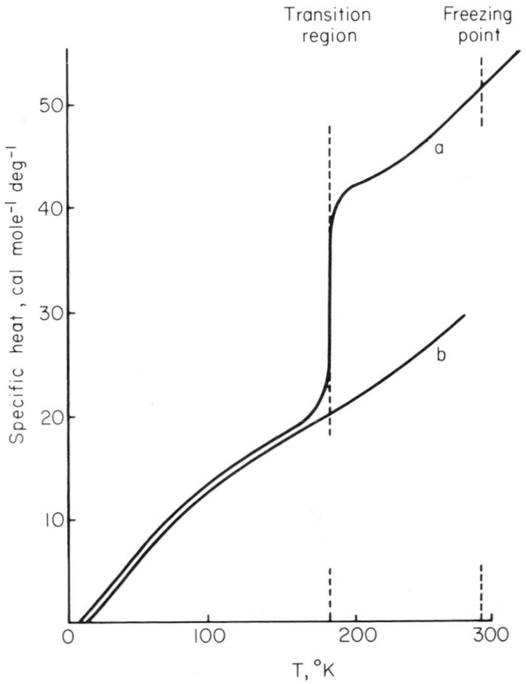

FIG. 11. The specific heat of glycerol: (a) crystal, (b) liquid, supercooled liquid, and glass. (After Simon and Lange (1926).)

A measurement of the latent heat shows that liquid glycerol at the melting point has an entropy about 15 e.u. greater than that of the solid. The difference at lower temperatures may then be estimated by integrating the specific heat curves in Fig. 11, the values obtained being shown in Fig. 12. We see that, below the transition region, the entropy difference between the glassy and crystalline states remains substantially constant at about 5 e.u. Because the glassy state is not in internal equilibrium, the entropy difference between the two states remains finite down to absolute zero.

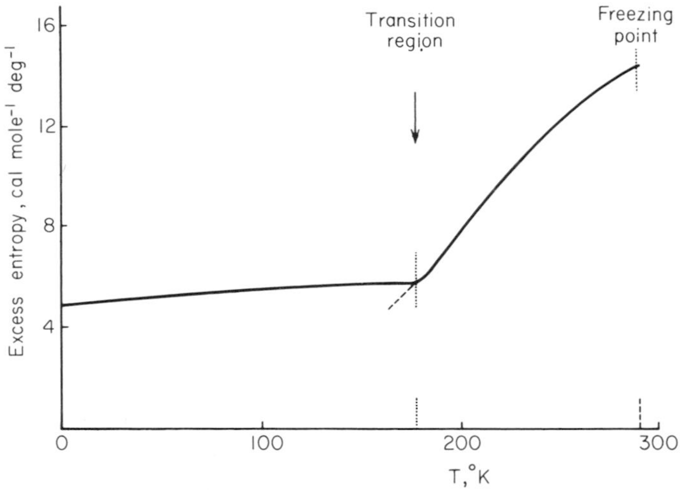

FIG. 12. The full line shows the amount by which the entropy of supercooled glycerol exceeds that of crystalline glycerol, calculated from the results of Fig. 11. The dashed line shows the entropy difference calculated from the results of Fig. 13. (After Simon (1956).)

The relevance of reaction rate to conditions of internal equilibrium is underlined by the experiments of Oblad and Newton (1937), who used the method of mixtures to measure the heat content of glycerol which had been previously allowed to rest at a temperature T for periods of up to 168 hr. Above the transition region, the heat content was found to be independent of the time of resting, as would be expected. However, below the transition region, the heat content decreased considerably if the glycerol was rested for long periods. periods. After 5 hours at 174°K, the heat content was 271 cal mol^{-1}, but after 72 hr, it fell to 190 cal mol^{-1}. Thus, at temperatures only a few degrees below the transition region, a period of several days is required for the glass to take up its equilibrium

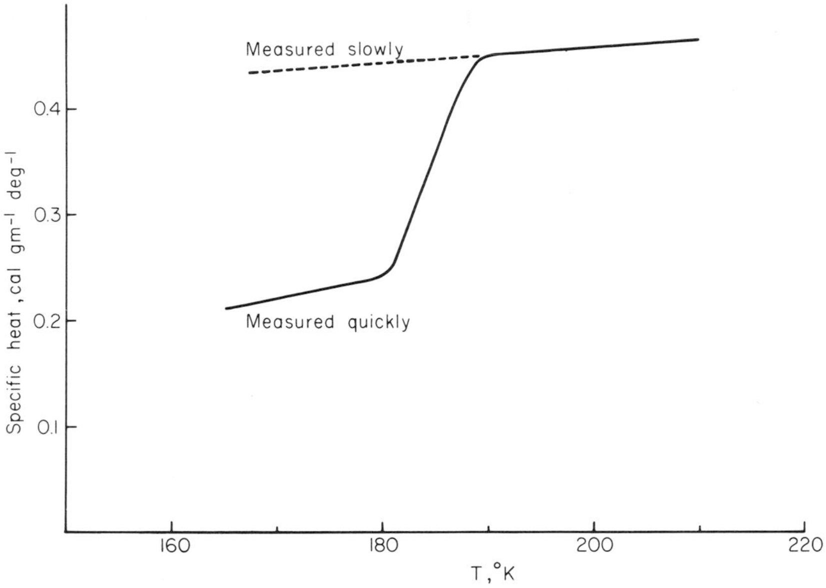

FIG. 13. The specific heat of supercooled glycerol. (After Oblad and Newton (1937).)

condition, and it was not practical to extend the measurements to any lower temperatures.

The above experiment shows that normal calorimetric measurements on glycerol below the transition region do not indicate the true equilibrium values of the thermodynamic functions. Figure 13 shows the specific heats obtained by normal techniques, while the dotted line shows the specific heats deduced from the heat contents obtained by Oblad and Newton for well-rested specimens. If we now calculate the entropy difference between the liquid and solid states using the equilibrium specific heats given by Oblad and Newton, we obtain the values of ΔS shown by the dashed line in Fig. 12. Although the line only extends a few degrees below the transition region, there is little doubt that, if the measurements could be extended to lower temperatures, the curve would extrapolate to zero at absolute zero in accord with the third law for systems in internal equilibrium.

Other quantitative measurements of frozen-in disorder have been made by Eastman and Milner (1933), who studied a solid solution containing 27.2% AgCl and 72.8% AgBr. The entropy at 298.16°K was estimated from the relation

$$S_{\text{sol}} = 0.272 S_{\text{AgCl}} + 0.728 S_{\text{AgBr}} + H/298.16, \tag{7.1}$$

where H is the heat of mixing, and S_{AgCl} and S_{AgBr} are the molar entropies of AgCl and AgBr obtained by integrating their specific heats. The value of the entropy so obtained was 26.06 ± 0.10 e.u. On the other hand, a direct integration of the specific heat of the solution gave an entropy of only 24.99 ± 0.10 e.u. The difference between these two values of the entropy arises because the disordered arrangement of the halogen atoms persists as a frozen-in state down to absolute zero. This state has an excess entropy of mixing which does not vanish, because the system is not in internal equilibrium. The value of this excess entropy, $-R(0.272 \times \ln 0.272 + 0.728 \ln 0.728) = 1.16$ e.u., accounts for the observed entropy discrepancy to within the experimental error.

The third law does no more than state that the entropy vanishes in systems in complete internal equilibrium. It does not say how the entropy behaves in a system which is not in complete internal equilibrium. In a glassy material, some disorder is frozen-in, and the entropy remains finite at absolute zero. On the other hand, the entropies of the metastable forms of the allotropic solids discussed in Section IV and VI do go to zero. How does the third law distinguish between these two types of behavior?

When discussing internal equilibrium, we must keep in mind the distinction between frozen-in and metastable states. In a metastable state, the atoms are located in well-defined potential wells, all of the same form, and are therefore stable with respect to small displacements. The system is in a clearly defined thermodynamic state, and is completely determined by the thermodynamic functions p, V, T, etc. In contrast, the atoms in a glass are in the random conditions in which they happen to have been frozen in at the transition region. Any small displacement will therefore tend to move the system to some different state. A glass has no tendency to maintain its configuration, and its entropy is not a unique value of the thermodynamic variables.

We can now differentiate between the behavior of frozen-in and metastable states by making use of the concept of aspects used in the definition of the third law. The entropy of a glass, such as supercooled glycerol, arises from a vibrational aspect and from an aspect associated with the configurational disorder of the atoms. The vibrational aspect remains in equilibrium down to the lowest temperatures, its specific heat is very similar to that of the crystalline solid (Fig. 11), and its entropy tends to zero. On the other hand, the configurational aspect becomes frozen-in and contributes a residual entropy at absolute zero.

The position with respect to metastable allotropic solids is rather different. Quite generally, we would expect that a system consisting of a simple solid with two allotropic forms α and β in equilibrium would have at least three aspects: the vibrational aspects of the allotropes α and β, and an aspect arising from the entropy of mixing. Just as in the case of a glass, we expect the entropy of the vibrational aspects to go to zero, but an entropy of mixing to be frozen in. However, the nature of metastable equilibrium is such that it is often possible to obtain one allotrope completely isolated from its other forms. Thus, we can make measurements on, say, gray tin completely free of white tin, and vice versa. Hence, the specific heat measurements on allotropic solids described in Sections IV and VI relate only to the simple vibrational aspect of the particular allotrope in question. That is, we are dealing only with the entropy of a vibrational aspect; this vanishes in accord with the third law, as we expect, and Eq. (4.2) and (4.3) are quite valid.

VIII. Nuclei and Entropy

The atomic nuclei contribute to the entropy of a substance in several ways. The disordered orientation of magnetic moments gives rise to a magnetic entropy. If an element has two or more isotopes, an entropy of mixing must be associated with them. We now consider the behavior of the entropy arising from these and other nuclear aspects of a system. Nuclear aspects are generally so weakly coupled to the rest of the system that we can regard them as being virtually independent of the other aspects, but we discuss some important exceptions in Section VIII,C.

A. Nuclear Spin

Many atomic nuclei have a spin and an associated magnetic moment, the spin being characterized by a quantum number I. At high temperatures, these moments are free to take up any one of $(2I + 1)$ orientations, and there is an associated entropy of magnitude $R \ln(2I + 1)$. According to the third law, this contribution must vanish at sufficiently low temperatures, and internuclear forces eventually produce an ordered arrangement of the spins. However, these forces are usually due to the magnetic interactions between the nuclei, and are so small that an appreciable orientation of the nuclei occurs only at temperatures of the order of 10^{-5}°K.

It follows that a normal specific heat measurement, and its extrapolation to zero, does not take account of the nuclear spin entropy. Hence, if the calorimetric entropy of a gas is to agree with the statistical entropy, we must ignore the nuclear spin states when computing the partition function. In fact, as the coupling between the spins and the other aspects of a system is generally very weak, we can usually ignore the spin aspect completely when discussing the other aspects. Thus, our previous conclusions regarding the entropy of simple systems are usually unaffected if the nuclei have spins. For example, when a solid melts, or a liquid vaporizes, there may be a nuclear spin entropy associated with each phase, but this is usually the same in each phase, and therefore has no effect on the thermodynamics of the transition. Similarly, we may ignore the effects of nuclear spin in a chemical reaction such as $A + B \rightleftharpoons C + D$. Each nuclear species remains unchanged during the reaction and will contribute a constant term of the form $R \ln(2I + 1)$ to the entropy. Thus, the total spin entropy remains constant, and does not affect the thermodynamics of the reaction.

B. Isotopic Mixing

Consider a system in which at least one of the elements is present in more than one isotopic form—say, for example, sodium chloride. The chlorine isotopes ^{35}Cl and ^{37}Cl will be distributed randomly, and there is an entropy of mixing corresponding to this disorder. According to the third law, any entropy of mixing must finally vanish, and we therefore expect that the isotopes will eventually take up some ordered arrangement. However, the energy differences involved are so small that ordering will only take place at a very low temperature. Nearly always, the substance is then in the solid phase, so that the disorder is frozen-in, and hence does not contribute to the calorimetric entropy. In fact, the only occasion when isotopic demixing is observed is in the phase separation of liquid mixtures of ^{3}He and ^{4}He (Section III).

It thus follows that calorimetric and statistical entropies will only agree if this entropy of mixing is not included in the statistical entropy. In fact, for nearly all purposes, it is permissible to ignore the entropy of mixing of the isotopes. Of course, if we wish to consider processes in which separation of the isotopes occurs, we must treat each isotope as a separate species, and include the mixing term. (The partition function of a substance with more than one isotope must also take account of the different values of the atomic mass and other parameters by using suitably weighted parameters; see, for example, Fowler and Guggenheim (1939).)

C. Nuclei and Symmetry

We have already noted that, if the nuclear aspects of a system are independent of the other aspects, we can generally discuss the behavior of the system without reference to the spins. However, the nuclear aspects are sometimes linked to the others by symmetry requirements, as in the case of certain diatomic molecules. We therefore briefly outline the effect of nuclear spin on the partition function of hetero and homonuclear molecules of the type AB and AA, where A and B are atoms of two different elements.

The partition function for the spin and rotational aspects of a heteronuclear molecule is given by

$$Z = \varrho_{\mathrm{A}}\varrho_{\mathrm{B}} \sum_{\varkappa=0123\cdots} (2\varkappa + 1) \exp[-\varkappa(\varkappa + 1)\theta_\tau/T], \tag{8.1}$$

where $\varrho_{\mathrm{A}} = 2I_{\mathrm{A}} + 1$ and $\varrho_{\mathrm{B}} = 2I_{\mathrm{B}} + 1$ are the statistical weights associated with each spin, and the term under the summation sign is the usual expression for the rotational partition function. For convenience, we abbreviate Eq. (8.1) to

$$Z = \varrho_{\mathrm{A}}\varrho_{\mathrm{B}} \sum_{0123} . \tag{8.2}$$

The effect of the spins is merely to introduce the additional factor $\varrho_{\mathrm{A}}\varrho_{\mathrm{B}}$ corresponding to an additional entropy of $(R \ln \varrho_{\mathrm{A}} + R \ln \varrho_{\mathrm{B}})$. This is the term discussed in Section VIII,A, where we showed that it may generally be ignored.

The case of a homonuclear molecule of the form AA is more complex, because the total wave function must be symmetric or antisymmetric, depending on whether the nuclei have even or odd mass. The symmetry of this total wave function is usually determined by the product of the spin and rotational wave functions. If each nucleus has a statistical weight ϱ, the two spins may be described by $\frac{1}{2}\varrho(\varrho + 1)$ symmetric wave functions and $\frac{1}{2}\varrho(\varrho - 1)$ antisymmetrical functions. The symmetric states have a total nuclear spin of $S = 1$ and are termed *ortho* states, while the antisymmetric states have a total nuclear spin $S = 0$ and are termed *para*. Let us suppose that we are dealing with a molecule consisting of two similar atoms of even mass number, so that the total wave function must be symmetric. It follows that the para states can only be combined with antisymmetrical rotational states, and ortho states with symmetrical rotation states. Hence, the partition function of a homonuclear diatomic molecule has the form

$$Z_{\text{homo}} = \left[\tfrac{1}{2}\varrho(\varrho \pm 1) \sum_{024}\right] + \left[\tfrac{1}{2}\varrho(\varrho \mp 1) \sum_{135}\right], \tag{8.3}$$

where the upper signs are to be taken if the total wave function is symmetric, and the lower signs if antisymmetric.

Equation (8.3) leads in general to different thermodynamic functions than does Eq. (8.2). However, it turns out that, for all substances *in the gaseous state* except the isotopes of hydrogen, $\theta_r \ll T$, so the quantization of the levels has little effect on the thermal motion, and the partition function assumes the classical value. In this case,

$$\sum_{135} = \sum_{024} = \tfrac{1}{2} \sum_{0123}, \tag{8.4}$$

and Eq. (8.3) reduces to

$$Z_{\text{homo}} = \tfrac{1}{2}\varrho^2 \sum_{0123}. \tag{8.5}$$

That is, the only effect of replacing Eq. (8.2) by (8.3) is to introduce a factor $\frac{1}{2}$. Hence, the rotational entropy of the homonuclear molecules is less than that of similar heteronuclear molecules by a constant term $R \ln 2$.

At first sight, this last term would appear very relevant to the calculation of the entropy of polyatomic gases, but in practice it does not matter very much. Suppose we wish to estimate the entropy of gaseous chlorine, which consists of a mixture of $^{35}Cl^{35}Cl$, $^{37}Cl^{37}Cl$, and $^{35}Cl^{37}Cl$; then the partition function must be constructed using suitably weighted functions for the masses and other molecular parameters. In addition, the rotational entropy of $^{35}Cl^{37}Cl$ will be greater than that of $^{35}Cl^{35}Cl$ and $^{37}Cl^{37}Cl$ by $R \ln 2$ per mole of $^{35}Cl^{37}Cl$. However, the $^{35}Cl^{37}Cl$ will also have an excess entropy in the solid state, again associated with the possibility of forming a different configuration by rotating the molecule through 180°. Therefore, this entropy will not affect the thermodynamics of, say, vaporization, as it is the same in both phases and may be ignored. Moreover, as this entropy will be frozen-in at low temperatures, it will not contribute to the specific heat. Thus, we shall obtain agreement between calorimetric and statistical entropies if we again ignore the presence of the isotopes, and use the partition function for heteronuclear molecules.

The position must be further considered if equilibrium between the ortho and para species only comes about very slowly. Hydrogen gas at low temperatures consists of a mixture of two virtually independent species, ortho and para; we must therefore write the partition function of each species separately, thus

$$Z_o = \tfrac{1}{2}\varrho(\varrho + 1) \sum_{135}, \qquad Z_p = \tfrac{1}{2}\varrho(\varrho - 1) \sum_{024}. \tag{8.6}$$

The derived thermodynamic quantities are then obtained by adding together the values for the ortho and para components derived from the separate partition functions (see, for example, Wilks (1961)). For all other gases, the ratio θ_r/T is small, so the partition functions always assume the high-temperature classical values. The thermodynamic functions are then independent of any ortho-para-type equilibrium.

Finally, we note a quite different way in which the symmetry of the nuclear wave function affects the entropy of a system. The strength of the magnetic interactions between the nuclear spins in liquid ^3He is such that they should produce an ordering of the spins at a temperature of the order of 10^{-5}°K. In fact, because the total wave function must be antisymmetrical, the value of the *electrostatic* exchange interaction also depends on the orientation of the spins. This exchange energy is much greater than the magnetic interaction, and produces an antiferromagnetic spin ordering below about 0.5°K. As the energy involved is still relatively small, this type of effect is observed only in ^3He, which is still a liquid at very low temperatures.

It is of interest to mention a rather unusual consequence of the exchange ordering in liquid ^3He, which results in the form of the melting curve being appreciably different from that of liquid ^4He described in Section III. The atoms in solid ^3He are much more localized than in the liquid, so the exchange forces are correspondingly smaller. It follows that spin ordering in the solid will only occur at a much lower temperature than in the liquid. Hence, below about 0.3°K, solid ^3He along the melting curve has a greater entropy than the adjacent liquid, and the melting curve in this region has a negative slope (Fig. 14). Eventually, of course, the slope of the curve must become zero at absolute zero, as the entropies of solid and liquid both tend to zero; for further details, see Wilks (1967).

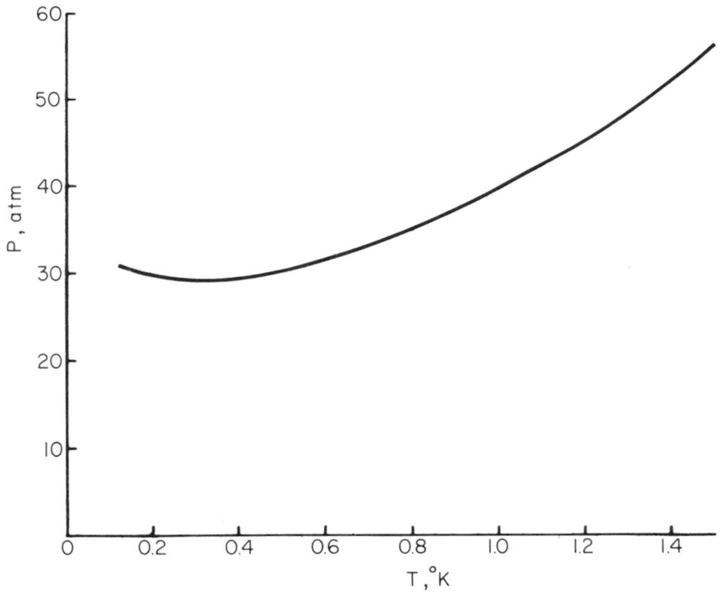

FIG. 14. The melting curve of liquid ^{3}He. (After Edwards *et al.* (1961).)

IX. The Third Law and the Entropy of Gases

In this and the next section, we describe how the third law provides information concerning the thermal motion in gases and solids.

A. Statistical and Thermal Entropies

As discussed in Section IV, the entropy of a gas may be obtained from thermal measurements and the third law, or from a statistical calculation, the two methods giving results in good agreement with each other. However, it sometimes happens that discrepancies are found between the statistical and thermal values. For example, Table IV gives figures which have been quoted for various gases, and we see that the discrepancies are significantly larger than the experimenal errors as typified by the values in Table III. These discrepancies are a valuable indication that the nature of the system in question has not been fully understood.

A discrepancy may arise in a number of ways. It may indicate that some of the thermal measurements are faulty, or that some arithmetical slip has occurred in the calculations. More fundamentally, there are at least three other possibilities. The statistical estimate of the entropy may be incorrect. This is unlikely if the molecules of the gas are fairly simple, but the calculation of the partition function of large organic molecules is more difficult. Again, symmetry requirements imposed by the nuclear spins may affect the form of the partition function. Secondly,

TABLE IV

CALCULATED AND MEASURED ENTROPIES OF GASES

	Temperature	Entropy (E.U.)		Reference
		Statistical	Thermal	
CO	B.P.	38.32	37.2	[a]
NO	B.P.	43.75	43.03	[b]
N_2O	B.P.	48.50	47.36	[c]
H_2O	298°K	45.10	44.29	[d]
D_2O	298°K	46.66	45.89	[e]
CH_3D	B.P.	39.49	36.72	[f]

[a] Clayton and Giauque (1932).
[b] Johnston and Giauque (1929).
[c] Blue and Giauque (1935).
[d] Giauque and Stout (1936).
[e] Long and Kemp (1936).
[f] Clusius *et al.* (1937).

the specific heat of the solid may not have been measured to sufficiently low temperatures before extrapolating to zero. For example, Table II shows that solid nitrogen undergoes a phase transition at 35.61°K; it is therefore necessary to measure the heat of transition at 35.61°K, and the specific heat to considerably lower temperatures, in order to get a reliable extrapolation to absolute zero. Thirdly, the solid phase may not have been in internal equilibrium, as discussed for the case of glycerol in Section VII. The discrepancies in Table IV arise in the values used for the entropy of the solid; but before considering these points in the next section, we first consider discrepancies associated with the partition function.

B. Hindered Rotation

A discrepancy between the values of the thermal and statistical entropies of a gas may have a constant value, or may depend on the temperature at which the comparison is made. Provided that the specific heat of the gas has been measured correctly, any such dependence on temperature indicates that the partition function used to describe the gas is incorrect.

As an example, we consider the entropy of dimethylacetylene $H_3C—C \equiv C—CH_3$. It is usual to assume that both methyl groups are capable of free rotation about the axis of the $C\equiv C$ triple bond. The evaluation of the partition function is then more involved than for simple diatomic molecules, but leads to an entropy in good agreement with the calorimetric one (Osborne *et al.*, 1940; Yost *et al.*, 1941). On the other hand, if we make a similar assumption about ethane $H_3C—CH_3$, namely, that each methyl group is free to rotate about the C—C bond, the calculated entropy does not agree with that obtained from thermal data (Kemp and Pitzer, 1937).

The behavior of ethane has been explained by postulating that the rotation of the methyl group is restricted by internal potential barriers. Pitzer and Gwinn (1942) discussed the situation when the methyl group rotates in a potential well of the form $\frac{1}{2}V_0(1 - \cos 3\phi)$, where ϕ is the azimuthal angle, and V_0 some constant. This potential leads to a different partition function and different thermodynamic quantities than those for simple rotation. The value of V_0 is derived from the difference between the statistical and calorimetric entropies at one temperature, and is then used to calculate the entropy at other temperatures. As the entropies obtained in this way agreed with the calorimetric ones, there was good ground for thinking that the treatment was essentially correct.

The first studies of "hindered rotation" were based primarily on comparisons of the thermal and statistical entropies. The principal features of the theoretical treatment have subsequently been confirmed by other techniques which permit more detailed studies of the molecular motion. These techniques include infrared and Raman spectroscopy, inelastic slow-neutron scattering, and measurements of ultrasonic relaxation. For a summary of such measurements on ethane, see Strong and Brugger (1967), who also describe a neutron scattering experiment in detail. Other instructive studies of hindered rotation are given in discussions of the entropy of methylchloroform (Pitzer and Hollenberg, 1953), of methylalcohol (Ivash *et al.*, 1955), and of octafluoropropane (Pace and Plaush, 1967).

X. Thermal Motion in Solids

We now discuss discrepancies between the thermal and statistical entropies (like those shown in Table IV) where the magnitude of the discrepancy does not depend on the temperature at which the entropy of the gas is evaluated. These discrepancies usually arise because of an incomplete understanding of the thermal motion in the solid state. It is, however, only quite recently that a fairly clear picture of the details of the thermal motion in molecular solids has begun to appear. We now give a brief review of some of the main ideas involved.

A. Frozen-In States

The behavior of glycerol (Section VII) shows that, if a substance is built up of large irregular molecules, a considerable amount of disorder may be frozen-in in the solid state. This frozen-in entropy is not reflected in the specific heat, so a discrepancy arises between the statistical and thermal entropies of the gas. Another example of a frozen-in state is that produced by the freezing-in of the disorder of solid solutions, also discussed in Section VII.

Quite generally, we may say that internal forces in a system will tend to produce an ordered arrangement, and a reduction in entropy, below a certain transition temperature T_i. More particularly, if the energy of interaction between two molecules has the value E_i, then

$$T_i \sim E_i/k, \tag{10.1}$$

where k is Boltzmann's constant. However, if the interaction is weak, the temperature T_i may be so low that no molecular rearrangement can take place. The motion of the molecules in a solid is restricted by internal potential barriers, rather as parts of a molecule of a gas are affected in hindered rotation (Section IX). If these internal barriers have a height E_b, then the molecules will be unable to change their orientation below a temperature T_b, where $T_b \sim E_b/k$. It follows that entropy of disorder will be frozen-in at low temperatures whenever $E_b > E_i$, that is, in solids where the intermolecular forces tending to produce orientation are weak, but where the barriers opposing molecular motion are relatively high.

Another classic example of frozen-in disorder is the behavior of ice. The angle between the two oxygen–hydrogen bonds in the water molecule is approximately 105°, while X-ray measurements indicate the positions of the oxygen atoms in ice. Hence, by making certain plausible assumptions, it is possible to deduce the number of possible positions for a hydrogen atom. The various arrangements correspond to a configurational entropy of $R \ln 3/2$ or 0.81 e.u., and, if this is frozen in, it accounts for the difference observed in Table IV. For the details of this approach, we refer the reader to the the original paper (Pauling, 1935), and to a confirmatory experiment using neutron diffraction techniques (Wollan *et al.*, 1949) (see also Flubacher *et al.* (1960), Leadbetter (1965)).

B. Molecular Rotation in Solid Hydrogen

In contrast to the frozen-in states discussed above, there are systems in which molecular rotation appears to be maintained down to absolute zero. We begin by comparing the statistical and thermal entropies of hydrogen gas.

As mentioned in Section VIII,C, the hydrogen molecule exists in two species, ortho and para. If we compute the statistical entropy of pure para hydrogen, it is found to be in good agreement with that deduced from thermal measurements (Johnston *et al.*, 1950). In particular, solid para hydrogen is an apparently straightforward substance whose specific heat may be safely extrapolated to zero from about 10°K. However, the behavior of ortho-para mixtures is complicated by the presence of the ortho molecules.

We have already referred to the restrictions imposed on the rotational states of the hydrogen molecule by symmetry requirements. *Ortho* molecules have a symmetrical spin wave function, and therefore their rotational states can only be antisymmetrical. It follows that the lowest rotational level for an *ortho* molecule is not the ground state, but the first excited state. Hence, the molecule has sufficient energy to overcome the small barrier potentials E_b in the solid, and the molecules continue to rotate down to the lowest temperatures, as indicated by magnetic resonance techniques (e.g., Reif and Purcell (1953)).

The specific heat of solid ortho-para hydrogen shows a large anomaly in the region of 1.5°K, the form of which depends on the *ortho* concentration (Fig. 15). Measurements show that this anomaly corresponds to an entropy change of $R \ln 3$ per mole of *ortho* hydrogen. Studies using nuclear magnetic resonance techniques show that the *ortho* molecules are rotating both above and below the anomaly, as we would expect from the previous discussion. The first excited rotational state is threefold degenerate, and this degeneracy is removed below the anomaly by an interaction between the molecules which depends on their relative orientation. Hence, the associated change in the entropy is $R \ln 3$.

We are now in a position to compare the statistical and thermal entropies of a gaseous mixture of *ortho* and *para* hydrogen. The statistical entropy is derived as in Section VIII,C, and includes the nuclear spin entropy. The specific heat measurements are extrapolated to zero from about 1°K, and include the above-mentioned term of $R \ln 3$ per mole of ortho

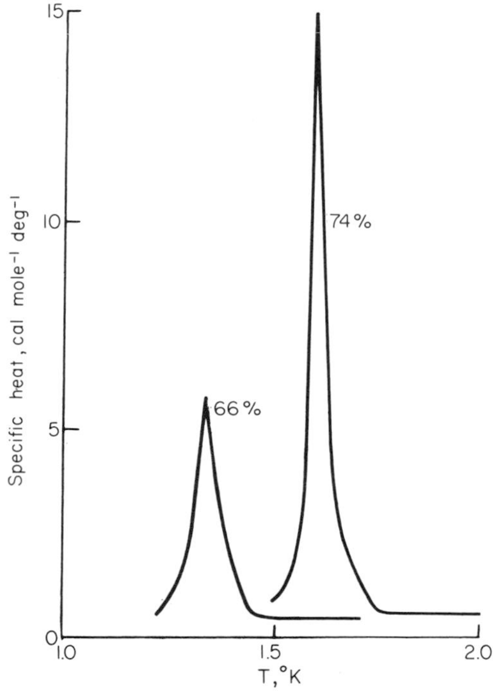

FIG. 15. The specific heat of solid hydrogen containing 66% and 74% of *ortho* molecules. (After Hill and Ricketson (1954).)

hydrogen. In addition, at 1°K, each molecule has a nuclear spin $S = 1$, which contributes an entropy of $R \ln 3$ per mole of *ortho* hydrogen, and there is an entropy of mixing $R[x \ln x + (1 - x) \ln (1 - x)]$ per mole of hydrogen, where x is the fraction of ortho. This entropy of mixing will remain frozen in, while the spin degeneracy will only be removed at extremely low temperatures. Hence, the measured thermal entropy must be augmented by both these terms. It is then in good agreement with the statistical entropy. Thus, solid ortho hydrogen gives a particularly striking verification of the third law. Even though the molecules are still rotating in an excited state at absolute zero, this motion makes no contribution to the entropy.

C. Molecular Libration and Rotation

We have just given examples of systems in which the molecular motion is completely frozen in, and in which rotation continues down to absolute zero. Very often, however, more complex intermediate situations are possible. If the barriers opposing rotation are relatively high, the rotational degrees of freedom may be limited to a vibratory torsional motion, termed a *libration*. For example, it appears clear that the ammonium group in solid NH_4Cl is librating at temperatures both above and below the characteristic lambda-type anomaly in the specific heat. Studies of the infrared spectrum, of neutron diffraction, and of the change in entropy show that this anomaly is associated with the ordering of the directions of the axes of libration (see Mellor (1964a)).

We now discuss how entropy measurements, together with the third law, have been used to gain information on these more complex modes of motion. Each particular solid must be considered individually, using as many techniques as possible, in order to obtain a full picture of the thermal motion. It is thus becoming apparent that the situation is not always as simple as has sometimes been assumed in the past. We consider a few typical examples.

The thermal and statistical entropies of partially deuterated methane CH_3D given in Table IV differ by 2.77 e.u. (Clusius *et al.*, 1937). In contrast, similar measurements on normal methane CH_4 gave a thermal entropy in good agreement with the statistical entropy, the spin entropy being ignored in the usual way. The discrepancy in the case of CH_3D has been plausibly explained in the past by noting that, in a completely ordered arrangement, the molecules should take up positions with the deuterium atoms in a particular one of the four hydrogen sites. If rotational motion becomes frozen in at a relatively high temperature, then we would expect a residual entropy of $R \ln 4$ or 2.75 e.u., which is close to the observed discrepancy. However, more recent experiments imply that the situation is not so straightforward.

Measurements below 10°K show that, below about 8°K, solid CH_3D exhibits a specific heat considerably in excess of the Debye form assumed by Clusius *et al.* (Colwell *et al.*, 1963, 1965). Hence, the discrepancy between the thermal and statistical entropies is considerably less than $R \ln 4$. Similar measurements on solid CH_4 also show an excess specific heat at lower temperatures, so that reevaluated thermal entropy is actually *greater* than the previously calculated statistical entropy! Colwell *et al.* point out that this behavior may be understood by (a) assuming that the molecules librate down to very low temperatures, and (b) taking into account the fact that the torsional motions will be influenced by the symmetry of the spin wave functions. That is, the allowed levels of libration will depend on the nuclear spin functions. Hence, it is necessary to include the spin entropy when comparing the thermal and statistical entropies in a way similar to that done for hydrogen. In this way, it seems possible to give a reasonable account of the behavior of all the deuterated methanes. For a more general discussion of nuclear spin entropy in molecules with two or more similar atoms, together with an application to solid ethane and diborane, see Hsien-Wen Ko and Steele (1969).

The discrepancies in Table IV concerning CO, NO, and N_2O have previously been explained by assuming that the molecules in a completely ordered state take up an arrangement of the form ABABABAB rather than random arrangements, such as ABABBAAB. Hence, if a disordered high-temperature arrangement becomes frozen-in, there will be an entropy discrepancy, which is of the same order as that observed. However, it now seems likely that further measurements are required; see, for example, the discussion of the behavior of CO by Gill and Morrison (1966). It is to be noted that the thermal and statistical entropies of HCl, which have quite dissimilar atoms, are in good agreement (Mellor, 1964b). Because the atoms are dissimilar, the interactions tending to produce an ordered arrangement are relatively strong, so that ordering occurs at a relatively high temperature where molecular rearrangement is still possible (the decrease in entropy being accompanied by a marked anomaly in the specific heat).

XI. The Third Law and Very Low Temperatures

A. The Unattainability of Absolute Zero

The third law implies that it is impossible to cool any system down to absolute zero (Nernst, 1912). The essential characteristics of a system which may be used to produce cooling are shown in Fig. 16. The entropy must depend substantially on some parameter other than the temperature,

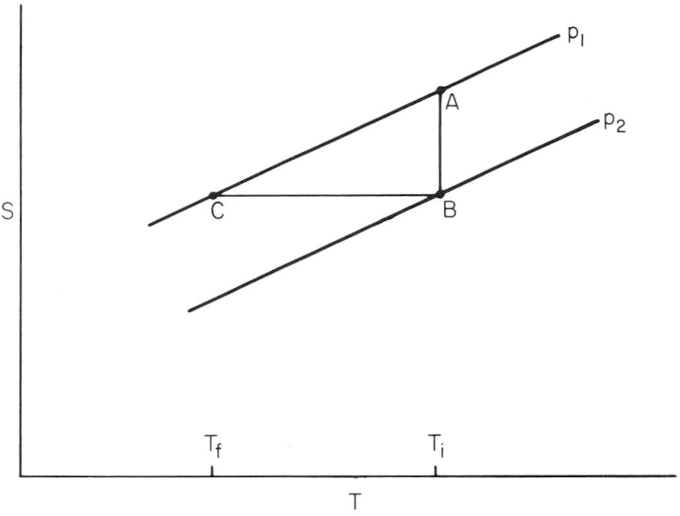

FIG. 16. Schematic entropy diagram of a system used to produce cooling. The parameter p is first varied isothermally so that the system moves from A to B; p is then restored to its original value by a reversible adiabatic process and the temperature falls to C.

and the two curves in the diagram show the entropy for different values of some external parameter p. (To fix our ideas, we may think of the system as a gas, and p_1 and p_2 two pressures such that $p_2 > p_1$.) The entropy is first reduced by varying the parameter p isothermally, thus tracing out the path AB. If the system is now isolated and p restored to its original value, the temperature falls. Thus, the isothermal compression of a gas followed by an adiabatic expansion results in cooling, as in the Claude process. The maximum cooling is obtained by a reversible process, that is, along the isentropic line BC. Similar arguments hold for the systems of paramagnetic salts used to attain temperatures below 1°K.

If the entropy curves of Fig. 16 could be extrapolated as in Fig. 17(a), there would be no difficulty in reaching absolute zero by a process of the type ABC shown in the diagram. However, according to the third law, the entropy curves of systems in internal equilibrium come together as in Fig. 17(b), and the process ABC does not lead to absolute zero. Absolute zero is only accessible by an *infinite* number of processes ABC. (Note that, although large entropy differences may persist down to absolute zero in frozen-in states, they are of no use for attaining low temperatures, as they cannot be varied by any external parameter.)

The unattainability of absolute zero may also be demonstrated more

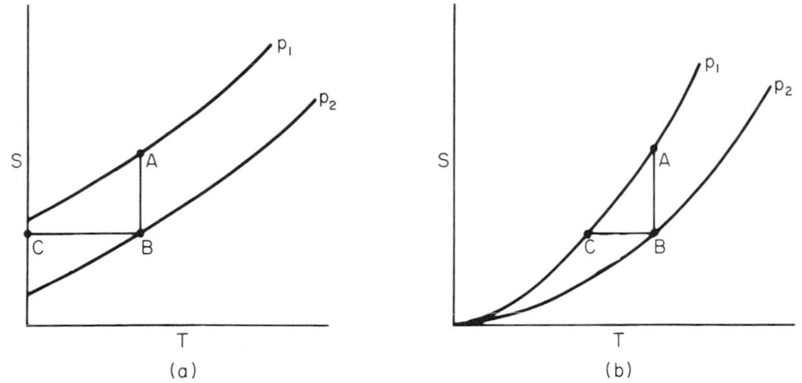

FIG. 17. Schematic entropy diagrams in which a cooling process *ABC* (a) leads to absolute zero, (b) does not lead to absolute zero.

formally by considering the change in a system consequent on varying some parameter adiabatically from a value α to a value β. Let us suppose that the system then passes adiabatically from a state with temperature T_1 and entropy S_1^α to state with a lower temperature T_2 and entropy S_2^β. According to the second law,

$$S_2^\beta \geq S_1^\alpha, \tag{11.1}$$

the equality being applicable if the process is reversible. By the third law

$$S_1^\alpha = \int_0^{T_1} \frac{C_\alpha \, dT}{T} \qquad \text{and} \qquad S_2^\beta = \int_0^{T_2} \frac{C_\beta \, dT}{T}, \tag{11.2}$$

where C_α and C_β are the specific heats corresponding to the values α and β of the variable parameter. If T_2 is to be zero, then S_2^β is also zero, so that, by Eqs. (11.1) and (11.2),

$$\int_0^{T_1} \frac{C_\alpha \, dT}{T} \leq 0. \tag{11.3}$$

This condition cannot be satisfied, as C_α is always a positive quantity; hence, the end temperature of the process cannot be absolute zero.

B. Negative Temperatures

It is well established that, under certain conditions, some systems behave, both statistically and thermodynamically, as if they were at a negative temperature. We briefly discuss the essential features of negative temperatures in order to show that the existence of such conditions does *not* imply the possibility of reaching absolute zero.

According to statistical mechanics, the equilibrium distribution of a number of localized particles between a set of energy levels ε_i is specified by Boltzmann's relation

$$n_i = A \exp(\beta\varepsilon_i), \tag{11.4}$$

where A and β are some constants. Most systems encountered in practice have an infinite number of higher levels, and a meaningful distribution is only obtained if the constant β is negative. However, this requirement is not inherent in the statistics. Let us consider a system in which each particle has access to only a limited number of levels, say four. The distribution of the particles between the levels is shown by the points in Fig. 18(a), and, according to Eq. (11.4), the curve through the points must be an exponential function. On supplying more energy to the system, some of the particles move to higher levels (Fig. 18b); the curve is still exponential, but with a lesser slope. If we continue to supply energy, there must eventually be more particles in the upper levels than the lower, as in Fig. 18(c). Statistically, this situation is quite admissible, provided that the curve is exponential. However, β is now positive, so that, if we still define temperature by the usual relation $T = -1/\beta k$, where k is Boltzmann's constant, then the system has a negative temperature.

Negative temperatures may be observed in nuclear spin systems, which have long spin–lattice relaxation times. If an experiment is conducted

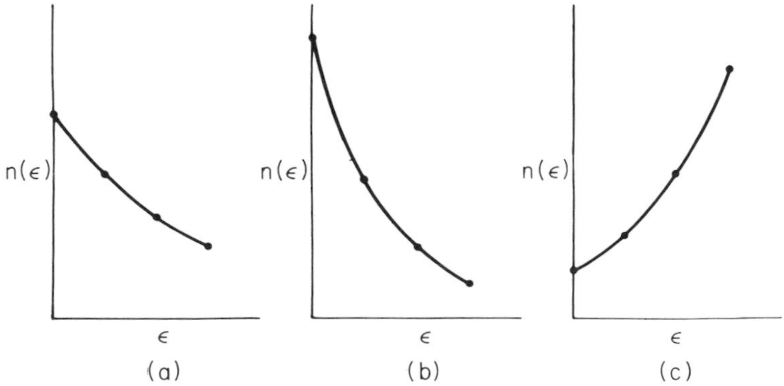

FIG. 18. Possible distributions of particles obeying Boltzmann statistics between four energy levels. In each case, the full line is an exponential function.

rapidly on such systems, the only levels sensed by the nuclei are the $2I + 1$ associated with different orientations of the nuclear spin I. For details of the original experiments, see Pound (1951), Purcell and Pound (1951), and Ramsey and Pound (1951); a short account is given by Wilks (1961).

The concept of negative temperature has been discussed by Ramsey (1956), who shows that the first and second laws of thermodynamics may be formulated so as to include systems characterized by a negative temperature (see also Bazarov (1964)). For our present purpose, it is sufficient to note that systems at a negative temperature are ones in which the population of the levels increases with the height of the level. Negative temperatures are really very hot, and the way to reach them is by supplying energy to populate the upper levels. Thus, the dividing line between positive and negative temperatures occurs when all the levels are equally populated, and, statistically, this corresponds to an infinite temperature. Therefore, systems do not pass through absolute zero when going from positive to negative temperatures. Finally, we note that, as negative absolute zero corresponds to all the particles in the top level, this will be just as inaccessible as the usual positive absolute zero.

C. Alternative Formulation of the Third Law

The first and second laws may be stated in a negative form by saying that it is impossible to construct perpetual motion machines of the first and second kind. Likewise, the third law is sometimes expressed by the statement that, "It is impossible to cool any substance to absolute zero in a finite number of processes."

We now consider the implications of this formulation. Suppose we cool a system, as in Section XI,A, by an adiabatic change of some external parameter from a value α to a value β, so that the system passes from a state with temperature T_1 and entropy S_1^α to a state with a lower temperature T_2 and entropy S_2^β. According to the second law,

$$S_1^\alpha = S^\alpha(0) + \int_0^{T_1} \frac{C_\alpha\, dT}{T},$$
$$S_2^\beta = S^\beta(0) + \int_0^{T_2} \frac{C_\beta\, dT}{T}, \tag{11.5}$$

where $S^\alpha(0)$ and $S^\beta(0)$ are the entropies at absolute zero for the particular values of the variable parameter. The lowest final temperature T_2 will be obtained with a reversible process; in this case, $S_2^\beta = S_1^\alpha$ and

$$S^\beta(0) + \int_0^{T_2} \frac{C_\beta\, dT}{T} = S^\alpha(0) + \int_0^{T_1} \frac{C_\alpha\, dT}{T}. \qquad (11.6)$$

Let us suppose that the temperature T_2 is zero; it then follows that

$$\int_0^{T_1} \frac{C_\alpha\, dT}{T} = S^\beta(0) - S^\alpha(0). \qquad (11.7)$$

Equation (11.7) then defines the value of T_1 which corresponds to a final temperature of zero. It has a real solution only if the right-hand side is positive. Hence, if we assert that it is impossible to reach absolute zero from any temperature T_1, it follows that

$$S^\alpha(0) \geq S^\beta(0).$$

If we now attempt to use the same transition in the reverse direction to reach a temperature $T_1 = 0$ from an initial temperature T_2, a similar argument shows that $S^\alpha(0) \leq S^\beta(0)$. Hence, it follows that $S^\alpha(0) = S^\beta(0)$. That is, the entropy of a system at absolute zero does not depend on any external parameter.

The restriction on a system implied by the unattainability principle is that entropy *differences* between different states of a system disappear at absolute zero. It is not necessary that the entropies themselves vanish. For example, Fig. 19 shows a schematic entropy–temperature diagram in which only entropy differences tend to zero; no finite number of processes such as ABC can lead to absolute zero. At first sight, the result that entropy *differences* should vanish appears significantly different from the statement that the entropy of each aspect should vanish. In fact, as we discussed in Section V, the point is of little importance, as each statement leads to the same practical consequences.

It is sometimes claimed that the unattainability formulation of the third law is preferable to that given previously, as it provides a precise statement of the law without any qualifying clause concerning the condition of internal equilibrium. That is, it states that absolute zero is unattainable, independent of whether or not the system is in internal equilibrium. In fact, the unattainability formulation offers little advantage. We have seen that one of the main applications of the law is to indicate whether the solid phase of a system is in internal equilibrium. Therefore, the concept of internal equilibrium must always be borne in mind when applying the third law, and this is best achieved by the standard formulation.

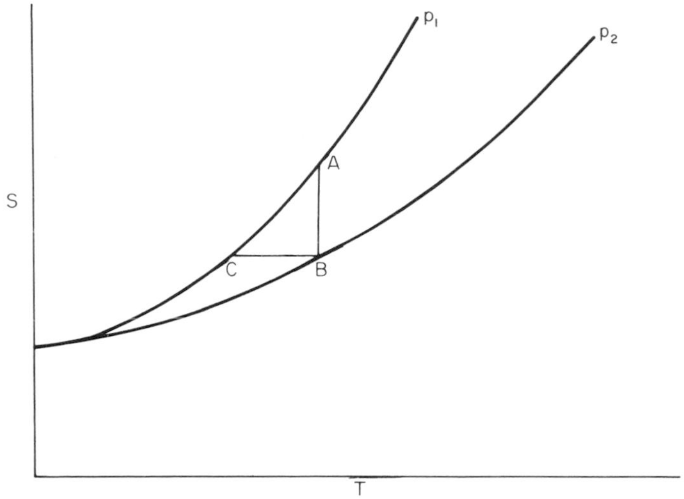

FIG. 19. Schematic entropy diagram of a system in which the entropy does not vanish, but in which absolute zero is unobtainable.

D. THE LOWEST TEMPERATURES

The third law implies that we can never attain the temperature of absolute zero. We now briefly consider to what extent this limitation restricts experimental work at low temperatures.

The method of reaching the lowest temperatures involves the adiabatic demagnetization of magnetic material, whose entropy depends on both temperature and magnetic field.* Figure 20 shows an entropy–temperature curve of a typical paramagnetic salt used to obtain temperatures below 1°K. At about 1°K, the entropy of the lattice is very small, and the entropy arises mainly from the random orientation of the electron spins. The figure shows that this entropy at first decreases quite slowly with falling temperature, and then, at some lower temperature, decreases rapidly, due to cooperative interactions between the dipoles. The actual temperature at which this drop occurs depends on the strength of the interaction forces, and, for the salts generally used, lies between 0.1 and 0.001°K.

* Temperatures down to a few millidegrees absolute may now be obtained with a helium dilution refrigerator (see, for example, Vilches and Wheatley (1967), Betts (1968)). The only method of reaching lower temperatures is by the adiabatic demagnetization of a nuclear spin system.

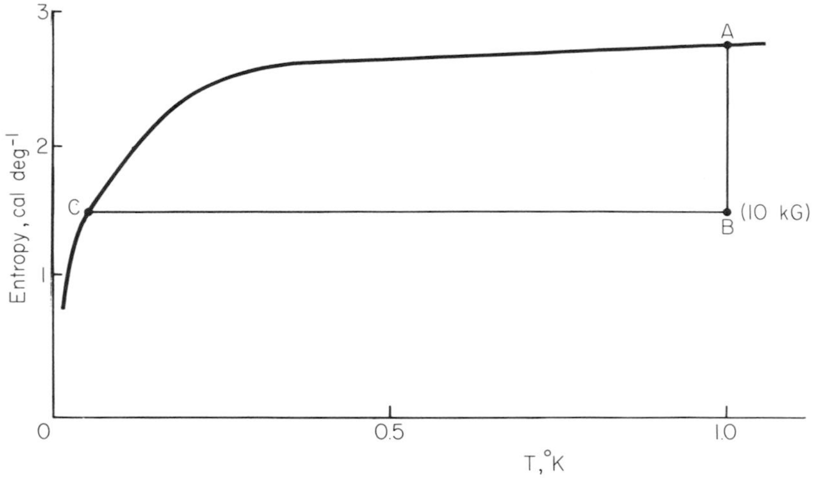

FIG. 20. The entropy of 1 mol of chromium methylamine alum as a function of temperature in zero magnetic field. *ABC* represents a magnetic cooling process. (After Gardner and Kurti (1954).)

To cool the salt from, say, the point *A* at 1°K, the entropy is reduced isothermally by applying a magnetic field and taking away the heat of magnetization, thus moving the system to point *B* on the diagram. A reversible adiabatic demagnetization then carries the system to point *C*, which corresponds to a much lower temperature. Using magnetic fields of the order of 20 kG, the final temperature will lie somewhere in the region where the entropy of the salt is decreasing rapidly, that is, between 0.1 and 0.001°K. Still lower temperatures are possible by the adiabatic demagnetization of nuclear spin systems, because the magnetic interactions between nuclei are generally so small that the spin entropy only falls toward zero at temperatures of the order of 10^{-5}°K. Such temperatures have been attained, although the technical difficulties are considerable because any thermal time constants become very long. For further details of adiabatic demagnetization techniques, see Kurti (1957), Mendoza (1961), Abel *et al.* (1965). These authors also discuss thermometry in the magnetic region, which is generally based on observations of the susceptibility of paramagnetic material, which may be related to the absolute thermodynamic scale by standard thermodynamic procedures.

The entropy of a system provides a useful indication of whether it is desirable to extend measurements of thermodynamic quantities to lower temperatures. If the entropy of a system is small, and may be accounted for by an extrapolation of the specific heat to absolute zero, then the other thermodynamic functions may also be extrapolated to absolute zero and no further experiments are needed. On the other hand, if the system still has an appreciable entropy, say, of the order of R, no such extrapolation is possible (except in the special case when the entropy is frozen-in), and further measurements are necessary. In this case, we can generally make measurements at lower temperatures by using the remaining entropy to cool the system as in Fig. 16. On the other hand, if the entropy is very small, not much cooling can be produced in this way, but all the other thermodynamic functions will be close to their values at absolute zero, and there is little interest in measuring them to lower temperatures. Therefore, the unattainability of absolute zero places little restriction on measurements of thermodynamic functions.

The situation regarding the kinetic properties of a system is somewhat different. By kinetic properties, we understand those characteristics of a system, such as the coefficients of viscosity, thermal conductivity, diffusion, etc., which describe its behavior under certain well-defined conditions representing various forms of departure from complete thermodynamic equilibrium. Just because kinetic processes refer to nonequilibrium conditions, the laws of thermodynamics are not directly applicable to them, and the entropy is not a good guide to the behavior of the kinetic coefficients. For example, the thermal conductivities of dielectric solids exhibit a prominent maximum at a temperature roughly equal to $\theta/30$, where θ is the Debye characteristic temperature; see, for example, Wilks (1967). Yet the lattice entropy in this region is already small and may be readily extrapolated to zero. Similarly, at the superconducting transition in metals, and at the zero-sound transition in liquid ^{3}He, the entropies of the associated aspects have already become quite small. Even in these cases, however, measurements are generally possible by making a suitable choice of temperature and frequency, and the principle of unattainability is again not very restrictive in practice.

Finally, it is worth noting that the Kelvin definition of the absolute temperature scale

$$Q_1/Q_2 = T_1/T_2$$

is not altogether appropriate in the region of very low temperature. We are accustomed to think of this scale at room temperatures as essentially a linear one, so that, at first sight, it might appear that a temperature shift of 1° is equally significant at all temperatures. In this case, the difference in temperature between 10^{-3} and 10^{-5}°K becomes trivial. However, from the point of view of statistical mechanics, it is more meaningful to say that two pairs of temperatures have equivalent differences if their ratios have the same value. That is, the difference between 10^{-3} and 10^{-5}°K is equivalent to that between 3 and 300°K. Hence, we should regard the absolute scale as extending *logarithmically* downward to absolute zero, giving an infinite region for exploration.

References

ABEL, W. R., ANDERSON, A. C., BLACK, W. C., and WHEATLEY, J. C. (1965). *Physics (N. Y.)* **1**, 337.

ABEL, W. R., ANDERSON, A. C., BLACK, W. C., and WHEATLEY, J. C. (1966). *Phys. Rev.* **147**, 111.

ALLEN, J. F., and MISENER, A. D. (1938). *Proc. Cambridge Phil. Soc.* **34**, 299.

BAZAROV, I. P. (1964). "Thermodynamics." Pergamon, New York.

BERMAN, R., and SIMON, F. E. (1955). *Z. Elektrochem.* **59**, 333.

BETTS, D. S. (1968). *Contemp. Phys.* **9**, 97.

BLACKMAN, M. (1955). "Encyclopedia of Physics" (S. Flügge, ed.) Vol. 7, P. I., p. 325. Springer, Berlin.

BLUE, R. W., and GIAUQUE, W. F. (1935). *J. Am. Chem. Soc.* **57**, 991.

BRIDGMAN, P. W. (1947). *J. Chem. Phys.* **15**, 92.

BUNDY, F. P., BOVENKERK, H. P., STRONG, H. M., and WENTORF, R. H. (1961). *J. Chem. Phys.* **35**, 383.

CLAYTON, J. O., and GIAUQUE, W. F. (1932). *J. Am. Chem. Soc.* **54**, 2610.

CLUSIUS, K., and FRANK, A. (1943). *Z. Elektrochem.* **49**, 308.

CLUSIUS, K., POPP, L., and FRANK, A. (1937). *Physica* **4**, 1105.

COLWELL, J. H., GILL, E. K., and MORRISON, J. A. (1963). *J. Chem. Phys.* **39**, 635.

COLWELL, J. H., GILL, E. K., and MORRISON, J. A. (1965). *J. Chem. Phys.* **42**, 3144.

CORAK, W. S., GARFUNKEL, M. P. SATTERTHWAITE, C. B., and WEXLER, A. (1955). *Phys. Rev.* **98**, 1699.

DE BRUYN OUBOTER, R., and BEENAKKER, J. J. M. (1961). *Physica* **27**, 219.

EASTMAN, E. D., and MCGAVOCK, W. C. (1937). *J. Am. Chem. Soc.* **59**, 145.

EASTMAN, E. D., and MILNER, R. D. (1933). *J. Chem. Phys.* **1**, 444.

EDWARDS, D. O., BAUM, J. L., BREWER, D. F., DAUNT, J. G., and MCWILLIAMS, A. S. (1961). *Proc. 7th Intern. Conf. Low Temp. Phys., Toronto, Ont., 1960* (G. M. Graham and A. C. Hollis-Hallett, eds.), p. 610. North-Holland Publ., Amsterdam.

EDWARDS, D. O., BREWER, D. F., SELIGMAN, P., SKERTIC, M., and YAQUB, M. (1965). *Phys. Rev. Letters* **15**, 773.

FAIRBANK, W. M., and WALTERS, G. K. (1958). *Nuovo Cimento Suppl.* 1, **9**, 297.

FLUBACHER, P., LEADBETTER, A. J., and MORRISON, J. A. (1960). *J. Chem. Phys.* **33**, 1751.

FOWLER, R. H., and GUGGENHEIM, E. A. (1939). "Statistical Thermodynamics." Cambridge Univ. Press, London and New York.

GARDNER, W. E., and KURTI, N. (1954). *Proc. Roy. Soc.* **A223**, 542.

GIAUQUE, W. F., and CLAYTON, J. O. (1933). *J. Am. Chem. Soc.* **55**, 4875.

GIAUQUE, W. F., and JOHNSTON, H. L. (1929). *J. Am. Chem. Soc.* **51**, 2300.

GIAUQUE, W. F., and POWELL, T. M. (1939). *J. Am. Chem. Soc.* **61**, 1970.

GIAUQUE, W. F., and STOUT, J. W. (1936). *J. Am. Chem. Soc.* **58**, 1144.
GIAUQUE, W. F., and WIEBE, R. (1928). *J. Am. Chem. Soc.* **50**, 101.
GILL, E. K., and MORRISON, J. A. (1966). *J. Chem. Phys.* **45**, 1585.
HILL, R. W., and RICKETSON, B. W. A. (1954). *Phil. Mag.* **45**, 277.
HSIEN-WEN KO, and STEELE, W. A. (1969). *J. Chem. Phys.* **51**, 4595.
IVASH, E. V., LI, J. C. M., and PITZER, K. S. (1955). *J. Chem. Phys.* **23**, 1814.
JOHNSTON, H. L., CLARKE, J. T., RIFKIN, E. B., and KERR, E. C. (1950). *J. Am. Chem. Soc.* **72**, 3933.
JOHNSTON, H. L., and GIAUQUE, W. F. (1929). *J. Am. Chem. Soc.* **51**, 3194.
KELLEY, K. K. (1929). *J. Am. Chem. Soc.* **51**, 1400.
KEMP, J. D., and PITZER, K. S. (1937). *J. Am. Chem. Soc.* **59**, 276.
KUPER, C. G. (1968). "Introduction to the Theory of Superconductivity," p. 22. Oxford Univ. Press (Clarendon), London and New York.
KURTI, N. (1957). *Nuovo Cimento Suppl.* 3, **6**, 1101.
LANGE, F. (1924). *Z. Physik. Chem.* **110**, 343.
LASAREW, B. G., and SUDOVSTOV, A. I. (1949). *Dokl. Akad. Nauk (SSSR)* **69**, 345.
LEADBETTER, A. J. (1965). *Proc. Roy. Soc.* **A287**, 403.
LEWIS, G. N., and RANDALL, M. (1961). "Thermodynamics" (revised by K. S. Pitzer and L. Brewer). McGraw-Hill, New York.
LONG, E. A., and KEMP, J. D. (1936). *J. Am. Chem. Soc.* **58**, 1829.
MELLOR, J. W. (1964a). "Supplement to Mellor's Comprehensive Treatise on Inorganic and Theoretical Chemistry," Vol. VIII, Suppl. I Nitrogen, (P. I) p. 384. Longmans, New York.
MELLOR, J. W. (1964b). "Supplement to Mellor's Comprehensive Treatise on Inorganic and Theoretical Chemistry,". Suppl. II, P. I (F, Cl, Br, I), p. 418. Longmans, New York.
MENDOZA, E. (1961). *In* "Experimental Cryophysics" (F. E. Hoare, L. C. Jackson, and N. Kurti, eds.), p. 165, Butterworths, London.
MISENER, A. D. (1940). *Proc. Roy. Soc.* **A174**, 262.
NERNST, W. (1906). *Kgl. Ges. Wiss. Gottingen* **1**, 1.
NERNST, W. (1912). *Ber. Ko. Preuss. Acad.* Feb. 1.
NEWTON, R. F., and DODGE, B. F. (1934). *J. Am. Chem. Soc.* **56**, 1287.
OBLAD, A. G., and NEWTON, R. F. (1937). *J. Am. Chem. Soc.* **59**, 2495.
OSBORNE, D. W., GARNER, C. S., and YOST, D. M. (1940). *J. Chem. Phys.* **8**, 131.
PACE, E. L., and PLAUSH, A. C. (1967). *J. Chem. Phys.* **47**, 38.
PAULING, L. (1935). *J. Am. Soc.* **57**, 2680.
PITZER, K. S., and GWINN, W. D. (1942). *J. Chem. Phys.* **10**, 428.
PITZER, K. S., and HOLLENBERG, J. L. (1953). *J. Am. Chem. Soc.* **75**, 2219.
POUND, R. V. (1951). *Phys. Rev.* **81**, 156.
PURCELL, E. M., and POUND, R. V. (1951). *Phys. Rev.* **81**, 279.
RAMSEY, N. F. (1956). *Phys. Rev.* **103**, 20.
RAMSEY, N. F., and POUND, R. V. (1951). *Phys. Rev.* **81**, 278.
REIF, F., and PURCELL, E. M. (1953). *Phys. Rev.* **91**, 631.
SCHRÖDINGER, E. (1952). "Statistical Thermodynamics," 2nd ed., p. 15. Cambridge Univ. Press, London and New York.
SIMON, F. E. (1927). *Z. Physik* **41**, 806.
SIMON, F. E. (1937). *Physica* **4**, 1089.
SIMON, F. E. (1956). "Year Book of the Physical Society," p. 1. Phys. Soc., London.

SIMON, F. E., and LANGE, F. (1926). *Z. Physik* **38**, 227.
SIMON, F. E., and SWENSON, C. A. (1950). *Nature* **165**, 829.
STEPHENSON, C. C., and GIAUQUE, W. F. (1937). *J. Chem. Phys.* **5**, 149.
STRONG, K. A., and BRUGGER, R. M. (1967). *J. Chem. Phys.* **47**, 421.
TER HAAR, D. (1966). "Elements of Thermostatics," p. 291. Holt, New York.
VILCHES, O. E., and WHEATLEY, J. C. (1966). *Phys. Rev.* **148**, 509.
VILCHES, O. E., and WHEATLEY, J. C. (1967) *Phys. Lett.* **24A**, 440.
WIEBES, J., Niels-Hakkenberg, C. G., and Kramers, H. C. (1957). *Physica* **23**, 625.
WILKS, J. (1961). "The Third Law of Thermodynamics." Oxford Univ. Press, London and New York.
WILKS, J. (1967). "The Properties of Liquid and Solid Helium." Oxford Univ. Press (Clarendon), London and New York.
WILSON, A. H. (1957). "Thermodynamics and Statistical Mechanics," p. 230. Cambridge Univ. Press, London and New York.
WOLLAN, E. O., DAVIDSON, W. L., and SHULL, C. G. (1949). *Phys. Rev.* **75**, 1348.
YOST, D. M., OSBORNE, D. W., and GARNER, C. S. (1941). *J. Am. Chem. Soc.* **63**, 3492.
ZINOV'EVA, K. N. (1955). *Zh. Éksperim. i Teor. Fiz.* **29**, 899.

Chapter 7

Practical Treatment of Coupled Gas Equilibrium

MAX KLEIN

I. Introduction

In this chapter, we shall discuss the determination of the properties of a gaseous mixture capable of changing composition through chemical reaction. Our approach will differ somewhat from the usual textbook approach in the sense that we shall go beyond the development of the fundamental thermodynamic relations to discuss some of the practical problems which arise in the actual calculation of the thermodynamic properties and compositions for such mixtures.* We shall restrict our

* There are, however, several papers and reports in which this problem is discussed from the same practical computational point of view. Particularly approparaite are the

discussion to homogeneous gas-phase reactions, the interested reader being referred to the appropriate literature for extensions which include condensed phases (see Feldman *et al.*, 1969; Boll, 1960; Boll and Patel, 1961; Wilkins, 1963, pp. 142–148; Boynton, 1960; Dobbins, 1959; Stroud and Brinkley, 1969; Barnhard and Hawkins, 1963).

In the derivations that follow, the usual assumption is made that chemical reactions can be frozen at any point in their approach to equilibrium. In other words, our system can be in thermal and mechanical but not chemical equilibrium. The reasonableness of treating such systems within the framework of thermodynamics is based essentially on the fact that the functions which describe the properties of the system when it is not in chemical equilibrium reduce to those for complete thermodynamic equilibrium on substitution of the equilibrium compositions, no other changes being required (Zemansky, 1943, p. 316).

The analysis of the chemical-equilibrium problem ultimately results in a set of nonlinear equations which has to be solved. Since these equations cannot be solved in closed form, numerical procedures for their solution must be developed. These numerical procedures then form an integral part of the overall solution of the problem. For this reason, a portion of this chapter has been devoted to the discussion of appropriate numerical methods.

Of primary importance in chemical-equilibrium calculations is the selection of a subset of species concentrations to be a basis set (in the sense of vector analysis) (Brinkley, 1946, 1947, 1956; Kandiner and Brinkley, 1950) for the mathematical description of the problem. There will be a minimum number of concentrations which span the space of all species concentrations. The choice of particular species concentrations for this basis set is somewhat arbitrary. The species whose concentrations have been selected to be that set will be referred to as *reference* species, all other species being designated *derived species.** Since the number of reference species is intimately connected to the number of distinct atomic

works of Brinkely (1946, 1947, 1956), Kandiner and Brinkley (1950), Zeleznik and Gordon (1968), Huff *et al.* (1951), Gordon *et al.* (1959), Chao (1967), Gilmore (1955, 1959), Clasen (1965), Shapiro (1963, 1964), Shapiro and Shapley (1964), DeLand (1967), and Wilkins (1963). Some of the conference papers contained in Bahn and Zukoski (1960), and Bahn (1963) are also relevant to this chapter.

* These two classes of species are given various names by various authors. Among these names are atoms, elements, and components for reference species and products, species, and constituents for the derived species. There are also several authors who make no distinction between these two classes.

species, a certain simplicity is obtained if the atoms (and free electrons) are chosen as reference species. This particular choice of reference species is not always convenient from a computational point of view, however that problem is discussed below and simple matrix methods are presented for the transformation of species from derived to reference species and vice versa. In most of our discussion, we shall consider the gaseous mixture to consist of l chemical species made up of free electrons, $c - 1$ atomic species, and m $(= l - c)$ other species, the atomic species together with the electron being designated as reference species and all other constituents being designated derived species.

The chemical formula of any given species can be written symbolically as

$$S_i = \prod_{j=1}^{c} (P_j)_{\nu_{ij}}, \qquad i = 1, \ldots, l, \tag{1.1}$$

where the P_j refer to the formulae of the reference species only (since we take these to be the atoms and free electrons). It should be noted that the reference species have also been included among the S_i. For these species, however, (1.1) becomes an identity, since for them $\nu_{ij} = \delta_{ij}$, the Kronecker delta. The subscripts ν_{ij} represent the amount of reference species j contained in one molecule of species i.

A chemical reaction connecting the four species S_1, S_2, S_3, and S_4 can be written as

$$\alpha_1 S_1 + \alpha_2 S_2 = \alpha_3 S_3 + \alpha_4 S_4. \tag{1.2}$$

According to convention, the left-hand members S_1 and S_2 are called reactants while the right-hand members S_3 and S_4 are called products. In what follows, it will be convenient to replace the stoichiometric coefficients α_i by coefficients t_i whose magnitudes are equal to those of the corresponding α_i, but which are negative for reactants and positive for products. Equation (1.2) becomes

$$\sum_{i=1}^{4} t_i S_i = 0. \tag{1.3}$$

The coefficients α_i, and hence the t_i, are chosen so that (1.3) is balanced for every reference species. This balancing can be simply expressed;

$$\sum_{i=1}^{4} t_i \nu_{ij} = 0, \tag{1.4}$$

where j runs over all the reference species.

Equation (1.3) has been written for the single reaction (1.2) involving the four species S_i. Its generalization to simultaneous chemical reactions for arbitrary numbers of species is simply

$$\sum_i t_{ik} S_i = 0, \tag{1.5}$$

where k runs over all the reactions and i over all species. Equation (1.4) is essentially unchanged by the addition of the second subscript to form t_{ik}. It will be noted that t_{ik} is zero when the ith reaction does not contain the kth species. In other words, the matrix $\mathbf{T}$ (whose elements are the t_{ik}) is defined such that each column is associated with one species and each row with one reaction ordered in some arbitrary fashion. Each row-column position of such a matrix is then associated with a particular chemical reaction and a particular chemical species which takes part in that reaction.

The specification of the thermodynamic state of a heterogeneous mixture requires the specification of two thermodynamic variables, e.g., temperature and pressure, along with the specification of the composition of each phase. Since we shall restrict ourselves to the gas phase, it is only necessary to specify the pair of thermodynamic variables along with the composition for the single phase. The specification of the path by which this thermodynamic state was obtained is totally irrelevant to the specification of the state. For systems capable of undergoing composition change through chemical reactions, this is equivalent to stating that the particular chemical reactions by means of which the equilibrium composition was attained need not be stated. In fact, in describing any particular such state, one is free to choose any convenient set of possible reactions that includes all species of interest. The results of all discussions are then independent of such a choice. In the initial parts of what follows, we shall consider all molecules and ions, which we take as the derived species, to be built up from their constituent atoms and free electrons, which we take as the reference species. This has computational advantages, particularly for systems consisting of molecules containing small numbers of atoms. Practical considerations related to the numerical finiteness of computers will require the modification of this approach when the concentrations of the atomic species and electrons become extremely small. In any event, we shall always assume that each chemical reaction equation contains only one derived species, all other species in the equation being reference species, which, in our discussion, we shall take to consist of atoms and electrons.

The formation of all derived species from the reference species results in a simplification in the **T** matrix. In particular, a column that refers to a derived species will contain zeros except for that one row position corresponding to the chemical reaction for the formation of that particular derived species since that derived species cannot appear in any other reaction equation. Furthermore, this single nonzero matrix position will contain the number plus one. In what follows, we shall, therefore, redefine the **T** matrix to omit such columns, that is, such that only columns associated with the reference species are included, it being understood that $t_{ij} = +1$ is associated with the derived species itself. There will be no confusion on this latter point since we shall shortly explicitly insert $t_{ij} = +1$ for the derived species in all the working relationships. An example of such a **T** matrix, as redefined, appears in Appendix A.

Any extensive property of a mixture can be written as a weighted sum over the corresponding partial molal quantities for each species, each such quantity being weighted by the number of moles of the species. The Gibbs free energy is such an extensive property, and since the partial molal Gibbs free energy is just the chemical potential, the appropriate weighted sum takes the form

$$G = \sum_{i=1}^{l} n_i \mu_i, \tag{1.6}$$

μ_i being the chemical potential and n_i the number of moles of the ith species. At equilibrium, the Gibbs free energy must be a minimum with respect to all virtual variations consistent with any constraints on the system. In the absence of nuclear transformations, it is clear that any variation carried out must be such as to preserve the total number of reference species whether bound or free. This is a constraint and leads to conservation equations which can be written in the form

$$\sum_{i=1}^{l} \nu_{ij} n_i - \chi_j = 0, \qquad j = 1, \ldots c. \tag{1.7}$$

For reference species other than electrons, χ_j is the concentration of the jth reference species, whether bound or free, in units consistent with the n_i. For the electrons, the analogous conservation equation is most conveniently expressed in terms of net charge conservation, i.e., χ_j, in this case, represents the net overall charge of the gas. In particular, for a neutral gas, $\chi_j = 0$.

The problem with which we shall be occupied thus can be stated as being the determination of the composition variables n_i such that the free energy given by (1.6) is a minimum under all virtual variations subject to the constraints (1.7). There have developed two numerical approaches to the solution of this problem. In one of these, the problem is numerically attacked directly as stated and, in fact, composition variables are sought that result in a stationary value of G, subject to the stated constraints. This method is referred to as the direct minimization of the free energy. In the second approach, the formalism is allowed to proceed further before the numerical attack is mounted. Thus, formal variation of (1.6) is carried out subject to the constraints (1.7). There results a series of equations each of which connects composition variables with the composition-independent parts of the chemical potentials for those species appearing in the equation. Since these composition-independent terms can be combined to define an equilibrium constant, this second method is referred to as the equilibrium-constant method. We shall now proceed to derive the working equations for the two methods. In what follows, we shall formally carry along in the derivation terms containing various departures from the ideal gas. Having derived the required relations, we shall then specialize them to the ideal gas in discussing methods for solution. The detailed way in which the nonidealities are calculated is contained in Appendix B. Modifications in the methods of solution as required by these nonidealities will be indicated but not discussed in detail.

II. Direct Minimization of the Free Energy†

In principle, the equations associated with this method have been derived, namely (1.6) and (1.7). For an ideal gas, these can be written more explicitly in terms of the mole fractions of the species. Thus, the chemical potential for a constituent in a mixture can be written as

$$\mu_i(T, P)/RT = \mu_i^*(T, P)/RT + \ln x_i + \ln \gamma_i, \tag{2.1}$$

where x_i is the mole fraction of the species in the mixture, $\mu_i^*(T, P)$ is the chemical potential of the pure species i at the same temperature and pressure, and γ_i is the activity coefficient of species i in the solution. The activity coefficient γ_i contains all departures from ideality. For a

† See Zeleznik and Gordon (1968), Gilmore (1955, 1959), Clasen (1965), Shapiro (1963, 1964), Shapiro and Shapley (1964), DeLand (1967), Wilkens (1963), White *et al.* (1957, 1958), and White (1967).

real gas, γ_i depends on T, P, and the concentrations of the various constituents. For an ideal gas, $\gamma_i = 1$, in which case the concentration dependence reduces simply to the natural logarithm of the mole fraction. Of course μ_i^* can be written as

$$\mu_i^*(T, P)/RT = \mu_i^\circ(T)/RT + \ln P, \tag{2.2}$$

in which the standard state is taken as 1 atm (101325.0 $\mathrm{N/m^2}$) and $\mu_i^\circ(T)$ depends only on the temperature. In what follows, it will be convenient to write the free energy G in dimensionless form by dividing it by RT. Substitution of (2.1) into the dimensionless form of (1.6) then yields

$$G/RT = \sum_{i=1}^{l} n_i[\mu_i^*/RT) + \ln(n_i/\sum n_i) + \ln \gamma_i]. \tag{2.3}$$

The chemical-equilibrium problem is then solved by the method of the direct minimization of the free energy when the set of composition variables n_i is found that minimizes (2.3) subject to the c equations (1.7) being satisfied. It should be noted that, when real-gas effects are included, the γ_i can depend on the concentrations of all species. This will complicate certain of the numerical methods used in the solution for the direct minimization of the free energy (Boynton, 1963).

III. The Equilibrium-Constant Method

As we have already stated, this method starts with the formal variation of (1.6) subject to (1.7). The fact that n_i appear in both (1.6) and (1.7) indicates that the formal variation of n_i in (1.6) cannot be carried out independently. These variations can be made independent by the elimination of the proper number of variables. This can be accomplished by the method of Lagrange multipliers (Courant, 1937), in which a sum over the c equations (1.7), each multiplied by an unknown multiplier λ_j, is added to (1.6). When this is done and the variation carried out, there results

$$\delta G = \sum_{i=1}^{l} \left(\mu_i + \sum_{j=1}^{c} \lambda_j \nu_{ij}\right) \delta n_i + \sum_{j=1}^{c} \left(\sum_{i=1}^{l} \nu_{ij} n_i - \chi_j\right) \delta \lambda_j, \tag{3.1}$$

where the unknown Lagrange multipliers λ_j are defined so as to eliminate the proper number of coefficients of the δn_i. All remaining variations then become independent. The two sets of variations in (3.1) can then be carried out independently. This leads to the set of equations

$$\mu_i + \sum_{j=1}^{c} \lambda_j \nu_{ij} = 0, \qquad i = 1, \ldots, l,$$

and

$$\sum_{i=1}^{l} \nu_{ij} n_j - \chi_j = 0, \qquad j = 1, \ldots, c.$$

The first of these can be used to evaluate the Lagrangian multipliers λ_j and to determine their meaning. Since $\mu_i = -\sum_j \lambda_j \nu_{ij}$, it follows from (1.6) that

$$G = \sum_i \mu_i n_i = - \sum_{i,j} \lambda_j \nu_{ij} n_i.$$

But, from the second set of relations above,

$$\sum_i \nu_{ij} n_i = \chi_j,$$

so that

$$G = - \sum_j \lambda_j \chi_j$$

and $-\lambda_j$ can be considered to be the contribution of reference species j (*whether bound or free*) to the free energy. Since

$$-\lambda_j = (\partial G / \partial \chi_j)_{T,P,\chi_i, i \neq j},$$

it follows that $-\lambda_j$ is, in effect, the chemical potential of the reference species whether bound or free.

Equation (3.1) contains the l variations δn_i. The fact that these are subject to the c constraints (1.7) means that there must exist $l - c$ variables which can be varied independently. This is equivalent to the statement that there exists at least one set of $l - c$ chemical reactions for the attainment of the thermodynamic state. One such set of reactions can be obtained by considering each derived species to be a product and requiring it to be built up from its constituent reference species, the latter considered to be reactants. One obtains, thereby, $l - c$ chemical reactions for the derived species (along with c superfluous identity chemical reactions for the reference species). Clearly, any other set of reactions containing more than $l - c$ reactions must contain redundancies.

Having reduced the number of reactions to $l - c$, one can obtain an equal number of variables by defining a variable to go with each reaction. This is the sense of De Donder's (1936; also see Zemansky, 1943, p. 235) introduction of the degree-of-reaction variable. This variable is defined in terms of the change in the number of moles of a species produced by a particular chemical reaction. Thus, the change in the number of moles of species j as a result of the ith reaction is given by

$$(\delta n_j)_i = t_{ji}(\delta \xi_i),$$

where ξ_i is the degree of reaction of chemical reaction i. By definition, the same value of $\delta\xi_i$ must apply to each species in the ith reaction, the scale for each such species being given by t_{ji}. The total change in the number of moles of the jth species for the mixture can be calculated by summing these changes over all reactions. This yields

$$\delta n_j = \sum_i t_{ji}\,\delta\xi_i, \tag{3.2}$$

which can be used in (3.1) to demonstrate the independence of the variations $\delta\xi_i$. Thus, substitution of (3.2) in (3.1) yields

$$\begin{aligned}\delta G &= \sum_{i=1}^{l}\left(\mu_i + \sum_{j=1}^{c}\lambda_j\nu_{ij}\right)\sum_{k=1}^{l} t_{ik}\,\delta\xi_k + \sum_{j=1}^{c}\left(\sum_{i=1}^{l}\nu_{ij}n_i - \chi_i\right)\delta\lambda_j \\ &= 0 \\ &= \sum_{i,k}\mu_i t_{ik}\,\delta\xi_k + \sum_{i,j,k}\lambda_j\nu_{ij}t_{ik}\,\delta\xi_k + \sum_{j=1}^{c}\left(\sum_{i=1}^{l}\nu_{ij}n_j - \chi_j\right)\delta\lambda_j.\end{aligned}$$

But, according to the generalized form of (1.4), $\sum_i \nu_{ij}t_{ik} = 0$, so that

$$\delta G = \sum_{i,k}\mu_i t_{ik}\,\delta\xi_k + \sum_{j=1}^{c}\left(\sum_i \nu_{ij}n_j - \chi_j\right)\delta\lambda_j = 0.$$

The fact that the λ_j have disappeared from the first term on the right means that the variations have been uncoupled. Independent variation of the terms in this last equation then leads to the relation

$$\sum_i \mu_i t_{ij} = 0, \qquad j = 1, \ldots, l - c, \tag{3.3}$$

(the so-called equations of reaction equilibrium) as well as the c equations (1.7). The $l - c$ equations (3.3) along with the c relations (1.7) also completely specify the system. Equations (3.3) will form the basis for the derivation of the working equations of the equilibrium-constant method.

Let us again consider the derived species to be formed only from reference species. As previously mentioned, for such chemical reactions the coefficient of the derived species must be plus unity, so that (3.3) can be written as

$$\mu_i + \sum_{j=1}^{c} \mu_j t_{ij} = 0, \qquad i = 1, \ldots, l - c, \tag{3.4}$$

where i now refers only to derived species and j refers only to reference species and where there no longer remains the possibility for confusion with regard to the removal from the **T** matrix of the columns corresponding to the derived species.

Equations (3.4) are valid for all values of temperature and pressure for which the system is in thermodynamic equilibrium. The explicit dependence of these equations on composition for the ideal gas becomes clear on the substitution of (2.2) into (2.1), setting $\gamma_i = 1$, and substituting the result into (3.4). We choose, however, to carry along the activity coefficient for the present. Equation (3.4) then becomes

$$\begin{aligned}\frac{\mu_i{}^\circ}{RT} + \sum_j t_{ij} \frac{\mu_j{}^\circ}{RT} + \ln P + \sum_j t_{ij} \ln P + \ln x_i \\ + \sum_j t_{ij} \ln x_j + \ln \gamma_i + \sum_j t_{ij} \ln \gamma_j = 0.\end{aligned} \tag{3.5}$$

The first two terms depend only on the temperature. It is customary to combine such terms into an equilibrium constant K_i for the ith reaction (and hence for the ith derived species) by the definition

$$-\ln K_i = \mu_i{}^\circ/RT + \sum_j t_{ij}\, \mu_j{}^\circ/RT. \tag{3.6}$$

Substitution into (3.5) leads to

$$-\ln K_i + \ln P^{(\Sigma t_{ij}+1)} + \ln x_i + \sum_j \ln x_j^{t_{ij}} + \ln \gamma_i + \sum_j \ln[\gamma_j^{t_{ij}}] = 0,$$

which can be written as

$$K_i = \left[\prod_j \gamma_i^{t_{ij}}\right]\gamma_i \cdot P^{-\omega_i}\left[\prod_j x_i^{t_{ij}}\right] x_i, \tag{3.7}$$

where $\omega_i = -[\Sigma\, t_{ij} + 1]$ is the increase in the number of particles in the reaction in going from derived to reference species. In what follows, it will be convenient to introduce an activity coefficient for the reaction by defining

$$\gamma_i' = \left[\prod_j \gamma_j^{-t_{ij}}\right] \Big/ \gamma_i.$$

Equation (3.7) can then be written;

$$K_i\gamma_i' = P^{-\omega_i}\left[\prod_j x_j^{t_{ij}}\right]x_i. \tag{3.7a}$$

We shall now restrict ourselves to the ideal gas, taking $\gamma_j = 1$ as required. Real-gas effects can be included through the introduction of explicit expressions for γ_i' as described in Appendix B.

For the ideal gas, (3.7) can be rewritten so that the mole fraction of the derived species appears on the left and only reference species appear on the right. Thus, by transposition, (3.7) becomes, for $\gamma_i' = 1$

$$x_i = K_i P^{\omega_i} \prod_j x_j^{-t_{ij}} \tag{3.7b}$$

Since, for reactants, $t_{ij} = -\nu_{ij}$, (3.7b) can now be simply written in terms of the stoichiometric coefficients, i.e.,

$$x_i = K_i P^{\omega_i} \prod_j x_j^{\nu_{ij}}, \tag{3.8}$$

while the activity coefficient for the reaction becomes

$$\gamma_i' = \left[\prod_j \gamma_j^{\nu_{ij}}\right] \Big/ \gamma_i$$

Equation (3.8) will form the basis for computations within the framework of the equilibrium-constant method. Because we have chosen to write the chemical reactions entirely in terms of atoms and the electron as reference species, (3.8) contains ν_{ij} in an unambiguous manner with regard to sign. In what follows, we shall also continue to use t_{ij} on occasion in order to maintain complete generality.

Equation (3.8) is appropriate for calculations in which pressure is a thermodynamic state variable. If, for some reason, partial pressures are preferred over mole fractions, use can be made of the definition $P_j = x_j P$, where P_j is the partial pressure, and (3.8) is then written as

$$P_i = K_i \prod_j P_j^{\nu_{ij}}$$

Quite often, density is a more convenient state variable than pressure. Conversion of (3.8) to the corresponding density form requires the explicit use of the equation of state that connects pressure and density.

Thus, since P is expressed in bars (or atmospheres), this can be written as

$$x_i = K_i(P/P_0)^{\omega_i} \prod_j x_j^{\nu_{ij}},$$

where P_0 is understood to refer to the pressure at specified reference conditions (usually, $T = 273.15°K$ and $P = 1$ bar (or 1 atm)). The equation of state can be written $P = \varrho ZRT$, where Z is the compressibility factor. For the one-component ideal gas, Z is unity, with departures from unity being due entirely to nonideality. We shall use the same form for the equation of state for the reacting multicomponent gas. In that case, it is convenient to let $1/\varrho$ be the volume per mole of the reaction mixture, so that, for the ideal gas, Z becomes the number of moles, a quantity which depends on the thermodynamic state parameters. Naturally, for the multicomponent real gas, Z also contains the effect of nonideality. Now, the equations of state are $P = \varrho ZRT$ for the conditions of interest and $P_0 = \varrho_0 Z_0 R T_0$ at the reference state, so that

$$P/P_0 = \varrho ZT/\varrho_0 Z_0 T_0.$$

Equation (3.8) can then be written,

$$C_i = K_i(T/T_0)^{\omega_i}(\varrho/\varrho_0)^{\omega_i} \prod_j C_j^{\nu_{ij}},$$

where $C_j = x_j Z/Z_0$ is the concentration of species j in moles per mole of equilibrated gas at the reference conditions. It should be noted that the reference state is being used for purposes of scaling all results and not as a standard state in the usual chemical sense. The use of this particular reference state may be inconvenient in some systems, in which case, if solutions are still desired at specified densities, either other definitions of the variables subscripted zero should be used or calculations at the desired densities carried out in a pressure formulation for which no such reference state is needed with an iteration inserted so that the particular pressure that corresponds to the desired density is determined.

Since in what follows a reference state will always be used for scaling, the dropping of the Z_0 should result in no confusion. With this in mind, Z/Z_0 can be replaced by Z and the concentrations defined by $C_i = x_i Z$, it being understood that these are with respect to one mole in the reference state.

It is convenient to combine the factors that depend only on temperature by defining an equilibrium constant

$$\bar{K}_i = K_i(T/T_0)^{\omega_i}.$$

One obtains thereby

$$C_i = \bar{K}_i(\varrho/\varrho_0)^{\omega_i} \prod_j C_j^{\nu_{ij}} \tag{3.9}$$

For consistency, (1.7) should be expressed in the same units as (3.9). Thus, (1.7) becomes

$$\sum_{i=1}^{l} \nu_{ij} C_i - \chi_j = 0, \qquad j = 1, \ldots, c, \tag{3.10}$$

where χ_j is now in moles per mole of gas at the reference state. Since, for the electrons, (3.10) represent the net charge conservation equation (with positive and negative charges balancing each other), ν_{ij} for them is replaced by t_{ij} in (3.10).

Equations (3.9) and (3.10) are the working equations for the equilibrium-constant method. These equations constitute a set of l equations for the concentrations of all of the species (i.e., both the reference and derived species). As mentioned earlier, these l concentrations are completely determined once the concentrations of the c reference species have been specified. The problem can be viewed, therefore, as requiring the solution of the c simultaneous equations (3.10) subject to the $l - c$ conditions (3.9). The meaning of this view of the problem will become somewhat more transparent when expressions are derived below for the calculation of the derivatives of the concentrations. The validity of this view can be demonstrated on substituting (3.9) into (3.10). This yields the set of equations

$$\sum_{i=1}^{l} \nu_{ij} \bar{K}_i(\varrho/\varrho_0)^{\omega_i} \prod_j C_j^{\nu_{ij}} - \chi_j = 0, \qquad j = 1, \ldots, c, \tag{3.11}$$

which is a set of c equations in c unknowns, i.e., the c reference species.

Equations (3.11) are a set of highly nonlinear equations. The nonlinearity arises both from the exponents ν_{ij} on the unknown variables as well as from the appearance of products of the unknowns. The question of the existence of multiple solutions naturally arises in such cases. Such problems are beyond the scope of this chapter. The interested reader is referred to recent literature addressed to this problem (Shears, 1968).

We have now obtained the working equations for the two approaches to the problem. These are (1.7) and (2.3) for the free-energy-minimization method and (3.9) and (3.10) for the equilibrium-constant method. Both methods require as input the chemical potentials of the individual chemical species [see Eqs. (2.3) and (3.6)]. In the ideal-gas limit appropriate to our calculations, these and other thermodynamic properties of the individual species can, in principle, be simply calculated by summation over the species energy levels (see, e.g., Mayer and Mayer, 1940; Gordon and Robinson, 1960, 1963; Durand and Brandmaier, 1960; Wilkens, 1963, Chapter I; Hilsenrath *et al.*, 1964; Drellishak, 1964; Reidin and Ragent, 1962). In practice, there can be problems, however, as we shall mention below (see also, Evans, 1963).

IV. Transformations among Sets of Reference Species

Either of the two sets of equations obtained, i.e., (1.7) and (2.3) or (3.9) and (3.10), can, in principle, be solved for the concentrations of the various constituents in the mixture. These equations contain the assumption that there is available a complete specification of the system for input, i.e., the temperature and density of the mixture, the thermodynamic properties of the constituents, and the specification of the mixture itself (through the values assigned to the χ_j). We have formally assumed that all derived species are obtained directly from the chemical combining of the atomic species. Because of this, the solution is specified in terms of the concentrations of these atomic species. This will become somewhat clearer below. As mentioned earlier, the specification of a precise set of chemical reactions from which the species are derived is irrelevant to the specification of the thermodynamic state of the system. The particularization of the reactions is needed only for converting the thermodynamic formalism into a computational structure. It follows, therefore, that changing from one set of reactions to another produces no fundamental thermodynamic changes in any of the results obtained.

Now, a particular set of chemical reactions will be convenient, from a computational point of view, for only a limited set of thermodynamic conditions. Thus, for example, for chemical reactions in the system H_2O, H_2, O_2, H, and O, the use of the atomic species H and O as reference species, while thermodynamically correct, will not be convenient from a computational point of view at temperatures sufficiently low that little dissociation occurs. While Eq. (3.9) in terms of atomic reference species is correct for the reaction

$$2H + O \rightleftharpoons H_2O$$

under all conditions, computational difficulties occur at low temperatures, where the concentrations of the atomic species C_i become extremely small and where the equilibrium constants K_i (and hence $\bar{K}_i$) become extremely large. Despite the fact that the C_i obtained at low temperatures for H_2O by solving (3.9) must be of reasonable magnitude, it is possible, in the process of carrying out the multiplications required in the right-hand side of (3.9), for numbers to be produced whose magnitudes are outside the limits set by the computer design. This can often be avoided by changing the order of multiplication so that small numbers like the C_j alternately multiply large ones like $\bar{K}_i$ or by taking logarithms. A second problem is associated with the magnitude of K_i. While this too is a soluble problem, solutions are always artificial (e.g., by representing K_i as a product of factors).

A much more reasonable and more physical way to solve both these problems exists, namely the replacing of the arbitrarily chosen set of chemical reactions used by another set from which reference species can be defined whose magnitudes are within the limits set by the computer. For example, at low temperatures it is obvious that H_2O should be formed from *molecular* oxygen and *molecular* hydrogen. That is, the computational framework should be based on H_2 and O_2 as reference species, with the H_2O reaction becoming

$$H_2 + \tfrac{1}{2}O_2 \rightleftharpoons H_2O.$$

It should be noted that the coefficient of the derived species has been left equal to unity to be consistent with the definition of the **T** matrix.

In this section, we shall indicate how such transformations of reference species may be carried out. We shall merely describe how the elements of the **T** matrix are to be changed, the details required to make the transformation part of a computational program being left to the interested reader.

Equation (3.4) can be written as

$$\mu_i = -\sum_{j=1}^{c} t_{ij}\mu_j, \qquad i = 1, \ldots, l, \tag{4.1a}$$

where j refers to the reference species and i to the derived species. Equations (4.1a) include the c identity equations for the reference species since these also undergo transformation. In matrix notation, (4.1a) can be written

$$\boldsymbol{\mu}' = -\mathbf{T} \cdot \boldsymbol{\mu} \tag{4.1b}$$

with $\boldsymbol{\mu}'$ an l-element column vector and $\boldsymbol{\mu}$ a c-element row vector; $\mathbf{T}$ is, of course, an l-element row by a c-element column matrix.

In carrying out a transformation of reference species, it is essential that the number of final reference species be the same as the number of original reference species. In Brinkley's terminology (1946), this requires that the dimension of the vector space (i.e., the rank of the T matrix) be invariant under the transformation. This requires the equations connecting the new reference species with the old to contain as many equations as unknowns, i.e., the matrix associated with the transformation must be square. In other words, if double primes refer to the new reference species, there is a subset of Eqs. (4.1a) which can be written

$$\mu_j'' = -\sum_{k=1}^{c} t_{kj}\mu_k, \qquad j = 1, \ldots, c,$$

where the t_{kj} are now the elements of a square matrix. This can also be written in the vector notation of (4.1c). Thus,

$$\boldsymbol{\mu}'' = -\mathsf{T}_s \cdot \boldsymbol{\mu}, \tag{4.1c}$$

where the subscript s is meant to indicate the square matrix associated with the two sets of reference species. Since T_s is square, (4.1c) can formally be solved for $\boldsymbol{\mu}$ in terms of $\boldsymbol{\mu}''$ by multiplying by T_s^{-1} from the left. There results

$$\mathsf{T}_s^{-1} \cdot \boldsymbol{\mu}'' = -\boldsymbol{\mu}. \tag{4.1d}$$

Equation (4.1b) can now be written in terms of the new reference species by substituting (4.1d) for $\boldsymbol{\mu}$ in (4.1b). There results

$$\boldsymbol{\mu}' = \mathsf{T} \cdot \mathsf{T}_s^{-1} \cdot \boldsymbol{\mu}'' = \mathsf{T}_1 \cdot \boldsymbol{\mu}'',$$

with the elements of T_1 constituting the new coefficient matrix. It should be noted that the elements associated with the previous reference species will no longer be Kronecker deltas, whereas those associated with the new reference species will become Kronecker deltas. As will become clear below, these transformations must be accompanied with various changes in the identities of rows and columns in the original T matrix.

The transformation is not complete until the constants χ_j have been transformed to those appropriate to the new reference species. In matrix form, (3.10) can be written

$$\mathbf{C} \cdot \mathsf{T} = \boldsymbol{\chi}.$$

Multiplication by the square array T_s^{-1} from the right yields

$$\mathbf{C} \cdot \mathsf{T} \cdot \mathsf{T}_s^{-1} = \boldsymbol{\chi} \cdot \mathsf{T}_s^{-1}.$$

But $\mathsf{T} \cdot \mathsf{T}^{-1} = \mathsf{T}_1$, so that this becomes

$$\mathbf{C} \cdot \mathsf{T}_1 = \boldsymbol{\chi} \cdot \mathsf{T}_s^{-1},$$

and it is clear that the transformations of the χ_j constitute the elements of the vector equation

$$\boldsymbol{\chi}' = \boldsymbol{\chi} \cdot \mathsf{T}_s^{-1}.$$

In order to clarify the mechanics of the transformation, let us take, as an example, the set of reactions among the species CO_2, CO, O_2, C, and O. When the atoms C and O are taken as reference species, the elements of the T matrix are given by

$$\begin{array}{c} \\ CO_2 \\ CO \\ O_2 \\ C \\ O \end{array} \begin{array}{c} \begin{array}{cc} C & O \end{array} \\ \begin{pmatrix} 1 & 2 \\ 1 & 1 \\ 0 & 2 \\ 1 & 0 \\ 0 & 1 \end{pmatrix} \end{array}.$$

Suppose, now, that it is desired to transform to CO and O_2 as reference species. The transformation matrix T_s and its inverse are easily seen to be given by

$$\mathsf{T}_s = \begin{array}{c} \\ C \\ O \end{array} \begin{array}{c} \begin{array}{cc} CO & O_2 \end{array} \\ \begin{pmatrix} 1 & 0 \\ 1 & 2 \end{pmatrix} \end{array}; \qquad \mathsf{T}_s^{-1} = \begin{pmatrix} 1 & -\frac{1}{2} \\ 0 & \frac{1}{2} \end{pmatrix},$$

with the new matrix T_1 calculated from

$$\mathsf{T}_1 = \mathsf{T} \cdot \mathsf{T}_s^{-1} = \begin{pmatrix} 1 & 2 \\ 1 & 1 \\ 0 & 2 \\ 1 & 0 \\ 0 & 1 \end{pmatrix} \cdot \begin{pmatrix} 1 & -\frac{1}{2} \\ 0 & \frac{1}{2} \end{pmatrix}$$

$$= \begin{array}{c} \\ CO_2 \\ CO \\ O_2 \\ C \\ O \end{array} \begin{array}{c} \begin{array}{cc} CO & O_2 \end{array} \\ \begin{pmatrix} 1 & \frac{1}{2} \\ \hdashline 1 & 0 \\ 0 & 1 \\ \hdashline 1 & -\frac{1}{2} \\ 0 & \frac{1}{2} \end{pmatrix} \end{array},$$

where the shift of reference species is indicated by a shift in the location of the diagonal unit submatrix between the dashed lines.

To illustrate the transformation on the χ_j, suppose there is one mole of CO_2 present, so that, for C and O as reference species, $\chi_1 = 1$, $\chi_2 = 2$. Thus,

$$\boldsymbol{\chi}' = (1 \quad 2)\begin{pmatrix} 1 & -\frac{1}{2} \\ 0 & \frac{1}{2} \end{pmatrix} = \begin{pmatrix} 1 \\ \frac{1}{2} \end{pmatrix}$$

and, for CO and O_2 as reference species, the values $\chi_1 = 1$, $\chi_2 = \frac{1}{2}$ are appropriate.

Clearly, transformations of this kind can be carried out automatically within the framework of any computer program designed to solve the equations for the concentrations. We shall not describe how this might be done in practice. A word of caution is in order, however, and that is that, as should become clear in the next section, since the reference species are changed, these transformations must always be accompanied with a shift in the energy differences (i.e., heats of reaction) used to produce derived species from reference species.

V. Some Problems in the Calculation of Ideal-Gas Functions for the Individual Species

The only microscopic information that has thus far been included in the formalism is contained in (1.1) and is merely the labeling of each derived species in terms of its constituent reference species. The points of entry for the detailed microscopic properties of individual constituents have already been passed. These are (2.3) for the free-energy-minimization method and (3.6) for the equilibrium-constant method. For a pure component, the chemical potential is the same as the free energy per mole, so that in (2.2)

$$\frac{\mu^\circ(T)}{RT} = \frac{(G_T^\circ - E_0^\circ)}{RT} + \frac{E_0^\circ}{RT}$$

where G_T° is the ideal-gas Gibbs free energy for the species and E_0° is the reference energy at zero kelvin. The expression for the equilibrium constant, (3.6), becomes

$$\ln K_i = -\frac{(G_T^\circ - E_0^\circ)_i}{RT} + \sum_j \nu_{ij} \frac{(G_T^\circ - E_0^\circ)_j}{RT} - \frac{(E_0^\circ)_i}{RT} + \sum_j \nu_{ij} \frac{(E_0^\circ)_j}{RT}.$$

Both methods, therefore, require the ideal-gas Gibbs function relative to the energy of a reference state $E_0°$ for the individual species.

Since only energy differences have physical meaning, the choice of $E_0°$ values would appear to be entirely arbitrary. In a reacting system, however, the difference between $E_0°$ for reactant and products is physically meaningful and is, in fact, the reaction energy

$$\frac{\Delta E_0°}{RT} = \frac{(E_0°)_i}{RT} - \sum_j \nu_{ij} \frac{(E_0°)_j}{RT}$$

at absolute zero. For this reason, the choice must be made in a manner that is consistent among species. In most cases, taking $E_0°$ to be zero for the *reference* species avoids complications. Of course, all choices are valid provided only that they are consistent. It should perhaps be pointed out again that whatever reference point is taken for the energy must be examined carefully after any transformation to new reference species is carried out.

The quantity $(G_T° - E_0°)/RT$ can be calculated for each species using statistical mechanics and the details of the internal structure of that species. The Gibbs free energy for an ideal-gas species is related to the partition function through the relation

$$(G_T° - E_0°)/RT = -\ln Q,$$

where Q is the ideal-gas partition function, and contains translational and electronic contributions for all species (except for the electrons, for which only a translational part is appropriate). For molecular species, there are additional contributions from vibrational and rotational energies. A discussion of methods for the evaluation of such partition functions is outside the scope of this chapter and the interested reader is referred elsewhere for such details (Mayer and Mayer, 1940; Gordon and Robinson, 1960, 1963; Durand and Brandmaier, 1960; Evans, 1963; Wilkens, 1963, Chapter 1; Hilsenrath *et al.*, 1964; Drellishak, 1964; Reidin and Ragent, 1962). In this chapter we assume that the Gibbs free energies have been calculated properly and are available as input to the calculation. We shall, however, mention some problems associated with the evaluation of the partition functions since such problems are not always described in the literature.

Perhaps the most serious such problem has to do with the actual divergence of these partition functions. There are an infinite number of energy levels just below the ionization limit of an atom. These terms, if included in the summation, would cause the partition function to diverge. At moderate temperatures, which can be as high as several thousand kelvins, this causes no trouble since the contributions of successive energy levels, starting from the lowest, tends to drop off rapidly and goes through a broad minimum starting with a relatively small quantum number. The series can therefore be treated as an asymptotic series and cut almost anywhere after the contribution of successive levels has become negligible. As the temperature is raised, however, the breadth of this minimum narrows until ultimately there is no level at which the contributions of successive levels become small. As soon as this happens, the advantage of an obvious cutoff point is lost and, in fact, the partition function begins to depend strongly on the choice of cutoff. This is obviously a signal that the theory on which the calculation is based has become inadequate, and this is indeed the case. Care must be taken even when this problem does not appear to exist. Exact partition-function expressions are often approximated by series which are cut off in a manner appropriate for low temperatures. When these series are extended to high temperatures, they sometimes give convergent, albeit wrong, results due to the neglect of terms not needed at low temperatures but needed at higher temperatures.* These neglected terms contain the divergence problem.

An interesting way of handling these divergences has been devised by Woolley (1958) and studied in some detail by several others (Kilpatrick, 1953; Hill, 1955; Haar, 1960). In this approach, sometimes referred to as physical cluster theory, no distinction is made between a molecule whose constituent atoms are bound together by means of covalent bonds and a physical cluster of these atoms for separations and conditions for which the weaker van der Waals interactions are appropriate. Indeed, reflection leads one to the conclusion that any distinction which might be made between these would indeed be artificial.

The divergence that arises from the summing of the energy levels of atoms and atomic ions over highly excited states of the outer electron has been considered in a number of different ways in the literature. Clearly, the representation of the partition function of the atom in the mixture as a sum over energy levels of the isolated atom is an approximation to the actual many-body problem. At ordinary temperatures, this approximation is reasonable since the number of atoms in any but the

* For some important comments on possible pitfalls in ideal-gas calculations, see the work by Evans (1963).

lowest-lying energy levels is quite small. Using semiclassical arguments, it is easy to see that the Bohr radii for such levels are quite small, and that therefore the "paths" of the outer electrons of different atoms do not overlap. As the temperature is raised, the situation changes and the Bohr orbits become sufficiently large for there to be overlap even at ordinary densities. The problem has therefore become a many-body problem whose partition function can no longer be approximated by a product of one-particle functions.

Several workers (Denton, 1963; Unsöld, 1948; Megreblian, 1952, 1953; Drellishak *et al.*, 1963, 1964) have devised density-dependent cutoff methods which are essentially variations of an approach due to Urey (1924) and Fermi (1924), and also contained in some unpublished work of Bethe (1942). In these approaches, the energy levels are summed only through those quantum numbers for which the Bohr radius is less than some function of the average interparticle distance. The problem now becomes density- and temperature-dependent, the density determining the average interparticle distance and the temperature the Bohr radius of the outer electrons as well as the probability of there being close collisions.

There are thus two main approaches to the problem of calculating the ideal-gas partition functions of atoms and atomic ions, depending on the temperature range of interest and the ionization energy. Up to moderately high temperatures, the sums can be cut off at a relatively low-lying level based on the smallness of successive contributions. At high temperatures, a cutoff can be taken which depends on temperature and density and which may vary for a given atomic species from mixture to mixture.

Since the quantum number takes on discrete values, it is possible for the summation to take on discontinuities as a function of temperature and density (Drellishak *et al.*, 1963, 1964). These discontinuities are found where a change occurs in the final quantum number accepted under the cutoff criterion. A way to reduce this problem has been proposed by Woolley (1958) and independently by Gilmore (1967). In these approaches, the partition function is divided into a sum of two parts. One of these is the contribution due to states below some quantum number n_A which is always less than n_c, the density-dependent cutoff. The other is due to the contribution of the levels between n_A and n_c considered now to be the levels of one electron on an ionized atom. The levels between n_A and n_c are, in fact, considered to be hydrogenlike on the ionic core. This method has the advantage of allowing one to use tables of ideal-gas partition functions and a cutoff criterion which depends on the species only through the nuclear charge. It is therefore of considerable computational advantage. Woolley's method is designed particularly for treating the ionized gas as an ionic solution within the theoretical framework set up by Mayer (1950) and Poirier (1953).

It should be clear from the preceding discussion that there are several ways in which one can approach the divergence problems that arise in the sum over states associated with the calculation of the ideal-gas partition functions of the individual species in a chemically reacting mixture. It is imperative, no matter which of these methods is used, that the counting of states for a given species above the dissociation or ionization limit for its constituents be done in a manner that is consistent with the counting of the states for these constituents themselves. If this is not done properly, a portion of the phase space for the mixture will have been included at least twice in the statistical-mechanical development for the mixture thermodynamic properties. Thus, one might mistakenly include the states associated with the separated atoms as a molecular state with a van der Waals intermolecular potential and include the same states as free atoms with a correction for nonideality. A consistent way in which this problem can be avoided has been described by Haar (1960).

VI. The Thermodynamic Properties of the Mixture

Once the thermodynamic functions for the individual species have been calculated at the temperature of interest and once the energies $\Delta E_0^\circ/RT$ have been chosen, it is possible, in principle, to solve for the species concentrations using either the method of free-energy minimization [minimization of (1.6) subject to (1.7)] or the equilibrium-constant method [solution of (3.9) subject to (3.8)] and to determine the concentrations of all species at that temperature as a function of the most useful thermodynamic variable (e.g., pressure, density, etc.). Given these concentrations and the properties for the individual species, one can then calculate the thermodynamic properties of the ideal-gas mixture. In this section, we shall indicate how this can be done. Real-gas corrections to these expressions are included in Appendix B.

According to our definitions of concentration and compressibility factor, the compressibility factor for the ideal gas relative to that at standard conditions is simply a sum over species concentrations, the sum going over both the derived and reference species. Thus,

$$Z = PV/RT = \sum_{i=1}^{l} C_i, \tag{6.1a}$$

where, as stated earlier, C_i is the number of moles of species i relative to one mole in a reference state and Z is the compressibility factor, defined to include the number of moles, also with respect to a reference state. As stated earlier, the choice of the reference state is a matter of convenience in scaling these quantities.

The internal energy can be calculated from

$$E/RT = \sum_{i=1}^{l} C_i(E_i/RT), \tag{6.1b}$$

where $E_i = (E_T{}^\circ - E_0{}^\circ)_i + (E_0{}^\circ)_i$ and where $(E_T{}^\circ - E_0{}^\circ)_i$ is calculated from statistical mechanics.

The entropy is given by

$$S/R = Z\left[\sum_{i=1}^{l} x_i(S^\circ/R)_i - \sum_{i=1}^{l} x_i \ln x_i - \ln(P/P^\circ)\right], \tag{6.1c}$$

where P° is the pressure of the thermodynamic standard state. All other thermodynamic potentials can be written in terms of combinations of (6.1a–c).

Among the gas properties of interest are the derivatives of the thermodynamic potentials. These lead to such properties as the specific heat and sound velocity. Such properties are often obtained by numerical differentiation of tables of the thermodynamic potentials (Lewis and Neel, 1964; Curtis and Wohlwill, 1961; Landis and Nilsen, 1961, 1962). Even the best of such methods requires considerable caution in its application. It is necessary to be careful in handling the end points and care must be taken to ensure that the data points are *neither* too close together *nor* too far apart. Numerical differentiation is entirely unnecessary, however, since the derivatives can be obtained as the solutions of sets of c linear equations in c unknowns, c being the number of *reference* species in the mixture (Brinkley, 1956; Klein, 1961; Newman and Allison, 1966). Several authors (Fenter and Gibbons, 1961; Hochstim and Adams, 1962) have developed similar approaches in which, however, the derivatives are expressed in terms of the solutions of l linear equations in l unknowns, l being the total number of species in the mixture. The difference between c and l is generally quite large. For high-temperature air, for example, $l = 30$, while $c = 6$. It should be understood, however, that the reduction from l equations to c equations occurs only under ideal-gas conditions (i.e., $\gamma_i = 1$). When real-gas effects are included, it is always necessary to solve a set of l equations in l unknowns for the derivatives. We shall proceed to derive the linear equations for the derivatives within the framework of the equilibrium-constant method. The results obtained are, of course, valid in either approach.

The derivatives required can in all cases be written in terms of the derivatives of properties that are additive functions of the properties of the individual constituents. This means that it is only necessary to calculate the derivatives of functions of the form

$$Y = \sum_{i=1}^{l} C_i Y_i^\circ, \tag{6.2}$$

where Y_i° is an arbitrary ideal-gas property for the ith constituent.

Thus, for the specific heat at constant volume, one has

$$C_v/R = (\partial E/\partial T)_\varrho, \qquad \text{where} \quad E = \sum_{i=1}^{l} C_i E_i.$$

The difference in the specific heats is given by

$$\frac{C_p - C_v}{R} = \frac{1 + (T/Z)(\partial Z/\partial T)_\varrho}{1 + (\varrho/Z)(\partial Z/\partial \varrho)_T}, \qquad \text{where} \quad Z = \sum_{i=1}^{l} C_i,$$

and the ratio of specific heats is obtainable then from $\gamma = C_p/C_v$. The sound velocity can be written as

$$\frac{a}{a_0} = \gamma RTZ\left[1 + \frac{\varrho}{Z}\left(\frac{\partial Z}{\partial \varrho}\right)_T\right].$$

Useful in aerodynamic calculations (Hochstim and Adams, 1967) are the following:

$$\gamma_{P,\varrho} = \left(\frac{\partial \ln P}{\partial \varrho}\right)_S = \gamma\left[1 + \frac{\varrho}{Z}\left(\frac{\partial Z}{\partial \varrho}\right)_T\right],$$

$$\gamma_{T,\varrho} = \left(\frac{\partial \ln T}{\partial \ln \varrho}\right)_S = 1 + \frac{1 + (T/Z)(\partial Z/\partial T)_\varrho}{C_v/R} Z,$$

and

$$\gamma_{T,P} = \frac{1}{1 - (\partial \ln T/\partial \ln P)_S}$$

$$= \gamma\left\{1 + \frac{T(\partial Z/\partial T)_\varrho[1 + (T/Z)(\partial Z/\partial T)_\varrho]}{[1 + (\varrho/Z)(\partial Z/\partial \varrho)_T]C_v/R}\right\}^{-1}.$$

The above expressions contain derivatives of additive functions as in (6.2) which are either with respect to temperature at constant density or are with respect to density at constant temperature.

Let us first consider the derivatives at constant density, i.e., we shall consider the evaluation of the quantity $(\partial Y/\partial T)_\varrho$. Now, from (6.2),

$$(\partial Y/\partial T)_\varrho = \sum_{i=1}^{l} (\partial C_i/\partial T)_\varrho Y_i^\circ + \sum_{i=1}^{l} C_i(\partial Y_i^\circ/\partial T)_\varrho. \tag{6.3}$$

The second term contains the species ideal-gas function $(\partial Y_i^\circ/\partial T)_\varrho$, which can be computed directly from the partition function of the ith species. The C_i are obtained as solutions of the equilibrium equations. The second term is therefore simply a cumulative sum involving known quantities. Since the Y_i° which appear in the first term are also known, the problem of evaluating (6.3) reduces to the determination of the concentration derivatives. We shall do this for the ideal gas (i.e., $\gamma_i = 1$) only, the extension to conditions where this is not appropriate (i.e., $\gamma_i \neq 1$) being left to the interested reader. Taking the derivative of (3.9) yields

$$(\partial \ln C_i/\partial T)_\varrho = (\partial \ln \bar{K}_i/\partial T)_\varrho + \sum_{j=1}^{c} \nu_{ij}(\partial \ln C_j/\partial T)_\varrho, \tag{6.4}$$

where j runs over the reference species only. There are also the conservation equations (3.10),

$$\chi_j = \sum_{i=1}^{l} \nu_{ij}C_i = \sum_{i=1}^{l-c} \nu_{ij}C_i + \sum_{i=l-c+1}^{c} \nu_{ij}C_i, \qquad j = 1, \ldots, c.$$

Since $\nu_{ij} = \delta_{ij}$, the Kronecker delta for reference species, this can be written

$$\chi_j = \sum_{i=1}^{l} \nu_{ij}C_i + C_j, \qquad j = 1, \ldots, c, \tag{6.5}$$

where, again, j runs over the reference species only. Differentiating (6.5) and using (6.4), there results

$$B_j = -\sum_{k=1}^{c} A_{jk}(\partial \ln C_k/\partial T)_\varrho, \qquad j = 1, \ldots, c, \tag{6.6}$$

where

$$B_j = -\sum_{i=1}^{l} \nu_{ij}C_i(\partial \ln \bar{K}_i/\partial T)$$

and

$$A_{jk} = \sum_{i=1}^{l} \nu_{ij}\nu_{ik}C_i + \delta_{kj}C_k \tag{6.7}$$

and where

$$\left(\frac{\partial \ln K_i}{\partial T}\right)_\varrho = \frac{1}{T}\left[\frac{E_i}{RT} - \sum_{j=1}^{c} \nu_{ij}\frac{E_j}{RT}\right]$$

is a known quantity, with E_i the internal energy of the ith species. The c unknowns $(\partial \ln C_k/\partial T)_\varrho$, i.e., the derivatives of the reference species, constitute the unknown vector in (6.6). Having solved these equations, e.g., by inverting the matrix A, (6.4) can be used to calculate the derivatives of the ordinary species.

Let us now consider derivatives with respect to the density at constant temperature, i.e., $(\partial Y/\partial \varrho)_T$. Differentiation of (3.9) leads to the relations

$$(\partial \ln C_i/\partial \varrho)_T = \omega_i/\varrho + \sum_{j=1}^{c} \nu_{ij}(\partial \ln C_j/\partial \varrho)_T. \tag{6.8}$$

On the other hand, differentiation of (6.5) yields

$$0 = \sum_{i=1}^{l} \nu_{ij}C_i(\partial \ln C_i/\partial \varrho)_T + C_j(\partial \ln C_j/\partial \varrho)_T.$$

Combining the latter with (6.8) leads to

$$B_j = -\sum_{k=1}^{c} A_{jk}(\partial \ln C_k/\partial \varrho)_T, \qquad j = 1, \ldots, c, \tag{6.9}$$

where

$$B_j = \sum_{i=1}^{l} \nu_{ij}C_i\omega_i/\varrho$$

and where A_{jk} is again given by (6.7). Since the coefficient matrix A_{jk} is identical in both cases, one can use essentially the same computer program to solve (6.6) and (6.9). These solutions are then substituted into (6.4) and (6.8) to obtain the derivatives of the remaining species.

It should be noted that Eqs. (6.6) and (6.9) are each c linear equations in the derivatives of the c reference species. At this point, it should be clear that, fundamentally, the equilibrium-constant approach to the problem is one of solving c nonlinear equations in c unknowns.

As mentioned above, the ability to express the derivatives in terms of c equations in c unknowns is peculiar to the ideal-gas approximation. As soon as concentration-dependent terms are introduced in the activity coefficients, concentrations of derived species appear on the right-hand side of (3.9) and hence on the right-hand sides of (6.4) and (6.8). This, then, requires the solution of l linear equation in l unknowns for the evaluation of the concentration derivatives. It may still be possible, however, to use these ideal-gas expressions for the derivatives under conditions where the concentrations include real-gas effects. This can be done when $\gamma_i \neq 1$, but $(\partial\gamma_i/\partial\varrho)_T$ and $(\partial\gamma_i/\partial T)_\varrho$ are negligible for major constituents. This can even be extended to situations in which these derivatives are small but not negligible by developing additive expressions for corrections due to the contribution of these derivatives of the γ_i. These expressions could then be evaluated by an approximate numerical scheme.

VII. Numerical Methods

A. Introduction

We now turn to the central problem of this chapter, namely that of solving for the species concentrations. This can be done within the framework of either the free-energy-minimization method or the equilibrium-constant method (Naphtali, 1960; Harnett, 1960). These are, of course, entirely equivalent methods (Brinkley, 1960; Harnett, 1960, especially the last paragraph) and the decision to choose one over the other is mainly a matter of taste. For the purposes of this chapter, numerical methods will mean those methods appropriate to digital calculators. It is interesting to note that chemical-equilibrium equations have been solved using analog computers (Edwards and Rubin, 1958). Such methods are considered to be outside the province of this chapter, however.

Within the framework of these two methods, the literature can be further divided between numerical approaches that are specifically tailored to a particular chemical problem and those that are general-purpose approaches. Special-purpose approaches mainly predate the advent of high-speed electronic computers. At the time, the carrying out of involved algebraic operations was much to be preferred over long and tedious numerical calculations which had to be done on a slide rule or desk calculator. These special-purpose approaches, in the main, consist in the reduction of the c equations (3.11) to one or two equations by successive substitution. With the increasing availability of large-scale, high-speed computers whose main purpose is just the carrying out of tedious repititious numerical operations, the need for the development of such special-purpose schemes has disappeared.

Intermediate between these special-purpose schemes and general-purpose methods are a number of approaches (Hansen, 1968; Hochstim, 1960) which apply to systems for which the concentrations of those derived species that contain more than one reference species is small. These approaches are based on the fact that the neglect of such species serves to uncouple Eqs. (3.10), thereby converting the problem into that of solving a set of c *independent* nonlinear equations. Such approaches, where they are appropriate, can be useful in preliminary studies of chemical systems.

It is interesting to note that, from a numerical point of view, both the free-energy-minimization and the equilibrium-constant methods require the determination of a set of values of the concentrations of the reference species that minimizes some function of these concentrations. This is obvious for the free-energy-minimization method, for which the free energy itself is being minimized. In the equilibrium-constant method, on the other hand, the exact solution of a set of equations is required. From a numerical point of view, this can also be cast into a form requiring a minimization, namely the minimization of an error function that represents departures from the solution (Householder, 1953). Thus, the solution of an arbitrary set of equations

$$\psi_i(x_1, \ldots, x_n) = 0, \qquad i = 1, \ldots, n, \tag{7.1}$$

is equivalent to the minimization of the function

$$\phi = \sum_{i=1}^{n} | \psi_i |^2 \tag{7.2}$$

with respect to the parameters x_i, $i = 1, \ldots, n$. This replacement by a minimization problem is not unique, however. In particular, if α_{ij} are the elements of any positive-definite matrix α, then the solution of the set of equations (7.1) is also equivalent to the minimization of the function $\phi = \sum_{ij} \psi_i \alpha_{ij} \psi_j$ with both minima being identical. Equation (7.2) is thus seen to be the special case where $\alpha = I$, the unit matrix. Since both methods lead to a numerical minimization problem, it is possible to describe the solution of the equations appropriate to either in the same numerical mathematical language.

Nonlinear problems generally cannot be solved in closed form. In the present case, the fact that the problem is nonlinear can perhaps best be seen in Eqs. (3.11) of the equilibrium-constant method, where the unknowns, C_j, appear as products raised to powers (some of which are negative). Nonlinear equations of this kind will certainly not be soluble in closed form, particularly in such a way as to be capable of handling different chemical systems. It should be noted in this regard that each chemical system requires the use of a different T matrix and hence a different set of exponents ν_{ij} in (3.11). This, in turn, requires the solution of a different set of nonlinear equations. In the absence of a general closed-form method of solution, therefore, it is necessary to develop a purely numerical approach to the solution of the problem of minimizing ϕ in (7.2).

Such methods start with a guess at the set of variables $x_1, \ldots, x_n$ in (7.1). This guess is substituted into ϕ in (7.2) and generally found not to be the set of variables that minimizes ϕ. Procedures are then invoked to improve on the guess in a systematic way. If the improvement procedure is properly designed, that set of values of the variables $x_1, \ldots, x_n$ is ultimately determined that does in fact minimize the function ϕ within a preassigned tolerance.

Clearly, whether use is made of the free-energy-minimization method or the equilibrium-constant method, and, in fact, regardless of whatever numerical method is chosen, there must ultimately result a minimum value of ϕ. Since the function ϕ describes a surface in the n-dimensional space whose coordinate axes are the concentrations of the reference species, all numerical methods which can be devised for finding the minimum of ϕ must be geometrically equivalent to starting from some initial point on this ϕ surface and searching for the surface minimum. For this reason, all such methods are referred to as search methods. The function being minimized [i.e., ϕ in (7.2)] is referred to as the objective function.

Nonlinear problems that are fundamentally different from each other make use of different objective functions. As will be seen below, it can also be true that different numerical approaches to the *same* nonlinear problem can be based on the use of different objective functions. In the latter case, it should be obvious that the n-dimensional surfaces corresponding to the different objective functions should have the same coordinates (but not necessarily the same value) for the objective function at their respective minima. This can usually be shown to be trivially true since these objective functions most often differ from each other by functions that vanish at the location of the minimum. Even when this is not so, they, of necessity, differ by functions whose surfaces also have minima at this same location.

There are a number of objective functions which have been widely used in the numerical methods associated with the direct free-energy-minimization and equilibrium-constant methods. Consider the direct free-energy-minimization method. Here, the objective function can simply be taken to be the free energy of the mixture. In another numerical approach to this method (Naphtali, 1959, 1961), constants times the sum of the squares of the left-hand sides of (1.7) are added to the free energy and the total function taken as the objective function. Since the left-hand sides of (1.7) vanish at the solution, this addition to the free energy has no effect on the solution. In fact, since these added quantities are positive away from the solution, they tend to increase the magnitude of the objective function away from the solution over the value it would have if it were simply the free energy. As a result, they tend to magnify the depth of the minimum in the objective function over that obtained using the free-energy minimum alone. Although other objective functions are possible, they will not be included in this discussion.

While the equilibrium-constant method can likewise be stated in more than one way according to the choice of objective function,* we shall restrict ourselves to that based directly on satisfying the relevant equations. Thus, the set of equations (3.11) will be satisfied if and only if the proper values are used for the reference species. In the spirit of Eq. (7.2), it is possible to define a set of functions of the concentrations of the reference species

$$\psi_i(C_1, \ldots, C_c) = \sum_{i=1}^{l} \nu_{ij} \bar{K}_i(\varrho/\varrho_0)^{\omega_i} \prod_j (C_j^{\nu_{ij}}) - \chi_j .$$

The problem of solving (3.11) is then equivalent to the minimization of the objective function

$$\phi = \sum_{i=1}^{c} |\psi_i|^2 .$$

It should be noted that the addition of the Gibbs free energy to this ϕ results in yet another possible objective function and, in fact, is essentially

* A method based on the introduction of the degree-of-reaction variable has been developed by Park (1967). With this approach, the number of equations reduces to the number of reference species in a natural way (see especially p. 13).

one of the objective functions described for the direct free-energy minimization method.

It should be obvious that the same search method can, in principle, be used for all minimization problems. For this reason, part of what follows contains descriptions of and references to general search methods (i.e., not specific to the chemical-equilibrium problem). In practice, however, it is often found that search methods need to be empirically tailored to the particular nonlinear problem at hand. This is not to say, in the problem under discussion, that the search methods must necessarily be tailored to the particular chemical system of interest. Rather, these methods must be adjusted to handle the chemical-equilibrium problem in a manner different from an adjustment made for some other nonlinear problem. This says, in essence, that variations of conditions within the same nonlinear problem (e.g., the variation due to changing the chemical system in the equilibrium problem) will result in objective-function surfaces which are much more nearly alike than are the surfaces associated with objective functions for quite different nonlinear problems.

Each chemical problem might be expected to have an optimum numerical method of solution. The chances of there being such an optimum numerical method of solution for the wide range of chemical problems associated with all possible variations in the T matrix is quite small, however. We shall, in fact, assume that no such general optimum method exists and shall aim at the description of general search methods which should have a high probability of converging to the correct answer for almost all chemical systems. It should be pointed out, however, that there can also be considerable advantage to making small adjustments on the search method for each particular chemical system where such adjustments are feasible. A particular method with which the author has considerable experience and which has been found to converge rapidly in a manner which is independent of the initial guesses will be discussed below in some detail.

A general philosophical point relating to search methods needs to be emphasized, that is, that there are no restrictions on the procedures that can be devised for going from point to point on the ϕ surface, provided only that a proper test is used for the determination of when the surface minimum is reached. In particular, the precise method for the development of the path to the minimum need have no relationship to the chemistry of the problem and, for that matter, none to its mathematics. In fact, it is not even necessary to follow a point having a particular value of ϕ by one with a smaller value of ϕ, if this turns out to be convenient.

As might be expected, however, search methods have been developed in which a sequence of points is traced out on the surface in such a way that the ϕ associated with a given point is indeed guaranteed to be less than the ϕ associated with the preceding point. The ability to do this guarantees convergence to the answer in the sense that successively decreasing ϕ ultimately leads to its minimum value. This guarantee is sometimes obtained at the cost of slow convergence, however.

The starting point has no more *a priori* physical significance than a point used in any other iteration. It follows, therefore, that the particular method chosen for determining initial guesses need have no relation to the chemistry or mathematics of the problem, just as for other iteration points. Many of the methods as described in the literature require starting from points that satisfy the mass balance equations (3.10). Such a restriction is entirely unnecessary, however, since a proper test for the solution will automatically guarantee that the equations (3.10) are satisfied at the accepted solution. In what follows, we shall have little to say about methods for choosing starting values.

While it is generally not possible rigorously to prove convergence for a search method except in the vicinity of the minimum, empirically it is found that most search methods do converge even from points quite far removed from the solution. The ease with which this can be done for a particular method, or equivalently, the sensitivity of the method to the choice of a starting point, depends very strongly on the method itself.

It should be clear from the preceding discussion that central to the solution of nonlinear minimization problems is the proper choice of a criterion by means of which one determines that the solution has been found. Criteria for solutions that are appropriate to the problems of interest here are obvious. Despite this, they are not always used. Clearly, every surface minimum, regardless of the problem, is characterized by the requirement that the derivatives of the objective function with respect to the unknown variables are each less than some arbitrarily small value. This criterion would appear to be a natural one for the direct free-energy-minimization method. In the case of the equilibrium-constant method, there is the further requirement that the equations (3.9) and (3.10) themselves must be simultaneously satisfied (or equivalently, that ϕ in (7.2) is sufficiently small). An intermediate method can also be developed by using the direct free-energy-minimization method of solution but requiring that (3.3) be satisfied at the solution, the μ_i being given by (1.8) (Naphtali, 1959, 1961). Despite these rather obvious criteria, however, search methods all too often use as a criterion for solution the condition that changes in all the unknowns from one iteration to the next become vanishingly small at the minimum.* Although the latter condition generally yields the same results as the former, it is not guaranteed to do so.

B. Search Methods

In recent years, an entire literature has developed that deals with nonlinear search methods.† This literature consists mainly of descriptions of various methods, with occasional reports of experience with particular problems. There is very little in the way of general proofs either for optimization procedures or for convergence except, perhaps, for points near the solution where linearization is valid. We shall only describe the general characteristics of nonlinear search methods, leaving the reader the option of going to the literature for more detailed reviews or for the details of specific methods.

The decision to choose one particular search method over all others will generally depend, at least in part, on such nonmathematical criteria as relative speed of attaining the solution and ease of programming. It will also depend on how often a particular problem is to be solved and for how many different chemical systems. As computing machines become faster, the need for a simply programmable method tends to outweigh considerations of machine speed. On the other hand, the availability of generalized subprograms for the complicated mathematical and logical details of a method tends to reduce the programming time required.‡ It is clear from this that criteria for the choice of one search method over another depend, to a large extent, on the details of the computing facility available for the solution of the problem. In the following, we shall describe a number of different search methods in general terms and shall include alternatives that might serve to reduce computing time. The coverage of these methods is not meant to be exhaustive. More complete

* A representative sampling of these include: Stroud and Brinkley (1969), Hilsenrath *et al.* (1959), Horton and Menard (1969), Palmer and Mahoney (1966).

† Reviews of this literature are contained in the following: Spang (1962), Wolfe (1962), Curtin (1968), Wilde and Beightler (1967).

‡ A number of authors either list complete computer programs in their publications or promise to make such programs available. Among these are: Horton and Menard (1969), Huber (1962), Stroud and Brinkley (1969).

discussions will be found in the several review articles included among the references which deal with nonlinear search methods.

The problem of finding the minimum of a multidimensional surface is a special case within the general class of problems of finding extreme values. Any method developed for the express purpose of finding minima of surfaces can be applied, with at most minor modifications, to the problem of finding surface maxima and vice versa. The latter are more natural within the framework of the disciplines of economics and operations research. This correspondence was specifically taken advantage of in an approach developed at the Rand Corporation in which the equations appropriate to the method of the direct minimization of the free energy were solved by means of linear programming techniques originally developed for the solution of operations research problems. This approach is discussed quite lucidly in a series of Rand reports and will not be considered here.

A nonlinear problem in a set of unknowns can always be converted to a linear problem in a set of deviations from the current values of the unknowns. In order to do this rigorously, it is necessary to make the assumption that such deviations are negligible compared to the current values. This can be done in (3.9) near $C_j^{(k)}$, the kth iterate for the reference species. By taking $C_j^{(k+1)} = C_j^{(k)} + \delta C_j$ and neglecting products of the δC_j, there results a set of linear equations for the δC_j. These can be solved for the δC_j and a new guess $C_j^{(k+1)}$ obtained. This method is rigorous very near the solution point. For points far from the solution, however, there is not even the guarantee that the corrections δC_j will be in the proper direction. In spite of this, the method has been used successfully (Chu, 1958).

Another method for linearizing nonlinear problems is to take a Taylor expansion about the current values of the unknowns and neglect terms involving products and squares of deviations. This method differs from the simple expansion in that the first derivatives at the current guess point appear in the linear equations. This latter method of linearization is essentially the Newton–Raphson method for the solution of sets of nonlinear equations. We shall indicate several ways in which this method has been applied to the chemical-equilibrium problem. A common variation of this approach makes use of a Taylor expansion of the first derivatives about the minimum. In this approach, the linear terms in the deviations contain the second derivatives of the objective function (Chu, 1958). The latter approach is often referred to as the Gauss–Newton method.

Search methods for the development of paths along the surface to the minimum can also be referred to as iterative methods. This general class of numerical methods starts with a guess value, which is substitued into the equations to be solved. On the basis of the result of this substitution, a new value is determined which then becomes the guess value for a repeat of this procedure. This process is continued until, given a convergent method, a solution is obtained. Iterative methods can be written in terms of a function of the variables, which gives a prescription for calculating the next guess from a given guess point. In particular, this function defines a sequence of operations

$$\mathbf{x}^{(k+1)} = \Gamma(\mathbf{x}^{(k)}). \tag{7.3}$$

The functions Γ can be viewed as operators which, when operating on a guess value $\mathbf{x}^{(k)}$, produce the next guess value $\mathbf{x}^{(k+1)}$. This operation can also be given a geometric interpretation by writing the sequence

$$\mathbf{x}^{(k+1)} = \mathbf{x}^{(k)} + \lambda\boldsymbol{\mu}^{(k)}, \tag{7.4}$$

where $\boldsymbol{\mu}^{(k)}$ is a unit vector in the direction of the next guess with λ the magnitude of the step in that direction. Equation (7.4) is generally used to describe descent methods, $\boldsymbol{\mu}$ then always being taken as a vector in the general direction of the minimum.

Iteration methods are generally described as being either of the Newton–Raphson type or of the descent type, although the distinction is not always clear (Zeleznik and Gordon, 1968). In the Newton–Raphson approach, $\Gamma(x)$ is obtained from a truncated Taylor series about the solution. For example, suppose that the surface minimum occurs at the point $\mathbf{x}^{(0)}$ with $\mathbf{x}^{(1)}$ a point close by. It follows, then, that $\psi(\mathbf{x}^{(0)}) = \psi(\mathbf{x}^{(1)}) + \mathsf{J}(\mathbf{x}^{(1)})(\mathbf{x}^{(0)} - \mathbf{x}^{(1)})$, where $\mathsf{J}(\mathbf{x})$ is the matrix of the partial derivatives of $\psi(\mathbf{x})$, i.e., $J_{ij} = \partial\psi_i(\mathbf{x})/\partial x_j$. If $\mathbf{x}^{(0)}$ is at the minimum, it follows that $\psi(\mathbf{x}^{(0)}) = 0$, so that, solving for $\mathbf{x}^{(0)}$, there results

$$\mathbf{x}^{(0)} = \mathbf{x}^{(1)} - \mathsf{J}^{-1}(\mathbf{x}^{(1)})\psi(\mathbf{x}^{(1)}).$$

If in a sequence of iterations the kth point is sufficiently near the minimum, it follows that the solution can be obtained as the $(k+1)$th point by solving the equation

$$\mathbf{x}^{(k+1)} = \mathbf{x}^{(k)} - \mathsf{J}^{-1}(\mathbf{x}^{(k)})\psi(\mathbf{x}^{(k)}). \tag{7.5}$$

While this is only rigorously correct at the solution, it is used in the

Newton–Raphson approach at all points on the surface as a means of determining a sequence of points $\mathbf{x}^{(k+1)}$, $\mathbf{x}^{(k+2)}$, ..., etc. which defines a path to the minimum. It should be noted that, for the Newton–Raphson method, the Γ operator is given by

$$\Gamma(\mathbf{x}) = \mathbf{x} - \mathsf{J}^{-1}(\mathbf{x})\psi(\mathbf{x}). \tag{7.6}$$

Substitution of (7.6) into (7.3) and comparison with (7.4) shows that there is a descent version of the Newton–Raphson search method with the direction vector given by

$$\lambda\boldsymbol{\mu} = -\mathsf{J}^{-1}(\mathbf{x})\psi(\mathbf{x}).$$

At the solution, (7.6) becomes $\Gamma(\mathbf{x}^{(k+1)}) = \mathbf{x}^{(k+1)}$, so that $\mathbf{x}^{(k+2)} = \mathbf{x}^{(k+1)}$ and the differences between successive iterations eventually vanish. As has been mentioned, this can be used as a basis for a criterion for solution. Thus, a solution is said to have been reached when the change between successive values of all components of the vector $\mathbf{x}$ become less than some fraction of their current values. Unfortunately, this is not a necessary and sufficient condition for a solution. Thus, while we have shown this property to hold at the solution (necessary condition), it is possible that for a particular search method a point on the surface can satisfy this condition away from the solution. Thus, satisfying this criterion does not guarantee a solution (sufficient condition).

Mention has been made of a version of the Newton–Raphson method (often referred to as the Gauss–Newton method) that makes use of second derivatives (Vinturella, 1968). This can be written in vector notation in a form similar to (7.5). For $\mathbf{x}^{(k+1)} - \mathbf{x}^{(k)}$ small, the matrix of first derivatives $\mathsf{J}(\mathbf{x})$ can be expanded about the point $\mathbf{x}_k$ to terms linear in the difference $\mathbf{x}^{(k+1)} - \mathbf{x}^{(k)}$. This results in the relation

$$\mathsf{J}(\mathbf{x}^{(k+1)}) = \mathsf{J}(\mathbf{x}^{(k)}) + \mathsf{H}(\mathbf{x}^{(k)}) \cdot (\mathbf{x}^{(k+1)} - \mathbf{x}^{(k)}),$$

where $\mathsf{H}(\mathbf{x})$ is the matrix of second derivatives, generally referred to as the Hessian matrix. If the assumption is made that $\mathbf{x}^{(k+1)}$ is the solution point, it follows that $\mathsf{J}(x^{(k+1)}) = 0$ and the above relation becomes

$$0 = \mathsf{J}(\mathbf{x}^{(k)}) + \mathsf{H}(\mathbf{x}^{(k)}) \cdot (\mathbf{x}^{(k+1)} - \mathbf{x}^{(k)})$$

or

$$\mathsf{H}(\mathbf{x}^{(k)})\mathbf{x}^{(k+1)} = \mathsf{H}(\mathbf{x}^{(k)})\mathbf{x}^{(k)} - \mathsf{J}(\mathbf{x}^{(k)}),$$

from which

$$\mathbf{x}^{(k+1)} = \mathbf{x}^{(k)} - \mathsf{H}^{-1}(\mathbf{x}^{(k)})\mathsf{J}(\mathbf{x}^{(k)}),$$

quite analogous to (7.5). In this case, $\Gamma(\mathbf{x})$ is given by

$$\Gamma(\mathbf{x}) = \mathbf{x} - \mathsf{H}^{-1}(\mathbf{x})\mathsf{J}(\mathbf{x}).$$

In what follows, we shall discuss the Newton–Raphson method only in the form (7.5).

Many variations of the Newton–Raphson search algorithm [i.e., (7.5)] have been devised through variation of the definition of $\mathsf{J}(\mathbf{x})$. These alternate methods are mainly used when the calculation of $\mathsf{J}(\mathbf{x})$ and its inverse either involves excessive computing time or is overly complex. The simplest variation involves a simple iteration. In that case, $\mathsf{J}(\mathbf{x}) = \lambda\mathsf{I}$, I the unit matrix, and the algorithm for the choice of successive values becomes

$$\Gamma(\mathbf{x}) = \mathbf{x} - \lambda\psi(\mathbf{x}),$$

which at the solution becomes $\Gamma(\mathbf{x}^{(k+1)}) = \mathbf{x}^{(k)}$. As will be seen below, this approach falls within a category of methods which we have designated direct search methods. An appropriate value of λ can generally be found empirically through monitoring the behavior of the $\mathbf{x}$ vector in successive iterations. The value of λ used will depend strongly on the nature of the function $\psi(\mathbf{x})$. For example, consider the one-dimensional problem $\psi(x) = 0$. If $\psi(x)$ is such that $x^{(1)} < x^{(0)}$ implies $\psi(x) > 0$, then clearly $\lambda > 0$ is required. If, on the other hand, $x^{(1)} > x^{(0)}$ implies $\psi(x) < 0$, $\lambda < 0$ is required. These requirements would easily be obtained empirically.

A more complicated variation of the Newton–Raphson method involves replacing the "tangent" matrix $\mathsf{J}(\mathbf{x})$ by a "secant" matrix $\mathsf{J}(\mathbf{x})$ (Wolfe, 1959). In this approach, the partial derivatives in the J matrix are replaced by finite differences based on neighboring solution points. For an n-dimensional ϕ-space (i.e., for n unknowns), the secant approximation requires having in hand $n + 1$ neighboring values. The method thus does not really get under way until the objective function has been evaluated at $n + 1$ points. The procedure can be started by choosing, somewhat arbitrarily, these $n + 1$ points, evaluating the function, calculating the elements of the matrix J, and proceeding to invert J as in the Newton–Raphson method. Since the derivatives of the function do not have to be evaluated, computing time is reduced. Since the secants make use of stored evaluations of the objective function at the $n + 1$ points, the programming is also simplified. Since the method still requires a matrix inversion for determining $\mathbf{J}^{-1}(\mathbf{x})$ from $\mathbf{J}(\mathbf{x})$, the saving in machine time will be modest. A dramatic reduction in computational time can be had by means of a variation on the secant version of the Newton–Raphson method in which each component x_j of the unknown vector $\mathbf{x}$ is treated separately, thus eliminating the need for matrix inversion. This approach is associated with the name of Wegstein (1958) and will be described below among the direct search methods.

An interesting variation of the Newton–Raphson method has recently been described (Vinturella, 1968). In this approach, the coordinate system in parameter space is rotated into the space spanned by the eigenvectors of $\mathbf{J}$. In that space, $\mathbf{J}$ can be written in diagonal form. The advantage of this approach is that, since the new "parameters" (i.e., the eigenvectors of $\mathbf{J}$) are orthogonal, they do not interact with each other, so that the minimization can be carried out independently for each.

A number of methods have been developed for the solution of chemical-equilibrium problems which make use of a Newton–Raphson search method. The two most widely used are one due to Brinkley (1947) and a second due to Huff *et al.* (1951); Gordon *et al.* (1959); Chao (1967). The relationship between them is discussed in some detail in the review of Zeleznik and Gordon (1968). Each method is sufficiently well described in the literature to preclude the need for a description here. Brinkley's approach is, of course, built around the solution of a set of equations whose number is the same as the number of reference species. It is interesting to note that the method of Huff *et al.*, (1951) as originally described, considers both the reference species and the derived species on an equal footing, thus requiring the solution of l equation for the l composition variables. Since l can become prohibitively large for matrix techniques, methods for reducing this requirement are of interest. Zelesnik and Gordon indicate that, as expected, the method can be converted to one in which only the reference species are required, thereby reducing the problem to that of solving c equations in c unknowns, where $c < l$ is always the case. They indicate, however, that this can result in convergence problems unless the reference species are carefully chosen so as to be major constituents or, if not, unless the initial guesses are good. This problem does not appear to arise when all l equations are solved, presumably since these *always* contain the major constituents.

C. Direct Search Methods

Search methods fall into two categories on the basis of the ease with which they can be applied to new problems. This can be of considerable practical importance in the chemical-equilibrium problem. For example, after programming an ideal-gas calculation, one might wish to add in real-gas effects by relaxing the restriction $\gamma_i = 1$. For some search methods, this can pose a major reprogramming problem, while for others, the conversion is rather simple. These two classes of search methods can be described, roughly, in terms of the extent to which they depend on the content of the equilibrium equations.

Search methods of the general Newton–Raphson type require considerable detail since they call for the calculation of first (and sometimes second) derivatives of the objective function in addition to the evaluation of the objective function itself. Reprogramming for such methods can call for considerable effort. On the other hand, search methods exist that at each step require only that the objective function be evaluated. Such methods we shall designate direct search methods since, in the case of the most obvious one, the minimum in the objective-function surface is sought by a direct search in variable space. The usual definition of direct search methods is much more narrow than this. We shall, in fact, include among our direct search methods certain ones that are variations of approaches not normally defined as being of the direct search type, e.g., variations of the Newton–Raphson method. This underlines the arbitrariness of these classifications.

There are several advantages to be had from the use of direct search methods. Since only the objective function is evaluated, a minimum of reprogramming effort is required when the objective function is altered. Furthermore, the simplicity of the search aspect allows for simple initial programming, thereby reducing the time required for going from the development of the problem to the working computer program. Experience has also shown that direct search methods can be successful for problems that are poorly conditioned for the other methods (Hooke and Jeeves, 1961; Wilde and Beightler, 1967, p. 307; Wegstein, 1958). An example might be a problem for which the matrix of the derivatives of the objective function with respect to the unknowns is ill-conditioned in some part of variable space even far removed from the solution. Since that matrix plays a key role in the Newton–Raphson method, there can be trouble with it if an iteration happens to come near such regions of ill-conditioning.

Perhaps the simplest of the direct search methods is one that merely involves testing the objective function in a stepwise variation of the unknown parameters (perhaps on a grid), accepting only those changes that reduce that function. The pattern search method of Hooke and Jeeves (1961) is an improvement over this simplistic approach. Their approach makes use of information obtained but usually ignored in a simple stepwise search on a grid. In their direct search method, two kinds of multidimensional steps on the ϕ-surface are defined—an *exploratory* step which, when successful, is followed by a *pattern* step. For the former, the unknown parameters are varied individually, while for the latter, the parameters are all changed simultaneously. Exploratory steps proceed as follows. The objective function is evaluated at the location of the initial guess $\mathbf{x}^{(0)} = (x_1^{(0)}, x_2^{(0)}, \ldots, x_n^{(0)})$ and the result stored. Using a prescribed step size for $x^{(1)}$, i.e., Δ_1, the parameter $x_1^{(0)}$ is increased to $x_1^{(0)} + \Delta_1$ and the objective function evaluated. Should this lead to a reduction in the objective function, $x_1^{(0)}$ is replaced by $x^{(1)} = x_1^{(0)} + \Delta_1$, so that the new guess point becomes $(x^{(1)}, x_2^{(0)}, \ldots, x_n^{(0)})$ and the same procedure is followed for $x_2^{(0)}$ using Δ_2, the prescribed step size for x_2. If, on the other hand, replacing $x_1^{(0)}$ by $x_1^{(0)} + \Delta$ does not reduce the objective function, a change of direction is attempted, i.e., $x_1^{(0)} - \Delta_1$ is tried in place of $x_1^{(0)}$. When this too does not lead to a reduction in ϕ, Δ_1 is replaced by $\alpha\Delta_1$, $\alpha < 1$, and the process is repeated from the start. This is continued, each new Δ_1 being decreased when there is no success, until there is either a decrease in the objective function for some value of Δ_1 or until Δ_1 becomes smaller than a preassigned limit. In either case, the same process is carried over to x_2. In the former case, $x_1^{(0)}$ is changed by the current value of Δ_1 in the direction indicated by the decrease in ϕ, while in the latter case, $x_1^{(0)}$ is used as $x_1^{(1)}$ and the guess point is therefore not changed. This process is applied to each parameter in turn until each of $x_1, \ldots, x_n$ has been varied. There are now two possibilities: either the new guess point $\mathbf{x}^{(1)} = (x_1^{(1)}, \ldots, x_n^{(1)})$ is identical with the initial guess point $\mathbf{x}^{(0)} = (x_1^{(0)}, \ldots, x_n^{(0)})$ or these two points differ in at least one parameter. In the former case, the surface minimum has probably been reached. At this point, a *proper* test for solution is made to see if, indeed, the surface minimum has been reached. When $\mathbf{x}^{(1)}$ differs from $\mathbf{x}^{(0)}$ in at least one parameter, a change vector $\mathbf{x}^{(1)} - \mathbf{x}^{(0)} \equiv \mathbf{\Delta}_p$ has been determined which, when drawn from the initial guess point $\mathbf{x}^{(0)}$, points in the *general* direction of the minimum. Hooke and Jeeves suggest that, on the average, this change vector indicates the direction to the minimum sufficiently well enough for the objective function to be still smaller at $\mathbf{x}^{(2)} = \mathbf{x}^{(1)} + \mathbf{\Delta}_p$.

The step $\mathbf{x}^{(2)} = \mathbf{x}^{(1)} + \mathbf{\Delta}_p$ is called a pattern step. Of particular importance is the fact that all parameters are varied simultaneously in a pattern step, whereas in an exploratory step, each of the n parameters is varied at least once. The time required for a complete exploratory step for n unknowns can obviously be much more than n times that required for a pattern step.

Hooke and Jeeves choose to restrict themselves to one pattern step after a successful exploratory step. They then follow this single pattern step by a new exploratory step. Clearly, one might modify their method by making the number of consecutive pattern steps after a successful exploratory step a variable or even by continuing with pattern steps until there ceases to be a reduction in the objective function.

A number of further variants of the pattern search method are possible. One might, for example, choose to vary each component step of the exploratory phase from the original guess point $\mathbf{x}^{(0)}$ so that $\mathbf{x}^{(1)} = \mathbf{x}^{(0)} + \mathbf{\Delta}$, where $\mathbf{\Delta} = (\Delta_1, \Delta_2, \ldots, \Delta_n)$ as calculated. This does not hold for the method as we have outlined it above since each component x_i is changed on a successful decrease in ϕ. Thus, in other words, $\mathbf{\Delta}_p \neq \mathbf{\Delta}$. One might, furthermore, devise variations on the pattern step itself by replacing the vector $\mathbf{\Delta}_p$ as determined by the exploratory step by a different change vector $\mathbf{\Delta}_p'$ for use in the pattern step. One might, for example, determine $\mathbf{\Delta}_p'$ as some kind of average over the $\mathbf{\Delta}_p$ vectors used in several of the preceding pattern steps or one might simply take a fraction of Δ_p.

Another approach that, according to our definition, can be included among the direct search methods is actually an independent-parameter version of a variation on the secant approximation to the Newton–Raphson method. The method is also referred to as the Wegstein method (Wegstein, 1958; Jeeves, 1958). The basic algorithm for one dimension is available among the basic library routines at many computer installations. The method has been used extensively by the author and his collaborators in the production of thermodynamic tables of atmospheric gases.* This method has been found to produce rapid convergence even when the initial guess values differ from the answers by more than factors of 10^{+6} or 10^{-6}. When coupled with a method for automatically converting from one set of reference species to another where necessary, the method should produce rapid convergence entirely independent of initial guesses and should therefore be a truly automatic approach. We shall describe

* The following reports, containing such tables, were published under the sponsorship of the Arnold Engineering and Development Center, Tullahoma, Tennessee: Hilsenrath and Klein (1963–1966) and Hilsenrath *et al.* (1966).

the method in some detail in one dimension but shall not specifically relate it to the actual Newton–Raphson method. The extension to multi-dimensional problems in the independent-parameter approximation will then be shown to be straightforward.

This method can best be described in terms of the iteration (7.3). It should be noted that (7.3) can also be applied to the solution of equations of the form $\psi(\mathbf{x}) = 0$ since one can take $\Gamma(\mathbf{x})$ in the form

$$\Gamma(\mathbf{x}^{(k)}) = \mathbf{x}^{(k)} + \alpha\psi(\mathbf{x}^{(k)}),$$

where α is nonzero. With this form for $\Gamma(\mathbf{x}^{(k)})$, (7.3) reduces to the identity $\mathbf{x}^{(k+1)} = \mathbf{x}^{(k)}$ at the solution where $\psi(\mathbf{x}^{(k)}) = 0$. In the Wegstein method, which we shall describe in one dimension, (7.3) is modified by superimposing on it an in–out averaging algorithm. Thus, rather than taking $x^{(k+1)}$ as given by (7.3) for the next guess, an average between the value $x^{(k)}$ into the iteration and $x^{(k+1)}$ out of the iteration is taken to obtain as the next guess

$$\bar{x}^{(k+1)} = x^{(k+1)} - q(x^{(k+1)} - x^{(k)}). \tag{7.7}$$

This produces a value given by $x^{(k+1)}$ as calculated in (7.3) minus a fraction q of the calculated change between $x^{(k)}$ and $x^{(k+1)}$. It is easy to show that, in the secant approximation, $q = -m/(m - 1)$, where m is the slope of the secant between $x^{(k-1)}$ and $x^{(k)}$. This enables q to be calculated at each iteration step. We are, however, much more interested in the advantages of using q as a fixed empirical quantity.

The use of (7.3) by itself will lead to one of the following behavior patterns for successive values of $x^{(k)}$ (i.e., as functions of k, the iteration number):

1. The values of $x^{(k)}$ oscillate and converge.
2. The values of $x^{(k)}$ oscillate and diverge.
3. The values of $x^{(k)}$ converge monotonically.
4. The values of $x^{(k)}$ diverge monotonically.

It is also possible for the character of the behavior pattern to change as n progresses. Generally, however, one particular kind of behavior should predominate. Each of the above behavior patterns for (7.3) can be improved through the use of (7.7) with fixed q in the sense that convergence can be speeded up. This can be done, in each case, by means of the following ranges of q values:

1. Oscillatory convergence: $0 < q < 0.5$.
2. Oscillatory divergence: $0.5 < q < 1$.
3. Monotonic convergence: $q < 0$.
4. Monotonic divergence: $q > 1$.

The reason, in each case, follows from the use of q in (7.7) as an in–out averaging parameter. Thus, for the case where successive values oscillate with a trend toward the answer superimposed on this oscillation, it is clear that $x^{(k+2)}$ will lie between $x^{(k)}$ and $x^{(k+1)}$. This is what is meant by oscillation. Each value will, however, on the average be closer to $x^{(k+1)}$ than to $x^{(k)}$. This, in turn, is what is meant by convergence. Thus, q must be so chosen that $\bar{x}^{(k+1)}$ lies between $x^{(k)}$ and $x^{(k+1)}$ but closer to $x^{(k+1)}$. This is clearly the case only for $0 < q < 0.5$. Examination of the other ranges listed for q values shows them also to be designed to produce values of $\bar{x}^{(k+1)}$ in the general direction of the solution from $x^{(k+1)}$. Wegstein (1958), with several examples, illustrates the damping effect this method has on oscillations, its ability to change divergent behavior into convergent behavior, and even to speed up convergence where there is monotonic convergence. It should be noted that $q = 1.0$ causes the iteration averaging scheme (7.7) to become $\bar{x}^{(k+1)} = x^{(k)}$, which destroys the ability to change x from one iteration to the next and hence causes a false locking in on an answer.

The extension of the Wegstein method to multidimensional problems is particularly simple. It is merely necessary to state such problems in such a way that the search method can be independently applied to each variable. In an approach used extensively by the author and his collaborators (Hilsenrath and Klein, 1963–1966; Hilsenrath *et al.*, 1966), n equations (7.7) were written, one being written for each variable, with each equation having its own q parameter. In this approach, the algorithm

$$C_j^{(k+1)} = \left[\chi_j \Big/ \sum_{i=1}^{l} |\, \nu_{ij} \,| C_j^{(k)}\right] C_j^{(k)}, \qquad j = 1, \ldots, c-1, \tag{7.8a}$$

was taken for (7.3) for all species except the electrons, the superscripts referring to iteration number. A different scheme had to be used for the electrons since for them $\chi_j = 0$. The algorithm

$$C_j^{(k+1)} = -\sum_{i=1}^{l-1} \nu_{ij} C_i^{(k)}, \qquad j = c, \tag{7.8b}$$

was taken for the electrons, the sum extending over all species *except the electrons themselves*. It should be noted that, according to our definitions, ν_{ij} for $j = c$ is positive for negative ions and negative for positive ions, so that the former subtract from the sum while the latter add to it, as required. Also note that (7.8b) is merely (3.10) for the electrons with $\chi_j = 0$.

The iteration process proceeds as follows. Initial guesses for the concentrations of the reference species are substituted into (3.9), thereby producing initial guesses for the derived species. The method has been found to be sufficiently independent of these initial guesses for the reference species to allow the same set of initial guesses to be used for all problems. The initial guesses thus calculated for the derived species are substituted into (7.8a) and (7.8b), yielding interim new guesses for the reference species. These are then substituted into (7.7) and new guesses obtained for the reference species. These are used in (3.9) to start the next iteration. This process is continued until the proper criterion for solution is satisfied. It should be noted that, according to (3.10),

$$\chi_j \Big/ \sum_{i=1}^{l} | \nu_{ij} | C_i = 1$$

at the solution, in which case (7.8a) becomes $C_j^{(k+1)} = C_j^{(k)}$, as required.

For the various systems studied by the author and his collaborators, the search method was found to be insensitive to the choice of each of the q_j except for a small sensitivity associated with the q_j of the electron concentration. This lack of sensitivity, particularly for the atomic species, persisted over an extremely large range of conditions. These included temperatures and densities for which the reference species were present in trace amounts, with the major constituents being molecular species; those for which the reference species dominated; as well as those for which the reference species were again present in trace amounts, with atomic ions being the main species. In all such cases, the value $q_j = 0.5$ was used for all atomic reference species.

The electron concentration behaved somewhat differently as a function of iteration number, as might have been expected, since (7.8b) is of quite a different form than is (7.8a). For the electron concentration, there tended to be oscillatory divergence with $q \cong 0.8$ being required. Somewhat more rapid convergence was obtained by starting each problem with $q = 0.5$ for the electron concentration and increasing the value of q slightly with each iteration so as to reach $q \cong 0.8$ after a small number of iterations. It is important to point out that this scheme (i.e., $q = 0.5$ for all species except the electrons and a slowly increasing q value for the latter), once adopted, was never modified regardless of the chemical system studied.

A number of precautions common to iteration methods had to be taken. A ceiling was placed on the relative magnitude by which the C_j were allowed to change in one iteration. Thus, for $| C_j^{(k+1)} - C_j^{(k)} | / C_j^{(k)} > 100$ or $< 1/100$, the algorithms $C_j^{(k+1)} = 100 C_j^{(k)}$ and $C_j^{(k+1)} = C_j^{(k)}/100$ were used. In practice, these restrictions were found to operate only for current guess points quite far removed from the solution. These ceilings, in effect, guide the iteration point into the neighborhood of the solution. They play an important role in making the problem independent of the initial guess values, as they undoubtedly also would if applied to other search methods.

Appendix A. Illustrations of the Matrices Associated with the Stoichiometry of Reacting Gaseous Mixtures

A number of matrices having to do with the stoichiometry of coupled chemical reactions in gaseous mixtures have been defined in the text. A complete understanding of these matrices and of their relationships to the chemistry of coupled chemical reactions is absolutely essential here. Without such an understanding, much of what appears in this chapter becomes unintelligible. For this reason, we include in this appendix the matrices associated with a typical set of coupled reactions. We have purposely chosen a set of reactions that includes electron attachment and detachment (i.e., ionization) in order to illustrate our treatment of the electron as a reference species. This treatment differs, in some respects, from standard chemical treatments, particularly with regard to notation.

The set of reactions with which we shall be concerned in this appendix are reactions among the following thirteen species: N_2, O_2, N_2O, NO_2, N^+, O^+, O_2^+, O_2^-, N^{2+}, O^{2+}, N, O, and e^-, where e^- refers to the electron. Equation (1.1) of the text requires these to be written N_2, O_2, N_2O, NO_2, Ne_{-1}^-, Oe_{-1}^-, $O_2e_{-1}^-$, $O_2e_{+1}^-$, Ne_{-2}^-, Oe_{-2}^-, N, O, and e^-, where a negative subscript for the electron indicates the absence of an electron and a positive subscript the presence of an extra electron. It should be noted that our notation differs from standard notation only in that the usual practice of writing ionic charge as a superscript on the chemical symbol has been replaced by specific reference to the presence or absence of the electron treated as a chemical species.

This notation leads to the following ν_{ij} matrix elements (N, O, and e^- the reference species):

		N	O	e^-
N_2		2		
O_2			2	
N_2O		2	1	
NO_2		1	2	
Ne^-_{-1}	(N^+)	1		−1
Oe^-_{-1}	(O^+)		1	−1
$O_2e^-_{-1}$	(O_2^+)		2	−1
$O_2e^-_{+1}$	(O_2^-)		2	1
Ne^-_{-2}	(N^{2+})	1		−2
Oe^-_{-2}	(O^{2+})		1	−2
N		1		
O			1	
e^-				1

It should be pointed out that this set of ν_{ij} matrix elements is associated with the chemical symbols of the various species and so is permanently associated with these species, *regardless of the definition of reference species.* The elements of the various **T** matrices, on the other hand, depend very strongly on the choice of reference species.

According to usual practice and notation, the chemical reactions to which we shall restrict this discussion might be written, with atoms and the electron as reference species,

$$2N \rightleftharpoons N_2$$
$$2O \rightleftharpoons O_2$$
$$2N + O \rightleftharpoons N_2O$$
$$N + 2O \rightleftharpoons NO_2$$
$$N \rightleftharpoons N^+ + e^-$$
$$O \rightleftharpoons O^+ + e^-$$
$$2O \rightleftharpoons O_2^+ + e^-$$
$$2O + e^- \rightleftharpoons O_2^-$$
$$N \rightleftharpoons N^{2+} + 2e^-$$
$$O \rightleftharpoons O^{2+} + 2e^-$$
$$N \rightleftharpoons N$$
$$O \rightleftharpoons O$$
$$e^- \rightleftharpoons e^-.$$

The T matrix associated with Eq. (1.5) of the text was a square matrix and included both a row and a column for each species. The reactions above define such a matrix as follows:

	N_2	O_2	N_2O	NO_2	N^+	O^+	O_2^+	O_2^-	N^{++}	O^{++}	N	O	e^-
N_2	1										−2		
O_2		1										−2	
N_2O			1								−2	−1	
NO_2				1							−1	−2	
N^+					1						−1		1
O^+						1						−1	1
O_2^+							1					−2	1
O_2^-								1				−2	−1
N^{2+}									1		−1		2
O^{2+}										1		−1	2
N											1		
O												1	
e^-													1

where a blank position is used to indicate a zero matrix element.

Somewhat better symmetry can be obtained, in the sense that reference species always appear on the same side of the equation, if the ionization reactions are written

$$N - e^- \rightleftharpoons N^+$$

$$O - e^- \rightleftharpoons O^+$$

$$O_2 - e^- \rightleftharpoons O_2^+$$

$$N - 2e^- \rightleftharpoons N^{2+}$$

$$O - 2e^- \rightleftharpoons O^{2+}$$

It should be noted that this has no effect on the matrix element t_{ij}.

The second form of the T matrix defined in the text uses the fact that the column under a derived species contains an entry only for the row associated with that same derived species, with that entry always being $+1$. With this $+1$ understood (and, more important, supplied when needed), the T matrix can be written in terms of the reference species only. This results in the following matrix:

	N	O	e^-
N_2	−2		
O_2		−2	
N_2O	−2	−1	
NO_2	−1	−2	
N^+	−1		1
O^+		−1	1
O_2^+		−2	1
O_2^-		−2	−1
N^{2+}	−1		2
O^{2+}		−1	2
N	1		
O		1	
e^-			1

It should be noted that, because atoms and the electron are used as reference species, these matrix elements correspond exactly to the ν_{ij}. That this is not so for other choices of reference species is obvious on examination of the matrix elements in the text in the discussion of the matrix manipulations required for transformation of reference species. As stated above, such manipulations do not affect the ν_{ij}.

Appendix B. The Introduction of Real-Gas Effects

The expressions derived for the calculation of the species concentrations and those developed for the calculation of the properties of the mixture need to be modified to include real-gas effects. In principle, these modifications should include both "internal" effects, i.e., modifications of the internal energy levels of individual species and "external" effects, i.e., those due to interactions between particles. For particle densities of interest here, however, changes in the internal energy levels of individual species will be negligible. As a result, the thermodynamic properties of individual species can be written as the sum of an ideal-gas part (calculated as indicated in the text) and a part due to particle interactions involving the species. In this appendix, we shall be concerned with the development of certain correction terms to the ideal-gas expressions developed in the text.

According to the formalism developed for the ideal gas, the inclusion of expressions for the species activity coefficients will automatically extend the calculations of the species concentrations to include real-gas effects. Such expressions need to be included in Eq. (2.3) for the free-energy- minimization method and in (3.7a) for the equilibrium-constant method. The extension of the calculation of the thermodynamic properties of the mixture to include real gases is then completed on adding to the relations (6.1) *et seq.* developed for the mixture properties of an ideal-gas appropriate terms for real-gas effects and using, in the resulting expressions, concentrations calculated using real-gas activity coefficients.

I. Real-Gas Effects on the Calculation of the Species Concentrations

The chemical potential of a species in a gas mixture can be written

$$\mu_i(T, P, \mathbf{x})/RT = \mu_i^\circ(T)/RT + \ln P + \ln x_i + \Sigma\, (\Delta_k\mu_i)/RT,$$

where $\mathbf{x}$ indicates a dependence on the concentrations of *all* species and where each $\Delta_k\mu$ is a different additive effect. An activity coefficient can be defined for each of these effects through the relation

$$\ln \gamma^{(k)} = \Delta_k\mu/RT,$$

leading to an expression for the chemical potential

$$\frac{\mu_i(T, P, \mathbf{x})}{RT} = \frac{\mu_i^\circ(T)}{RT} + \ln P + \ln x_i + \ln\Big[\prod_k \gamma_i^{(k)}\Big].$$

The expression (2.3) on which the calculation of the concentrations in the free-energy-minimization method is based then need only be modified by taking $\gamma_i = \prod_k \gamma_i^{(k)}$ including as many of the k effects as desired. The modification of (3.7a) is equally straightforward, merely requiring the definition of a γ' for each of the k effects. It should be noted that these effects can include such diverse items as quantum effects, higher virial coefficients, etc. In short, the formalism includes *all* effects that produce additive terms in the chemical potential.

The effect of second virial coefficients can be included immediately on writing down the appropriate correction to the chemical potential. Thus,

$$\frac{\Delta\mu_i}{RT} = \frac{2}{V} \sum_{k=1}^{l} C_k B_{ik}, \tag{B.1}$$

so that

$$\ln \gamma_i = \frac{2}{V} \sum_{k=1}^{l} C_k B_{ik}$$

This can be substituted directly into (2.3) for the free-energy-minimization method. For the equilibrium-constant method, it is necessary to calculate

$$\gamma_i' = \left[\prod_j \gamma_j^{\nu_{ij}}\right] \Big/ \gamma_i$$

for the ith chemical reaction. Thus,

$$\gamma_i = \exp\left[\frac{2}{V} \sum_{k=1}^{l} C_k B_{ik}\right],$$

so that

$$\frac{\prod_j \gamma_j^{\nu_{ij}}}{\gamma_i} = \exp\left[\frac{2}{V} \sum_j \nu_{ij} \sum_{k=1}^{l} C_k B_{jk} - \frac{2}{V} \sum_{k=1}^{l} C_k B_{jk}\right]$$

and

$$\gamma_i' = \exp\left[\frac{2}{V_0} \frac{\varrho}{\varrho_0} \sum_j \nu_{ij} \sum_{k=1}^{l} C_k B_{jk} - \frac{2}{V_0} \frac{\varrho}{\varrho_0} \sum_{k=1}^{l} C_k B_{ik}\right], \quad \text{(B.2)}$$

where V_0 is the volume of one mole at standard conditions. It should be noted that this γ_i' should be substituted into (3.9) as a factor multiplying $\bar{K}_i$.

The limiting-law Debye–Hückel correction to the calculation of the species concentrations can also be included in a straightforward manner. In this case,

$$\frac{\Delta\mu_i}{RT} = \frac{-\pi^{1/2}(\sum_s Z_s^2 C)^{1/2}}{(DkT)^{3/2}} \left(\frac{\varrho}{\varrho_0}\right)^{1/2} \left(\frac{N_0}{V_0}\right)^{1/2} Z_i^2$$

where N_0 is the number of particles per mole at standard conditions, V_0 the corresponding volume, D the dielectric constant of the mixture, and Z_s the ionic charge of each ion of type s. It follows then that

$$\ln \gamma_i = \left[\frac{-\pi^{1/2}}{(DkT)^{3/2}} \left(\frac{\varrho}{\varrho_0}\right)^{1/2} \left(\frac{N_0}{V_0}\right)^{1/2} Z_i^2 \left(\sum_s Z_s^2 C_s\right)^{1/2}\right]. \quad \text{(B.3)}$$

Substitution of this expression into (2.3) guarantees the inclusion of the limiting-law Debye–Hückel correction. In the equilibrium-constant method, this correction is inserted by means of the additional factor

$$\gamma_i' = \frac{\pi_j \gamma_j^{\nu_{ij}}}{\gamma_i}$$

$$= \exp\left[\frac{-\pi^{1/2}}{(DkT)^{3/2}}\left(\frac{\varrho}{\varrho_0}\right)^{1/2}\left(\frac{N_0}{V_0}\right)^{1/2}\left(\sum_s Z_s^2 C_s\right)^{1/2}\left(\sum_j \nu_{ij} Z_j^2 - Z_i^2\right)\right]. \tag{B.4}$$

Where the reactions are written in terms of complete ionization to ions and electrons, this can be simplified slightly by making use of the identity

$$\sum_j \nu_{ij} Z_j^2 = - Z_i$$

for that case. This follows from the fact that, for such reactions, $\nu_{ij} = 0$ except for the electrons, while for the electrons, $\nu_{ij} = -Z_i$, the ionic valence, and $Z_i = 1$, the electronic valence.

Woolley (1959) has developed additional corrections for various other effects including third virial coefficients, the detailed calculation of the dielectric constant, and the finiteness of the ionic core.

At this point, the advantages of the direct search method should be apparent. We have defined direct search methods to include all those for which only the objective function needs to be evaluated. The addition of γ_j' factors in (3.7a) as complex as (B.2) and (B.4) or of $\ln \gamma_j'$ terms in (2.3) of the complexity of (B.1) and (B.3) would require much more complex reprogramming in methods involving the use of derivatives than is required in methods for which only the objective function is evaluated. Furthermore, the former requires some additional mathematical analysis in these cases, whereas the latter approach does not.

II. Real-Gas Effects on the Calculation of the Properties of the Mixture

In part I of this appendix, we showed how the species concentrations can be calculated including various real-gas effects. These are the concentrations that must now be used in the expressions for the thermodynamic properties of the mixture in terms of the appropriate properties of the individual species. These expressions, however, are not simply those derived in the text for the ideal-gas mixture. In addition to these ideal-gas terms, there must be included terms that specifically refer to the real-gas effects on the properties themselves. Thus, the compressibility factor must now be written

$$Z = PV/RT = Z^* + \sum_k \Delta_k Z,$$

where Z^* is the ideal-gas value, i.e., $Z^* = \sum_{i=1}^{l} C_i$, and the $\Delta_k Z$ are expressions for the various real-gas effects. Thus, for the effect of the second virial coefficient,

$$\Delta Z = \frac{1}{V_0} \frac{\varrho}{\varrho_0} \sum_{\alpha=1}^{l} \sum_{\beta=1}^{l} C_\alpha C_\beta B_{\alpha\beta}, \tag{B.5}$$

where $B_{\alpha\beta}$ is the second virial coefficient, which describes the interaction between species α and species β. For the limiting-law Debye–Hückel effect,

$$\Delta Z = \frac{-\pi^{1/2}}{3} \frac{\varepsilon^3}{(DkT)^{3/2}} \left(\frac{\varrho}{\varrho_0}\right)^{3/2} \left(\sum_{\alpha=1}^{l} Z_\alpha^2 C_\alpha\right)^{3/2} \left(\frac{N_0}{V_0}\right)^{1/2}, \tag{B.6}$$

where Z_α is the ionic charge of species α.

The internal energy needs now to be computed using

$$E/RT = E^*/RT + \sum_k \Delta_k(E_k/RT),$$

where E^* is the ideal-gas value and, for the effect of the second virial coefficient,

$$\Delta \frac{E}{RT} = \frac{-1}{V_0} \left(\frac{\varrho}{\varrho_0}\right) \sum_{\alpha=1}^{l} \sum_{\beta=1}^{l} C_\alpha C_\beta T \frac{dB_{\alpha\beta}}{dT}, \tag{B.7}$$

while for the Debye–Hückel effect, for the correction for the internal energy is three times (B.6), i.e., three times that for the compressibility factor.

The entropy is to be calculated from

$$S/R = S^*/R + \sum_k \Delta_k(S/R)$$

where S^* is the ideal-gas value, where the second virial effect, is given by

$$\Delta \frac{S}{R} = \frac{-1}{V_0} \left(\frac{\varrho}{\varrho_0}\right) \sum_{\alpha=1}^{l} \sum_{\beta=1}^{l} \left[C_\alpha C_\beta B_{\alpha\beta} + T \frac{dB_{\alpha\beta}}{dT}\right], \tag{B.8}$$

and the Debye–Hückel limiting-law effect is given by (B.6).

Corrections are easily derived for the various properties that depend on derivatives of these properties by differentiation of (B.5)–(B.8) as required.

Acknowledgments

The author is pleased to acknowledge his indebtedness to the various members of the National Bureau of Standards staff who were responsible for the development of many of the methods on which much of this chapter is based. These include Dr. Charles W. Beckett, Mr. Joseph Hilsenrath, and Dr. Harold W. Woolley. The author is also indebted to the U.S. Air Force and particularly the Arnold Engineering Development Center, Tullahoma, Tennessee for financial support and for providing the specific need on which the motivation for working in this area of research was based. He is also indebted to Dr. W. H. Evans of the National Bureau of Standards for a careful reading of the manuscript and many helpful suggestions.

References

The list of articles and books is meant to be extensive and sufficient for most purposes but is not meant to be exhaustive. It should, however, present an excellent starting point for anyone interested in compiling an exhaustive list of works on problems associated with the calculation of the thermodynamic properties and compositions of chemically reacting mixtures. The list can be expanded considerably merely by adding items which are contained as references in the members of the list but which are not included in the list itself.

Certain of the technical reports listed below are available from central sources as follows:

A. Reports with numbers preceded by the letters RM are available from the Rand Corporation, 1700 Main Street, Santa Monica, California 90406.

B. Reports with numbers preceded by the letters AD are available either from DDC, Headquarters, Cameron Station, 5010 Duke Street, Building 5, Alexandria, Virginia 22314, or The Clearing House for Federal Scientific and Technical Information (CFSTI), Springfield, Virginia 22151.

C. Reports with numbers preceded by the letters PB are available from CFSTI, Springfield, Virginia 22151.

D. Reports with the letters NACA or NASA along with the designated number are available either from the indicated source or from The NASA Scientific and Technical Information Facility, P.O. Box 33, College Park, Md. 20740.

E. All other reports should be requested either from the publication office of the originating agency or, by author, title and number, from CFSTI, Springfield, Virginia 22151.

Bahn, G. S. (1963). *Kinet. Equilibria Performance High Temp. Syst. Proc. Conf., 2nd,* 1962. Gordon and Breach, New York.

Bahn, G. S., and Zukoski, E. E. (1960). *Kinet. Equilibria Performance High Temp. Syst. Proc. Conf., 1st, 1959.* Butterworth, London and Washington, D. C.

Barnhard, P., and Hawkins, A. W. (1963). *Kinet. Equilibria Performance High Temp. Syst. Proc. Conf., 2nd, 1962,* p. 235. Gordon and Breach, New York.

Bethe, H. A. (1942). OSRD Rep. 369 (PB27307).

Boll, R. H., and Patel, H. C. (1961). *J. Eng. Power* **83**, 451.

Boll, R. H. (1961). *J. Chem. Phys.* **34**, 1108.

Boynton, F. P. (1960), *J. Chem. Phys.* **32**, 1880.

Boynton, F. P. (1963). *Kinet. Equilibria Performance High Temp. Syst. Proc. Conf., 2nd, 1962,* especially pp. 193–198. Gordon and Breach, New York.

Brinkley, S. R. (1946). *J. Chem. Phys.* **14**, 563.

Brinkley, S. R. (1947). *J. Chem. Phys.* **15**, 107.

Brinkley, S. R. (1956). High speed aerodynamics and jet propulsion. *In* "Combustion Processes" (B. Lewis, R. N. Pease, and H. S. Taylor, eds.), Vol. II, p. 64. Princeton Univ. Press, Princeton, New Jersey.

Brinkley, S. R. (1960). *Kinet. Equilibria Performance High Temp. Syst. Proc. Conf., 1st, 1959,* p. 73. Butterworths, Washington, D. C.

Chao, K. C. (1967). Instructional Monogr. TD-1-67 obtainable from Technol. Utilization Officer, NASA, Lewis Res. Center, Cleveland, Ohio.

Chu, S. T. (1958). *Jet Propul.* **28**, 252.

Clasen, R. J. (1965). Rep. RM 4345 PR.

Courant, R. (1937). "Differential and Integral Calculus," 2nd ed., Vol. II, p. 188. Nordeman, New York.

Curtin, J. F. (1968). Tech. Note SAD205. Dept. of Supply, Australian Defense Sci. Serv., Weapons Res. Establ., Salisbury, South Australia.

Curtis, M. W., and Wohlwill, H. E. (1961). Tech. Rep. 7102-0013-MU-00. Space Technol. Lab., Los Angeles, California.

De Donder, Th. and van Bysselberghe, P. (1936). "Thermodynamic Theory of Affinity." Stanford University Press, Palo Alto, California.

DeLand, E. C. (1967). Rep. RM 5404 PR.

Denton, J. C. (1963). Ph. D. Thesis, Dept. of Mech. Eng., Texas A. and M. Univ., College Station, Texas.

Dobbins, T. W. (1959). Rep. WADC TR-59-757, available as AD 232465 or PB 157176.

Drellishak, K. S. (1964). Rep. AEDC-TDR-64-22. Arnold Eng. and Develop. Center, Tullahoma, Tennessee. AD 428210.

Drellishak, K. S., Knopp, C. F., and Cambel, A. B. (1963). Rep. AEDC-TDR-63-146. Arnold Eng. and Develop. Center, Tullahoma, Tennessee. AD 414708.

Drellishak, K. S., Aeschliman, D. P., and Cambel, A. B. (1964). Rep. AEDC-TDR-64-12. Arnold Eng. and Develop. Center, Tullahoma, Tennessee. AD 427839.

Durand, J. L., and Brandmaier, H. E. (1960). *Kinet. Equilibria Performance High Temp. Syst. Proc. Conf., 2nd, 1962,* p. 115. Butterworth, London and Washington, D. C.

Edwards, A. G., and Rubin, A. I. (1958). Tech. Reps. 2501 and 2529. Feltsman Res. and Eng. Labs., Picatinny Arsenal, Dover, New Jersey. AD 202072 and AD 201271 respectively.

Evans, W. H. (1963). *Kinet. Equilibria Performance High Temp. Syst. Proc. Conf., 2nd, 1962,* p. 151. Gordon and Breach, New York.

Feldman, H. F., Simons, W. H., and Bienstock, D. (1969). *U. S. Bur. Mines Rep. Invest.* **RI 7257.**

Fenter, F. W., and Gibbons, H. B. (1961). Rep. RE-IR-14. Link-Temco-Vought Co., Dallas, Texas (rep. written at the former Chance Vought Res. Center).

Fermi, E. (1924). *Z. Phys.* **26**, 54.

Gilmore, F. R. (1955). Rep. RM 1543.

Gilmore, F. R. (1959). Rep. RM 2328.
Gilmore, F. R. (1967). Rep. DASA 1971-1. Lockheed, Palo Alto Res. Lab., Palo Alto, California. AD65054.
Gordon, J. S., and Robinson, R. (1960). *Kinet. Equilibria Performance High Temp. Syst. Proc. Conf., 1st, 1959*, p. 30. Butterworth, London and Washington, D. C.
Gordon, J. S., and Robinson, R. (1963). *Kinet. Equilibria Performance High Temp. Syst. Proc. Conf., 2nd, 1962*, p. 39. Gordon and Breach, New York.
Gordon, S., Zeleznik, F. J., and Huff, V. N. (1959). NASA TN-132.
Haar, L. (1960). *Kinet. Equilibria Performance High Temp. Syst. Proc. Conf., 1st, 1959*, p. 35. Butterworth, London and Washington, D. C.
Hansen, F. C. (1968). NACA Rep. TN 4150.
Harnett, J. J. (1960). *Kinet. Equilibria Performance High Temp. Syst. Proc. Conf., 1st, 1959*, p. 185. Butterworth, London and Washington, D. C.
Hill, T. L. (1955), *J. Chem. Phys.* **23**, 617.
Hilsenrath, J., and Klein, M. (1963). Rep. AEDC-TDR-63-161. Arnold Eng. and Develop. Center, Tullahoma, Tennessee. AD 339389.
Hilsenrath, J., and Klein, M. (1964). Rep. AEDC-TDR-63-162. Arnold Eng. and Develop. Center, Tullahoma, Tennessee. AD 432210.
Hilsenrath, J., and Klein, M. (1965). Rep. AEDC-TR-65-58. Arnold Eng. and Develop. Center, Tullahoma, Tennessee. AD 612301.
Hilsenrath, J., and Klein, M. (1966). Rep. AEDC-TR-66-65. Arnold Eng. and Develop. Center, Tullahoma, Tennessee. AD 630461.
Hilsenrath, J., Klein, M., and Sumida, D. Y. (1959). In. *Thermodyn. Transp. Prop. Gases Liquids Solids, Pap. Symp. Therm. Prop., 1959*, ASME or McGraw-Hill, New York.
Hilsenrath, J., Messina, C. G., and Evans, W. H. (1964). Rep. AFWL TDR-64-44. Res. and Technol. Div., Air Force Weapons Lab., Air Force Systems Command, Kirtland Air Force Base, New Mexico. AD 606163.
Hilsenrath, J., Messina, C. G., and Klein, M. (1966). Rep. AEDC-TR-66-248. Arnold Eng. and Develop. Center, Tullahoma, Tennessee. AD 644081.
Hochstim, A. (1960). *Kinet. Equilibria Performance High Temp. Syst. Proc. Conf., 1st, 1959*, p. 39. Butterworth, London and Washington, D. C.
Hochstim, A. R., and Adams, B. (1962). *Progr. Int. Res. Thermodyn. Transp. Prop. Pap. Symp. Thermophys. Prop., 2nd, 1962*. p. 228. ASME, New York.
Hooke, R., and Jeeves, T. A. (1961). *Comm. ACM* **8**, 212.
Horton, T. E., and Menard, W. A. (1969). Tech. Rep. 32-1350. Jet Propul. Lab., California Inst. of Technol., Pasadena, California.
Householder, A. S. (1953). "Principles of Numerical Analysis." McGraw-Hill, New York.
Huber, R. A. (1962). Rep. AERE-R5793. Chem. Eng. Div., AERE, Harwell, Berkshire, England.
Huff, V. N., Gordon, S., and Morrell, V. E. (1951). NACA Rep. 1037.
Jeeves, T. A. (1958). *Comm. ACM* **1**, No. 8, p. 9.
Kandiner, H. J., and Brinkley, S. R. (1950). *Ind. Eng. Chem.* **42**, 850.
Kilpatrick, J. (1953). *J. Chem. Phys.* **21**, 274.
Klein, M. (1961). Unpublished work.
Landis, F., and Nilsen, E. (1961). Rep. 1921. Pratt and Whitney Co., Hartford, Connecticut.

Landis, F., and Nilsen, E. (1962). *Progr. Int. Res. Thermodyn. Transp. Prop. Pap. Symp. Thermophys. Prop., 2nd, 1962,* p. 218. ASME, New York.

Lewis, C. H., and Neel, C. A. (1964). Reps. AEDC-TDR-64-36 (AD 600469) and AEDC-TR-64-113 (AD 601241). Arnold Eng. Develop. Center, Tullahoma, Tennessee.

Mayer, J. E. (1950). *J. Chem. Phys.* **18**, 1426.

Mayer, J. E., and Mayer, M. G. (1940). "Statistical Mechanics." Wiley, New York.

Megreblian, R. V. (1952). Rep. No. 7. Guggenheim Jet Propul. Center, California Inst. of Technol., Pasadena, California.

Megreblian, R. V. (1953). Ph. D. Thesis, California Inst. of Technol., Pasadena, California.

Naphtali, L. M. (1959). *J. Chem. Phys.* **31**, 263.

Naphtali, L. M. (1960). *Kinet. Equilibria Performance High Temp. Syst. Proc. Conf., 1st, 1959,* p. 184. Butterworth, London and Washington, D. C.

Naphtali, L. M. (1961). *Ind. Eng. Chem.* **53**, 387.

Newman, P. A., and Allison, D. V. (1966). NASA Rep. TN-3540.

Palmer, A. S., and Mahoney, D. (1966). Rep. 66354. Royal Aircraft Estab., Ministry of Aviation, Farborough Hants, England.

Park, S. D. (1967). NASA-TN-D-4106.

Poirier, J. C. (1953). *J. Chem. Phys.* **21**, 965, 972.

Reidin, M., and Ragent, B. (1962). Rep. ARL 62-358. Vidya, Inc., Palo Alto, California. AD 278 589.

Shapiro, N. Z. (1963). Rep. RM 3677 PR.

Shapiro, N. Z. (1964). Rep. RM 4205 PR.

Shapiro, N. Z., and Shapley, L. S. (1964). Rep. RM 3935 1 PR.

Shears, D. B. (1968). *J. Chem. Phys.* **48**, 4144.

Spang, H. A., III (1962). *SIAM* (*Soc. Ind. Appl. Math.*) *Rev.* **4**, 343.

Stroud, C. W., and Brinkley, K. L. (1969). NASA Rep. TN D-5391.

Unsöld, A. (1948). *Z. Astrophy.* **24**, 355.

Urey, H. C. (1924). *Astrophys. J.* **59**, 1.

Vinturella, J. B. (1968). Ph. D. Thesis, Dept. of Chem. Eng., Tulane Univ., New Orleans, Louisiana.

Wegstein, J. (1958). *Comm. ACM* **1**, No. 6, p. 9.

White, W. B. (1967). *J. Chem. Phys.* **46**, 4171.

White, W. B., Johnson, S. M., and Dantzig, G. B. (1957). Rep. P-1059 Revised. The Rand Corporation, Santa Monica, California.

White, W. B., Johnson, S. M., and Dantzig, G. B. (1958). *J. Chem. Phys.* **28**, 751.

Wilde, D. J., and Beightler, C. S. (1967). "Foundations of Optimization." Prentice-Hall, Englewood Cliffs, New Jersey.

Wilkins, R. L. (1963). "Theoretical Evaluation of Chemical Propellants" (Prentice-Hall Int. Ser. in Space Technol.). Prentice-Hall, Englewood Cliffs, New Jersey.

Wolfe, P. (1962). Rand Rep. R 401 PR. The Rand Corp., Santa Monica, California.

Wolfe, P. (1959). *Comm. ACM* **2**, No. 12, p. 12.

Woolley, H. W. (1958). *J. Res. Nat. Bur. Stand.* **61**, 469.

Zeleznik, F. J., and Gordon, S. (1968). *Ind. Eng. Chem.* **60**, 27; also published in "Applied Thermodynamics," p. 303. Amer. Chem. Soc. Publ., Washington, D. C.

Zemansky, M. (1943). "Thermodynamics," 2nd ed. McGraw-Hill, New York.

Chapter 8

Equilibria at Very High Temperatures

H. KREMPL

I. Introduction

The gaseous state of matter generally can be treated as a (hot) plasma because the number of free electrons never becomes exactly zero. Within the scope of this chapter, we must restrict ourselves to classical non-relativistic and nondegenerate plasmas. They are represented by the conditions $kT \leqq mc^2$ and $\lambda_e \leqq n_e^{-1/3}$, where λ_e is the thermal de Broglie wavelength and n_e the electron density.

Some problems in this field are not yet understood completely and are treated in contradictory ways by different authors. Therefore, we have to choose the way which seems to us to be the best one, weighing the different conceptions against one another and using experimental results when they are available. If we have to choose between a derivation which is more exact and one which is more concrete, we will prefer the latter. In most cases, a semiquantitative analysis offers the utmost physical transparency and brevity without losing too much in accuracy and generality. The applied total pressure imposed by the surrounding atmosphere is used as a boundary condition because it is experimentally more relevant for the case of a noncompressed plasma.

II. The Ideal Plasma

An ideal plasma implies an ideal hot gas in thermal equilibrium, consisting of n_e free electrons and n_ξ ions per unit volume. Ions are characterized by charge numbers ξ, and the case of $\xi = 0$ is given to neutral atoms. The equation of state then reads, for a monatomic one-element plasma,

$$p = p_e + \sum_{\xi=0}^{Z} p_\xi = kT\left(n_e + \sum_{\xi=0}^{Z} n_\xi\right). \tag{2.1}$$

The generalization to a multielement system is trivial, from Dalton's law. Also, molecules can easily be included when required. As indicated, the summation over ξ begins with $\xi = 0$ (in case negative ions are present, with $\xi = -1$) and ends with $\xi = Z$, the atomic number of the element under consideration. Here, we have used the particle density instead of the more familiar number of moles and the volume because it is nearly impossible to enclose a plasma in a vessel of definite volume with isothermal walls. By definition, the entire plasma must be neutral, which can be expressed by the condition of quasineutrality

$$p_e = \sum \xi p_\xi = kT \sum \xi n_\xi. \tag{2.2}$$

Using the notation familiar in electrochemistry, the "strength" of a plasma may be described by the ionic strength, as defined by

$$n_e^* = (n_e + \sum \xi^2 n_\xi)/2 = n_e(1 + \langle\xi\rangle)/2, \tag{2.3}$$

where the abbreviation $\langle\xi\rangle = \sum \xi^2 n_\xi / n_e$ is the mean charge number, which is equal to ξ if only one ionic species is present. In a singly ionized plasma, $\xi = 1$ and $n_e^* = n_e$.

Numerically, we find as orders of magnitude $n_e = 10^3$ cm^{-3} in the interstellar HI region at 100°K and $n_e = 10^6$ cm^{-3} in air. Values of n_e of order from 10^{10} to 10^{20} cm^{-3} may be reached in special plasma furnaces, (Finkelnburg and Maecker 1956, Maecker 1963) as shown in Fig. 1.

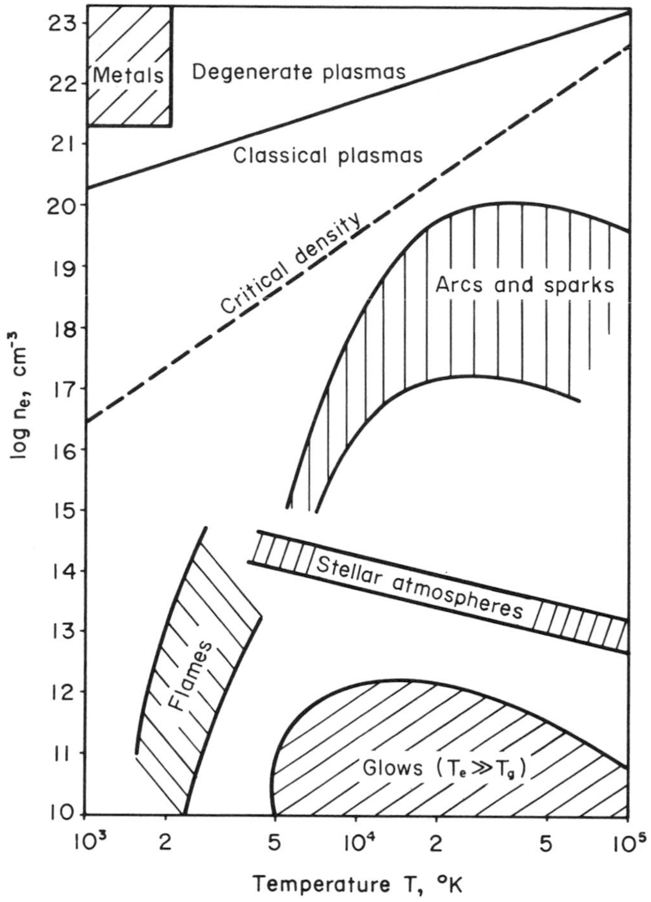

FIG. 1. Various natural and laboratory plasmas, shown on a T–n_e plot.

A gas being heated indefinitely passes through successive states of ionization which can be expressed on a density basis by

$$n_0 \xrightarrow[\chi_0]{T_0} n_1 \xrightarrow[\chi_1]{T_1} n_2 \cdots n_\xi \xrightarrow[\chi_\xi]{T_\xi} n_{\xi+1} \cdots \xrightarrow[\chi_{Z-1}]{T_{Z-1}} n_Z \qquad (2.4)$$

with increasing T, where T_ξ, the ionization temperature, denotes the temperature at which $n_{\xi+1} = n_\xi$. The degree of ionization, which is defined by

$$x_\xi = n_{\xi+1}/(n_\xi + n_{\xi+1}), \qquad (2.5)$$

is then 0.5.

Anticipating the results below, we have $T_\xi \propto \chi_\xi$, where χ_ξ is the ionization energy, or the energy required to remove one electron from the ion of charge ξ. Since $\chi_{\xi+1} \geqq 1.5\chi_\xi$ is valid for all elements of low ξ ($T_\xi < 10^5$°K), the equilibria will not overlap considerably. Thus, each ion species exists only in a small temperature range. We shall use the three-particle approximation unless otherwise stated, which means that, at the relevant temperature T_ξ, only the species with the densities n_e, n_ξ, and $n_{\xi+1}$ are coexistent. Then, we have the conservation equation [see Eqs. (2.1) and (2.2)]

$$p \approx (\xi + 1)p_\xi + (\xi + 2)p_{\xi+1} \tag{2.6}$$

$$p_e \approx \xi p_\xi + (\xi + 1)p_{\xi+1}. \tag{2.7}$$

In this approximation, definition (2.5) becomes identical with the expression based on the more familiar mole fraction $x_\xi = p_{\xi+1}/\sum p_\xi$. Since $x_\xi \geqq x_{\xi+1}$, the ionization degree of the atoms x_0 is a measure of the

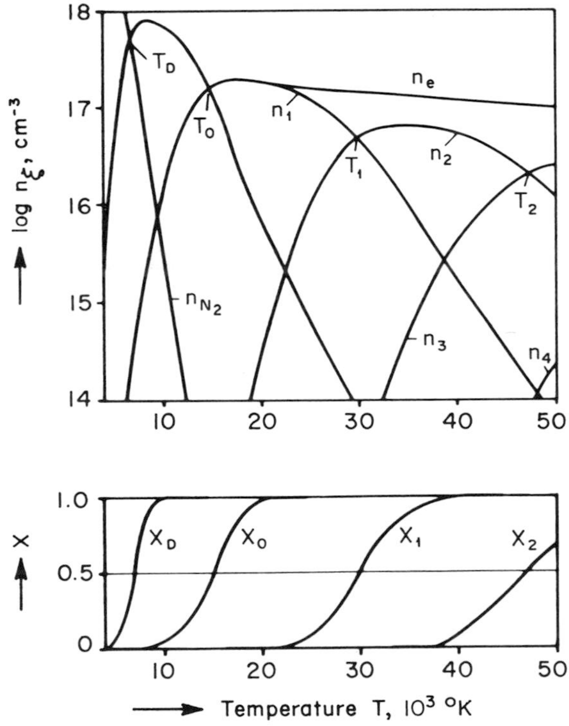

FIG. 2. Composition of nitrogen at atmospheric pressure versus temperature.

plasma strength. In terms of electrochemistry, we call a plasma with $x_0 \ll 1$ a weak plasma, one with $x_0 \sim 1$ a strong plasma, and one with $x_0 = 1$ a fully ionized plasma. The corresponding temperature conditions are $T < T_0$, $T \sim T_0$, and $T > T_0$. The mean charge number $\langle \xi \rangle$ may be useful, too.

In the following, we shall often use the term "relevant temperature" with respect to charge number ξ, as we did in connection with the three-particle approximation. We use the temperatures at which the maxima of n_ξ and $n_{\xi+1}$ occur as the lower and upper limits, respectively, of this temperature. These maxima occur as shown in the n_ξ–T curve in Fig. 2. In this range, x_ξ varies appreciably with the temperature. In the case $\xi = 0$, the lower limit $T = 0$ may be chosen; or, in dealing with diatomic gases, the dissociation temperature T_{D} is selected.

III. Internal Partition Function and Internal Energy*

A. Diatomic Molecules

We consider an oscillating rotator of moment of inertia $I = \mu r_{\mathrm{e}}^2$, frequency $\omega^2 = k_{\mathrm{e}}/\mu$. We further assume that rotational energy $\mathscr{E}_j$, vibrational energy $\mathscr{E}_v$, and electronic energy $\mathscr{E}_t$ are completely independent, and have for the partition function (ter Haar, 1967, p. 65)

$$\begin{aligned} Q_{\mathrm{AB}}(T) &= Q_j Q_{\mathrm{v}} Q_t \\ &= \{2kTI/\hbar^2\sigma_j\}[1 - \exp(-\hbar\omega/kT)]^{-1} \sum_{t=1}^{\infty} g_t \exp(-\mathscr{E}_t/kT). \end{aligned} \quad (3.1)$$

At temperatures where the dissociation $\mathrm{AB} \rightarrow \mathrm{A} + \mathrm{B}$ is perceptible, we can expand the second bracket in a series and retain the first term, for $\hbar\omega \ll kT$ (Artmann, 1965, 1968). In the same way, the sum can be replaced by its first term, which becomes g_1, because here and in the following all energies are related to the normal state of the molecule, where $\mathscr{E}_1 = 0$. The symmetry factor is $\sigma_j = 1$ for $\mathrm{A} \neq \mathrm{B}$. In the case of a homonuclear molecule $\mathrm{A} = \mathrm{B}$, $\sigma_j = 2$, since the higher symmetry of the AA molecule must be taken into account. Thus, we find

$$Q_{\mathrm{AB}}(T) \approx (2g_1^{\mathrm{AB}}/\sigma_j)(I/\hbar^3\omega)(kT)^2 \propto \mu^{3/2}T^2. \quad (3.2)$$

* See Münster (1956), Aston and Fritz (1959), and Godnew (1963).

The T^2 dependence of $Q_{AB}(T)$ gives rise to the well-known high-temperature limit of the heat capacity,

$$C_{int}/R = (1/T^2)\, \partial^2 (\ln Q_{AB})/\partial^2(1/T) = 2, \tag{3.3}$$

which corresponds to four degrees of freedom, two vibrational and two rotational ones.

Now, let us compare this simple expression with the exact value of Q_{AB} (Fig. 3). Presuming that all vibrational and rotational states of the ground state are known from spectroscopic data, we have to sum all Boltzmann factors of those states. Here, the sequence of the states may be expanded in a power series, the coefficients accounting for the anharmonicity of the vibration and the nonrigidity of the rotation. The first excited electron level $t = 2$, being a small correction to that of the ground state, may be treated in the classical way as shown above. Higher elec-

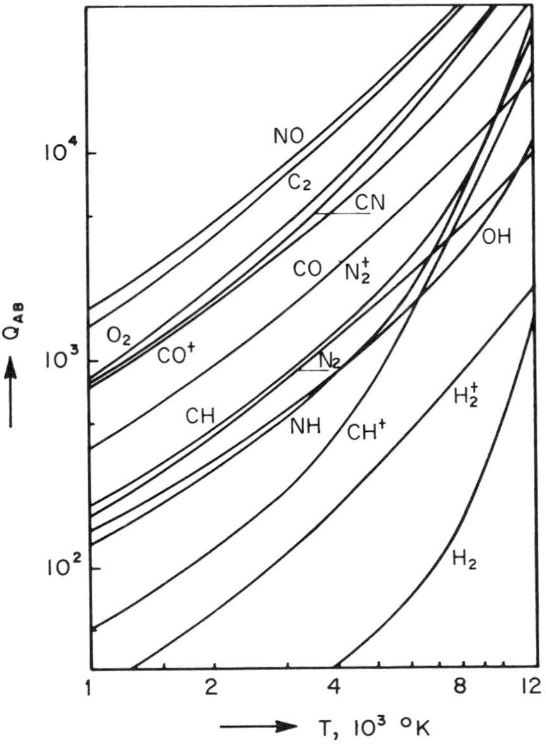

FIG. 3. Partition function of various diatomic molecules versus temperature (Bögershausen and Schneider, 1971).

tronic states, $t > 2$, may be neglected (Burhorn and Wienecke, 1959, 1960). An analysis of $Q_{AB}(T)$ by Krauss (1969) points to a limiting law, valid near the dissociation temperature ($\varkappa_D \approx 0,5$):

$$Q_{AB}(T) = aT^2, \tag{3.4}$$

where a is a constant, depending on the molecule under consideration. The a factor is 2–3 times larger then that found in the classical expression, Eq. (3.2). It seems that at high temperatures the internal energies are decoupled. Further, it can be assumed that molecules having similar molecular constants μ, I, and ω will have a value a of the same order of magnitude. The aT^2 behavior of the partition function fails for molecules and radicals of the type XH (i.e., molecules with extremly low reduced mass).

B. Atoms and Ions

1. *Truncation of the Partition Function**

With respect to molecules, we need not discuss the convergence of partition functions because molecules at elevated temperatures are mostly dissociated. However, suppose that, in accord with Scheibe (Scheibe *et al.*, 1952), molecules are hydrogenlike, then there could be Rydberg series of energy levels with statistical weights approaching infinity as $g_t = 2t^2$, and the sum might diverge.

In omitting these higher energy levels and confining ourselves to the convergent part of the sum, we are using a rather formal approximation. At moderate temperature and not too low pressure, this procedure is also used for ions (Riewe and Rompe, 1938; Ebeling 1968), by summing up to a quantum number of about $t_\Delta = 5$, where t_Δ is not critical. At extreme conditions (low pressure and high temperature), the omitted part of the sum, even truncated according to a suitable physical model, becomes comparable with its initial terms. Therefore, we try to submit the individual particle to a boundary condition. One should no longer use energy levels of an isolated particle (Unsöld, 1948; Ecker and Weizel, 1956; Margenau and Lewis, 1959). This procedure is needlessly laborious, as the shift of each term depends on both the temperature and

* See Margenau and Lewis (1959), Drawin and Felenbok (1965), and Ecker anp Kröll (1966).

the ionic strength. In order to reduce the calculations to a tolerable level, we shall use a cutoff procedure applied to the ion under consideration, which we assume to be hydrogenlike for $t > t_H$. The energy levels and statistical weights, according to the hydrogenlike rule, are now

$$\mathscr{E}_t = (\xi + 1)^2 \chi_H/(1 - t^{-2}), \quad g_t = 2t^2 \quad \text{for} \quad t \geqq t_H, \tag{3.5}$$

whereby t_H is the lowest quantum number for which the observed energy levels can be expressed to a sufficient approximation by the following familiar expression for the tth orbit of a hydrogenlike ion (that is, speaking in a classical sense, the orbit number at which the orbiting electron does not penetrate the inner shells):

$$a_t = a_0 t^2/(\xi + 1) \tag{3.6}$$

(a_0 is the first Bohr radius).

Now, we assume, somewhat artificially, that each orbit with a radius $a_t \leqq a_{t_\Delta}$ is undisturbed by the neighboring particles and orbits $a_t > a_{t_\Delta}$ are to be canceled by the surrounding particles. The boundary radius a_{t_Δ} may be choosen as one of the following characteristic lengths:

(a) The Landau length $\varrho_L = e^2/kT$, which means that an orbit whose binding energy is $\mathscr{E}_t' = \chi_H/t_\Delta^2$ should be removed if $\mathscr{E}_t' \geqq kT$. ($\mathscr{E}_t'$ is counted from the ionization limit).

(b) The mean distance of the electrons $d_e = (3/4\pi n_e)^{1/3}$.

(c) The Debye radius $\varrho_D = (kT/8\pi e^2 n_e^*)^{1/2}$.

(d) $t_\Delta^2 = \chi/\Delta\chi$, i.e., the partition function should be cut off at the so-called reduced ionization potential.

This last procedure is often used in the literature, although there is no evidence that $\Delta\chi$ is correlated to t_Δ. Setting respectively each of these lengths equal to a_{t_Δ}, we get at once, with the help of Eq. (3.6), the upper quantum number t_Δ at which the sum of states must be truncated:

$$\begin{aligned} &\text{(a)} \quad t_\Delta^2 = (\xi + 1)\varrho_L/a_0 \\ &\text{(b)} \quad t_\Delta^2 = (\xi + 1)d_e/a_0 \\ &\text{(c)} \quad t_\Delta^2 = (\xi + 1)\varrho_D/a_0 \\ &\text{(d)} \quad t_\Delta^2 = (\xi + 1)\chi_H/\Delta\chi_\xi. \end{aligned} \tag{3.7}$$

Since, in the case of an ideal plasma, $\varrho_L < d_e \ll \varrho_D$, it follows that the assumption (a) yields the lowest and (c) the highest value of the partition function (Fig. 4).

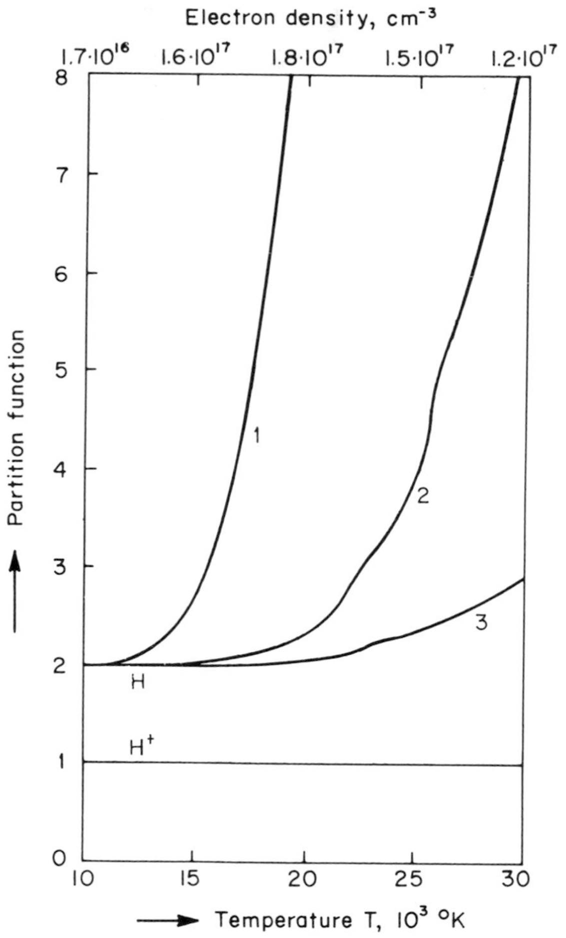

FIG. 4. Partition function of hydrogen versus temperature for various cutoff principal quantum numbers t_Δ: (1) $t_\Delta = 40$; (2) $t_\Delta = 15$; (3) $t_\Delta = 6$; upper scale $n_e(T)$ at $p = 1$ bar (Diermeier, 1969).

A decision on the basis of experiment as to the best approximation is not yet possible. Case (d) is added for ions, where the hydrogen-likeness is not yet reached at the quantum number t_Δ. The value of $\Delta\chi$ will be evaluated in Section V, C.

2. *Calculation of the Partition Function*

Tables of atomic energy levels (Moore, 1949, 1952, 1958) always become incomplete before the cutoff limit is reached. Since these missing levels, having high statistical weights contribute notably to the partition function, it is advisable to separate the latter into three parts: (1) the statistical weight of the ground state, (2) a sum which contains all low-lying levels up to a quantum number $t_{\mathrm{H}} - 1$, and (3) a part which sums up the hydrogenlike terms.

We can convert these parts into an elementary integral,

$$\sum_{t_{\mathrm{H}}}^{t_\Delta} 2t^2 \exp(-\mathscr{E}_t/kT) \approx 2 \exp(-\mathscr{E}_{t_\Delta}/kT) \int_{t_{\mathrm{H}}}^{t_\Delta} t^2 \, dt, \tag{3.8}$$

assuming that energies of all levels in this sum are equal to the reduced

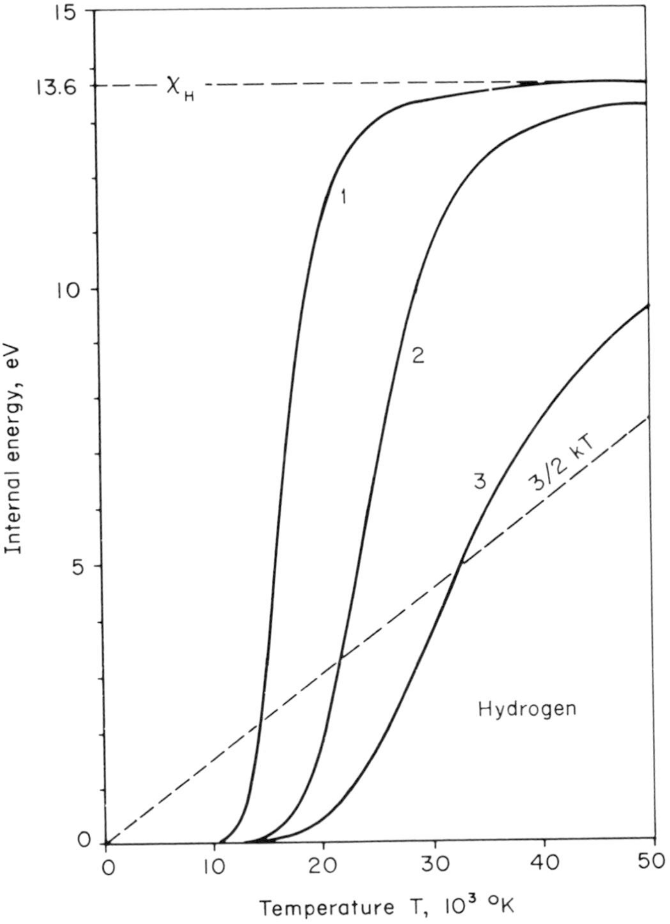

FIG. 5. Internal energy of atomic hydrogen versus temperature for various cutoff principal quantum numbers t_Δ: (1) $t_\Delta = 40$; (2) $t_\Delta = 15$; (3) $t_\Delta = 6$ (Diermeier, 1969).

ionization energy. Now, neglecting the lower limit of the integral, the resulting partition function reads

$$Q_\xi(T, t_\Delta) = g_1 + \sum_{t=2}^{t_{\mathrm{H}}-1} g_t \exp(-\mathscr{E}_t/kT) + (2t_\Delta{}^3/3) \exp(-\mathscr{E}_{t_\Delta}/kT). \quad (3.9)$$

For the calculation of the second term, a method given by Schlender and Traving (1965) is recommended, where the sum is reduced by introducing a few equivalent levels and weights. Also, the tables of Drawin and Felenbok (1965) are useful. Now, the internal energy due to excitation can be easily calculated using the statistical relation $E_\xi = kT^2(\partial \ln Q_\xi/\partial T)$, which is demonstrated in Fig. 5.

IV. Transition Equilibria and Characteristic Temperature

A. Dissociation

The dissociation equilibrium $\mathrm{AB} \rightleftharpoons \mathrm{A} + \mathrm{B} - \chi_\mathrm{D}$ is governed by the law of mass action (ter Haar, 1967, p. 64)

$$\frac{n_\mathrm{B}}{n_\mathrm{AB}} n_\mathrm{A} = \left(\frac{2\pi\mu kT}{h^2}\right)^{3/2} \left(\frac{Q_\mathrm{B} Q_\mathrm{A}}{Q_\mathrm{AB}}\right) \exp\left(-\frac{\chi_\mathrm{D}}{kT}\right), \quad (4.1)$$

where the first parentheses represent the reduced translational partition function and the second ones the internal partition function ratio; the energy χ_D again is counted from the ground state of the particles (Figs. 6 and 7). Let us now introduce the degree of dissociation in a rather unusual way,

$$x_\mathrm{D} = n_\mathrm{B}/(n_\mathrm{B} + n_\mathrm{AB}), \quad (4.2)$$

with $n_\mathrm{B} = n_\mathrm{A}$, and the independent variable $n_\mathrm{A} = p_\mathrm{A}/kT$ which, from the spectroscopic point of view, is the easiest parameter to determine. Now, Q_A and Q_B may be replaced by the statistical weights $g_1{}^\mathrm{A}$ and $g_1{}^\mathrm{B}$ and Q_AB by the approximation. Eq (3.2). We find then

$$\frac{p_\mathrm{B}}{p_\mathrm{AB}} p_\mathrm{A} = \frac{x_\mathrm{D}}{1 - x_\mathrm{D}} p_\mathrm{A}$$

$$= \left(\frac{\mu}{2\pi}\right)^{3/2} \frac{\sigma_j}{2} \frac{g_1{}^\mathrm{A} g_1{}^\mathrm{B}}{g_1^\mathrm{AB}} \frac{\omega}{I} (kT)^{1/2} \exp\left(-\frac{\chi_\mathrm{D}}{kT}\right). \quad (4.3)$$

In this equation, the reduced mass μ is the quantity with the highest

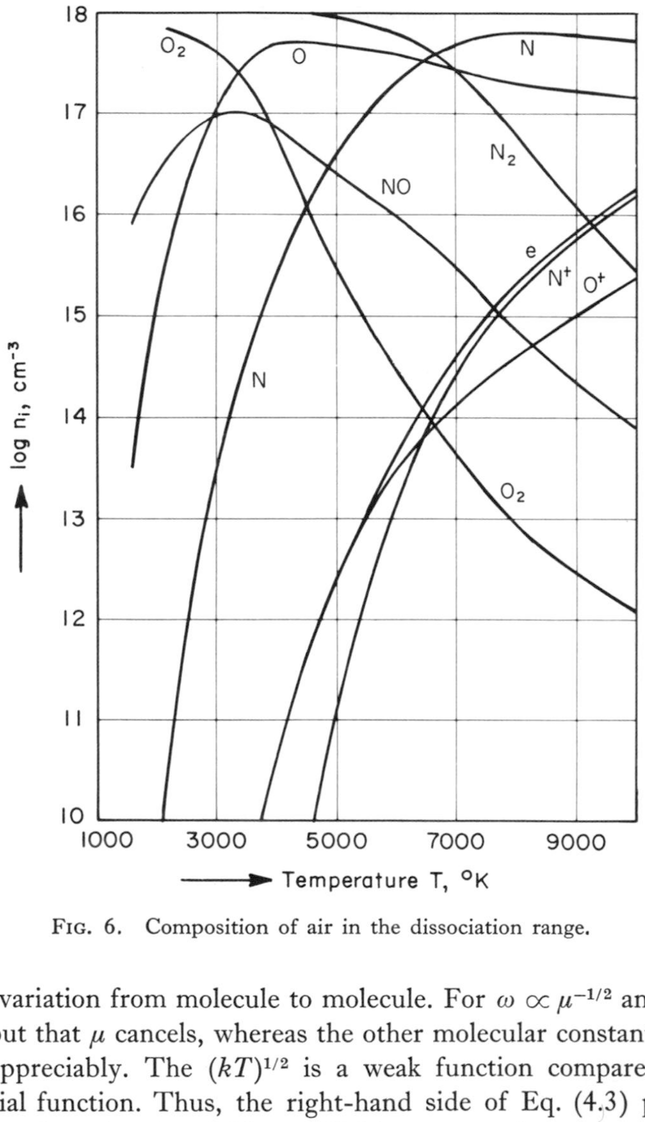

FIG. 6. Composition of air in the dissociation range.

range of variation from molecule to molecule. For $\omega \propto \mu^{-1/2}$ and $I \propto \mu$, it turns out that μ cancels, whereas the other molecular constants do not change appreciably. The $(kT)^{1/2}$ is a weak function compared to the exponential function. Thus, the right-hand side of Eq. (4.3) primarily depends on the reduced variable χ_D/kT. Now, comparing various diatomic molecules at the corresponding state $x_D = 0.5$, we find, by a more detailed examination (Krauss, 1970, 1969),

$$\chi_D/kT \approx 14. \tag{4.4}$$

For the evaluation of a gas with many constituents, we refer to Frie (1967), as well as similar texts.

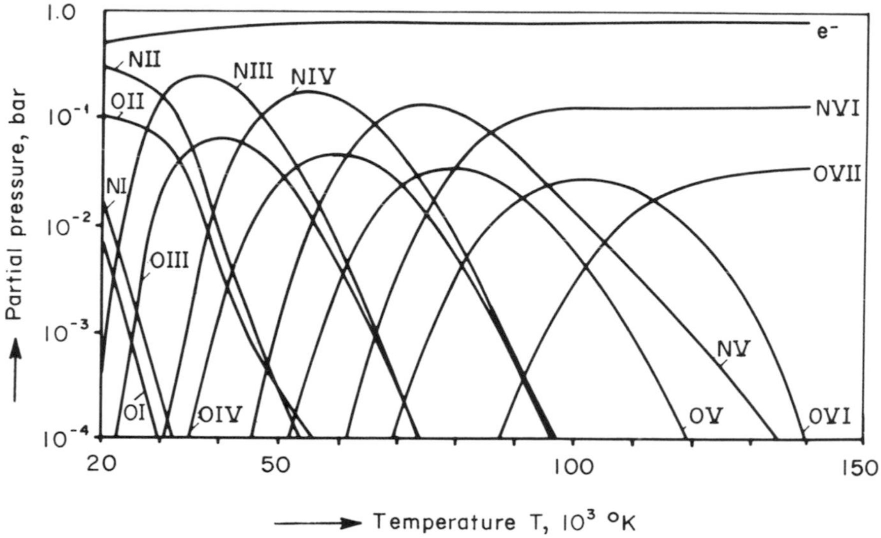

FIG. 7. Composition of air in the ionization range (Neumann, 1962).

B. IONIZATION*

1. *Reduced Saha Equation and Ionization Temperatures*

The ionization equilibrium $A_\xi \rightleftharpoons A_{\xi+1} + e - \chi_\xi$ with $\xi = 0, 1, 2, \ldots, Z - 1$ is governed by Saha's equation (ter Haar, 1967, p. 65), which is written on a pressure basis as

$$\{x_\xi/(1 - x_\xi)\}p_e = (2\pi m/h^2)^{3/2}(2Q_{\xi+1}/Q_\xi)(kT)^{5/2} \exp(-\chi_\xi/kT)$$
$$\equiv K_\xi(T, n_e) \tag{4.5}$$

where

$$x_\xi/(1 - x_\xi) = p_{\xi+1}/p_\xi. \tag{4.6}$$

In order to render the variables p_e and kT dimensionless, we introduce $\chi_\xi = z^2\chi_H$ and $r_\xi = a_0/z$. Thus, we replace the charge distribution by an effective point charge ze acting upon the electron to be detached; r_ξ is the first Bohr radius. We use for $r_\xi = a_0$ and χ_H the usual expressions $a_0 = \hbar^2/me^2$ and $\chi_H = me^4/2\hbar^2$, and find from Eq. (4.5) in the

* See Unsöld (1948) and Lochte-Holtgreven (1968).

approximation $Q_\xi = g_\xi$:

$$\frac{x_\xi}{1-x_\xi}\,\frac{p_e r_\xi^3}{\chi_\xi} = \left(\frac{1}{4\pi}\right)^{3/2} \frac{2g_{\xi+1}}{g_\xi} \left(\frac{kT}{\chi_\xi}\right)^{5/2} \exp\left(\frac{-\chi_\xi}{kT}\right). \tag{4.7}$$

Now, the dimensionless variables are the reduced entropy χ_ξ/kT and the reduced electron pressure p_e divided by $\chi_\xi/r_\xi^3 = z^5\chi_H/a_0^3$; the quantity $\chi_H/a_0^3 = 147$ Mbar is the so-called atomic unit pressure, i.e., the pressure at which the electronic shell structure collapses (Hund, 1961). If we approximate $2g_{\xi+1}/g_\xi = 1$, Eq. (4.7) can be recognized as the reduced Saha equation. It enables us to estimate the dissociation temperature T_ξ as follows. At $x_\xi = 0.5$, p_e extends from $p/3$ to $\approx p$ ($\xi = 0$ to $\xi = Z$) and can therefore be considered as a constant, because χ_ξ/kT depends logarithmically on p_e. Now, accepting an error of ± 1 on χ_ξ/kT, Eq. (4.7) gives with p expressed in bars

$$\chi_\xi/kT_\xi \approx 10 - 2\log p \tag{4.8}$$

in the range 8 eV $< \chi_\xi <$ 20 eV, or, more conveniently, $T_\xi \approx 1180\chi_\xi^{[\mathrm{eV}]}$ for $p = 1$ bar ≈ 1 atm. An examination of the ionization energy sequence of various molecules shows that, in any case, $(\chi_{\xi+1} - \chi_\xi) > 5$ eV, which corresponds to a temperature interval $(T_{\xi+1} - T_\xi) > 6000$°K. In this rather large region, the A_ξ ion is predominant, as can be seen from Fig. 8. Since n_ξ drops rapidly beyond this interval, we may consider T_ξ as the lower and $T_{\xi+1}$ as the upper limit of the region of existence of the A_ξ ion with the density n_ξ.

2. *The Common Electron Pressure*

*a. Interstage Coupling.** The Z Saha equations are linked together by the common electron pressure, Eq. (2.2), produced by all ionization stages and connected with the total pressure by Eq. (2.3). Combining these $Z + 2$ equations, one finds easily

$$\frac{p_e}{p - p_e} = \sum_{\xi=1}^{Z} \xi \prod_{m=0}^{\xi-1} \frac{K_m}{p_e} \Bigg/ \left\{1 + \sum_{\xi=1}^{Z} \prod_{m=0}^{\xi-1} \frac{K_m}{p_e}\right\}. \tag{4.9}$$

In the three-particle approximation, the ionization equilibria are considered as mutually decoupled, which can be expressed by

$$p_e/kT = \xi n_\xi + (\xi + 1) n_{\xi+1} \tag{4.10}$$

* See Krempl (1962).

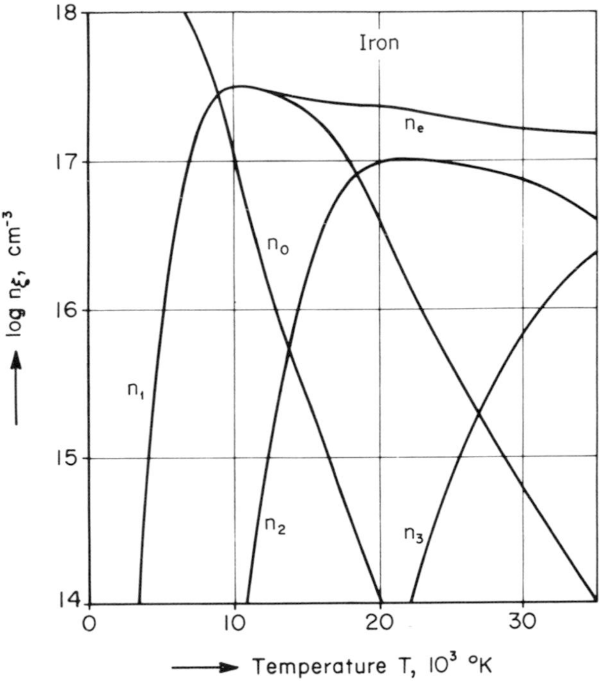

FIG. 8. Composition of iron versus temperature at atmospheric pressure.

or, using Eq. (2.5) and Eq. (2.6), by

$$p_e/p = (\xi + x_\xi)/(\xi + 1 + x_\xi). \tag{4.11}$$

(*i*) *Weak plasma.* The relevant temperatures for a weakly ionized plasma are $T < T_0$, where $x_0 \ll 1$, and therefore, according to Eq. (4.11), $p_e/p = x_0$. Then, Eq. (4.6) becomes

$$p_e^2 = pK_0(T), \tag{4.12}$$

which gives rise to a steep slope in the function $p_e = p_e(T)$ nearly like $\exp(-\chi_0/kT)$.

(*ii*) *Strong plasma.* This term denotes a plasma where x_0 is of the order of 0.5, corresponding to $p_e = p/3$ and thus $T \approx T_0$. Combining Eq. (4.11) with Eq. (4.6), we find easily

$$p_e^2/(p - 2p_e) = K_0(T). \tag{4.13}$$

Thus, the slope of $p_e(T)$ becomes flatter compared to that of Eq. (4.12).

(*iii*) *Fully ionized plasma.* At temperatures $T > T_0$, the electron pressure no longer varies perceptibly with temperature. Therefore, we can express $p_e(T)$ by a sequence of pairs of parameters $\xi = 1, 2, \ldots, Z - 1$, i.e.,

$$T = T_\xi \qquad \text{and} \qquad p_{e\xi} = p(2\xi + 1)/(2\xi + 3), \tag{4.14}$$

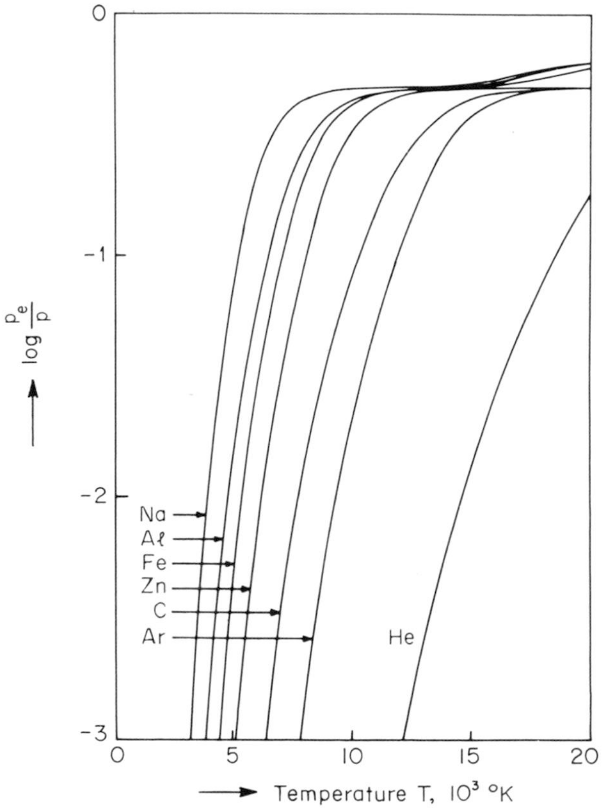

FIG. 9. Electron pressure produced by various elements at atmospheric pressure.

where Eq. (4.11) with $x_\xi = 0.5$ has been used. These distinct p_e values may also be used to a fairly good approximation in the whole range of applicability of the ξth Saha equation, thus getting a linearized approximation with respect to x_ξ. The application of Eq. (4.11) to Eq. (4.14) is shown by Fig. 9. Combining these equations with Eq. (2.5), we find

$$p_\xi = (p - p_e)(1 - x_\xi), \qquad p_{\xi+1} = (p - p_e)x_\xi. \tag{4.15}$$

*b. Interelement Coupling.** In plasma experiments, one takes care to produce a plasma of the highest purity possible. In spectrochemical light sources such as low-current arcs, flames, and sparks, the plasma contains a rather large number of elements. Depending on the ionization energy, each element contributes more or less to the total electron pressure, by which the ionization equilibria of all elements present are linked together. To deal with the "interelement effect," we presume a multielement plasma consisting of the elements $A = B, C, D \ldots$. The temperature of the plasma should be sufficiently low for us to write the partial pressure of A as $p_A = p_0^A + p_1^A$, the higher ionization terms $p_2^A, p_3^A, \ldots$ being negligible. Then, the mole fraction X_A ($\Sigma X_A = 1$) and the degree of ionization can be expressed respectively by

$$X_A = (p_0^A + p_1^A)/(p - p_e) \qquad \text{and} \qquad x_0^A = p_1^A/(p_0^A + p_1^A). \tag{4.16}$$

Due to the quasineutrality, $\Sigma p_1^A = p_e$, we find, combining these two equations,

$$\sum_A x_0^A X_A = p_e/(p - p_e). \tag{4.17}$$

Now, we eliminate x_0 from Eq. (4.6) and have finally

$$p_e/(p - p_e) = \sum_A \{p_e/(K_0^A + 1)\}^{-1} X_A, \tag{4.18}$$

which is the generalization of Eq. (4.13) to multielement plasmas. As can be seen easily, Eq. (4.18) is an equation of the $(\nu + 1)$th degree in p_e, with ν the number of elements present. The most instructive case is that of a two-element system, where a rather easily ionizable element A, say a metal, is mixed with a gas B of high ionization energy, e.g., a permanent gas (see Fig. 10). Denoting $X_A = X$, we then have $X_B = 1 - X$. At low temperature, where B is not yet ionized, the second B term of the sum can be neglected; further, as $T \to 0$, we get $p - p_e \to p$ and $p_e/K_0^A + 1 \to p_e/K_0^A$. Thus, we have for the low-temperature behaviour of p_e

$$p_e^2 = pXK_0^A. \tag{4.19}$$

We see that, at a given temperature, $p_e \propto \sqrt{X}$, just like the square-root dependence of the concentration in Ostwald's dilution law.

At intermediate temperatures, we should expect a tangent to an inflection point, or both a maximum and a minimum in the p_e–T curve

* See Boumans (1966).

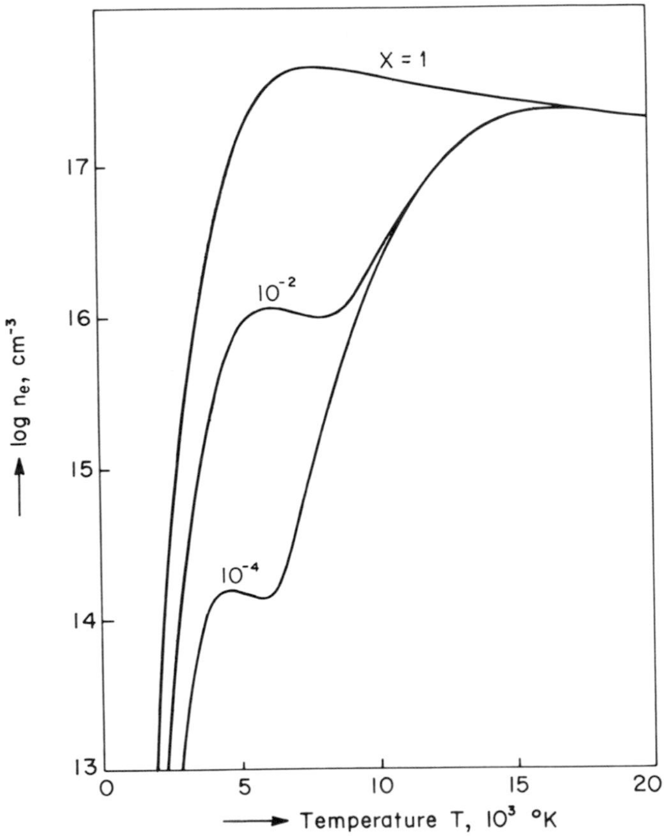

FIG. 10. Common electron density produced by mixture at various mole fractions $X = [Na]/([Na] + [Ar])$ versus temperature.

according to the properties of a polynomial of third degree except for $X = 1$ and $X = 0$, which represent the case of pure A or pure B. If we now increase the temperature into a region of multifold ionization, the interelement effect is weakened more and more, because $p_e \rightarrow p$, and therefore p_e no longer varies much with the element composition.

*c. Practical Evaluation.** Now, we have accounted for the excitation of high electronic levels in the partition function up to a density-dependent truncation of the summation and the lowering of the ionization potential $\Delta\chi$, which also depends strongly on the electron pressure.

* See Olsen (1962), Neumann (1962), Ruthe and Neumann (1965), Bögershausen *et. al.* (1967).

For this, we use the estimate made above and improve the equilibrium constants K_ξ used in Eq. (4.9) and Eq. (4.18). Then, the electron pressure is recalculated. The old and the new values are compared and, if agreement is fairly good, the solutions are retained. Otherwise, further iteration is carried out until self-consistent results are obtained.

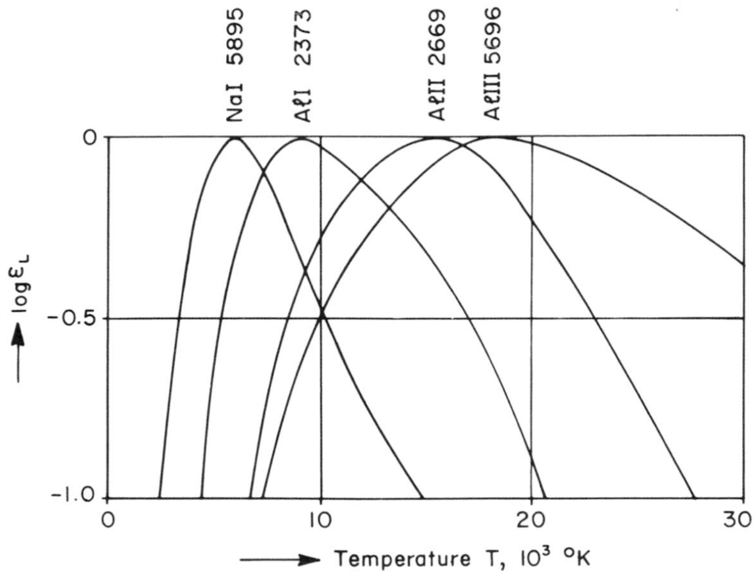

FIG. 11. Relative population of electronic states of various elements with various excitation energy.

C. Excitation and Line Emission*

In thermal equilibrium, the population of a level with the exciting energy $\mathscr{E}_t$ is given by the Boltzmann expression

$$n_{\xi t} = (g_{\xi t}/Q_\xi)n_\xi \exp(-\mathscr{E}_t/kT). \tag{4.20}$$

With increasing temperature, the population first rises due to the Boltzmann factor, but at higher temperature, the population decreases because n_ξ becomes small due to the increasing ionization $n_\xi \to n_{\xi+1}$ (Diermeier and Krempl, 1966) (Fig. 11). The balance between these two effects occurs in the vicinity of the ionization temperature T_ξ. The exact value

* See Maecker and Peters (1954), Fowler (1964), Griem (1964), Neumann, W. (1967), Richter (1968), and Zwicker (1968).

of this so-called "norm-temperature" T_ξ^* can be evaluated by putting $dn_{\xi t}/dT = 0$, with the help of the Saha equation (Larenz, 1951). For $\xi = 0$, one finds by straightforward calculation the degree of ionization at T_0^*

$$x_0^* = [(\mathscr{E}_t/kT^*) - 1]/[(\chi_0/kT^*) + 2.5], \tag{4.21}$$

which is of the order of 0.5. Inserting x_0^* into Eq. (4.6), we get a transcendental equation containing T_0^* implicitly.

Linearizing this equation by suitable approximation and expanding the result at any ξ (Krempl, 1962), we have

$$T_\xi^* = 950\chi_\xi\{1 - (0.33\mathscr{E}_t)/\chi_\xi + 0.37 \log(\chi_\xi/10) - 0.14 \log(p_e^* g_\xi/g_{\xi+1})\}^{-1}, \tag{4.22}$$

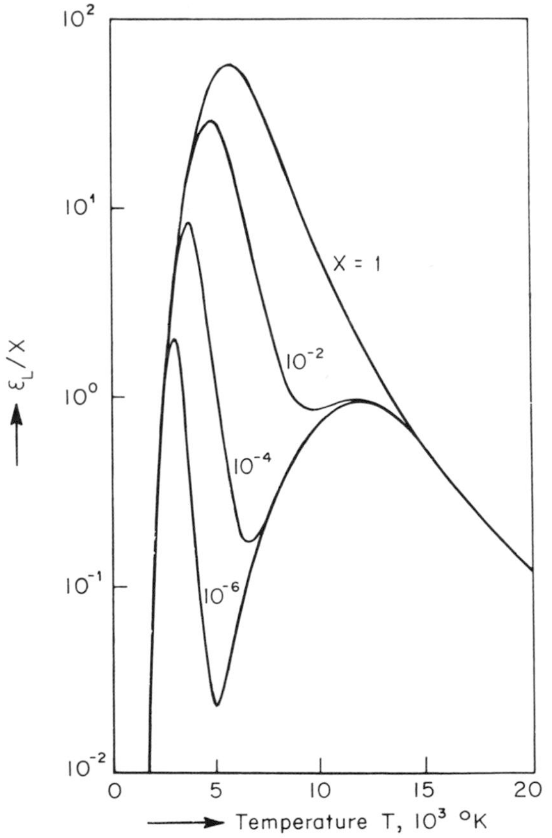

FIG. 12. Influence of the dilution $X = [\mathrm{Na}]/([\mathrm{Na}] + [\mathrm{Ar}])$ on the emission coefficient of the line NaI 5895 (Diermeier and Krempl, 1966).

where χ_ξ is to be expressed in eV and $p_e{}^*$ in bar (atm). Equation (4.14) may be used to approximate $p_e{}^*$.

In a two-element plasma, where the electron pressure yields a double-peaked p_e–T curve, as seen in Fig. 12, we expect to obtain two maxima in the n–T curve, where the emission coefficient ε_L of a spectral line is plotted versus temperature. This quantity is proportional to the population $n_{\xi t}$ in an optically thin plasma, where no absorption takes place. If A is the Einstein transition probability, we have the familiar relation

$$\varepsilon_L = (A/4\pi)h\nu n_{\xi t}. \tag{4.23}$$

Observing ε_L by spectroscopic methods, we may find a distinct point in a temperature-inhomogeneous plasma where a maximum of ε_L is found. Then, we are able to assign to this maximum the norm temperature of the line under examination. With the help of a few lines, a temperature field can be mapped out very exactly.

V. Plasma at Higher Densities†

In order to obtain a convergent partition function, it is necessary to introduce a limiting force. This limitation does not belong to the perfect-gas concept; it follows from geometric reasons. In the present section, we deal with electrical forces arising when the separation of the particles becomes small enough. For this concept, the additional energy resulting from these forces should not be greater than the thermal energy.

In treating the equilibrium properties of a plasma, we assume that quantities such as internal energy, free energy, and pressure are the sum of contributions arising from two sources: (1) a perfect, classical gas of neutrals, electrons, and ions acting together as kernels; and (2) electrostatic interaction between classical point-charge particles, which we shall denote by the index C.

A. Fundamental Lengths and Critical Concentrations

To deal with the problem of interaction, we introduce three characteristic radii around the test particle α surrounded by field particles β:

† See Harris (1959), Zeleznik and Gordon (1965), and Ebeling (1968).

(1) The scattering amplitude or Landau length

$$\varrho_L = e^2/kT. \tag{5.1}$$

In the region $r > \varrho_L$, the Boltzmann factor can be linearized (Fowler, 1964).

(2) The average mean distance

$$d = (3/4\pi n)^{1/3} \qquad \text{with} \quad n = n_e + \sum_0^Z n_\xi \tag{5.2}$$

of all particles, including electrons. For $r > d$, the occupation number of a unit region is large enough so that the discontinuous ensemble of particles may be represented sufficiently accurately by a continuous probability density distribution, as given by the laws of statistical mechanics (Ecker, 1955), in particular, the Boltzmann population.

(3) The Debye shielding radius

$$\varrho_D = \left(kT/4\pi \sum_{\beta=e_1}^{Z} n_\beta e_\beta{}^2\right)^{1/2} = \{kT/4\pi e^2 n_e(1 + \langle\xi\rangle)\}^{1/2}. \tag{5.3}$$

For field particles outside the Debye sphere, the interaction force with the test particle becomes zero.

According to Ecker and Kröll (1963), there exists an upper limit of the particle density for applying the linearized Poisson–Boltzmann equation, as used in the Debye–Hückel theory of the strong electrolyte (Davis, 1967, p. 405), if d becomes equal to ϱ_L. Thus, we find (cf. Fig. 1)

$$n_{cr} = (3/4\pi)(kT/e^2)^3. \tag{5.4}$$

B. Average Micropotential

Our calculation starts with the considerations of Ecker and Kröll (1963) concerning the average micropotential between a test charge α and its surrounding field particles β. The authors found that the established Debye shielding potential, which is inside the Debye sphere $r/\varrho_D < 1$,

$$\psi_1(r) = \psi_\alpha + \psi_{\beta,1} = (e_\alpha/r)\exp(-r/\varrho_D) \approx (e_\alpha/r) - (e_\alpha/\varrho_D) \tag{5.5}$$

is no longer valid at high densities $n > n_{cr}$. In this case, fluctuation becomes too great for a statistical treatment of the problem. In particular, the linearized Poisson–Boltzmann equation, which yields Eq. (5.5), is no longer applicable. In this case, the potential ψ_β around the test ion originating from the surroundings n_β is determined by the nearest neighbor. Therefore, the second term of Eq. (5.5) should be replaced by $c_2 e_\alpha/d$, whereas the first Coulomb term attached to the origin of r cannot be influenced;

$$\psi_2(r) = \psi_\alpha + \psi_{\beta,2} = (e_\alpha/r) - (c_2 e_\alpha/d). \tag{5.6}$$

The constant c_2 can be found by joining the two cases steadily at $n = n_{\rm cr}$. Setting $\psi_1 = \psi_2$ we have

$$c_2 = (d/\varrho_{\rm D})_{\rm cr} \approx 1. \tag{5.7}$$

If we denote $c_1 = 1$ and $c_2 = 0$ for $n < n_{\rm cr}$ and $c_1 = 0$ and $c_2 = (d/\varrho_{\rm D})_{\rm cr}$ for $n > n_{\rm cr}$, we can write the potential produced by the field particles

$$\psi_\beta = \psi_{\beta,1} + \psi_{\beta,2} = -(c_1 e_\alpha/\varrho_{\rm D}) - (c_2 e_\alpha/d). \tag{5.8}$$

To find the Coulomb energy $E_{\rm c}$, we should insert $e_\alpha \psi_\beta$ in the correlation function and calculate the thermodynamic function Davis (1967). For brevity, we exploit the fact known from electrostatics that the energy of the electrical interaction of a system of charged particles is half the sum of the products of each charge with its potential in the field of all other charges. Therefore, the Coulomb interaction energy can be calculated ($N = nV$)

$$E_{\rm C} = (1/2) \sum_\alpha e_\alpha n_\alpha \psi_\beta V. \tag{5.9}$$

Inserting Eq. (5.8), we find easily, using the definition of Eq. (5.3),

$$-E_{\rm c}/NkT = (1/8\pi)[(c_1/\varrho_{\rm D}^3) + (c_2/\varrho_{\rm D}^2 d)](V/N). \tag{5.10}$$

The expression for the Helmholtz free energy can be found by integrating the Gibbs–Helmholtz equation $A/T = \int (E/T^2)\, dT$. The integration constant in the case $n < n_{\rm cr}$ is $A_{\rm c} = 0$ for $T \to \infty$, because of the ideal-gas behavior at high temperature. For $n > n_{\rm cr}$, it is given by the requirement that both branches of $A_{\rm c}$ join smoothly at $n = n_{\rm cr}$. We find

$$-A_{\rm c}/NkT = (1/\pi)\{(c_1/12\varrho_{\rm D}^3) + (c_2/8\varrho_{\rm D}^2 d)\}(V/N) + \text{const.} \tag{5.11}$$

Once the free energy has been determined in this form, all other thermodynamic functions can be calculated using the usual thermodynamic relations.

The equation of state for the system is obtained from the thermodynamic relation $P = -(\partial A/\partial V)_{T,N_\beta}$ and can be written in the form

$$-p_c V/NkT = (1/24\pi)[(c_1/\varrho_D^3) + (c_2/\varrho_D^2 d)](V/N). \quad (5.12)$$

Using now $\mu_\beta = (\partial A_c/\partial N_\beta)_{T,V,N_\alpha}$ for the chemical potential, we find the activity coefficient due to Coulomb forces f_β for particles of charge β

$$-\mu_\beta = -kT \ln f_\beta = (e_\beta^2/2)[(c_1/\varrho_D) + (c_2/d)]. \quad (5.13)$$

A more rigorous derivation of the free-energy expansion was developed originally by Berlin and Montroll (1952) for the case of strong electrolytes. Evaluating the partition function, they found a somewhat smaller critical density and therefore a different constant in A_c. In addition, they have a small logarithmic term. For better comparison, we shall write the free-energy expansion for a strong plasma, $n_0 = 0$, in the general form

$$A_c/V = An^x + Bn^y + C \ln n + D. \quad (5.14)$$

In our case, we have $x = 3/2$ and $y = 4/3$, with $B = 0$ if $n < n_{cr}$ and $A = 0$ if $n > n_{cr}$. The logarithmic term, which has also been found by Vedenov and Larkin (1959), should be neglected in our consideration because it is small and vanishes for a symmetrical plasma, where $\sum n_\beta e_\beta^3 = 0$. The values 3/2 for x and 2 for y are valid in the whole density range. It should be noted that, for short-range forces, $x = 2$ and $B = C = D = 0$ in the van der Waal approximation.

C. Ionization Equilibrium and Deviations from Ideal Behavior

To take into account the Coulomb interaction in the Saha equation, the partial pressure p_β should be replaced by the corresponding activities $f_\beta p_\beta$. Inserting the activity coefficients on the right side of Eq. (4.5) we have to replace χ_ξ by the reduced ionization energy $\chi_\xi - \Delta\chi_\xi$. The correction $\Delta\chi_\xi$ can easily be calculated by

$$\begin{aligned}\Delta\chi_\xi &= kT(-\ln f_{\xi+1} + \ln f_\xi - \ln f_e) \\ &= e^2(c_1/\varrho_D + c_2/d)(\xi + 1) \quad (5.15)\end{aligned}$$

Figure 1 shows that practically all laboratory plasmas lie within the range of the Debye approximation (for numerical values, of c_1 and c_2, see Ecker and Weizel, 1957, 1958; Ecker and Kröll, 1966; Neumann, 1967; and Richter, 1968).

References

ARTMANN, J. (1965). *Z. Phys.* **183**, 65.

ARTMANN, J. (1968). *Z. Angew. Phys.* **25**, 104.

ASTON, J. G., and FRITZ, J. J. (1959). "Thermodynamics and Statistical Thermodynamics." Wiley, New York.

BERLIN, T. H., and MONTROLL, E. M. (1952). *J. Chem. Phys.* **20**, 75.

BÖGERSHAUSEN, W., HINGSAMMER, J., and KREMPL, H. (1960). *Ber. Bunsenges. Phys. Chem.* **71**, 64.

BÖGERSHAUSEN, W., and SCHNEIDER, J. (1971). "Diplom Arbeit." Technische Universität München.

BOUMANS, P. W. J. (1966). "Theory of Spectrochemical Excitation." Hilger and Watts, London.

BURHORN, F., and WIENECKE, R. (1959). *Z. Phys. Chem. (Leipzig)* **212**, 105.

BURHORN, F., and WIENECKE, R. (1960). *Z. Phys. Chem. (Leipzig)* **213**, 37.

DAVIS, H. T. (1967). "Physical Chemistry," Vol. 2. Academic Press, New York.

DIERMEIER, R. (1969). Thesis. Technische Hoschschule München.

DIERMEIER, R., and KREMPL, H. (1966). *Z. Phys.* **200**, 239.

DRAWIN, H. W., and FELENBOK, P. (1965). "Data for Plasmas in Local Thermodynamic Equilibrium." Gauthier-Villars, Paris.

EBELING, W. (1968). *Physica* **38**, 378.

ECKER, G. (1955). *Z. Phys.* **140**, 274.

ECKER, G., and KRÖLL, W. (1963). *Phys. Fluids* **6**, 62.

ECKER, G., and KRÖLL, W. (1966). *Z. Naturforsch. A* **21**, 2012 and 2023.

ECKER, G., and WEIZEL, W. (1956). *Ann. Phys.* **17**, 126.

ECKER, G., and WEIZEL, W. (1957). *Z. Naturforsch. A* **12**, 859.

ECKER, G., and WEIZEL, W. (1958). *Z. Naturforsch. A* **13**, 1093.

FALKENHAGEN, H., and EBELING, W. (1963). *Monatsber. Deut. Akad. Wiss. Berlin* **5**, 615.

FINKELNBURG, W. and MAECKER, H. (1956). Elektrische Bögen und thermisches Plasma, *In* "Handbuch der Physik." (S. Flügge, ed.) Vol. 22. Springer, Berlin,

FOWLER, R. G. (1964). Radiation from gas discharges. *In* "Discharge and Plasma Physics" (S. C. Haydon, ed.). Armidale, New York.

FOWLER, R. H. (1966). "Statistical Mechanics," p. 267. Cambridge Univ. Press, New York and London.

FRIE, W. (1967). *Z. Phys.* **201**, 269.

GODNEW, I. N. (1963). "Berechnung Thermodynamischer Funktionen aus Moleküldaten." Deut. Verlag. Wiss., Berlin.

GRIEM, H. R. (1964). "Plasma Spectroscopy." McGraw-Hill, New York.

HARRIS, G. M. (1959). *J. Chem. Phys.* **31**, 1211.

HUND, F. (1961). "Theorie des Aufbaues der Materie." Teubner, Stuttgart.

KRAUSS, L. (1970). *Z. Naturforsch.* **25a**, 724.

KRAUSS, L. (1969). *Z. Phys.* **223**, 39.

Krempl, H. (1962). *Z. Phys.* **167**, 302.

Larenz, R. W. (1951). *Z. Phys.* **129**, 327.

Lochte-Holtgreven, W. (1968). Evaluation of Plasma parameters. *In* "Plasma Diagnostics" (W. Lochte-Holtgreven, ed.). North-Holland Publ., Amsterdam.

Maecker, H. (1963). "Discharge and Plasma Physics" (S. C. Haydon, ed.). Univ. New England, Armidale, New York.

Maecker, H., and Peters, T. (1954). *Z. Phys.* **139**, 448.

Margenau, H. and Lewis, M., (1959). *Rev. Mod. Phys.* **31**, 569

Moore, C. E. (1949). "Atomic Energy Levels," Vol. I. Nat. Bur. of Stds., Washington, D. C.

Moore, C. E. (1952). "Atomic Energy Levels," Vol. II. Nat. Bur. Stds., Washington, D. C.

Moore, C. E. (1958). "Atomic Energy Levels," Vol. III. Nat. Bur. Stds., Washington, D. C.

Münster, A. (1956). "Statistische Thermodynamik." Springer, Berlin.

Neumann, Kl.-K. (1962). *Ber. Bunsenges. Phys. Chem.* **66**, 551.

Neumann, W. (1967). *In* "Ergebnisse der Plasmaphysik und der Gaselektronic" (R. Rompe and M. Steenbeck, eds.) Vol. 1. Akademie Verlag, Berlin.

Olsen, H. N. (1962). Determination of properties of an optically thin argon Plasma. *In* "Temperature, Its Measurement and Control in Science and Industry," Vol. 3, p. 593. Reinhold, New York.

Richter, J. (1968). *In* "Plasma Diagnostics." (W. Lochte-Holtgreven, ed.) North-Holland Publ., Amsterdam.

Riewe, K. H., and Rompe, R. (1938). *Z. Phys.* **111**, 79.

Ruthe, R., and Neumann Kl.-K. (1965). *Ber. Bunsenges. Phys. Chem.* **69**, 414.

Scheibe, G., Brück, D., and Dörr, F. (1952). *Chem. Ber.* **85**, 867.

Schlender, B., and Traving, G. (1965). *Z. Astrophys.* **61**, 92.

ter Haar, D. (1967). "Physical Chemistry," Vol. 2. Academic Press, New York.

Unsöld, A. (1948). *Z. Astrophys.* **24**, 355.

Unsöld, A. (1948). "Physik der Sternatmosphären." Springer, Berlin.

Vedenov, A., and Larkin, A. (1959). *Zh. Eksp. Teor. Fiz.* **36**, 1133.

Zeleznik, F. J., and Gordon, S. (1965). *Can. J. Phys.* **44**, 877.

Zwicker, H. (1968). Evaluation of plasma parameters in optical thick plasmas. *In* "Plasma Diagnostics" (W. Lochte-Holtgreven, ed.). North-Holland Publ., Amsterdam.

Chapter 9

High Pressure Phenomena

ROBERT H. WENTORF, JR.

I. Introduction

In this chapter, we shall be primarily concerned with pressures above about 20 kilobars (kbar). One kilobar is 10^9 dyn cm^{-2}. This choice of pressure range is arbitrary and is based on limitations of space—pressure influences the behavior of matter in so many areas of study that only the strong or unusual effects can be sketched here. The reader who is deeply interested in certain areas may consult the references listed at the end of the chapter.

At this time, about 1000 persons in 250 laboratories in the world use high pressure as a main tool of their researches. Many of the basic techniques had been developed before about 1950; however, then, only a few scientists regarded pressure as a thermodynamic variable of overwhelming interest. Chief among these was Bridgman (1964a,b); his collected papers were published in 1964 by Harvard University Press and still guide many workers. Modern high-pressure work appears in diverse journals, and the High-Pressure Data Center at Brigham Young University, Provo, Utah, collects, organizes, and provides information on nearly all the published work.

A few remarks about the general effects of high pressures may be appropriate. One notes that pressures in the range 30–100 kbar, not too difficult to achieve with modern techniques, compress most solids by at least a few per cent and thereby change the average interatomic spacings several times more than does cooling from 300°K to 1°K. The work done on the material by compressing it is not so large, only a few kilocalories, but this is often sufficient to make a new crystalline form stable. If a chemical reaction takes place with a volume change of 10 cm^3 mol^{-1}, then, at 43 kbar, the corresponding free-energy change is 10 kcal mol^{-1}, sufficient to affect many equilibria. Often, a particular structural arrangement of atoms will persist at pressures for which it is not thermodynamically stable; this often occurs in materials with high melting temperatures and may make difficult the determination of an equilibrium (thermodynamically consistent) phase diagram. However, at the same time, this persistence effect provides many interesting materials for study or use at 1 atm, e.g., diamond.

II. Techniques

A. Apparatus for Producing and Measuring High Static Pressures

1. *Piston and Cylinder*

A straight cylindrical piston may be driven by a hydraulic press to compress material contained in a strong cylinder, as shown in Fig. 1. Many details of sealing the piston, supporting the cylinder, etc. are described by Bridgman (1952, 1964a,b). A specimen may be immersed in a pressure-transmitting fluid, but most ordinary liquids freeze or become very viscous at 20–30 kbar; exceptions are mixtures of pentanes, or of *n*-pentane with isoamyl alcohol (Jayaraman *et al.*, 1967a). Pressures above 40 kbar may be reached with solid pressure-transmitting media such as AgCl or pyrophyllite, but the finite shear strength of the medium and its friction against the cylinder walls generate pressure gradients; the specimen may be sheared and the pressure on it is not quite so definite.

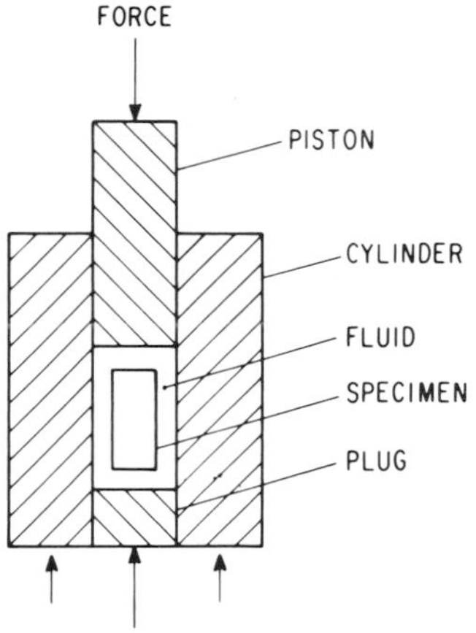

FIG. 1. Simple cylindrical piston and cylinder.

Above about 40–50 kbar, unsupported pistons tend to fail and cylinders tend to rupture radially or axially, even when made of the strongest available materials such as cobalt-cemented tungsten carbides. Extra support must be given to these parts.

TABLE I

PROPERTIES OF MATERIALS USEFUL IN HIGH-PRESSURE CONSTRUCTION

Material	Average strength (kbar)		Average Young's modulus (kbar)
	Compression	Tension	
Steel, 18-8 stainless	5–10	5–10	2000
Steel, 4340 or maraging	15–25	15–20	2000
Cobalt-cemented tungsten carbide:			
3% Co	60	11	6500
6% Co	45	14	5600
13% Co	40	22	5300
Beryllium copper	12	12	1200
Glass fibers, laid parallel in resin	6	10	380
Boron fibers, laid parallel in resin	10	20	4000
Single graphite fibers	—	25	4500
Sapphire	40	—	3500
Diamond	120+	—	13000
Pyrophyllite	1	0.1	800
NaCl	0.3	—	400

Table I lists some of the properties of materials for high-pressure construction. One should note that an elastic material which supports a stiffer material tends to bulge around the stiffer material and impose some tensile stress on the stiffer material. One might also note that the metallic materials can suffer some plastic flow at high stresses and this prolongs their service lives, brief though they may be.

Figure 2 shows a successful design by Boyd (1962) which has been used for high-temperature (1850°C) studies of silicate systems. Note

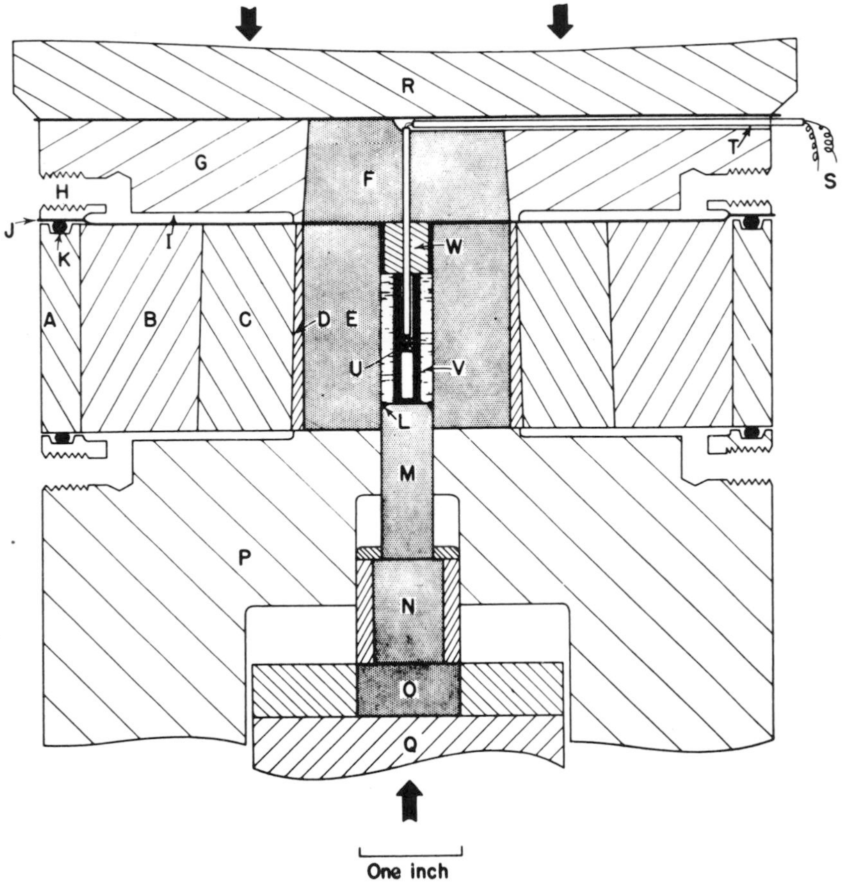

FIG. 2. Internally heated piston and cylinder of Boyd (1962). *A*, soft steel safety ring; *B*, *C*, *D*, *G*, steel binding rings; *E*, cemented carbide cylinder; *F*, cemented carbide end plate; *L*, steel piston sealing ring; *M*, *N*, *O*, cemented carbide pistons; *P*, steel bridge; *Q*, 1000-ton ram; *R*, 800-ton ram; *S*, thermocouple leads; *U*, sample; *V*, talc sample cell with graphite heating element; *W*, stainless steel power lead.

the extensive support from all directions. The internal furnace is heated electrically; insulation is talc, surrounded by a lead sheath for low wall friction.

In some designs, the piston friction is low enough to permit close estimation, within 2%, of the pressure by measuring the piston force and area; Kennedy *et al.* (1962, 1967) have thereby made careful measurements of the pressures for certain phase transformations used as pressure references. For example, bismuth at 25°C undergoes phase changes at 25.3 and 26.8 kbar. These changes are marked by changes in density and also electrical resistivity which are easily detected outside the pressure vessel. Further remarks about pressure reference standards are found in Section II,C of this chapter.

2. *Tapered Piston and Cylinder*

A tapered piston can support a higher tip pressure than a cylindrical one before disastrous failure. The blunter the piston, the greater the attainable pressure—Bridgman called this "the principle of massive support."

One may also support the flanks of the piston so that the allowable shear stress in the material is never exceeded, and thereby reach even higher tip pressures. In principle, any finite pressure can be sustained by a properly supported tapered piston, but practical problems arise when one attempts to generate this pressure, or when one considers that only a small portion of the total force on the piston appears as tip pressure—the bulk of the force is consumed as flank support. Thus, the highest static pressures demand extra-large presses or extra-small working volumes.

A successful example of a tapered piston apparatus is the "belt" design (Hall, 1960, 1966). A cross section of the highly stressed region is shown in Fig. 3; one half is shown at 0.001 kbar pressure; the other half, at, say 60 kbar, to illustrate the compressions involved.

The tapered pistons and cylinder are made of cemented tungsten carbide and are set into strong, tapered steel rings. A compressible gasket, made of two layers of pyrophyllite (a soft stone which has high friction against metal) and a thin, soft steel cone, separates the piston flank from the tapered wall of the cylinder. This gasket seals the variable gap as the piston advances and retreats, and provides flank support and electrical insulation for the piston. The working volume in the cylinder can be heated electrically. Pyrophyllite around it provides good thermal and electrical insulation to 2000°C or so. The current flows from one piston through a ring, radially inward via a disk to the furnace heater tube, through the heater tube, and out the other piston. The stone disk on the piston face furnishes thermal insulation. The temperature gradients in the heater tube can be adjusted by selecting the materials and dimensions of the disks and furnace tube. Heating currents of several hundred amperes are common. Slender wires for thermocouples, etc., may be led out through the gaskets.

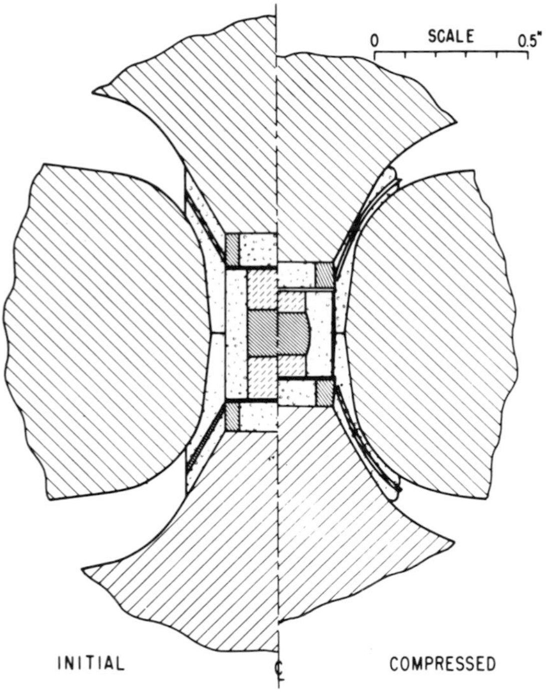

FIG. 3. Belt apparatus. One side shown before compression; the other side shown at about 60 kbar.

The pressure inside the chamber is usually estimated from the piston force by means of a calibration curve which is obtained by noting the piston force necessary to effect phase transformations in reference substances. The curve depends somewhat on the amount and compressibility of the material in the chamber because this affects the relative loads on the piston tip and flank. The calibration is performed at 25°C; however, the experiments may be done at higher temperatures. Thermal expansion tends to increase the pressure, but the conversion of hot materials into denser forms usually predominates, so that the pressure falls somewhat; e.g., hot pyrophyllite in the range 30–100 kbar changes into a mixture of kyanite and coesite. Pistons driven by a constant force will move inward to compensate for the densification, but then the flank gasket carries more load and the tip pressure falls. Thermal or piston force cycles also affect the pressure in ways which must be ascertained for each particular case.

Several variations of this basic design have been made using slightly different piston shapes, gasketing materials, etc. By making the piston relatively long and narrow with more flank support, Bundy (1963) was able to reach 150–200 kbar in a reaction volume which could be heated briefly to over 3000°C.

3. *Anvil Designs*

a. Simple Opposed Anvils. The "principle of massive support" ultimately leads one to what are known as "Bridgman anvils," shown in Fig. 4. The sample is in the form of a thin disk between two opposed anvil working faces; frictional forces prevent its complete extrusion. Often, an outer ring made of some high-friction material is used to confine a more fluid central region; e.g., a pyrophyllite or nickel ring may be used to confine the soft AgCl or liquids. Typical ratios of sample diameter to thickness are 15 or 30 or more. This type of device is easy to make and use and has proven to be one of the most versatile in modern work.

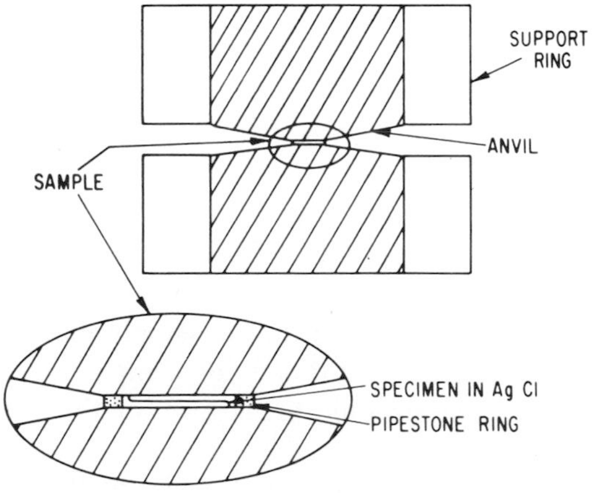

Fig. 4. Simple opposed "Bridgman anvils" with pipestone confining ring.

In the sample wafer, consider a ring of width dr in mechanical equilibrium so that the pressure-difference force, $-2\pi rt\,dp$, is balanced by the frictional force, $2\mu 2\pi rp\,dr$, where r is the radius, t is the thickness, p is the local pressure, and μ is the local coefficient of friction between the material and the piston face (see Fig. 5). Then we have

$$-2\pi rt\,dp = 2\mu 2\pi rp\,dr \tag{2.1}$$

or

$$p = p_0 \exp[-2\mu(r - r_0)/t], \tag{2.2}$$

and the pressure may fall off exponentially with radius at incipient extrusion. For a typical value of $\mu = 0.5$, we see that $p = 10^5 p_0$ for $(r_0 - r)/t = 11.3$; i.e., a ratio of sample radius to thickness of 11.3 should suffice to contain 100 kbar.

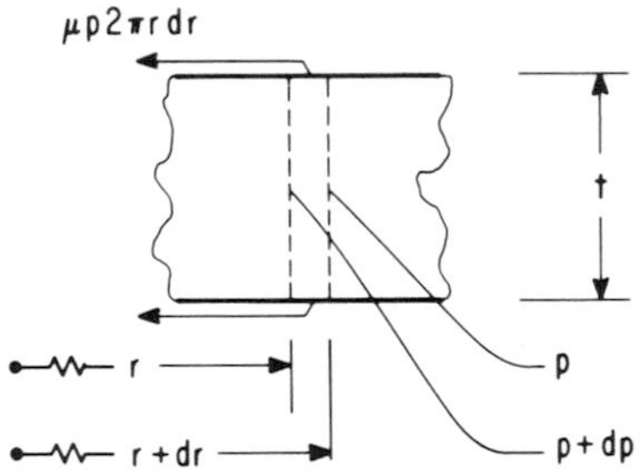

FIG. 5. Section of confining gasket between opposed anvils.

The disk of compressed material tends to become lens-shaped at higher pressures because the central regions of the anvils are deformed more. The thinning of the edges permits a higher pressure gradient there. Thus, certain combinations of anvil and sample material may result in a substantially uniform pressure over a large portion of the center of the sample and a very sharp pressure gradient at the edges; other combinations may produce undesirable pressure distributions (Jackson and Waxmann, 1963; Myers *et al.*, 1963, Connell, 1966). Clearly, the sample pressure is rarely equal to piston force divided by sample area, and separate calibrations must be used. Some sample materials may change their friction characteristics with flow or temperature, and it is not uncommon to have a catastrophic breakdown of the pressure seal, i.e., an explosive extrusion.

Various strong materials may be used for the anvils: high-cobalt alloys, or cemented tungsten carbides (Dachille and Roy, 1962). Even small diamonds have been used with great success for optical spectra or X-ray diffraction of compressed matter up to 150 kbar (Weir *et al.*, 1962; Whatley and Van Valkenburg, 1966; Bassett and Takahashi, 1964; Bassett *et al.* 1967; Connell, 1966). A collimated beam of X-rays passes through one diamond, is scattered by the sample, and passes out through the other diamond. In another variation using X-rays, the sample is contained between tungsten carbide anvils with a confining ring of powdered boron; the beam passes in and out through the boron in the gap between the anvils (Jamieson and Lawson, 1962).

Such anvil devices may be cooled to low temperatures or heated up to about 500°C without serious deterioration in performance if the proper construction materials are used. It is even possible to fit tiny furnaces into the wafer; the long axis of the furnace lies parallel with the anvil surface (Ringwood and Major, 1967).

Drickamer and his co-workers have extended the pressure range of Bridgman anvils by adding support to the tapered flanks. Figures 6 and 7 illustrate the method. The support is provided by the pyrophyllite wafers; the best piston flank angle was found by trial and error. The average pressure on the anvil tip plus flank area is usually below about 15 kbar.

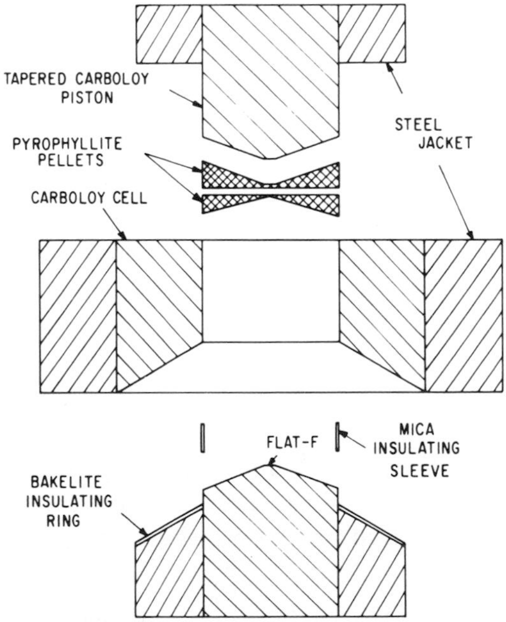

FIG. 6. Very-high-pressure anvil apparatus of Drickamer (1962).

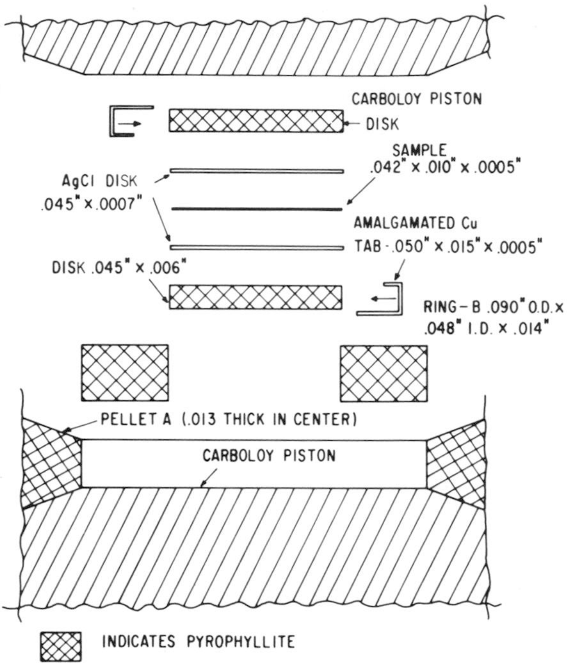

FIG. 7. Central high-pressure region of Drickamer anvils. Dimensions are in thousandths of an inch.

For the highest pressures, the anvils are prestressed to a moderate pressure—this leaves them with dished faces which are reground flat—and then used once more before being discarded. Electrical resistance studies have been made to 500 kbar (Balchan and Drickamer, 1961a,b 1962). Wafers of NaCl allow optical studies to 150 kbar (Fitch *et al.*, 1957); wafers of LiH and B permit X-ray studies to 450 kbar (Perez-Albuerne *et al.*, 1964). For Mössbauer studies to about 200 kbar, the pyrophyllite of the lateral supporting wafers is replaced by lead sectors to reduce stray radiation, and the sample disk is prepressed from a combination of B, LiH, and the substance under study (DeBrunner *et al.*, 1966). The actual high-pressure region is only about 2 mm in diameter and 0.3 mm thick, but this is sufficient for many interesting studies between 77 and 500°K. The pressures are estimated by X-ray measurement of the lattice constants of well-behaved materials whose compressions are known from shock-wave data.

b. Polyhedral Anvil Devices. These may be regarded as a set of anvils whose working faces define a polyhedron. The material of the polyhedron is compressed as the anvils are driven toward a common center. Compressible gaskets are placed between adjacent anvil flanks, or gaskets may be formed spontaneously as the material of the polyhedron, purposely made too large, oozes out into the gaps between the anvils. The simplest form is a tetrahedron with four rams having triangular faces; a version due to Hall (1958, 1966) is shown in Fig. 8. Pressures up to about 100 kb are attainable in such apparatus; it has been used for high-temperature work with an internal furnace, and for X-ray work at high temperatures and pressures—the beam may enter through a normal gap or through a special hole in one anvil and the diffracted rays are observed along the gaps between anvils (Hall, 1966; Barnett and Hall, 1964). Other popular forms of this type of apparatus include the cube, with six square-faced anvils, octahedron, dodecahedron, etc. The use of more than six anvils does not seem to confer any special advantages to offset the greater complexities. As with most other apparatus using solid pres-

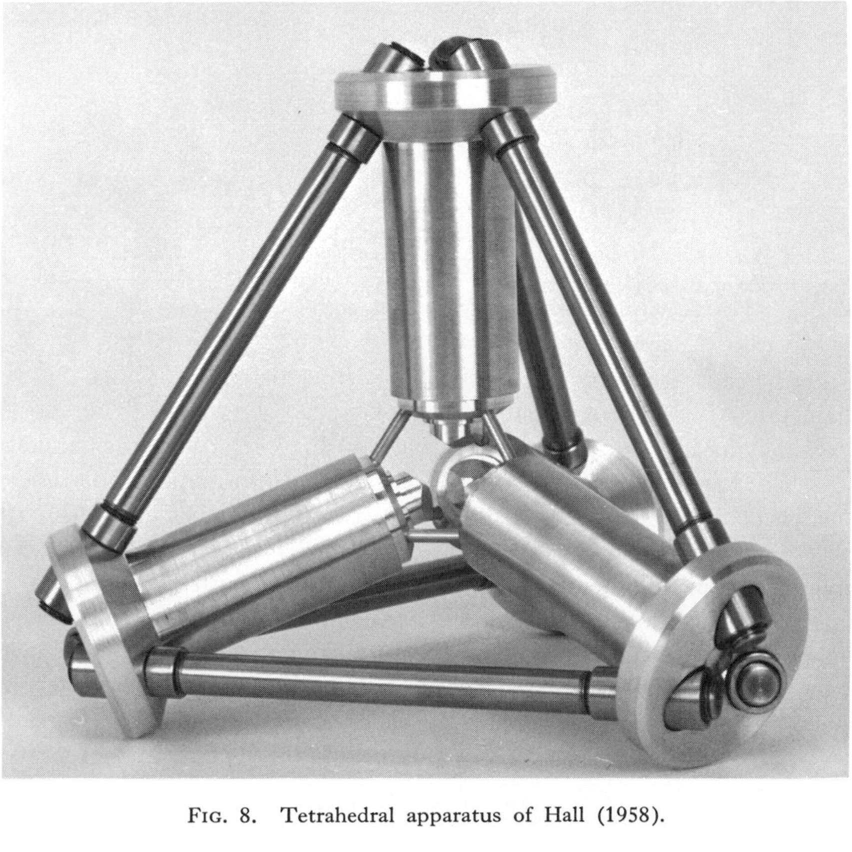

FIG. 8. Tetrahedral apparatus of Hall (1958).

sure-transmitting media, the internal pressure must be estimated by means of a calibration curve, except when the volume of a reference substance inside can be continuously monitored by means of X-ray diffraction.

Lees (1966) has given an extended description of the stresses involved in the various parts of a tetrahedral anvil device and their effects on the specimen.

B. Dynamic High-Pressure Generation

This method produces the highest laboratory pressures, over 1000 kb, but they usually last only a few microseconds. The pressures are produced by the rapid acceleration of matter, driven by detonating explosives, compressed gases, electromagnetic forces, etc. A large body of literature exists on the various methods of producing, measuring, and interpreting the phenomena (Deal, 1962; Duvall and Fowles, 1963; McQueen, 1964; Doran and Linde, 1966).

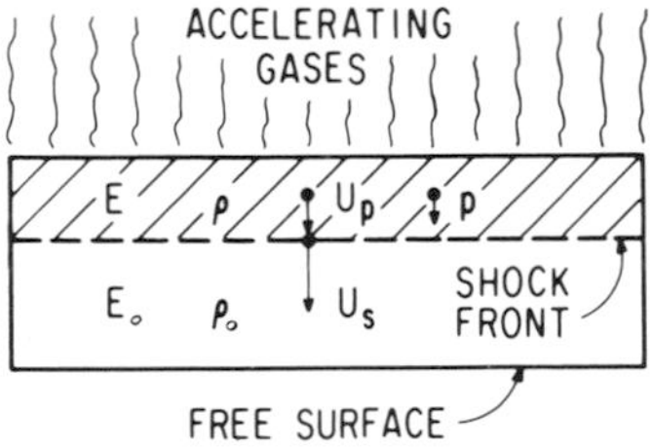

Fig. 9. Plane shock wave passing through a slab of material.

Let us consider a simple plane shock wave generated in a disk whose diameter is large enough that edge effects may be ignored. The shock wave may have been generated by a block of high explosive; the accelerating gases at high pressure compressed and set into motion the shaded portion of the disk, as shown in Fig. 9. Let this portion have velocity U_p, and let it be at pressure p and density ϱ. Between the moving compressed region and the material originally at rest with density ϱ_0 is a sharp boundary called the shock front which moves along at velocity U_s, and across which the velocity and pressure change abruptly. (The shock front develops and persists because a compression wave travels faster in the compressed material. At low shock pressures, an elastic precursor wave may run ahead of the shock front.) By conservation of matter,

$$\varrho_0 U_s = \varrho(U_s - U_p). \tag{2.3}$$

By conservation of momentum,

$$\varrho_0 U_s U_p = p. \tag{2.4}$$

By conservation of energy,

$$U_s \varrho_0 E_0 + p U_p = U_s \varrho_0 E + \tfrac{1}{2} \varrho_0 U_s U_p^2, \tag{2.5}$$

where E_0 is the original and E the final internal energy per gram of material. We note that p is really the stress in the direction of motion.

From the foregoing equations, one concludes, among other things, that

$$E - E_0 = \tfrac{1}{2} p (V_0 - V), \tag{2.6}$$

where $V = 1/\varrho$ and $V_0 = 1/\varrho_0$. This is called the Hugoniot equation.

The velocities involved are of the order of a few km sec^{-1} and they can be measured to within 1% or less by a number of methods using high-speed cameras, flash gaps, contact pins, electronic equipment, etc., but a detailed description is outside the scope of this chapter. It may be sufficient to point out that the shock-front velocity may be determined by the closure times of pin contacts or the flashing times of gas-filled gaps set at different path lengths in the material.

The measurement of U_p is more difficult. Consider the situation when the shock front of Fig. 9 has reached the far end (free surface) of the plate. For an instant, the entire plate moves along at velocity U_p, but then, the compressed material is free to expand because no pressure confines it at the free surface. The free surface then reaches an even higher velocity as the material expands, and a rarefaction wave travels back toward the high-pressure driving gas side. One can measure this free-surface velocity by suitable contact pins, etc. It turns out that the free-surface velocity is almost exactly twice U_p.

Knowing U_s, U_p, and ϱ_0, one can find p, V, and E by Eqs. (2.3)–(2.5). Each experiment yields one value for p, V, and E. A number of experiments at different pressures yields a relationship for E as a function of p and V, and, from these data and suitable assumptions which are too lengthy to describe here, one may calculate temperature and estimate the p–V–T equation of state of the material. One notes that, in general, the final T is higher for larger $V_0 - V$.

The material at the shock front is first rapidly compressed up to the yield point of the material. Above this yield point, whose value depends on the pressure and rate of loading, the material shears, relaxes, and behaves as a fluid, and the dynamic pressure above this level approaches an ordinary hydrostatic pressure (Fowles, 1961). Careful measurement of U_p or free-surface velocities (Allen and McCrary, 1953; Jones *et al.*, 1962) have shed more light on this relaxation process; it can be described in terms of activation energies (McQueen, 1964). Obviously, weak shocks in strong materials will not display the ideal simple behavior upon which Eqs. (2.3)–(2.6) are based; the energy dissipated by plastic flow will then be more important.

When the compressed material undergoes a phase change, at least one more wavefront will be produced in the material and appear at the free surface (Alder, 1963; McQueen, 1964). Usually, some time is required for the change to occur, and one's chances of observing its effects are greater with thicker slabs (longer travel paths) of material. Some phase changes, usually those involving extensive rearrangements of atoms, may be too sluggish to occur in the few microseconds available. For example, the meteoritic glass maskelynite is a product of low–temperature shock waves in lime-soda feldspars (Milton and DeCarli, 1963; Stoffler, 1966). There is, of course, no guarantee that the shocked material will proceed from one equilibrium phase to the next; unstable intermediate phases may form and persist.

The highest dynamic pressures are reached by accelerating a "driver" plate with, say, high explosive and allowing this plate to strike a specimen plate a few centimeters away. Then, one is accelerating solid with solid instead of solid with gas, and pressures comparable to those in the earth's core, 2000 kbar, may be reached.

Hamann (1966) has discussed some of the techniques available for producing extra-high shock pressures as well as measuring many electrical and optical effects in the shocked media. One might also remark here that extremely high, brief magnetic fields may be produced by cylindrically converging shocks in a conductor carrying an axial magnetic field. Keeler and Mitchell (1967) describe the electrical conductivities of some insulators under shock compression, and Doran and Linde (1966) review a number of electrical and crystal imperfection phenomena.

C. Measurement of Pressure and Temperature

The estimation of pressure in shock experiments follows most directly from the experimental data, provided that there are no phase changes, etc. to complicate the picture. Quartz transducers may be used for lower pressures; Fowles (1967) has described the behavior of *X*-, *Y*-, and

Z-cut quartz in shock compression. The densities of simple well-behaved metals such as Ag, Rh, Pd, or Ni as a function of pressure can be determined to 1% or better. The high density and small compression mean that the shock produces only a small temperature rise, and this reduces the error in determining isotherms from the shock data.

Turning now to static experiments, if one can measure the compression of such a well-behaved metal, e.g., by X-ray diffraction at high pressure, then one has a good estimate of the pressure. The accuracy is mainly limited by the accuracy of measuring lattice spacings. The shock data are better at higher pressures, i.e., upward from 100 kbar, but the interpolation to lower pressures is easy. Furthermore, quite accurate extrapolations from low pressure to about 100 kbar can be made by using elastic-constant data obtained in the 10-kbar range (Anderson, 1966).

The compression of NaCl has been used as a pressure reference, and various workers are coming to agree on the p–V curve (Decker, 1966; Weaver *et al.*, 1971; Perez-Albuerne and Drickamer, 1965; Christian, 1957; Altshuler *et al.*, 1960, 1961). An uncertainty of about 0.1% in the measurement of the lattice-spacing of NaCl corresponds to an uncertainty of about 2 kbar in the range 0–100 kbar, and the uncertainty about the position of the p–V curve adds about 2% more. There seems to be a phase change at about 300 kbar (Bassett *et al.*, 1968).

At pressures up to about 30–40 kbar, where some substances are still fluid and stresses are not severe, pressures may be estimated from force and area after allowing for friction. Carefully designed pistons can develop about 60 kbar (Kennedy *et al.*, 1967) and have been used to determine the transitions in reference substances such as Bi, Tl, Ba, etc. More complicated pistons have been used to 100 kb (Vereshchagin *et al.*, 1966).

The range 40–150 kbar currently presents the greatest difficulties; it is a little high for simple piston and cylinder, and a little low for shock measurements. Most work in this region is reported on the basis of calibration curves based upon transitions in reference substances. A listing is given in Table II. Ultimately, these pressures will be more definitely fixed; in the meantime, experimenters have a common basis for reference.

The determination of transition pressures in reference substances by shock-wave methods is difficult because of the short times available for the transition to occur and the effects of crystal orientation relative to a plane shock wave. The 25°C transition in Fe, for example, was long taken to be 131 kbar, but recent work indicates that the pressure may be some-

TABLE II

PRESSURE REFERENCE TRANSITIONS[a]

Substance	Transition	Pressure (kbar)
Hg	Freeze, 0°C	7.569
Bi	I–II	25.50
Tl	II–III	36.7 ± 0.3
Cs	II–III	42.5 ± 1
	III–IV	43.0 ± 1
Ba	I–II	55 ± 2
Bi	III–V	77 ± 3
Sn	Resistance	100 ± 6
Fe	α–ε	126 ± 6
Ba	Resistance	130 ± 10
Pb	Resistance	$1.2–1.6 \times 10^2$

[a] Adopted as temporary working standards at NBS Conf. on High Pressure, Gaithersburg, Maryland, October, 1968. All values at 25°C except as noted. Values above 100 kbar are more uncertain.

what lower, perhaps as low as 120 kbar, and the transition may be spread out over several kb (Mao *et al.*, 1967; Wong *et al.*, 1968).

It should be borne in mind that variations in loading and materials may produce variations in the true pressure inside an apparatus which has been calibrated in separate experiments. These effects are particularly pronounced at pressures over 100 kbar because here most materials are not so compressible and a small local change in volume corresponds to a large local change in pressure. Furthermore, the shear strengths of the usually solid media inevitably support pressure gradients.

Probably today most high-pressure workers would be pleased if they knew their pressures above 40–50 kbar to an accuracy of 2–5%, and an uncertainty of twice this is not unexpected. An excellent discussion of the general problem of pressure calibration is given by Decker *et al.* (1971).

The measurement of temperature at high pressure is straightforward up to about 700°K when the entire apparatus is heated or cooled in a bath or oven. Difficulties arise when internal furnaces are used. The problems of providing optical paths have not made blackbody radiation methods popular. Most temperatures are therefore estimated by thermocouples.

Most of the common thermocouple metals behave differently at high pressures than at 1 atm. In general, the thermocouple indicates a temperature which is somewhat too low. Figures 10 and 11 show the estimated corrections to be added for Pt–Pt 10% Rh and chromel–alumel couples according to Hanneman and Strong (1965, 1966). These are based on observations of diverse phenomena such as melting, diffusion, and phase transformations. The thermal (Johnson) electrical noise power

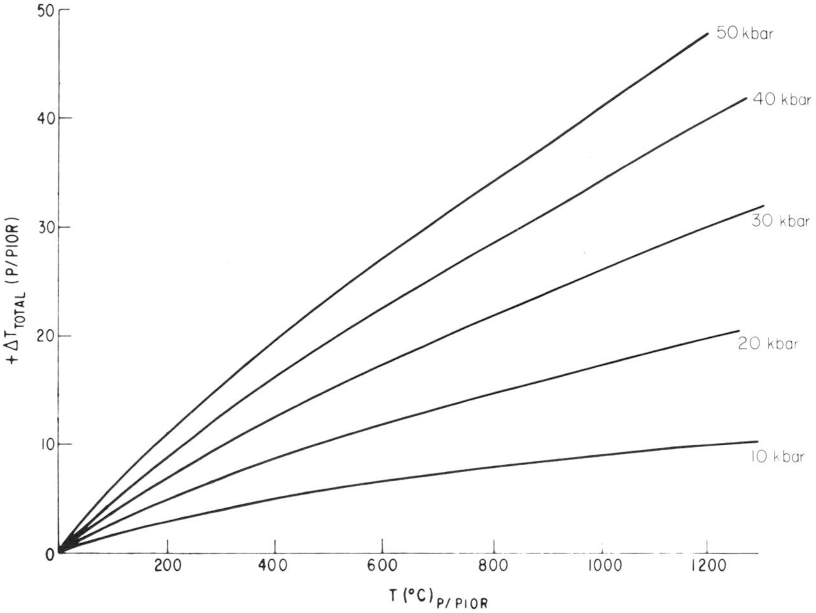

FIG. 10. Temperature correction to be added to Pt–Pt 10%Rh thermocouple on account of pressure.

developed in a resistor is proportional to absolute temperature but is independent of pressure, and has been used to estimate thermocouple corrections which are in general agreement with those of Fig. 10 and 11 (Wentorf, 1968).

One must always be wary of chemical attack of thermocouples. Furthermore, above 2000°C, most thermocouples become unreliable. Temperatures there may be estimated from heating-power inputs (Wentorf, 1959; Bundy, 1963). Resistance thermometry is not often used, because of spurious changes due to mechanical strain.

For general discussions of the problems of measuring pressure and

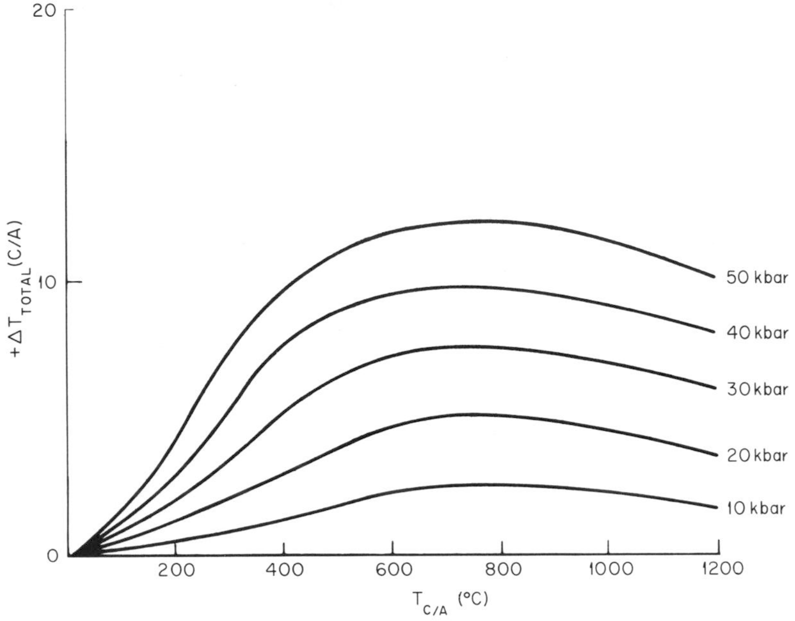

FIG. 11. Temperature correction to be added to chromel–alumel thermocouple on account of pressure.

temperature, one may consult Giardini and Lloyd (1963), Strong (1962), Cohen *et al.* (1966), Lees (1966), Lloyd (1971), and Bell and England (1967).

The measurement of temperature in shocked materials is difficult to do directly because thermocouples behave erratically and optical methods don't work too well with opaque substances (Hamann, 1966). Blackbody emission has been used to estimate temperatures in transparent materials and on the free surface of copper. Otherwise, temperatures in shocked materials are calculated by an iterative procedure based on reasonable assumptions for the behavior of the specific heat (Deal, 1962). However, slight differences in the calculations from one laboratory to the next may result in significant differences in the calculated p–V isotherms. As examples of the temperatures involved, polymethylmethacrylate shocked to about 2000 kbar comes up to about 10,000°K; water reaches about 1050°K at 130 kbar; tin starting at 300°K arrives at about 700°K at 200 kbar. The initial temperatures of material to be shocked may be changed, within limits set by thermal decomposition and cryogenic capabilities.

III. Observations

A. Compressibility

1. *Gases and Liquids*

The compressibility of gases and liquids carries some theoretical interest. The subject is experimentally somewhat limited by the freezing of most gases at 30 kbar or less at ordinary temperatures, and at high temperatures, the pressures are limited by the strengths of available materials. High-pressure hydrogen weakens most strong metals (Bridgman, 1964a,b). Hamann (1957) has discussed the compression of gases and liquids; Bridgman's data indicate that most liquids shrink by 20–30% by 12 kbar; those with hydrogen bonding shrink less. Other excellent summaries are those of Vodar and Saurel (1963) and Steele and Webb (1963). Tsiklis (1968) gives further information on the compression of technically important gases and liquids.

2. *Solids*

The compressibilities of solids may be measured by several methods. The shock-wave method, discussed earlier, is suitable for very high pressures and moderate temperatures; in this way, the p–V behavior of most of the likely materials in the earth has been estimated to 1000–2000 kbar, and data on many metals and alloys are available.

At somewhat lower pressures, X-ray diffraction techniques have provided much information on both the compression and the crystal structure of the material (Jamieson and Lawson, 1962; Jamieson and Olinger, 1968; Mao *et al.*, 1967; Perez-Albuerne *et al.*, 1964; Perez-Albuerne and Drickamer, 1965; Lynch and Drickamer, 1966; Barnett *et al.*, 1966; McWhan, 1967). Another interesting technique employs the change in inductance of a coil wound closely around the specimen inside the pressure vessel. The sample is buried in a soft material such as AgCl to minimize distortions (Giardini *et al.*, 1964; Giardini and Samara, 1965).

As a guide, one might note that most solids shrink about 10–40% upon being compressed to 100 kb (Bridgman, 1964 a, b; Hamann, 1957), some more, some less, e.g., Fe 6%, K 50%, NaCl 21%.

When the pressures are low enough that simple pistons suffice, one may measure compression by the piston displacement, after making suitable corrections for cylinder dilation, etc. Bridgman (1964a,b) used this method with great success and it has been extended to cryogenic temperatures as well (Stewart, 1962, 1965; Bernardes and Swenson, 1963).

Many glasses are found to be denser, e.g., 10%, after exposure to high pressures than they were before. MacKenzie (1963) has studied this phenomenon in some detail. The shear which the specimen suffers affects its final density; temperature and time also play a role.

On the other hand, alpha quartz may persist at 25°C to pressures of 150 kbar (McWhan, 1967).

3. *Theories and Extrapolation Methods*

For simple ionic solids, the main factor affecting the estimation of the compressibility is the representation of the repulsive forces—these are not quite a simple power law and seem softer than an exponential (Tosi and Arai, 1966); in metals, the electronic terms predominate, but are not easy to estimate (Austin, 1966; Stokes, 1966; Swenson, 1966).

Extensive energy-band calculations have been performed for a large number of important semiconductor materials (Herman *et al.*, 1968). For more complicated molecules or atomic bonding systems, very few calculations from first principles have been made. At sufficiently high pressures for any solid, the electrons would be shared among all the atoms according to the Thomas–Fermi metallic model (March, 1955; Ross and Alder, 1967; Knopoff, 1963).

It turns out that accurate extrapolations of compressibility and elastic constants can be made to high pressures if one has accurate knowledge of the elastic constants at moderate pressures (Anderson, 1966). Other interesting properties of solids can be deduced from a study of vibration spectra and anharmonicity (Plendl, 1961; Plendl and Gielisse, 1962). A detailed theoretical calculation for the compressibility of sodium metal has been given by Pastine (1968). Two parameters were based on experimental data to 22 kbar; the predictions are probably accurate to a few per cent at 100 kbar, and the density of sodium might be used as a pressure indicator.

B. Transport Properties

1. *Viscosity and Plastic Flow*

As recommended earlier, the viscosities of most liquids rise sharply with pressure (Hamann, 1957; Bridgman 1964a,b), and if the liquid does not actually crystallize at a high pressure, owing to the slowness of organizing large molecules, it certainly becomes essentially solid. In the same way, many elastic or plastic materials of high molecular weight become very stiff and brittle at high pressures.

If one takes a simple rate-theory view of viscosity, then the local volume increase involved in the slipping of one molecule past another results in a pV^* term which is added to the ordinary activation energy, and which becomes appreciable at high pressures. Naturally the larger the molecule, the larger is V^*.

High pressures do not seem to affect the viscosities or shear strengths of most metals to a great extent, but what is noteworthy is that most metals at high pressures become extremely plastic and can survive enormous deformations without breaking. This property is particularly valuable when one wishes to shape normally brittle metals such as Be, or when extremely large changes in shape are required of an ordinary metal (Pugh and Ashcroft, 1963; Bridgman, 1964b; Beresnev *et al.*, 1963). Evidently, the flaws which would lead to fracture at low pressures are healed up or never grow when the ambient pressure becomes comparable to the shear strength.

One may also recall the fluidlike state produced by intense shock waves—this has been put to use for making complicated shapes in thin sheets without tearing, or welding dissimilar metals (Shewmon and Zackay, 1961; Kiker, 1966).

Bobrovsky (1966) and Martynov *et al.* (1967) have given reviews of the field of high-pressure metal flow and its practical applications.

2. *Diffusion*

Diffusion in liquids is generally studied at pressures below those included in the scope of this chapter. However, the subject carries great interest, particularly from the points of view of geology and crystal growth. Hydrothermal syntheses of many minerals have elucidated many geological puzzles, and quartz and other technically important crystals are regularly synthesized in large sizes from supercritical water solutions (e.g., see Gilman (1963)).

Diffusion at high pressures in solids has been studied in some detail because it helps shed light on the atomic processes involved (Lazarus and Nachtrieb, 1963; Goldstein *et al.*, 1965; Hanneman *et al.*, 1965). In most solids, the local dilation necessary for the diffusion jump makes the process sensitive to pressure, as with viscosity; i.e., one may write

$$D = A \exp[(-E^* - pV^*)/RT], \tag{3.1}$$

where D is the diffusion coefficient, A is some sort of frequency factor which usually has only a small temperature and pressure dependence, E^* is an activation energy at negligible pressure, p is pressure, and V^* is an activation volume, the local dilation.

Equation (3.1) gives a good approximation to the observations for most systems. For many metals, V^* is typically about half the atomic volume. If the diffusion proceeds interstitially, then the local dilation is small and V^* is found to be nearly zero. If the diffusion proceeds by vacancy motion, one may regard V^* as the sum of two terms, the volume for vacancy formation plus the volume for vacancy motion (Huebner and Homan, 1963). In ionic solids, diffusion can be measured by ionic conductivity. Impurities play a role here, in that a positive ion vacancy in, say, NaCl, is associated with a compensating divalent ion; this association depends on both pressure and temperature, and hence so does V^* (Abey and Tomizuka, 1966).

Ordinarily, one considers diffusion as being driven by a concentration or temperature gradient. But sometimes a pressure gradient will do, as in AgI (Schock and Katz, 1967). Between diamond anvils at 20–40 kbar, one can observe I_2 collecting in the high-pressure central region and metallic Ag (reduced by light) in the low-pressure edge regions. The AgI decomposes and the more compressible iodine moves to the higher-pressure region.

3. *Thermal and Electrical Conductivity*

Strong (1962) and Cohen *et al.* (1966) have written good discussions of high-temperature phenomena in high-pressure apparatus. They and Fujisawa *et al.* (1968) have given some values for thermal conductivity at high pressures. Usually, the effect of pressure is not great, pressure tending to increase conductivity. When a phase change is involved, the change may often be detected by a differential method, comparing the substance to a nearby standard (Claussen, 1963).

Electronic electrical conduction is usually not a simple affair and depends on carrier density and scattering, crystal structure and perfection, etc. Its use as a detector of phase transformations is popular, but theoretical attempts to explain what is going on are difficult (Paul, 1963;

Falicov, 1965; Schirber, 1965; Landwehr, 1965; Goree and Scott, 1966). When phase changes are not involved, the stiffening of the lattice usually reduces scattering somewhat. Balchan and Drickamer (1961a, 1962) have examined the behavior of many metals to pressures of about 500 kbar.

High-pressure studies on several important semiconducting materials have helped to elucidate their band structures, e.g.: silicon and germanium (Paul and Warschauer, 1963; Fritzsche, 1965), GaAs (Lees *et al.*, 1967), GaSb (Kosicki *et al.*, 1968). Jayaraman *et al.* (1967b) have looked at the effect of pressure on the band-gap and *p–n* junction characteristics in Si and GaAs. Most of these studies are at pressures below about 60 kb, so that there is little chance of phase changes to denser, metallic forms. Samara and Drickamer (1962a,b) have examined the collapse of many semiconductors with the open, zinc blende type of structure to denser forms. For example, Ge at about 200 kb has the white tin structure and a resistivity about that of aluminum.

Some of the poorer insulators also yield to compression. Selenium and iodine, when compressed to a few hundred kilobars, first become semiconducting, but the activation energy for conduction falls steadily with increasing pressure and finally becomes zero; i.e., the substance becomes metallic. The pressure at which this occurs is about the same pressure at which the band gap would be expected to vanish on the basis of extrapolated optical measurements (Drickamer, 1963) and is about 230 kbar for iodine and 130 kbar for Se. Several other examples are discussed by Samara (1967).

Organic semiconductors also show interesting changes in resistivity under pressure (Batley and Lyons, 1966). In the range to 60 kbar, for example, copper phthalocyanine shows two new crystalline phases and corresponding discontinuities in activation energy. The mobility and sign of the carriers also appear to change (Vaisnys and Kirk, 1966). Drickamer (1967) and his colleagues have exposed a number of large organic molecules to pressures in the range 200–400 kbar. At 80°K, the resistivity decreases reversibly with pressure, but at 300°K, several compounds show irreversible increases in resistivity which are associated with the formation of new intermolecular bonds.

The phenomena in supercritical water solutions have been explored in great detail by Franck and his co-workers (Maier and Franck, 1966). Pure water at 98 kbar and 1000°C becomes a fairly good conductor (Holzapfel and Franck, 1966), in accordance with the decrease in volume associated with partial ionization.

Hamann (1966) has reported many findings on changes in electrical conductivity, dielectric constant, and ionization in strongly shocked aqueous solutions and other substances.

Usually, the superconducting state is slightly less dense than the normal state (Koehnlein, 1967). However, many of the new metallic phases produced by pressure turn out to be superconductors, e.g., Si, Ge (Buckel and Wittig, 1965), InSb (Bommel *et al.*, 1963), and especially phosphorous (Wittig and Matthias, 1968). The existence of these superconductors has assisted the understanding of superconductivity as a whole.

C. Phase Transformations

There exist thousands of phase transformations which may be produced in elements or compounds by the application of pressure. A goodly number of these have been catalogued and discussed (Klement and Jayaraman, 1967). Space permits only a few examples and comments here.

In general, of course, pressure favors the stability of the denser phase. If we take pressure and temperature as the running variables, then the slope of the equilibrium line between two phases depends on the changes in volume and entropy. It is not necessary that the denser phases have the lower entropy, as we know from water–ice. Often, directional bonding between atoms or molecules is involved, and the number of possible arrangements of the energy of the system changes in ways which are difficult to predict. Thus, we find phase boundaries running every which way for diverse reasons. An experimental complication is the difficulty of measuring pressure and temperature accurately, for reasons mentioned earlier.

Phase transformations may be studied by many different techniques, depending upon the information being sought. The most common methods use changes in volume, resistivity, dielectric constant, permeability, thermal conductivity, structure (as determined by X-rays or neutron diffraction), thermoelectric power, magnetic resonance, sound velocity, etc. Particularly useful are spectroscopic methods because they can tell one about the local environments of atoms. Whatley and Van-Valkenburg (1966) have given us a good summary of high-pressure optical methods. Whalley (1966, 1967) has reviewed techniques for the results obtained by studying refractive and dielectric properties. X-ray and Mössbauer studies have been mentioned earlier in the discussion of apparatus. Magnetic resonance work yields information about local magnetic and electric fields and about diffusional or relaxation processes (Benedek, 1963a,b). Far-infrared studies tell one about metal–halogen bonds (Postmus *et al.*, 1967).

1. *Liquid–Solid Phase Changes*

The melting temperatures for most solids increase with pressure, but naturally there are exceptions. Rowlinson (1967) has discussed several current theories of melting and finds them inadequate. Most of them are based on Grüneisen's equation of state and the Lindemann concept, according to which melting occurs when crystal vibrations exceed a critical amplitude. One should properly consider both phases in any treatment of their equilibria, and the state of the liquid should not be neglected.

Even more difficult is the explanation of apparent maxima in melting temperature for various elements such as carbon (graphite), cesium, rubidium, barium, etc. for pressures in the 10–100-kbar range (Klement and Jayaraman, 1967). Cesium has recently been studied intensively (Jayaraman *et al.*, 1967c). The solid shows three phase transformations, a dimpled melting point maximum at about 24 kbar corresponding to the I–II phase change, bcc–fcc, and a melting-point minimum at about 42.5 kbar corresponding to the II–III change in the solid. Phase III is believed to be a collapsed form of II, an electron leaving 6s and entering 5d, with an accompanying increase in resistivity. Phase IV forms at about 44 kbar; its structure is not yet known for certain. The melting temperatures of III and IV rise with pressure. The melting-point maxima can be explained on the basis of the liquid density increasing with pressure, probably as the result of the 6s-5d electronic collapse; this idea is supported by measurements of the electrical resistivity of the liquid, which shows a marked (sixfold) increase with pressure from 10 to 50 kbar.

So far, there is no evidence for a solid–liquid critical point in any substance; at extremely high pressures, it is expected that the difference between solid and liquid will gradually fade away as the increasing compression permits a closer approach to the Thomas–Fermi state of ions embedded in a sea of free electrons.

Usually, one can extrapolate melting-point curves to moderate pressures, e.g., 40 kbar, by using volume and entropy changes obtained at 1 atm, assuming that no phase changes occur in the solid. For examples with alkali halides, see Pistorius (1967). One must take experimental care that the system is pure; for example, small amounts of water may have large effects on "melting" of silicates (Boettcher and Wyllie, 1967).

2. *Solid–Solid Phase Changes*

There seem to exist more kinds of solids than liquids, and the phase-diagram literature is correspondingly richer. However, one must be cautious in interpreting solid phase diagrams because it is often difficult to decide whether thermodynamic equilibrium has been reached. This is particularly true for refractory materials, cold materials, strongly anisotropic materials, or materials with high shear strengths. One may wait a long time and still not have the equilibrium phase ("A diamond is forever"). This phenomenon can be helpful if one wishes to examine or use at low pressures materials prepared at high pressures, but the phase which is recovered at low pressure is not necessarily the same phase which existed at the high pressure, as work with Ge has shown.

Or one may arrive at phase A by one pressure–temperature path and at phase B by another path. One or neither phase may be the thermodynamically stable one. Which phase forms may depend on structural factors which kinetically favor certain paths from certain precursors. Electromotive-force measurements would offer a method of deciding about the relative stability of different phases, but suitable electrolytes or electrodes are not always available, and little work has been done in this area so far.

Tosi and Arai (1966) have discussed the stability of various types of solids under pressure, from a theoretical standpoint. A general criterion is that the crystal must be stable against infinitesimal shear stresses, i.e., the elastic shearing constants must be greater than zero. It appears that simple theories are usually inadequate, and many-body forces must be taken into account, even for simple structures as alkali halides.

From the experimental point of view, the fact that solids are involved raises several difficulties. Usually, the phase transformation does not appear to be perfectly sharp, but instead seems to be spread out over a range of pressure. Part of the spread may lie in hysteresis or lack of followthrough in the apparatus itself; part of the spread often lies in the substance under study. Nuclei of the new phase usually must form and grow; the movements of the interphase boundaries may be hampered by impurities, grain boundaries, etc.; the volume changes involved may shield unchanged portions from the external pressure. The addition of shearing strain often accelerates solid–solid phase changes (Dachille and

Roy, 1960). The overall temperature rise due to the shear is negligible, but rearrangements are facilitated at the many sites which possess unusually high local strain energy.

As an example of the subtle problems which may be encountered in determining solid–solid phase boundaries, one may read the papers of Jamieson and Olinger (1968) on the transition of TiO_2 to the alpha lead dioxide form, or of Miller *et al.* (1966) on CdS and MnS, or of Evans (1967) on selective nucleation of the high-pressure ices.

Solid–solid phase transitions may also be greatly accelerated by using some sort of solvent or liquid catalyst to provide a fluid intermediate phase. Small amounts of water are extremely effective with most oxide and silicate systems at moderate temperatures (600–1200°C). Sometimes, the fluid phase has special characteristics that influence which solid phase crystallizes from it; for example, some carbon solvents known as catalysts (such as molten Fe or Ni) produce diamond or graphite more or less according to which phase is thermodynamically stable; other solvents for carbon (such as AgCl) strongly favor (metastable) graphite even in the presence of diamond seed crystals (Wentorf, 1966; Strong and Hanneman, 1967).

Although the available reaction times in shocked materials are brief, rarely over 10 μsec, this corresponds to several million crystalline vibrations, and many instances of phase transformations are known in shocked material. The shear near the shock front probably assists some transformations. One may cite the transformation of graphite to diamond (DeCarli and Jamieson, 1961), and of BN to denser forms (DeCarli, 1967; Coleburn and Farbes, 1968). A hexagonal kind of diamond may also form by shock (Hanneman *et al.*, 1967). Especially well studied is the transformation in iron at around 130 kb; the change to a nonferromagnetic form produces inductive effects (Wong *et al.*, 1968). Demagnetization of yttrium-iron garnet may be studied similarly (Shaner and Royce, 1968).

A general rule is that the phases which are stable at higher pressures feature higher coördination numbers, i.e., each atom or molecule has more nearest neighbors. For example, many zinc blende structures (fourfold coordination) such as ZnO, Ge, InSb, CdS, etc. go over to sixfold coordinated structures. Further examples of great geological interest are the changes from quartz to stishovite (rutile structure), and olivines to spinels (Ringwood and Major, 1967). Coordination-number changes occur in many other structures, including metals and salts. Naturally, there is an increasing tendency to approach a metallic state which may actually be reached with heavier atoms.

Another general rule is that crystals composed of heavier atoms may serve as prototypes for the high-pressure phases of lighter atoms. For example, the germanates often serve as high-pressure models for the silicates (Reid *et al.*, 1967), and white tin serves as a model for high-pressure Ge and Si. This trend has assisted the search for an identification of high-pressure phases. Evidently, there is no overall trend toward simpler structures at high pressures.

3. *Magnetic Phase Transformations*

Pressure may affect the cooperative alignment of spins in a solid in a continuous manner as the interatomic distances decrease, or in a discontinuous manner when a phase change occurs. The change in magnetic state can be followed with coils, inductors, etc., or by the Mössbauer effect, discussed later. Usually, the changes in nonmetals are better understood than in metals.

Leger *et al.* (1967) have reported on the changes of Curie temperature in Fe, Co, Ni, and six Fe–Ni alloys, up to about 60 kb. Iron and Co are hardly affected; the Curie temperature for Ni rises about 0.32°K $kbar^{-1}$; the largest changes are shown by 30–30% Ni alloys, about -3°K $kbar^{-1}$. Graham (1968) has measured the decrease in magnetization during application of shock pressures of up to 300 kbar to invar and a 3% Si–Fe alloy. Invar displays very little magnetism at 120 kbar; the Si alloy behavior is complicated by a phase change (probably related to the formation of the nonferromagnetic hcp form of pure iron at high pressures) spread out from 150 to 225 kbar. The Fe–Rh alloys are especially interesting, and studies up to 25 kbar and 400°C (Wayne, 1968) have helped the progress toward understanding these alloys.

Even more complicated magnetic behavior is shown by the rare earths Gd, Tb, Dy, and Ho under pressure (McWhan and Stevens, 1967). Their Curie temperatures fall with pressure, -0.5 to -1.5°K $kbar^{-1}$, but at higher pressures, 35–70 kbar, depending on the metal, two distinct ordering temperatures appear; they are associated with the transformation to a samarium type of structure. Kafalas *et al.* (1968) report on the magnetic transitions in spinels.

D. Chemical Reactions at High Pressures

Usually, the larger the reacting molecule, the lower are the pressure and temperature at which a reaction can be studied, on account of the large increases in melting point, viscosity, or energies of the activated states which very high pressures produce. Thus, the literature on high-pressure chemical reactions shrinks rapidly as the pressures rise above 20 kbar or so. Pressures of even 100 kbar add compression energies which are small compared with chemical bond energies, so the main effects of high pressure are in displacing equilibria or affecting activated states and not in extensive alterations of molecular structure. Pressures of 10^3–10^5 kbar may be high enough for some molecules to lose their chemical identity; the normally localized bonding electrons would be free to roam through the entire mass of reacting material, which would resemble a high-density plasma.

Weale (1967) has reviewed and summarized the basic principles, techniques, and results of studies on high-pressure chemical reactions in the range 1–15 kbar. This covers a lot of ground, most of which is outside the scope of this chapter. Of great interest so far have been cyclization and polymerization reactions. These proceed with an overall increase in density, and in the case of radical-addition polymerizations (e.g., polystyrene, methyl methacrylate, vinyl acetate), the activated state is also denser than the starting materials, so that pressure increases the rate. Pressure also affects the branching of the polymer chains, e.g., polyethylene prepared at 3.5 kbar has less branching than that prepared at 1 kbar.

High-pressure studies have also been of great value in deciding which of several possible reaction mechanisms predominate. It is possible nowadays to estimate the changes in volume demanded by various paths and select the most likely ones from the observed effects of pressures to about 15 kbar on reaction rates. (Weale, 1967; Whalley, 1962).

Turning now to pressures above 20 kbar, one finds a review of high-pressure inorganic chemistry by Hall (1966), which also includes a discussion of techniques. In one of the rare very-high-pressure organic-chemistry papers, Bengelsdorf (1958) reported on the trimerization of aromatic nitriles, at pressures in the range of 30–50 kbar to triazines. Soulen and Silverman (1963) studied the polymerization of $(PNCl_2)_3$ at pressures up to 70 kbar; higher pressures favored formation of a more thermally stable polymer. The polymerization of CS_2 was first noted by Bridgman, studied by several workers, and fairly well completed by Butcher *et al.* (1963). Pressures above about 30 kbar are needed at temperatures of 200–250°C; the polymer appears to be linear. It has also been reported that COS polymerizes at high pressures, but so far CO_2 has not yielded.

At pressures in the range 95–150 kbar, many organic compounds pyrolyze to carbon—eventually, as diamond at very high temperatures (3000°C) (Wentorf, 1965). However, at intermediate temperatures (1500°C), some variations occur. Paraffins or polyethylene seem to lose hydrogen and collapse to soft, white solids whose only crystalline content is diamond; naphthalene or anthracene form graphite instead of diamond, but the addition of some diamond-forming hydrocarbon such as fluorene produces diamond. Evidently, molecular structure plays a role at these high pressures. The results of Drickamer (1967) and his colleagues with large ring compounds also illustrate structural effects. A new form of carbon from the Ries crater which is related to graphite has been described by El Goresey and Donnay (1968).

Dense forms of BN can be prepared by catalyst-solvent methods (Wentorf, 1961), by higher pressures in the range 130 kb (Bundy and Wentorf, 1963), or by shock compression (DeCarli, 1967).

Often, extremely high partial pressures of volatiles such as oxygen, water, sulfur, etc. can be maintained in high-pressure apparatus, and in this way, interesting compounds may be prepared at high temperatures. Thus, one has the synthesis of CrO_2 at 60 kbar and 1500°C (DeVries, 1967; Fukunaga *et al.*, 1968). De Vries and Roth (1968) describe the formation of $PbCrO_3$ under similar conditions. Silverman (1964) has prepared various magnesium and calcium silicides, bismuth sulfides, etc.; Munson (1968) describes IrS_2 and $NiAs_2$ in the pyrite structure. Rooymans (1967) and Albers and Rooymans (1965) describe the syntheses of many spinels and chalcogenides. The large molecules which may exist in S or Te can combine with each other to form helices at high pressures (Geller, 1968). Special magnesium borides may be prepared (Filonenko *et al.*, 1968). Certain titanate spinels decompose to ilmenite structures (Akimoto and Syono, 1967). High pressures may favor stoichiometry by making lattice deficiencies more costly in free energy, as in the preparation of stoichiometric FeO by Katsura *et al.* (1967).

E. Miscellaneous Observations

1. *Biology*

Pressures of above about 5 kbar are usually lethal to living organisms, partly because of the freezing of the aqueous medium and destruction of cells and partly because of the denaturing effect of pressure on proteins. However, below 2 kbar, pressure affects cell morphology (Landau,

1961; Libby *et al.*, 1965), protein and nucleic acid synthesis (Pollard and Weller, 1966), mitosis (Marsland, 1965), etc. Reviews in this field have been given by Johnson *et al.* (1954) and Heden (1964).

2. *Mössbauer Studies*

Mössbauer techniques for high pressure were mentioned in Section II,A,3a (DeBrunner *et al.*, 1966). Drickamer and Fung (1968) have summarized much of the work at Illinois at pressures up to about 220 kbar. To put it briefly, the Mössbauer spectra are affected by the s-electron density, the magnetic field, and the electric field gradient, all at the nucleus. Pressure affects the s-electron density directly and also indirectly by affecting the d electrons. These effects are noted for iron and dilute alloys by Pipkorn *et al.* (1964) and Edge *et al.* (1965). In ionic compounds, both the s-electron densities and the electric field gradient (quadrupole splitting) can be altered (Champion *et al.*, 1967; Vaughan and Drickamer, 1967a). An unexpected effect is the reversible reduction of ferric to ferrous ion at high pressures. The behavior of iron in phosphate and silicate glasses depends on whether it occupies tetrahedral or octahedral sites. The behavior of the covalent ferrocene is described by Vaughan and Drickamer (1967b).

A magnetic field at the nucleus splits the emission line into four, and magnetic behavior can thus be followed. For example, the hcp form of iron above 130 kbar is found to be nonferromagnetic and the field in Ni rises until about 60 kbar; then it falls. Antiferromagnetic CoO has been studied by Coston *et al.* (1966). Obviously, Mössbauer effects at high pressure will aid the understanding of solids.

3. *Geology*

Geological studies form a natural field for high-pressure research. Gravity and rock densities tell us that the pressure in the crust and mantle increases at about 0.47 kbar/km depth. At the boundary between mantle and molten core (2900 km), the pressure is about 1370 kbar, slightly beyond the reach of current static apparatus, but well within range of shock pressures. Much of our information about the interior of the earth is based upon seismic studies of earthquake wave velocities. There are at least two interesting regions: the first, about 15 km under the oceans and 60 km under the continents, is called the Mohorovicic discontinuity; the wave velocity becomes abruptly higher just below it.

The second region is broader and displays a further increase in seismic velocity at depths of 400–500 km.

The deepest wells (8–9 km) and heat-flow measurements indicate that the temperature increases with depth; the temperature at the molten core must be at least about 2000°C, but the uncertainties of melting theories and the ignorance of the composition at this depth inhibit accurate estimates. At shallower depths, the melting curves of silicates (Boyd, 1962) set some limits.

The density distributions in the earth allowed by its observed mass and moment of inertia imply that the core has a density of about 9.6 at its exterior and 17.2 at its center and so is probably metallic, probably mainly iron. The mantle densities range between about 3.3 near the surface to 5.7 next to the core and suggest a rocky material. However, the chemical composition of the rock is fixed only within broad limits which are currently set by the ingenuity of men pressed to account for the natural observations.

Many interlocking alternate hypotheses have been proposed to explain the observed geological data, and they could not possibly be discussed here. The interested reader is referred to the excellent reviews by Newton (1966), Clark and Ringwood (1964), Katz (1966), Ringwood (1966), Birch (1964), Boyd (1967), and McQueen *et al.* (1964).

High-pressure laboratory work seems to have two main limitations compared with the earth itself. The first is the problem of time; men do not have millions of years and must extrapolate the relatively fast reactions of the laboratory. However, it may be that new mechanisms operate at lower temperatures and longer times. The second limitation is the relatively small ratio of volume to surface area in laboratory samples. Often, the encapsulating material reacts with the charge or allows leakage of gases. Hydrogen is especially difficult to contain; it plays some role in determining oxidation states and is a major constituent of water, H_2S, CH_4, etc.

In spite of these limitations, much has been learned about the behavior of rocks in the earth. One of the most fruitful fields has been mineral synthesis. Minerals such as diamond, pyrope garnet, jadeite, etc. are true high-pressure visitors to our low-pressure world of the surface; their examination (Meyer, 1967; Soga, 1967) and synthesis tell us something about the depths to, say, 300 km. Often, a mineral type has a range of stability depending on its composition: e.g., substitution of iron or nickel for magnesium lowers the pressure required for stability. Possibilities for fractionation exist (O'Hara and Yoder, 1967).

Even though thousands of natural minerals have been identified and classified, some new types were created in the laboratory, and men began to wonder how or whether they fitted into natural schemes. Silica can exist in two new forms: coesite, with a density of about 3.0 (Coes, 1962), and stishovite, with the rutile (sixfold coordination) structure and density 4.35 (Stishov and Popova, 1961; Holm *et al.*, 1967). Stishovite, which forms at about 110 kbar, might have been predicted by the analogy with germania.

A similar analogy between Mg_2GeO_4 spinel and Fe_2SiO_4 fayalite produced the spinel forms of Fe_2SiO_4, Co_2SiO_4, and Ni_2SiO_4 at pressures in the 60-kbar range (Ringwood, 1963). Akimoto *et al.* (1967) studied the melting curves of both olivine and spinel forms of Fe_2SiO_4. Further work at pressures up to about 150 kbar has showered us with various new forms which could be transformation products of olivines and pyroxenes (Ringwood and Major, 1966), feldspars (Ringwood *et al.*, 1967), nepheline (Reid and Ringwood, 1968), and ilmenites (Ringwood and Major, 1967). None of these new materials has been found in the field, but they offer possibilities for explaining the deeper seismic data because they are stiffer than their low-pressure forms.

Coesite and stishovite have been found in the field as minute inclusions in silica-rich rocks (sandstone, granite, etc.) which have been compressed by the impact of meteorites at the craters in Arizona and Ries, Bavaria. The diamonds found in some meteorites are also believed to have been produced by shock compression, and similar diamonds have been made in the laboratory by shock (De Carli and Jamieson, 1961). Other interesting materials have been found in these meteorite craters (French, 1966). Probably, the moon bears some exotic substances.

Sclar *et al.* (1967) have reported on the binding of water as hydroxyl at high pressures in certain magnesium silicates. The geological implications of this discovery are interesting for earthquakes and some types of vulcanism. Water often has pronounced effects on melting phenomena (Boettcher and Wyllie, 1967).

Recent shock-wave work on typical earth mineral systems indicate that, at pressures above about 500 kbar, the deep earth's behavior may not be explained on the basis of simple mixtures of oxides in their close-packed forms such as MgO, Al_2O_3, stishovite silica, FeO, etc. (Wang, 1968).

High-pressure studies on the iron–nickel system (Kaufman *et al.*, 1961) show that pressures of 50 kbar permit the fcc phase to be stable at lower temperatures and increase the supercooling required to form bcc.

This strongly implies a low-pressure cooling stage for nickel–iron meteorites.

One can see from this brief account that, during the past few years, high-pressure research has opened up more possibilities for geological processes than could have been imagined ten years ago. However, we still do not understand the earth's magnetic field, which presumably depends on properties such as viscosity and electrical conductivity (Khitarov and Slutsky, 1967). And what about continental drift? Probably, further work will serve to narrow our range of choices and lead to better understanding of the past, present, and future behavior of this planet and others.

General References

The following books or articles represent reviews or collections of reports of work in various areas of high-pressure research and help a reader to obtain an introduction to the field as well as detailed views on particular areas:

Accurate characterization of the high pressure environment. *NBS Symp. Gaithersburg, Maryland, October 1968.* (E. Lloyd, ed.). In press (1971).

"Advances in High Pressure Research" (R. S. Bradley, ed.), Vol. 1 (1967); Vols. 2, 3 (1969). Academic Press, London and New York.

Benedek, G. B. (1963). "Magnetic Resonance at High Pressure." Wiley (Interscience), New York.

Bridgman, P. W. (1949). "The Physics of High Pressure." Bell, London.

"Collected Experimental Papers of P. W. Bridgman." Harvard Univ. Press, Cambridge, Massachusetts, 1964.

Drickamer, H. G. (1965). The effect of high pressure on the electronic structure of solids. *Solid State Phys.* **17**, 1–89.

Hamann, S. D. (1957). "Physico-Chemical Effects of Pressure." Butterworths, London.

"High Pressure Measurement" (A. A. Giardini and E. C. Lloyd, eds.). Butterworths, London and Washington, D. C., 1963.

"High Pressure Physics and Chemistry" (R. S. Bradley, ed.). Academic Press, New York, 1963.

Irreversible effects of high pressure and temperature on materials. *ASTM Special Techn. Publ.* No. 374, Philadelphia, Pennsylvania, 1965.

Klement, Jr., W., and Jayaraman, A. (1967). Phase relations and structures of solids at high pressures (Rev. article). *Progr. Solid State Chem.* **3**, 289–376.

"Metallurgy at High Pressures and High Temperatures" (K. A. Gschneider, Jr., M. T. Hepworth, and N. A. O. Parlee, eds.). Gordon and Breach, New York, AIME, 1964.

"Modern Very High Pressure Techniques" (R. H. Wentorf, Jr., ed.) Butterworths, London and Washington, D. C., 1962.

"Physics of High Pressures and the Condensed Phase" (A. Von Itterbeek, ed.). Wiley, New York, 1965.

"Physics of Solids at High Pressures" (C. T. Tomizuka and R. M. Emrick, eds.). Academic Press, New York, 1965.

Progr. Very High Pressure Res., Proc. Intern. Conf., Bolton Landing, Lake George, N. Y., 1960 (F. P. Bundy, W. R. Hibbard, Jr., and H. M. Strong, eds.). Wiley, New York, 1961.

Solid State Phys. **13**, 81–143 (1962).

"Solids Under Pressure" (W. Paul and D. Warschauer, ed.), McGraw-Hill, New York, 1963.

'The Physics and Chemistry of High Pressures." Soc. of Chem. Ind., London 1963.

TSIKLIS, D. S. (1968). *In* "Handbook of Techniques in High-pressure Research and Engineering" (A. Bobrowsky, ed.) translated by A. Peabody. Plenum, New York.

WEALE, K. E. (1967). "Chemical Reactions at High Pressures." Spon, London.

WENTORF, JR., R. H. (1967). Modern very high pressure research (Rev. article). *Brit. J. Appl. Phys.* **18**, 865–882.

WHALLEY, E. (1967). High pressures. *Ann. Rev. Phys. Chem.* **18**, 205–232.

ZEITLIN, A. (1964). "Annotated Bibliography on High Pressure Technology." Butterworths, London and Washington, D. C.

SPECIAL REFERENCES

ABEY, A. E., and TOMIZUKA, C. T. (1966). *J. Phys. Chem. Solids* **27**, 1149.

AKIMOTO, S., and SYONO, Y. (1967). *J. Chem. Phys.* **47**, 1813.

AKIMOTO, S., KOMADA, E., and KUSHINO, I. (1967). *J. Geophys. Res.* **72**, 679.

ALBERS, W., and ROOYMANS, C. J. M. (1965). *Solid State Commun.* **3**, 417.

ALDER, B. J. (1963). *In* "Solids Under Pressure" (W. Paul and D. Warschauer, eds.), Chapter 13. McGraw-Hill, New York.

ALLEN, W. A., and MCCRARY, C. L. (1953). *Rev. Sci. Instr.* **24**, 165.

ALTSHULER, L., KULESHOVA, L., and PAVLOVSKII, M. (1960). *Zh. Eksperim. i. Teor. Fiz.* **39**, 16.

ALTSHULER, L., KULESHOVA, L., and PAVLOVSKII, M. (1961). *J. Exp. Theor. Phys.* **12**, 10.

ANDERSON, O. L. (1966). *J. Phys. Chem. Solids* **27**, 547.

AUSTIN, I. G. (1966). *Contemp. Phys.* **7**, 173.

BALCHAN, A., and DRICKAMER, H. (1961a). *Rev. Sci. Instr.* **32**, 308.

BALCHAN, A., and DRICKAMER, H. (1961b). *J. Chem. Phys.* **34**, 1948.

BALCHAN, A., and DRICKAMER, H. (1962). *In* "Modern Very High Pressure Techniques" (R. H. Wentorf, ed.), Chapter 2. Butterworths, London and Washington, D. C.

BARNETT, J., and HALL, H. T. (1964). *Rev. Sci. Instr.* **35**, 175.

BARNETT, J. D., BEAN, V. E., and HALL, H. T. (1966). *J. Appl. Phys.* **37**, 875.

BASSETT, W. A., and TAKAHASHI, T. (1964). Specific volume measurements of crystalline solids at pressures up to 200 kilobars by X-ray diffraction. *Am. Soc. Mech. Eng., New York*, November 1964, Paper 64-WA/PT-24.

BASSETT, W. A., TAKAHASHI, T., and STOOK, P. W. (1967). *Rev. Sci. Instr.* **38**, 37.

BASSETT, W. A., TAKAHASHI, T., MAO, H. K., and WEAVER, J. S. (1968). *J. Appl. Phys.* **39**, 319.

BATLEY, M., and LYONS, L. E. (1966). *Australian J. Chem.* **19**, 345.

BELL, P. M., and ENGLAND, J. L. (1967). *In* "Researches in Geochemistry." (P. H. Abelson, ad.), Vol. 2, pp. 619–638. Wiley, New York.

BENEDEK, G. B. (1963a). "Magnetic Resonance at High Pressure." Wiley (Interscience), New York.

BENEDEK, G. B. (1963b). *In* "Solids Under Pressure" (W. Paul and D. Warschauer, eds.), Chapter 9. McGraw-Hill, New York.

BENGELSDORF, I. S. (1958). *J. Am. Chem. Soc.* **80**, 4442.

BERESNEV, B. I., VERESHCHAGIN, L. F., RYABININ, Y. N., and LIVSHITS, L. D. (1963). "Some Problems of Large Plastic Deformation of Metals at High Pressures" translated by V. M. Newton. Macmillan, New York.

BERNARDES, N., and SWENSON, C. A. (1963). *In* "Solids Under Pressure" (W. Paul and D. Warshauer, eds.), Chapter 5. McGraw-Hill, New York.

BIRCH, F. (1964). *J. Geophys. Res.* **69**, 4377.

BOBROVSKY, A. (1966). Metal under high pressure. *New Scientist*, p. 840.

BOETTCHER, A. L., and WYLLIE, P. J. (1967). *Nature* **216**, 572–573.

BOMMEL, H. E., DARNELL, A. J., LIBBY, W. F., TITTMAN, B. R., and YENCHA, A. J. (1963). *Science* **141**, 714.

BOYD, F. R. (1962). *In* "Modern Very High Pressure Techniques" (R. H. Wentorf, Jr., ed.). Butterworths, London.

BOYD, F. R. (1967). Petrological problems in high pressure research. *In* "Researches in Geochemistry" (P. H. Abelson, ed.), Vol. 2, pp. 593–618. Wiley, New York.

BRIDGMAN, P. W. (1952). "The Physics of High Pressure." Bell, London.

BRIDGMAN, P. W. (1964a). "Collected Experimental Papers." Harvard Univ. Press, Cambridge, Massachusetts.

BRIDGMAN, P. W. (1964b). "Studies in Large Plastic Flow and Fracture." Harvard Univ. Press, Cambridge, Massachusetts.

BUCKEL, W., and WITTIG, J. (1965). *Phys. Letters* **17**, 187.

BUNDY, F. P. (1963). *J. Chem. Phys.* **38**, 631.

BUNDY, F. P., and WENTORF, R. H. (1963). *J. Chem. Phys.* **38**, 1144.

BUTCHER, E. G., WESTON, J. A., and GEBBIE, H. A. (1963). *Nature* **199**, 756.

CLARK, S. P., and RINGWOOD, A. E. (1964). *Rev. Geophys.* **2**, 35–88.

CHAMPION, A. R., DRICKAMER, H. G., and VAUGHAN, R. W. (1967). *J. Chem. Phys.* **47**, 2583–2591; *Proc. Natl. Acad. Sci. U. S.* **58**, 876.

CHRISTIAN, R. (1957). Lawrence Radiation Lab. Rept. UCRL-4900, May. Univ. of California, Berkeley, California.

CLAUSSEN, W. F. (1963). *In* "High Pressure Measurement" (A. A. Giardini and E. C. Lloyd, eds.), pp. 125–151. Butterworths, London and Washington, D. C.

COES, L. (1962). *In* "Modern Very High Pressure Techniques" (R. H. Wentorf, ed.), chapter 7. Butterworths, London and Washington, D. C.

COHEN, L. H., Klement, W. and Kennedy, G. C. (1966). *J. Phys. Chem. Solids* **27**, 179–186.

COLEBURN, N. L., and FARBES, J. W. (1968). *J. Chem. Phys.* **48**, 555–559.

CONNELL, G. A. N. (1966). *Brit. J. Appl. Phys.* **17**, 399.

COSTON, C. J., INGALLS, R. L., and DRICKAMER, H. G. (1966). *Phys. Rev.* **145**, 409.

DACHILLE, R., and ROY, R. (1960). Influence of displacive shearing stresses on the kinetics of reconstructive transformations effected by pressure in the range 0-100,000 bars. *Proc. Intern. Symp. Reactivity Solids, 4th, Amsterdam 1960*, p. 502. Elsevier, Amsterdam.

DACHILLE, F., and ROY, R. (1962). *In* "Modern Very High Pressure Technique" (R. H. Wentorf, ed.) Chapter 9, Butterworth's, London and Washington, D. C.

DEAL, W. E. (1962). *In* "Modern Very High Pressure Techniques" (R. H. Wentorf, ed.), Chapter 11. Butterworths, London and Washington, D. C.

DEBRUNNER, P., VAUGHAN, R. W., CHAMPION, A. R., COHEN, J., MOYZIS, J., and DRICKAMER, H. G. (1966). *Rev. Sci. Instr.* **37**, 1310.

DECARLI, P. S. (1967). *Bull. Am. Phys. Soc.* [2], **12**, 1127.

DECARLI, P. S., and JAMIESON, J. C. (1961). *Science* **133**, 1821.

DECKER, D. (1966). *J. Appl. Phys.* **37**, 5012.

DECKER, D. L., BASSETT, W. A., MERRILL, L., HALL, H. T., and BARNETT, J. D. (1971). High pressure calibration - A critical review. *NBS Symp. High Pressure, Gaithersburg, Maryland, October 1968*, E. Lloyd, ed. In press.

DEVRIES, R. C. (1967). *Mater. Res. Bull.* **2**, 999–1008.

DEVRIES, R. C., and ROTH, W. L. (1968). *J. Am. Ceram. Soc.* **51**, 72.

DORAN, D. G., and LINDE, R. K. (1966). *Solid State Phys.* **19**, 229–290.

DRICKAMER, H. G. (1963). *Phys. Chem. High Pressures Papers Symp., London, 1962*, p. 122–127. Soc. Chem. Ind., London.

DRICKAMER, H. G. (1967). *Science* **156**, 1183.

DRICKAMER, H. G., and FUNG, S. C. (1968). High pressure Mössbauer Studies. Rept. C00-1198-560. Univ. of Illinois, Urbana, Illinois [see also *Advan. High Pressure Res.* **3**, 1–38].

DUVALL, G. E., and FOWLES, G. R. (1963). *In* "High Pressure Physics and Chemistry" (R. S. Bradley, ed.), Chapter 9. Academic Press, New York.

EDGE, C. K., INGALLS, R., DEBRUNNER, P., DRICKAMER, H. G., and FRAUENFELDER, H. P. (1965). *Phys. Rev.* **138**, 729.

EL GORESEY, A., and DONNAY, G. (1968). *Science* **161**, 363.

EVANS, L. F. (1967). *J. Appl. Phys.* **38**, 4930–4932.

FALICOV, L. M. (1965). *In* "Physics of Solids at High Pressures" (C. T. Tomizuka and R. M. Emrick, eds.), pp. 30–45. Academic Press, New York.

FILONENKO, N. E., INANOV, V. I., FEL'DYUN, L. I., SOKHAR, M. I., and VERESHCHAGIN, L. F. (1968). *Soviet Phys. "Doklady" (English Transl.)* **12**, 833.

FITCH, R., SLYKHOUSE, T., and DRICKAMER, H. (1957). *J. Opt. Soc. Am.* **47**, 1015.

FOWLES, G. R. (1961). *J. Appl. Phys.* **32**, 1475.

FOWLES, G. R. (1967). *J. Geophys. Res.* **72**, 5729–5742.

FRENCH, B. M. (1966). *Science* **153**, 903.

FRITZSCHE, H. (1965). *In* "Physics of Solids at High Pressures" (C. T. Tomizuka and R. M. Emrick, eds.), pp. 184–195. Academic Press, New York.

FUJISAWA, H., FUJII, N., MIZUTANI, H., KANAMORI, H., and AKIMOTO, S. A. (1968), Tech. Repts. Inst. of Solid State Phys., Ser. A. No. 298, February. Univ. of Tokyo, Tokyo.

FUKUNAGA, OSAMU, and SAITO, SHINROKU (1968). *J. Am. Ceram. Soc.* **51**, 362–363.

GELLER, S. (1968). *Bull. Am. Phys. Soc.* **13**, 444.

GIARDINI, A. A., and LLOYD, E. C., eds. (1963). "High Pressure Measurement." Butterworths, London and Washington, D. C.

GIARDINI, A., POINDEXTER, E., and SAMARA, G. (1964). *Rev. Sci. Instr.* **35**, 713.

GIARDINI, A., and SAMARA, G. A. (1965). *J. Phys. Chem. Solids* **26**, 1523.

GILMAN, J. J. ed. (1963). "The Art and Science of Growing Crystals." Wiley, New York.

GIRIFALCO, L. A. (1964). *In* "Metallurgy at High Pressures and High Temperatures" (K. A. Gschneider, M. T. Hepworth, and N. A. D. Parlee, eds.), pp. 260–279. Gordon and Breach, AIME, New York.

Goldstein, J. I., Ogilvie, R. E., and Hanneman, R. E. (1965). *Trans. AIME*, **233**, 812.
Goree, W. S., and Scott, T. A. (1966). *J. Phys. Chem. Solids* **27**, 835.
Graham, R. A. (1968). *J. Appl. Phys.* **39**, 437.
Hall, H. T. (1966). *Progr. Inorg. Chem.* **7**, 1–38.
Hall, H. T. (1958). *Rev. Sci. Instr.* **29**, 267.
Hall, H. T. (1960). *Rev. Sci. Instr.* **31**, 125–131.
Hamann, S. D. (1957). "Physico-Chemical Effects of Pressure." Butterworths, London.
Hamann, S. D. (1966). *Advan. High Pressure Res.* **2**, 85–137.
Hanneman, R. E., and Strong, H. M. (1965). *J. Appl. Phys.* **36**, 523.
Hanneman, R. E., and Strong, H. M. (1966). *J. Appl. Phys.* **37**, 612.
Hanneman, R. E., Ogilvie, R. E., and Gatos, H. C. (1965). *Trans. AIME* **233**, 685, 691.
Hanneman, R. E., Strong, H. M., and Bundy, F. P. (1967). *Science* **155**, 995.
Heden, C. G. (1964). *Bacteriol. Rev.* **28**, 14.
Herman, F., Kartum, R. L., Kuglin, C. D., VanDyke, J., and Skillman, S. (1968). *Methods Comput. Phys.* **8**, 1–34.
Holm, J. L., Kleppa, O. J., and Westrum, E. F. (1967). *Geochim. Cosmochim. Acta* **31**, 2289–2307.
Holzapfel, W., and Franck, E. U. (1966). *Ber. Bunsenges. Physik. Chem.* **70**, 1105.
Huebner, R. P., and Homan, C. G. (1963). *Phys. Rev.* **129**, 1162.
Jackson, J., and Waxmann, M. (1963). *In* "High Pressure Measurement," (A. Giardini and E. Lloyd, eds.), pp. 39–58. Butterworths, London and Washington, D. C.
Jamieson, J. C., and Lawson, A. W. (1962). *J. Appl. Phys.* **33**, 776.
Jamieson, J. C., and Olinger, B. (1968). *Science* **161**, 893–895.
Jayaraman, A., Hutson, A. R., McFee, J. H., Coriell, A. S., and Maines, R. G. (1967a). *Rev. Sci. Instr.* **38**, 44.
Jayaraman, A., Sikorski, M. E., Irvin, J. C., and Yates, G. H. (1967b). *J. Appl. Phys.* **38**, 4454.
Jayaraman, A., Newton, R. C., and McDonough, J. M. (1967c). *Phys. Rev.* **159**, 527–533.
Johnson, F. H., Eyring, H., and Polissar, M. J., eds. (1954). "Kinetic Basis of Molecular Biology." Wiley, New York.
Jones, O., Nielsen, F., and Benedick, W. B. (1962). *J. Appl. Phys.* **33**, 3224.
Kafalas, J. A., Dwight, K., Menyuk, N., and Goodenough, J. B. (1968). *Solid State Res.* **4**, 19.
Katsura, T., Iwasaki, B., Kimura, S., and Akimoto, S. (1967). *J. Chem. Phys.* **47**, 4559.
Katz, S. (1966). *Trans. Am. Geophys. Union* **47**, 1.
Kaufman, L., Leyenaar, A., and Harvey, J. S. (1961). *Progr. Very High Pressure Res. Proc. Intern. Conf., Bolton Landing, Lake George, N.Y.*, 1960, pp. 90–106, Wiley, New York.
Keeler, R. N., and Mitchell, A. C. (1967). *Bull. Am. Phys. Soc.* [2], **12**, 1128.
Kennedy, G. C., and LaMori, P. (1962). *J. Geophys. Res.* **67**, 851.
Kennedy, G. C., Haygarth, J. C., and Getting, I. C. (1967). *J. Appl. Phys.* **38**, 4557.
Khitarov, N. T., and Slutsky, A. V. (1967). *J. Chem. Phys.* **64**, 7/8, 1, 85–1091.
Kiker, J. L. (1966). *J. Sci. Instr.* **43**, 269.

KLEMENT, JR., W., and JAYARAMAN, A. (1967). *Progr. Solid State Chem.* **3**, 289–376.
KNOPOFF, L. (1963). *In* "High Pressure Physics and Chemistry" (R. S. Bradley, ed.), Chapter 5. Academic Press, New York.
KOEHNLEIN, D. (1967). *Z. Physik* **208**, 142–158.
KOSICKI, B. B., JAYARAMAN, A., and PAUL, W. (1968). *Bull. Am. Phys. Soc.* **13**, 431.
LANDAU, J. V. (1961). *Exptl. Cell. Res.* **27**, 123.
LANDWEHR, G. (1965). *In* "Physics of High Pressures and the Condensed State" (A. van Itterbeek, ed.), Chapter 14. Wiley, New York.
LAZARUS, D., and NACHTRIEB, N. H. (1963). *In* "Solids Under Pressure" (W. Paul and D. Warschauer, eds.), Chapter 3. McGraw-Hill, New York.
LEES, J. (1966). *Advan. High Pressure Res.* **1**, 2–78.
LEES, J., WASSE, M. P., and KING, G. (1967). *Solid State Commun.* **5**, 521–523.
LEGER, J. M., SUSSE, C., and VODAR, B. (1967). *Solid State Commun.* **5**, 755–758.
LIBBY, W. F., SOLOMON, L., ZEEGEN, P., and EISERLING, F. (1965). *In* "Science in the Sixties", (D. L. Arm, ed.), pp. 176–179. Univ. of New Mexico, Albuquerque, New Mexico.
LLOYD, E. (1971). *NBS Symp. Accurate Characterization High Pressure Environment, Gaithersburg, October 1968*, E. Lloyd, ed. In press.
LYNCH, R. W., and DRICKAMER, H. G. (1966). *J. Chem. Phys.* **44**, 181.
MACKENZIE, J. D. (1963). *J. Am. Ceram. Soc.* **46**, 461.
MCQUEEN, R. G. (1964). *In* "Metallurgy at High Pressures and Temperatures" (K. A. Gschneider, Jr., M. T. Hepworth, and N. A. D. Parlee, eds.), pp. 44–132. Gordon and Breach, New York.
MCQUEEN, R. G., FRITZ, J. N., and MARSH, S. P. (1964). *J. Geophys. Res.* **69**, 2947.
MCWHAN, D. B. (1967). *J. Appl. Phys.* **38**, 347.
MCWHAN, D. B., and STEVENS, A. L. (1967). *Phys. Rev.* **154**, 438.
MAIER, S., and FRANCK, E. U. (1966). *Ber. Bunsenges. Physik. Chem.* **70**, 639.
MAO, H. K., BASSETT, W. A., and TAKAHASHI, T. (1967). *J. Appl. Phys.* **38**, 272.
MARCH, N. H. (1955). *Proc. Phys. Soc.* (*London*) **A68**, 726.
MARSLAND, D. (1965). *Exptl. Cell Res.* **38**, 592.
MARTYNOV, E. D., BERESNEV, B. I., RYABININ, Y. N. (1967). *Fiz. Metal. i Metalloved.* **24**, 3, 522–527.
MEYER, H. O. A. (1967). Mineral inclusions in diamond. Geophys. Lab., Ann. Rept. of Director, 1966–1967, Publ. No. 1499. Carnegie Inst. of Technol., Pittsburgh, Pennsylvania.
MILLER, R. O., DACHILLE, F., and ROY, R. (1966). *J. Appl. Phys.* **37**, 4913.
MILTON, D. J., and DECARLI, P. S. (1963). *Science* **140**, 670.
MUNSON, R. A. (1968). *Inorg. Chem.* **7**, 389.
MYERS, M., DACHILLE, F., and ROY, R. (1963). *Rev. Sci. Instr.* **34**, 401.
NEWTON, R. C. (1966). *Advan. High Pressure Res.* **1**, 195–258.
O'HARA, M. J., and YODER, H. S. (1967). *Scot. J. Geol.* **3**, 67–117.
PASTINE, D. J. (1968). *Phys. Rev.* **166**, 703.
PAUL, W. (1963). *In* "High Pressure Physics and Chemistry" (R. S. Bradley, ed.), Chapter 5. Academic Press, New York.
PAUL, W., and WARSCHAUER, D. (1963). *In* "Solids Under Pressure" (W. Paul and D. Warschauer, eds.), Chapter 8. McGraw-Hill, New York.
PEREZ-ALBUERNE, E. A., and DRICKAMER, H. G. (1965). *J. Chem. Phys.* **43**, 1381.
PEREZ-ALBUERNE, E., FORSGREN, K., and DRICKAMER, H. (1964). *Rev. Sci. Instr.* **35**, 29.

PIPKORN, D., EDGE, C. K., DEBRUNNER, P., DEPASQUALI, G., DRICKAMER, H. G., and FRAUENFELDER, H. (1964). *Phys. Rev.* **A135**, 1604.
PISTORIUS, C. W. F. T. (1967). *J. Chem. Phys.* **47**, 4870.
PLENDL, J. (1961). *Phys. Rev.* **123**, 1172–1180.
PLENDL, J., and GIELISSE, P. (1962). *Phys. Rev.* **125**, 828–832.
POLLARD, E. C., and WELLER, P. K. (1966). *Biochim. Biophys. Acta* **112**, 573.
POSTMUS, C., NAKAMOTO, K., and FERRARO, J. R. (1967). *Inorg. Chem.* **6**, 2194–2199.
POSTMUS, C., MARONI, V. A., FERRARO, J. R., and MITRA, S. S. (1968). *Inorg. Nucl. Chem. Letters* **4**, 269–274.
PUGH, H. L. D., and ASHCROFT, K. (1963). "The Physics and Chemistry of High Pressures," pp. 163–176. Soc. of Chem. Ind., London.
REID, A. F., and RINGWOOD, A. E. (1968). *Inorg. Chem.* **7**, 443–445.
REID, A. F., WADSLEY, A. D. and RINGWOOD, A. E. (1967) *Acta Cryst.* **23**, 736.
RINGWOOD, A. E. (1963). *Nature* **198**, 79.
RINGWOOD, A. E. (1966). *Advan. Earth Sci. Contrib. Intern. Conf., Cambridge, Mass., 1964*, P. M. Hurley, ed., pp. 357–399. M.I.T. Press, Cambridge, Massachusetts.
RINGWOOD, A. E., and MAJOR, A. (1966). *J. Geophys. Res.* **71**, 4448.
RINGWOOD, A. E., and MAJOR, A. (1967). *Nature* **215**, 1367–1368.
RINGWOOD, A. E., REID, A. F., and WADSLEY, A. D. (1967). *Acta Cryst.* **23**, 1093–1095.
ROOYMANS, C. J. M. (1967). High pressure studies of oxides and chalcogenides. Ph.D. Thesis, Amsterdam Univ., Netherlands (see also *Philips Res. Rept. Suppl.* **5**, (1968) and *Adv. High. Press. Res.* **2**, 1–95 (1969).
ROSS, M., and ALDER, B. J. (1967). *J. Chem. Phys.* **47**, 4129–4133.
ROWLINSON, J. S. (1967). *Nature* **213**, 440.
SAMARA, G. A. (1967). *J. Geophys. Res.* **72**, 671.
SAMARA, G. A., and DRICKAMER, H. G. (1962a). *J. Chem. Phys.* **37**, 1159.
SAMARA, G. A., and DRICKAMER, H. G. (1962b). *J. Phys. Chem. Solids* **23**, 457.
SCHIRBER, J. E. (1965). *In* "Physics of Solids at High Pressures" (C. T. Tomizuka and R. M. Emrick, eds.), pp. 46–67. Academic Press, New York.
SCHOCK, R. N., and KATZ, S. (1967). *J. Phys. Chem. Solids* **28**, 1985–1994.
SCLAR, C. B., CARRISON, L. C., and STEWART, O. M. (1967). *Am. Geophys. Union Meeting, Washington, D. C., April 1967*, Paper.
SHANER, J. W., and ROYCE, E. B. (1968). *J. Appl. Phys.* **39**, 492.
SHEWMON, P. G., and ZACKAY, V. F., eds. (1961). "Response of Metals to High Velocity Deformation." Wiley (Interscience), New York.
SILVERMAN, M. S. (1964). *Inorg. Chem.* **3**, 1041.
SOGA, N. (1967). *J. Geophys. Res.* **72**, 4227–4234.
SOULEN, J. R., and SILVERMAN, M. S. (1963). *J. Polymer. Sci. Pt. A* **1**, 823.
STEELE, W. A., and WEBB, W. (1963). *In* "High Pressure Physics and Chemistry" (R. S. Bradley, ed.), Chapter 4. Academic Press, New York.
STEWART, J. W. (1962). *In* "Modern Very High Pressure Techniques" (R. H. Wentorf, Jr., ed.), Chapter 10. Butterworths, London and Washington, D. C.
STEWART, J. W. (1965). *In* "Physics of High Pressures and the Condensed Phase" (A. Van Itterbeek, ed.), Chapter 5. North-Holland Publ., Amsterdam.
STISHOV, S. M., and POPOVA, S. V. (1961). *Geokhimiya* **10**, 837.
STOFFLER, D. (1966). *Contrib. Mineral. Petrology (Berlin)* **12**, 15.
STOKES, R. H. (1966). *J. Phys. Chem. Solids* **27**, 487.

STRONG, H. M. (1962). *In* "Modern Very High Pressure Techniques" (R. H. Wentorf, ed.), Chapter 5. Butterworths, London and Washington, D. C.
STRONG, H. M., and HANNEMAN, R. E. (1967). *J. Chem. Phys.* **46**, 3668.
SWENSON, C. A. (1966). *J. Phys. Chem. Solids* **27**, 39.
TOSI, M., and ARAI, T. (1966). *Advan. High Pressure Res.* **1**, 265–324.
VAISNYS, J. R., and KIRK, R. S. (1966). *Phys. Rev.* **141**, 641.
VAUGHAN, R. W., and DRICKAMER, H. G. (1967a). *J. Chem. Phys.* **47**, 1530.
VAUGHAN, R. W., and DRICKAMER, H. G. (1967b). *J. Chem. Phys.* **47**, 468.
VERESHCHAGIN, L. F., ZUBORA, E. B., BUIMORA, I. P., and BURDINA, K. P. (1966). *Doklady Akad. Nauk SSSR* **169**, 74–76 (*Soviet Phys. "Doklady"* (*English Transl.*) **11**, 585 (1967)).
VODAR, B., and SAUREL, J. (1963). *In* "High Pressure Physics and Chemistry" (R. S. Bradley, ed.), Chapter 3. Academic Press, New York.
WANG, C. Y. (1968). *Nature* **218**, 560–561.
WAYNE, R. C. (1968). *Phys. Rev.* **170**, 523.
WEALE, K. E. (1967). "Chemical Reactions at High Pressures." Spon, London.
WEAVER, J. S., TAKAHASHI, T., and BASSETT, W. A. (1971). P-V Relation for NaCl. *NBS Symp. High Pressure Environment, Gaithersburg, Maryland, October 1968*, E. Lloyd, ed. In press.
WEIR, C. E., VAN VALKENBURG, A., and LIPPINCOTT, E. (1962). *In* "Modern Very High Pressure Techniques" (R. H. Wentorf, ed.), Chapter 3. Butterworths, London and Washington, D. C.
WENTORF, R. H. (1959). *J. Phys. Chem.* **63**, 934.
WENTORF, R. H. (1961). *J. Chem. Phys.* **34**, 809.
WENTORF, R. H. (1965). *J. Phys. Chem.* **69**, 3063.
WENTORF, R. H. (1966). *Ber. Bunsenges. Phys. Chem.* **70**, 975–982.
WENTORF, R. H. (1968). Private communication.
WHALLEY, E. (1962). *Advan. Phys. Org. Chem.* **2**, 93.
WHALLEY, E. (1966). *Advan. High Pressure Res.* **1**, 143–189.
WHALLEY, E. (1967). High pressure. *Ann. Rev. Phys. Chem.* **18**, 205–232.
WHATLEY, L. S., and VAN VALKENBURG, A. (1966). *Advan. High Pressure Res.* **1**, 327–369.
WITTIG, J., and MATTHIAS, B. T. (1968). *Science* **160**, 994.
WONG, J. Y., LINDE, R. K., and DECARLI, P. S. (1968). *Nature* **219**, 714.

Chapter 10

Carathéodory's Formulation of the Second Law

S. M. BLINDER

I. Introduction

Classical thermodynamics rests on a phenomenological foundation of immense breadth and depth, encompassing a vast and diverse array of accumulated experience in chemistry, physics, biology, and engineering. The subject is most generally formulated in terms of four empirical laws—the zeroth through third laws of thermodynamics. Among these, the second law is undoubtedly the most profound and interesting conceptually. The second law had its origin in Carnot's analysis of the performance of steam engines (ca. 1824). Carnot's ideas were subsequently elaborated and extended—major contributions being made by Clausius and by Kelvin to make them applicable to a broad range of physical and chemical phenomena. This must, indeed, be numbered among the most remarkable episodes in the history of science.

Not surprisingly, there exist a number of alternative formulations of the second law. The classical formulations take as their starting point either the Kelvin–Planck or the Clausius principle. According to the former, you cannot continuously transform heat into work without something else happening. According to the latter, you cannot transfer heat from a cooler to a warmer body without something else happening. From either of these can be deduced two far-reaching mathematical generalizations:

2A. (Carnot's theorem). For all thermodynamic systems, the absolute temperature is an integrating denominator for the reversible heat differential. This implies the existence of a new function of state: the entropy.

2B. (Law of increasing entropy). Entropy increases monotonically in every irreversible adiabatic process.

Principles 2A and 2B comprise, in effect, a mathematical statement of the second law.

In either the Kelvin–Planck or the Clausius formulation, the derivation of the above mathematical principles is unreasonably lengthy and tenuous. In its course, certain idealized devices and processes must be evoked and extensive use made of engineering terminology and concepts. Although the arguments and conclusions are, without doubt, correct, the line of development is open to criticism on the following logical and aesthetic grounds:

1. The line of reasoning leading from the physical to the mathematical principles is excessively long.
2. There is no clear separation between the physical and mathematical content of the theory.
3. The laws governing the more commonly encountered physico-chemical systems ought to be derivable solely from phenomena occurring in such systems—not from the performance characteristics of steam engines!

It is understood, of course, that the historical development of a science will not generally proceed in a sequence of logical orderly steps. But once all the facts are known, the subject ought to be restructured with a view toward aesthetics and logical consistency.

Carathéodory (1873–1950) was a mathematician of Greek origin, working in Germany during the early years of this century. In 1909, following a suggestion by Born (see Born (1949)), he attempted to reformulate classical thermodynamics in accord with the criteria enumerated above. The outcome, a theory of considerable elegance and generality, has two principal ingredients: Carathéodory's theorem and Carathéodory's principle. The former is a purely mathematical result on geometrical attributes of linear differential forms. The latter is an abstract generalization based on the universal characteristics of irreversible processes. As a statement of the physical content of the second law, it is contrived so as to lead, in the most direct and transparent manner, to the entropy principles 2A and 2B.

II. Theory of Linear Differential Forms

Differential quantities of the type

$$dq = X(x, y, z)\,dx + Y(x, y, z)\,dy + Z(x, y, z)\,dz \tag{2.1}$$

play a central role in the Carathéodory formalism. These are known as linear differential forms or Pfaff differential expressions.* The case of three independent variables is especially singled out, since it contains sufficient generality for thermodynamic purposes, while allowing of geometrical representation in 3-space. The theory can also, without undue difficulty, be generalized to any number of variables. In application of the theory to thermodynamics, dq will represent an element of heat transfer in a *reversible* process. (We denote by dQ the element of heat in the general case.) The independent variables x, y, z can stand for P, V, T, composition variables, etc. Every point (x, y, z) accordingly represents a possible equilibrium state of a thermodynamic system while every curve in xyz-space represents a possible reversible process.

Every linear differential form in three (or more) variables belongs to one of three categories:

(a) dq is exact: there exists some function $F(x, y, z)$ whose total differential equals dq, i.e.,

$$dq = d[F(x, y, z)]. \tag{2.2}$$

It is known from experience, however, that such is not, in general, the case for the differential element of heat.

* For the requisite mathematical background, the reader is referred to Blinder (1966).

(b) dq is integrable: there exists some function $T(x, y, z)$ such that

$$dq/T(x, y, z) = d[S(x, y, z)]. \tag{2.3}$$

In such cases, $T(x, y, z)$ is known as an integrating denominator. (Alternatively, $1/T$ is called an integrating factor.) Division by $T(x, y, z)$, in this case, converts dq into the total differential of some function $S(x, y, z)$. Case (b) trivially includes case (a), say, when $T = \text{const}$.

(c) dq is nonintegrable: no integrating denominator of the form $T(x, y, z)$ exists.

The properties of differential expressions in three variables can be compactly formulated in terms of vector analysis (see Appendix A).

A differential equation of the type

$$dq = 0, \tag{2.4}$$

in which dq is given by (2.1), is known as a Pfaff (differential) equation. Solutions to (2.4) can be represented by curves in xyz-space. Each such solution curve corresponds evidently to a reversible adiabatic process. (A simple two-dimensional analog is one of the curves $PV^{\gamma} = \text{const}$, representing reversible adiabatics in an ideal gas.) Solution curves to (2.4) can, in concept, be constructed by point-by-point integration of the differential equation. In the general case, there should exist at least one solution curve between any two points in xyz-space (see Appendix B).

When every dq is integrable, however, *not* every pair of points can be so connected by a solution curve. This follows from the fact that solutions to an integrable Pfaff equation are inherently expressible in terms of a one-parameter family of *solution surfaces**:

$$S(x, y, z) = \text{const}, \tag{2.5}$$

obtained by explicit integration of dq in case (b). The only admissible solution curves to (2.4) are, accordingly, those which lie entirely within one of the solution surfaces (2.5). The situation is represented in Fig. 1. Evidently, if a differential expression (2.1) is integrable, there must exist points arbitrarily close to every point P which are inaccessible from it along any solution curve to the associated Pfaff equation (2.4).

* In general, for an integrable Pfaff equation in n independent variables, there exist solution hypersurfaces of dimensionality $n - 1$.

The converse of the last statement is Carathéodory's theorem:

> If, arbitrarily close to every point P, there exist points inaccessible from it along any solution curve to $dq = 0$, then dq possesses an integrating denominator.

A proof is outlined in Appendix C.

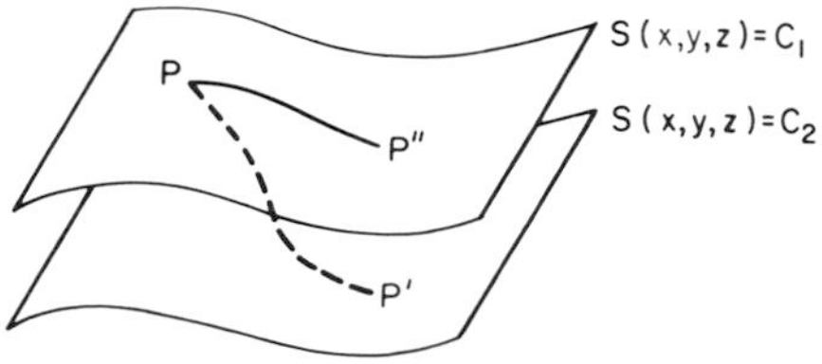

FIG. 1. Solution surfaces for integrable Pfaff equation. Points P and P'' are mutually accessible by Pfaff solution curves, while points P and P' are mutually inaccessible.

III. Carathéodory's Principle

Let us return now to the world of thermodynamics. The essential basis of the second law is the definite unidirectional tendency of all naturally occurring physical and chemical phenomena. Certain physical processes must accordingly be unrealizable, most obviously, those which entail the reversal of a spontaneous process. As Bertrand Russell put it most succinctly, "You cannot unscramble eggs." The Kelvin–Planck principle focuses attention on the impossibility of uncompensated conversion of heat into work. The Clausius principle focuses on the unidirectionality of heat flow. Carathéodory expressed the physical content of the second law in the following principle:

> Arbitrarily close to every equilibrium state of a thermodynamic system, there exist states unattainable from it by any adiabatic process.

Carathéodory's principle thereby proscribes a somewhat more general type of process than do either the Kelvin–Planck or Clausius statements of the second law.*

* Actually, assuming that $T > 0$, Carathéodory's principle can be derived from either the Kelvin–Planck or the Clausius principle. For $T < 0$—negative absolute temperatures, associated, for example, with an inverted Boltzmann distribution in nuclear-spin systems—the classical formulations of the second law must be amended, but Carathéodory's principle remains valid (see Ramsey (1956)).

Processes which do actually occur under given conditions are termed *spontaneous*. A more inclusive notion is that of *irreversible* processes. They comprise, in addition to spontaneous processes, those which could potentially occur, were an appropriate mechanism available. Since thermodynamic results must be independent of rate or mechanism, irreversibility is the more fundamental concept.

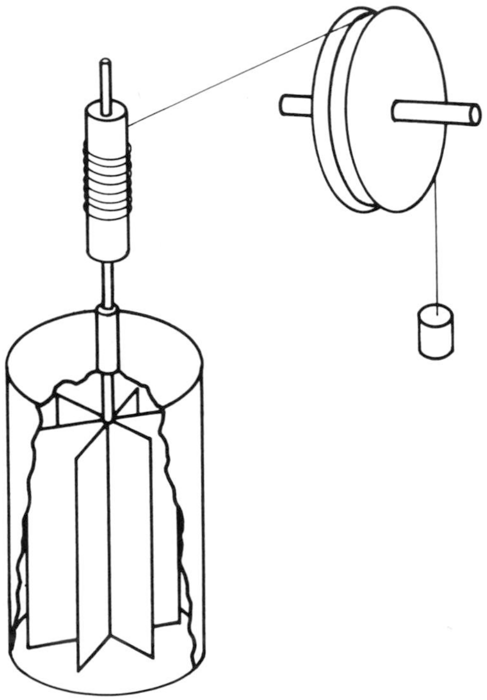

FIG. 2. Joule's paddle-wheel experiment.

A transparent demonstration of Carathéodory's principle is provided by Joule's paddle-wheel experiment, represented somewhat schematically in Fig. 2. This was, in fact, one of the methods employed by Joule to determine the mechanical equivalent of heat, which, of course, led to the *first* law of thermodynamics. Let the walls of the cylinder be thermally insulating, so that the fluid comprises a closed adiabatic (although not isolated) system. Now, if the weight were released from a given height, it would spontaneously fall, producing a clockwise rotation of the paddle wheel. Work would thereby be dissipated by friction in the fluid, causing a rise in temperature. The reverse process, in which the weight would *rise* spontaneously to convert heat into work while cooling the fluid, is contrary to all experience. It is therefore to be regarded as an impossible process. The paddle wheel will never, of its own accord, rotate counterclockwise, even infinitesimally. Therefore, for a given state of the system, that associated with a particular temperature of the fluid or elevation of the weight, there exist neighboring states—those in which the fluid is cooler or the weight more elevated—which are inaccessible by any adiabatic process.

Some insight into the limit of applicability of the second law can be gained by considering an idealized modification of the paddle-wheel experiment. Suppose the cylinder contained just 10 molecules and the apparatus were appropriately scaled such that individual molecular collisions with the vanes had perceptible effect. Under such circumstances, the pressure exerted on the vanes would be impulsive and widely fluctuating. It would, therefore, be quite conceivable for the paddle wheel to rotate in either direction, in apparent violation of Carathéodory's principle. Equilibrium states in the thermodynamic sense are, however, not definable for such fluctuating systems: thus, Carathéodory's principle is inapplicable, rather than violated. In any case, the second law emerges as a fundamentally statistical principle, pertaining only to systems of macroscopic extent. In contrast, the first law, being based on an inviolate mechanical principle, is fully valid even for a system of 10 molecules.

The full physical content of the second law is contained, most economically, in Carathéodory's principle. No further physical discussion is required to get to the entropy principles. In contrast, the classical formulations require, to reach the same conclusions, additional physical arguments which necessitate introduction of idealized devices and processes (Carnot cycles, etc.).

IV. Carnot's Theorem

The existence of an integrating factor for dq follows almost trivially from Carathéodory's principle specialized to the reversible case. As noted in Section II, Pfaff solution curves can be associated with reversible adiabatic processes. The inadmissibility of certain such processes corresponds, in mathematical terminology, to the existence of points, arbitrarily close to one another, which are mutually inaccessible by solution curves. Carathéodory's theorem thereby becomes applicable, leading to the conclusion that an integrating denominator for dq *does* exist for every thermodynamic system. It remains to determine the explicit form of the integrating denominator. This will be deduced first by an argument based on the ideal gas, which actually does not require any explicit reference to Carathéodory's principle. In Appendix D, the same result will be obtained by a more rigorous and correct procedure, which places no reliance on extrathermodynamic notions, such as the ideal gas.

By the first law, in a reversible process,

$$dq = dE + P\,dV. \tag{4.1}$$

Specializing to the case of an ideal gas,

$$dq = C_V(T)\,dT + (nRT/V)\,dV, \tag{4.2}$$

having noted that the internal energy is a function of temperature alone. Clearly, division by T separates variables T and V. Therefore, absolute temperature is an integrating denominator for dq*:

$$dS = dq/T. \tag{4.3}$$

This result, demonstrated thus far only for the ideal gas, can be shown to be of more general validity. Consider, accordingly, a thermodynamic system of whatever complexity, possibly several components distributed among several phases. Let an ideal-gas thermometer be inserted and let the composite system be adiabatically enclosed. Accordingly, heat can neither be gained nor lost by the composite system, but only exchanged between the (original) system and the thermometer. In an arbitrary infinitesimal process,

$$dq + dq' = 0, \tag{4.4}$$

where dq (for the original system) is given by an expression such as (2.1) and dq' (for the thermometer), is given by (4.2). Dividing by T, the temperature prevailing throughout the composite system,

$$dq/T + dq'/T = 0. \tag{4.5}$$

But, by (4.3), dq'/T, for the ideal gas, is an exact differential, say, dS'. Thus,

$$dq/T = -dS', \tag{4.6}$$

* Existence of an integrating denominator for (4.2) is independent of Carathéodory's theorem, since every differential expression in two variables is integrable.

showing that dq/T is also an exact differential, which can be denoted by dS ($= -dS'$). This completes the proof of Carnot's theorem, namely, that (4.3) applies to all thermodynamic systems.

V. Entropy Changes

Based on the idealization of reversible adiabatic processes, the existence of a new function of state, the entropy, has been deduced from the second law. Analogously, the first law gave, as a mathematical consequence, the existence of internal energy.

The second mathematical deduction from the second law is concerned with real, rather than hypothetical, processes—more particularly, with entropy changes in irreversible processes. In an arbitrary process AB, reversible or irreversible, the entropy change is given by a line integral:

$$\Delta S_{AB} = \int_A^B dS = \int_A^B dq/T, \tag{5.1}$$

where dq represents the (perhaps hypothetical) reversible heat element associated with the path of integration. But, since S is a function of state,

$$\Delta S_{AB} = S_B - S_A, \tag{5.2}$$

independent of the path between initial and final states. To evaluate the entropy difference between two states, it is necessary to conceptualize at least one reversible path between them, so that (5.1) can be applied. The associated process, since it is evoked solely for computational purposes, can be as highly idealized as desired. Once ΔS_{AB} is known, however, it applies equally well to any process, however irreversible, connecting the same two states. In reversible adiabatic process, $dq = 0$ at every stage, so that

$$\Delta S = 0. \tag{5.3}$$

It is very important to recognize that this does *not* apply to arbitrary adiabatic processes, in which $\Delta Q = 0$ in our notation, but only to reversible adiabatic processes, in which $\Delta q = 0$ as well. For irreversible adiabatic processes, in particular, $\Delta S \neq 0$.

As a consequence of (5.3), any two states which can be connected adiabatically and reversibly have equal entropy values. More generally, for a thermodynamic system specified by independent variables x, y, z, each manifold of states mutually accessible by reversible adiabatic processes defines an isentropic (constant-entropy) surface

$$S(x, y, z) = \text{const.} \tag{5.4}$$

This corresponds, of course, to a solution surface [cf. Eq. (2.5)] for the Pfaff equation associated with reversible adiabatic processes in the system.

As the constant in (5.4) runs over a continuous range of values, a one-parameter family of isentropic surfaces is defined. Since entropy is presumably a single-valued function of state, these must be discrete, nonintersecting surfaces. Through every point (x, y, z) there must consequently pass but one isentropic surface. This is an important consequence of the integrability of dq; analogous families of nonintersecting solution surfaces could not arise from a nonintegrable Pfaff equation. Moreover, by reasons of continuity, the family of surfaces (5.4) should be ordered monotonically. Thus, referring to Fig. 3, the entropy values should either increase or decrease in the order S', S, S''.

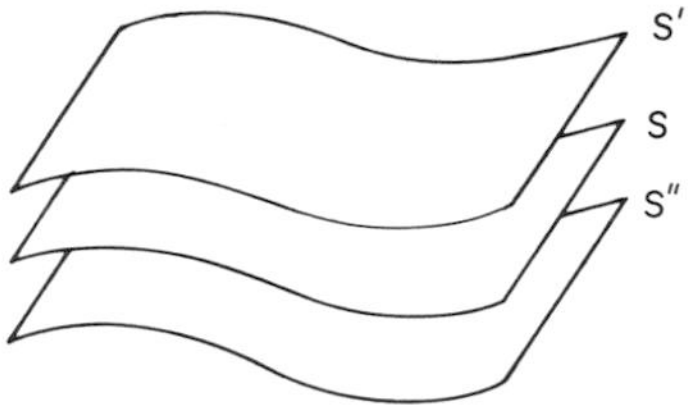

FIG. 3. Isentropic surfaces; S' is accessible, S'' inaccessible, from S by spontaneous adiabatic processes.

The classic example of an irreversible thermodynamic process is the flow of heat from a warmer to a cooler body. Consider two heat reservoirs, at temperatures T_1 and T_2 $(>T_1)$. Let them be in thermal contact with one another, but insulated from the surroundings, so that the overall process is adiabatic. Now, let a small increment of heat ΔQ flow from the warmer to the cooler reservoir. Although this clearly does not represent a reversible process, we assert nevertheless that ΔQ can be equated to an equal increment, Δq, of reversible heat. In justification, let it be idealized that neither reservoir undergoes any mechanical changes (or, at least, negligible ones) in the course of the process under consideration. Then, $\Delta W = 0$ and, by the first law, $\Delta Q = \Delta E$. Since the same reasoning would apply if the contemplated process were reversible, $\Delta q = \Delta E$.

An equal quantity of heat is therefore transferred, irrespective of whether the overall process is reversible or irreversible, so that ΔQ can be equated to Δq.

By virtue of the additivity of entropy [cf. Eq. (D.16)]

$$\Delta S = \Delta S_1 + \Delta S_2, \tag{5.5}$$

where S_1 and S_2 represent the entropies of the cool and warm reservoirs, respectively, and S is the entropy of the composite system. As the cool reservoir gains the increment of heat ΔQ at temperature T_1,

$$\Delta S_1 = \Delta Q/T_1. \tag{5.6}$$

As the warm reservoir gives up the same increment at T_2,

$$\Delta S_2 = -\Delta Q/T_2. \tag{5.7}$$

Thus, for the composite system,

$$\Delta S = \Delta Q\left(\frac{1}{T_1} - \frac{1}{T_2}\right) > 0, \tag{5.8}$$

showing that this particular irreversible adiabatic process is associated with a positive entropy increment.

VI. Law of Increasing Entropy

The second entropy principle requires the unrestricted statement of Carathéodory's principle. This powerful generalization on the adiabatic inaccessibility of certain thermodynamic states from others can be reexpressed in a suggestive way, making use of an observation to be developed in the following paragraph.

The manifold of states, for a thermodynamic system, having a given value of the entropy can be represented by a surface of the form (5.4). Now, any pair of states lying on the same isentropic surface ought, in principle, to be mutually accessible by some reversible adiabatic process. The latter can, of course, be represented by some Pfaff solution curve lying wholly within the surface. Now, every irreversible adiabatic process must of necessity be represented by a path between states on two different isentropic surfaces. Since all states on each surface are mutually accessible by reversible adiabatic processes, they become, in a sense, equivalent.

The following generalization emerges: If state P' is adiabatically accessible [inaccessible] from state P, then every state on the surface S' through P' is adiabatically accessible [inaccessible] from every state on the surface S through P. Accessibility of thermodynamic states has thereby been reexpressed in terms of accessibility of isentropic surfaces.

There should exist, arbitrarily close to every state, any number of states which *are* accessible by irreversible adiabatic processes. For example, a process in which work is dissipated by friction could, by appropriate variation of the mechanical parameters, be terminated arbitrarily close to the initial state. States which lie arbitrarily close to one another —while *not* on the same surface—can be termed as lying on *neighboring* surfaces. It can therefore be generalized:

> For every isentropic surface S, there exist neighboring surfaces S' which are *accessible* by irreversible adiabatic processes.

Using the same terminology, Carathéodory's principle can thereby be restated:

> For every isentropic surface S, there exist neighboring surfaces S'' which are *inaccessible* by irreversible adiabatic processes.

There consequently exist, for each surface S, neighboring surfaces of both types S' and S''. This must hold true even for the nearest-neighboring surfaces, those only infinitesimally removed from S. Even in the ultimate limit, neighboring isentropic surfaces corresponding to $S + dS$ and $S - dS$, respectively, one (S') is evidently accessible from S, while the other (S'') is inaccessible (see Fig. 3).

Irreversibility implies unidirectionality. Therefore, if S' were accessible from S by some irreversible adiabatic process, then S would, of necessity, be *inaccessible* from S' by any other irreversible adiabatic process.

In consequence, all neighboring surfaces of type S' must lie on one side of S, all neighboring surfaces of type S'', on the other side. In view of the assumed continuity of the entropy function, entropy must vary monotonically in traversing a family of surfaces. We arrive, therefore, at what might be termed "the principle of monotonic entropy change," namely: In all possible irreversible adiabatic processes, the entropy must either always increase or always decrease. To ascertain which of these two possibilities is indeed true, requires a further appeal to experiment. But to know ΔS for a single irreversible adiabatic process suffices to establish the universal principle.

It might be proposed that $\Delta S > 0$ for certain types of systems or adiabatic processes, while $\Delta S < 0$ for others. If this were true, however, then, by virtue of the additivity of entropy [cf. Eq. (D.16)], it would be possible to construct composite systems which violated Carathéodory's principle.

Most fundamentally, the sense of entropy change is determined by the natural direction of heat flow. Based on the conventional definition of absolute temperature, heat flows irreversibly from a higher to a lower temperature. Based on the adopted definition of entropy, such a process is associated with an increase in entropy [cf. Eq. (5.8)]:

$$\Delta S > 0. \tag{6.1}$$

This must therefore represent the universal principle for irreversible adiabatic processes. We have thus arrived at the law of increasing entropy.

One should clearly distinguish between the intrinsic content of the second law and consequences of the conventional definitions of temperature and entropy. The monotonicity of entropy belongs in the former category, the increasing property of entropy, in the latter.

Equation (6.1), in combination with Eq. (5.3), can be stated

$$\Delta S \geq 0. \tag{6.2}$$

Thus, in any adiabatic process, reversible or irreversible, the entropy is a never-decreasing function.

To make more concrete the geometrical properties of entropy surfaces contained in the foregoing results, refer again to the paddle-wheel experiment (Fig. 2). The schematic analog of an isentropic surface S might be the locus of points swept out by the weight in lateral, pendulum-like motions, which would evidently correspond to a reversible adiabatic process. Then, any surface situated *beneath* a given one, which could be realized in a spontaneous (hence, irreversible) adiabatic process, would belong in the category S'. But any surface situated *above* the given one, which is impossible to achieve spontaneously and adiabatically, would belong in the category S'' covered by Carathéodory's principle. In this example, entropy evidently decreases monotonically with elevation.

It is worthwhile to note that the conditions under which the law of increasing entropy was derived makes it applicable to *closed adiabatic**

* A closed adiabatic system can exchange neither heat nor matter with the surroundings. An isolated system can exchange neither work, heat, nor matter.

systems. Often, in the traditional treatments, it is erroneously concluded that the law applies only to *isolated* systems. It is clear, however, that the latter condition is unnecessarily restrictive.

VII. Nonadiabatic Processes

The law of increasing entropy, (6.1) or (6.2), applies specifically to adiabatic processes. It is, nonetheless, a principle of far-reaching applicability, since what is defined as the system can always be made sufficiently inclusive—by incorporation of essential elements of the surroundings—such that all contemplated processes become adiabatic. It is of interest, nevertheless, to deduce the analog of (6.1) or (6.2) for nonadiabatic processes—hence, for the individual components of adiabatically—isolated composite systems.

Consider, accordingly, a thermodynamic system undergoing an arbitrary infinitesimal process in which an element of heat dQ is transferred. Denote by dS the (differential) entropy change associated with the process. (To evaluate dS would, of course, require an alternative reversible path, but this is not essential to the following argument.) Assume, next, that those parts of the surroundings which participate in the contemplated process do so in reversible fashion. This constitutes no essential restriction, since there presumably exist innumerable alternative ways in which the original system can undergo the same process. Among these, there can almost certainly be idealized at least one in which the surroundings behave reversibly.

For the surroundings, let dq' and dS' represent, respectively, the differentials of heat and entropy. For the composite system, comprising the original system and essential surroundings, every process is, by supposition, adiabatic:

$$dQ + dq' = 0. \tag{7.1}$$

Since, for the surroundings, $dS' = dq'/T$, we have

$$dS' = -dQ/T. \tag{7.2}$$

By the additivity of entropy in composite systems,

$$dS_{\text{total}} = dS + dS'. \tag{7.3}$$

But, since the composite system is undergoing an adiabatic process,

$$dS_{\text{total}} \geq 0, \tag{7.4}$$

by the differential analog of (6.2).

It follows, therefore, after substituting (7.2) into (7.3), that

$$dS \geq dQ/T. \tag{7.5}$$

In integrated form,

$$\Delta S \geq \int dQ/T \tag{7.6}$$

for an arbitrary process undergone by the original system. This represents the desired generalization of the second entropy principle. Note that (7.6) reduces to (6.2) in adiabatic processes, when $dQ = 0$.

In contrast to the adiabatic case, entropy *can* decrease in an irreversible nonadiabatic process—crystallization from a melt for example. But, by (7.6), the entropy decrease will always be less than $\int dQ/T$.

Let (7.6) now be applied to an arbitrary *cyclic* process. Since entropy is a function of state, $\Delta S = \oint dS = 0$. The result is Clausius' inequality:

$$\oint dQ/T \leq 0. \tag{7.7}$$

Specialized to the limit of reversibility, (7.7) becomes

$$\oint dq/T = 0, \tag{7.8}$$

a result known as Clausius' theorem. Since the identical vanishing of a cyclic line integral implies the exactness of its integrand, (7.8) leads to Carnot's theorem. Clausius' inequality (7.7) thereby implies the full mathematical content of the second law.

VIII Conclusion

Carathéodory's approach to thermodynamics, in spite of its logical and aesthetic advantages, has not gained widespread favor among scientists and engineers. Its obvious pedagogical drawbacks are, first, the rather abstract nature of its physical basis, and, second, its heavy reliance on somewhat specialized mathematics. Whereas the Kelvin–Planck and Clausius principles are easily visualized, Carathéodory's principle, although asserting the impossibility of a fundamentally simpler type of process, is less readily associated with common experience. Carathéodory's principle is specifically contrived, however, to lead, most directly and efficiently, to the entropy principles 2A and 2B. It thus plays the formal role in thermodynamics of a minimal postulate of the second law, analogous, in this respect, to Hamilton's principle in mechanics or Fermat's principle in optics.

The conceptual and mathematical demands of the Carathéodory approach are becoming, with time, less and less of an obstacle to students of chemistry, physics, and engineering, in view of their continually increasing level of mathematical sophistication.* Much of this is, of course, due to the earlier introduction of quantum mechanics into chemistry and physics. It is not amiss, therefore, to suggest that the Carathéodory formulation represents a viable alternative to the steam engine in the presentation of the second law of thermodynamics.

Appendix A. Vector Formulation of the Theory of Differential Expressions

In terms of the displacement vector

$$\mathbf{r} = \mathbf{i}x + \mathbf{j}y + \mathbf{k}z \tag{A.1}$$

and the vector field

$$\mathbf{R}(\mathbf{r}) = \mathbf{i}X(\mathbf{r}) + \mathbf{j}Y(\mathbf{r}) + \mathbf{k}Z(\mathbf{r}), \tag{A.2}$$

a Pfaff differential in three variables, Eq. (2.1), can be expressed

$$dq = \mathbf{R}(\mathbf{r}) \cdot d\mathbf{r}. \tag{A.3}$$

When dq is exact [case (a)], (A.3) represents the differential of a scalar function $F(\mathbf{r})$ (which plays the role of a potential). This is true if and only if

$$\mathbf{R}(\mathbf{r}) = \text{grad}\, F(\mathbf{r}). \tag{A.4}$$

But a necessary (and sufficient) condition for a vector field $\mathbf{R}(\mathbf{r})$ to be

* Around the turn of the century, trigonometry was, not uncommonly, a subject taken up by mathematics majors in their senior year of college.

representable as the gradient of a scalar function is that it be irrotational, i.e.,

$$\text{curl}\,\mathbf{R} = 0. \tag{A.5}$$

The scalar components of (A.5), namely,

$$\partial Y/\partial x - \partial X/\partial y = 0, \quad \textit{et cyc.}, \tag{A.6}$$

are equivalent to the Euler reciprocity conditions for an exact differential.

Even when the vector field $\mathbf{R}(\mathbf{r})$ is *not* irrotational, there may still exist a scalar function $G(\mathbf{r})$ such that

$$\text{curl}(G\mathbf{R}) = 0. \tag{A.7}$$

This would imply (when $G \neq 0$) the existence of an integrating factor for dq [case (b)]. Now,

$$\text{curl}(G\mathbf{R}) = G\,\text{curl}\,\mathbf{R} - \mathbf{R} \times \text{grad}\,G. \tag{A.8}$$

Taking the scalar product with $\mathbf{R}$,

$$\mathbf{R} \cdot \text{curl}(G\mathbf{R}) = G\mathbf{R} \cdot \text{curl}\,\mathbf{R}, \tag{A.9}$$

the second term in curl $(G\mathbf{R})$ being normal to $\mathbf{R}$. The vanishing of curl $(G\mathbf{R})$ when $G \neq 0$ therefore implies

$$\mathbf{R} \cdot \text{curl}\,\mathbf{R} = 0. \tag{A.10}$$

This integrability condition, expanded out in scalar notation, reads

$$X\Big(\frac{\partial Z}{\partial y} - \frac{\partial Y}{\partial z}\Big) + Y\Big(\frac{\partial X}{\partial z} - \frac{\partial Z}{\partial x}\Big) + Z\Big(\frac{\partial Y}{\partial x} - \frac{\partial X}{\partial y}\Big) = 0. \tag{A.11}$$

It is easily shown that every differential expression in two variables is integrable, for, when, say, $Z = 0$ and X and Y are independent of z, (A.11) is fulfilled.

When (A.10) or (A.11) is *not* fulfilled, i.e.,

$$\mathbf{R} \cdot \text{curl}\,\mathbf{R} \neq 0, \tag{A.12}$$

then the differential expression dq is nonintegrable [case (c)].

Appendix B. Solution Curve for Nonintegrable Pfaff Equation

The differential expression

$$dq = y\,dx + dy - dz \tag{B.1}$$

does not satisfy (A.11), so that the corresponding Pfaff equation

$$y\,dx + dy - dz = 0 \tag{B.2}$$

is nonintegrable. It will be shown that two arbitrary points, P and P', can be connected by a solution curve to (B.2). Starting at the origin (0, 0, 0), a solution curve can be constructed of three linear segments: (*i*) $(0, 0, 0) \to (d, 0, 0)$, along which $y = 0$, $dy = 0$, $dz = 0$, so that (B.2) is satisfied. (*ii*) $(d, 0, 0) \to (d, b, b)$, along which $dx = 0$, $dy - dz = 0$, so that the differential equation is again satisfied. The last relation implies the linear path $y - z =$ const. But the constant must be zero, since the line passes through $(d, 0, 0)$ $(y = 0, z = 0)$. Thus, (d, b, b) lies on the same linear segment. (*iii*) $(d, b, b) \to (a, b, c)$, in which $dy = 0$, $b\,dx - dz = 0$. The last equality integrates to $bx - z =$ const. Since the line must pass through (d, b, b), the constant is $bd - b$. Thus, when $x = a$ on this last linear segment, $ba - z = bd - b$ or $z = b(1 - \mathrm{d} + \mathrm{a}) \equiv c$. Now, if two arbitrary points (x, y, z) and (x', y', z') can each be reached from (0, 0, 0) by solution curves of (B.2), they must be accessible, one from the other, by solution curves. There must exist, of course, innumerable other solution curves between these two points.

Appendix C. Proof of Carathéodory's Theorem

Following is a somewhat simplified proof for the three-dimensional case.* We require first a result pertaining to solution curves lying upon arbitrary surfaces. Let the equation for a surface be

$$F(x, y, z) = 0. \tag{C.1}$$

Between (C.1) and a Pfaff equation

$$X(x, y, z)\,dx + Y(x, y, z)\,dy + Z(x, y, z)\,dz = 0 \tag{C.2}$$

* An alternative proof based on an analytical approach is given by Buchdahl (1949).

one of the three variables can be eliminated. The result is a Pfaff equation in two variables, say,

$$U(u, v)\, du + V(u, v)\, dv = 0, \tag{C.3}$$

which is always integrable, irrespective of whether (C.2) is integrable or not. Therefore, through every point (x, y, z) on the surface (C.1) there passes one unique solution curve to (C.3). But solution curves to (C.3) are likewise solution curves to (C.2). Consequently, from among the solution curves to (C.2) lying on an arbitrary surface, one and only one passes through each point on the surface.

Let it be given now that, arbitrarily close to every point P, there exist innumerable points which are inaccessible from P by solution curves of a given Pfaff differential equation (Fig. 4). Through P, an arbitrary

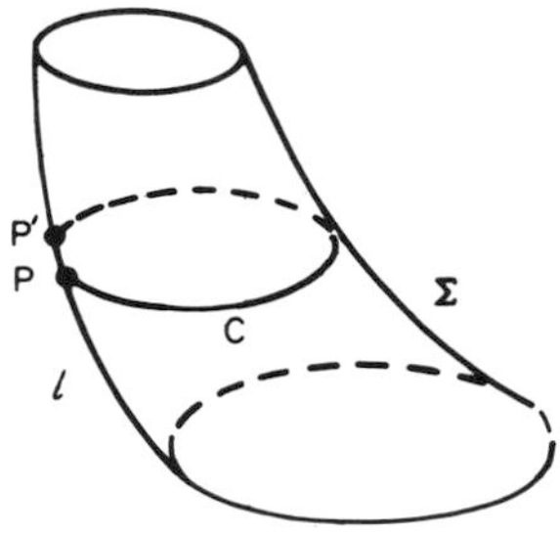

FIG. 4. Proof of Carathéodory's theorem.

curve l, *not* a solution curve, is passed. Along every such curve there will also lie innumerable inaccessible points arbitrarily close to P.* Now pass through l a cylinderlike closed surface Σ. By the result given in the preceding paragraph, there will be a unique solution curve, denoted by C, lying in Σ and passing through P. Denote the next intersection of C with l by P'. If P' is distinct from P, it can be made to lie arbitrarily close to P by suitable deformation of the surface Σ. But this contradicts the original supposition, since all points P' along l arbitrarily close to P would then be attainable from P by a solution curve. Evidently, P and P'

* To verify this, let Q represent some point inaccessible from P. Pass through Q and l some surface Σ'. Then, according to the preceding paragraph, a unique solution curve, say C', on Σ' will pass through Q. Let C' intersect l at Q'. Now, since Q is inaccessible from P but Q' is accessible from Q, Q' must be inaccessible from P. But, when $X(x, y, z)$, $Y(x, y, z)$, and $Z(x, y, z)$ are continuous functions of x, y, and z, then Q' can be made to lie as close as we please to P by choosing Q close enough to P.

must coincide and C must be a closed curve. But then, by continuous deformation of Σ, a solution *surface* containing C can be generated. Let the above procedure be repeated for two points Q and Q', inaccessible from P, which lie arbitrarily close to P on opposite sides of the adiabatic surface through P. That surface can thereby be cordoned off by two inaccessible surfaces arbitrarily close. There must therefore exist a family of mutually inaccessible solution surfaces of the form (7.4).

To complete the proof, the last result must be shown to imply the existence of an integrating factor for the differential expression in (C.2). Along any infinitesimal segment of a curve entirely within a solution surface, both $dq = 0$ and $dS = 0$. Along an infinitesimal segment *not* within a solution surface, $dS \neq 0$ and, in general, $dq \neq 0$. It is thus suggested that dq and dS are, in some way, proportional. Consider segments between points in two neighboring solution surfaces $S(x, y, z) = c$ and $S(x, y, z) = c + dc$. Then $dS = dc$, irrespective of the point of crossing. But dq *will* depend in general on the point (x, y, z) of crossing. This shows that the proportionality between dq and dS is of the form

$$dq = T(x, y, z)\, dS, \tag{C.4}$$

equivalent to the existence of an integrating denominator for dq. This completes the proof of Carathéodory's theorem for the case of three variables. Extension of this result to any number of independent variables introduces no difficulties.

Appendix D. Rigorous Proof of Carnot's Theorem

It was demonstrated in Section IV that, by virtue of Carathéodory's principle and theorem, an integrating factor for dq exists for every thermodynamic system. It will now be shown more rigorously that (1) the integrating denominator is a universal function for all thermodynamic systems in mutual thermal equilibrium, and (2) that it can be identified with the ideal-gas absolute temperature. In addition, the additivity property of entropy and the notion of a thermodynamic temperature scale will emerge.

In order to carry through this derivation, we must consider at least two systems in mutual thermal equilibrium. Each can be of the most general type, with any number of independent variables. Sufficient generality is achieved, however, for the case of three variables, say,

x, y, z. The generality of the derivation can further be extended by specifying the temperature on an arbitrary empirical scale. The empirical temperature, designated θ, might, for example, be based on the Celsius scale or even on the height, in centimeters, of the column in a mercury thermometer.

Let a composite system be constructed of two subsystems: subsystem 1 specified by the variables $x_1 = \theta_1, y_1, z_1$; subsystem 2, by the variables $x_2 = \theta_2, y_2, z_2$. Assume that the two subsystems interact only via heat transfer, so that their variables suffice as well to describe the composite system. Assume further that the subsystems have achieved thermal equilibrium at a common empirical temperature: $\theta_1 = \theta_2 = \theta$.

For an arbitrary infinitesimal reversible process,

$$dq = dq_1 + dq_2, \tag{D.1}$$

where dq is the net heat transfer to the composite system and dq_1 and dq_2 are the corresponding quantities for the respective subsystems. According to what has been shown, each of these elements of heat is integrable. Denoting by Θ the integrating denominator and by $d\Sigma$ the exact differential, we write

$$\begin{aligned} dq_1 &= \Theta_1(\theta, y_1, z_1)\, d\Sigma_1(\theta, y_1, z_1) \\ dq_2 &= \Theta_2(\theta, y_2, z_2)\, d\Sigma_2(\theta, y_2, z_2) \\ dq &= \Theta(\theta, y_1, z_1, y_2, z_2)\, d\Sigma(\theta, y_1, z_1, y_2, z_2). \end{aligned} \tag{D.2}$$

(We are anticipating, of course, the connection between Θ and T, Σ, and S.) Equation (D.1) can, accordingly, be expressed

$$\Theta\, d\Sigma = \Theta_1\, d\Sigma_1 + \Theta_2\, d\Sigma_2. \tag{D.3}$$

The Σ's are, of course, functions of state in their respective systems—for example,

$$\Sigma_1 = \Sigma_1(\theta, y_1, z_1). \tag{D.4}$$

It is convenient to utilize Σ_1 and Σ_2 directly as independent variables (say, in place of y_1 and y_2). On this basis, we obtain, after dividing (D.3) through by Θ,

$$d\Sigma = \frac{\Theta_1(\theta, \Sigma_1, z_1)}{\Theta(\theta, \Sigma_1, \Sigma_2, z_1, z_2)}\, d\Sigma_1 + \frac{\Theta_2(\theta, \Sigma_2, z_2)}{\Theta(\theta, \Sigma_1, \Sigma_2, z_1, z_2)}\, d\Sigma_2. \tag{D.5}$$

From the form of (D.5), it is evident that Σ must represent a function of Σ_1 and Σ_2 alone:

$$\Sigma = \Sigma(\Sigma_1, \Sigma_2) \tag{D.6}$$

independent of z_1, z_2 (plus any additional variables) and, in particular, of θ.

The ratios Θ_1/Θ and Θ_2/Θ must therefore contain no dependence on either z_1 or z_2. This implies that Θ is independent of *both* z_1 and z_2, which, in turn, implies that Θ_1 is independent of z_1, and Θ_2 of z_2. Therefore, dependence on variables other than θ, Σ_1, and Σ_2 need not be considered further.

The two ratios must also be independent of θ. This is accomplished, in the most general way, by integrating denominators of the following form:

$$\begin{aligned} \Theta_1(\theta, \Sigma_1) &= KT(\theta)F_1(\Sigma_1), \\ \Theta_2(\theta, \Sigma_2) &= KT(\theta)F_2(\Sigma_2), \end{aligned} \tag{D.7}$$

and

$$\Theta(\theta, \Sigma_1, \Sigma_2) = KT(\theta)F(\Sigma_1, \Sigma_2), \tag{D.8}$$

where $T(\theta)$ is a universal function of the empirical temperature and K is an arbitrary constant. Since the composite system has, in fact, no greater generality than either of the subsystems, (D.8) should be symmetrical in structure to each of (D.7). Therefore, Θ and F should contain the variables Σ_1 and Σ_2 only in combination as a function of Σ [cf. (D.6)] and, in place of (D.8) we have*

$$\Theta(\theta, \Sigma) = KT(\theta)F(\Sigma). \tag{D.9}$$

* This can be demonstrated more formally as follows. Substituting (D.7) and (D.8) in (D.3), $F(\Sigma_1, \Sigma_2)\, d\Sigma = F_1(\Sigma_1)\, d\Sigma_1 + F_2(\Sigma_2)\, d\Sigma_2$. Writing the total differential of $d\Sigma$ and equating coefficients of $d\Sigma_1$ and $d\Sigma_2$, $F(\Sigma_1, \Sigma_2)\, \partial\Sigma/\partial\Sigma_1 = F_1(\Sigma_1)$, $F(\Sigma_1, \Sigma_2) \times \partial\Sigma/\partial\Sigma_2 = F_2(\Sigma_2)$. Differentiating the first relation wrt Σ_2 and the second wrt Σ_1,

$$\frac{\partial F}{\partial \Sigma_2}\,\frac{\partial \Sigma}{\partial \Sigma_1} + F\,\frac{\partial^2 \Sigma}{\partial \Sigma_2\, \partial \Sigma_1} = \frac{\partial F_1}{\partial \Sigma_2} = 0$$

$$\frac{\partial F}{\partial \Sigma_1}\,\frac{\partial \Sigma}{\partial \Sigma_2} + F\,\frac{\partial^2 \Sigma}{\partial \Sigma_1\, \partial \Sigma_2} = \frac{\partial F_2}{\partial \Sigma_1} = 0.$$

Subtracting the first equation from the second,

$$\frac{\partial F}{\partial \Sigma_1}\,\frac{\partial \Sigma}{\partial \Sigma_2} - \frac{\partial F}{\partial \Sigma_2}\,\frac{\partial \Sigma}{\partial \Sigma_1} = 0.$$

But the left-hand side represents the functional derivative $\partial(F, \Sigma)/\partial(\Sigma_1, \Sigma_2)$, and its vanishing implies that a functional relationship exists between F and Σ.

Substituting (D.7) and (D.9) into the corresponding members of (D.2), we obtain

$$dq_1 = KT(\theta)F_1(\Sigma_1)\,d\Sigma_1,\ dq_2 = KT(\theta)F_2(\Sigma_2)\,d\Sigma_2,\ dq = KT(\theta)F(\Sigma)\,d\Sigma. \tag{D.10}$$

Entropy is next introduced, defining its differentials by

$$dS_1 = KF_1(\Sigma_1)\,d\Sigma_1,\quad dS_2 = KF_2(\Sigma_2)\,d\Sigma_2,\quad dS = KF(\Sigma)\,d\Sigma. \tag{D.11}$$

Equations (D.10) hence assume the form

$$dq_1 = T(\theta)\,dS_1,\qquad dq_2 = T(\theta)\,dS_2,\qquad dq = T(\theta)\,dS. \tag{D.12}$$

Formal expressions for the entropy are obtained from the indefinite integrals of (D.11):

$$\begin{aligned} S_1 &= K\int F_1(\Sigma_1)\,d\Sigma_1 + \text{const} \\ S_2 &= K\int F_2(\Sigma_2)\,d\Sigma_2 + \text{const} \\ S &= K\int F(\Sigma)\,d\Sigma + \text{const.} \end{aligned} \tag{D.13}$$

Substituting (D.12) into (D.1),

$$T(\theta)\,dS = T(\theta)\,dS_1 + T(\theta)\,dS_2. \tag{D.14}$$

Dividing by $T(\theta)$,

$$dS = dS_1 + dS_2 = d(S_1 + S_2), \tag{D.15}$$

and, by appropriate choice of additive constants in (D.13),

$$S = S_1 + S_2. \tag{D.16}$$

The important conclusion is therefore reached that entropy is an additive function: the entropy of a system equals the sum of the entropies of its component parts. This has been shown, in particular, for systems undergoing reversible processes but, since S is a function of state, it must be true in general.

Although, for each system, the particular form of Θ and S depends on $KF(\Sigma)$, this factor being transferrable from one to the other, the product $dq = \Theta\,d\Sigma = T\,dS$ is invariant to the functional form of $F(\Sigma)$

or the value of K. This is, of course, a necessary consequence of the fact that, if one integrating denominator exists, then an infinite number must exist. Alternative integrating denominators differ by factors which are arbitrary functions of Σ. One can, in particular, arbitrarily choose

$$KF_1(\Sigma_1) = KF_2(\Sigma_2) = KF(\Sigma) = 1, \qquad \text{(D.17)}$$

in which case, Θ becomes T and Σ becomes S.

Among the infinite number of integrating denominators for dq, there evidently exists one (to within an arbitrary multiplicative constant) which is a function of the empirical temperature alone. This is the choice which will evidently secure the greatest simplicity in thermodynamic formulas. Moreover, for all systems, of whatever composition, in mutual thermal equilibrium (equal values of θ), that integrating denominator is a universal function of θ, namely, $T(\theta)$. This is a far from casual result, considering that the integrating denominators were initially set up as distinct functions, each of at least three independent variables [cf. (D.2)].

The functional form of $T(\theta)$ depends, of course, on the particular empirical temperature scale. However, the numerical values of the function are entirely independent of this choice. It is possible, therefore, to define a temperature scale based entirely on thermodynamic concepts, independent of the properties of any particular thermometric substance (e.g., the mercury thermometer), even of the universal limiting behavior of all substances (e.g., the ideal-gas thermometer). Accordingly, we define a thermodynamic temperature equal to the universal integrating denominator for dq, i.e.,

$$dS = dq/T(\theta). \qquad \text{(D.18)}$$

It remains only to specify the size of the degree, since $T(\theta)$ still remains undefined to within a multiplicative constant K. This will, at the same time, establish the scale of entropy. It is obvious, comparing (D.18) with (4.3), that, by the appropriate choice, the thermodynamic temperature can be made to coincide exactly with the ideal-gas absolute temperature scale

$$T = T(\theta). \qquad \text{(D.19)}$$

This defines the so-called Kelvin temperature scale, its readings being denoted by °K. In practically all applications, the Kelvin and absolute scales are employed synonymously and interchangeably.

Acknowledgment

I should like to express my appreciation to The Macmillan Company for permission to include some short passages and especially Figs. 2 and 4 herein from my book "Advanced Physical Chemistry" (1969).

General References

Carathéodory's Formulation of Thermodynamics

Original Reference (in German)

Born, M. (1921). *Physik. Z.* **22**, 218, 249, 282.

Carathéodory, C. (1909). *Math. Ann.* **67**, 355.

Carathéodory, C. (1925). *Sitzber. Preuss. Akad. Wiss. Physik. Math. Kl.* **39**.

Lande, A. (1926). *In* "Handbuch der Physik" (S. Flügge, ed.), Vo. IX, Chapter IV, p. 281. Springer, Berlin.

Recent Discussions and Simplifications

Buchdahl, H. A. (1949). *Am. J. Phys.* **17**, 41, 44, 212.

Buchdahl, H. A. (1954). *Am. J. Phys.* **22**, 182.

Buchdahl, H. A. (1955). *Am. J. Phys.* **23**, 65.

Buchdahl, H. A. (1958). *Z. Physik* **152**, 425.

Buchdahl, H. A. (1960). *Am. J. Phys.* **28**, 196.

Landsberg, P. T. (1964). *Nature* **201**, 485.

Sears, F. W. (1963). *Am. J. Phys.* **31**, 747.

Sears, F. W. (1966). *Am. J. Phys.* **34**, 665.

Turner, L. A. (1960). *Am. J. Phys.* **28**, 781.

Turner, L. A. (1961). *Am. J. Phys.* **29**, 40, 71.

Turner, L. A. (1962). *Am. J. Phys.* **30**, 506.

Zemansky, M. W. (1966). *Am. J. Phys.* **34**, 914.

Textbooks

Bazarov, I. P. (1964). "Thermodynamics." Pergamon, New York.

Chandrasekhar, S. (1957). "Stellar Structure," Chapter I. Dover, New York.

Kestin, J. (1966). "A Course in Thermodynamics," Chapter 10. Ginn (Blaisdell), Boston.

Kirkwood, J. G., and Oppenheim, I. (1961). "Chemical Thermodynamics." McGraw-Hill, New York.

Wilson, A. H. (1960). "Thermodynamics and Statistical Mechanics." Cambridge Univ. Press, London and New York.

Special References

Blinder, S. M. (1966). *J. Chem. Ed.* **43**, 85.

Born, M. (1949). "Natural Philosophy of Cause and Chance," p. 38. Oxford Univ. Press, London and New York.

Buchdall, H. A. (1949). *Am. J. Phys.* **17**, 44.

Ramsey, N. F. (1956). *Phys. Rev.* **103**, 20.

Author Index

Numbers in italics refer to the pages on which the complete references are listed.

E

F

L

M

N

O

P

Q

R

S

T

Subject Index

O

P

Q

R

S

T

U

V